utledge Studies of Societies in Transition

he Economics of Soviet Break-up
rt van Selm

stitutional Barriers to Economic Development
land's incomplete transition
lited by Jan Winiecki

he Polish Solidarity Movement
volution, democracy and natural rights
rista Maria Cirtautas

rviving Post-Socialism
cal strategies and regional response in Eastern Europe and the former Soviet Union
lited by Sue Bridger and Frances Pine

nd Reform in the Former Soviet Union and Eastern Europe
lited by Stephen Wegren

nancial Reforms in Eastern Europe
policy model for Poland
nhaya L. Gupta and Robert Lensink

e Political Economy of Transition
portunities and limits of transformation
ef van Brabant

ivatizing the Land
ral political economy in post-communist socialist societies
ited by Ivan Szelenyi

kraine
te and nation building
ras Kuzio

een Post-Communism?
vironmental aid, innovation and evolutionary political economics
kael Sandberg

ganisational Change in Post-Communist Europe
nagement and transformation in the Czech Republic
Clark and Anna Soulsby

litics and Society in Poland
nces Millard

perimenting with Democracy
gime change in the Balkans
ffrey Pridham and Tom Gallagher

verty in Transition Economies
ted by Sandra Hutton and Gerry Redmond

rk, Employment and Transition
rian Smith, Al Rainnie and Adam Swain

vironmental Problems of East Central Europe: Second edition
ted by F. W. Carter and David Turnock

Environmental Problems o
East Central Europe

East Central Europe has many attractive landscapes
value. However, the rapid industrial development un
many pollution problems, which impacted negatively
living standards. In this new edition, the progress ma
solve the environmental problems described in the firs
attempts to bring environmental legislation into line
norms are also described.

Environmental Problems of East Central Europe looks
modern farming, water supplies, waste management a
These topics are placed within economic, social and po
ing on a clean environment must be reconciled with
safeguarding of jobs. European Union assistance, civi
environmental NGOs are also discussed.

All of these matters are considered within the context
area, as well as by each individual country. All the form
to the west of the Soviet Union are included, now wit
of Czechoslovakia and Yugoslavia broken up into sev
former Soviet Union is also reviewed, with particular ref

Environmental Problems of East Central Europe pr
date reference material, with a vast amount of suppor
mental conditions and the functioning of civil society, a
The environment is being taken seriously by them al
the Rio sustainability agenda in general and the EU
particular. The book reveals that East Central Europe
in some respects has a higher biodiversity than Wester
is enormous waste and inefficiency in energy use, peo
ively little and the East therefore has some lessons
managing on the basis of a 'fair share' of the earth's

F. W. Carter, who died in May 2001, was formerly F
Social Sciences at the School of Slavonic and East Eu
College London. **David Turnock** is Reader in Geog
Leicester.

Rc

Environmental Problems of East Central Europe

Second edition

Edited by F. W. Carter and David Turnock

London and New York

First published in 1993 in hardback by Routledge
Paperback edition published in 1996

Second edition published 2002
by Routledge
11 New Fetter Lane, London EC4P 4EE

Simultaneously published in the USA and Canada
by Routledge
29 West 35th Street, New York, NY 10001

Routledge is an imprint of the Taylor & Francis Group

Typeset in 9.5/12pt Galliard by Graphicraft Limited, Hong Kong
Printed and bound in Great Britain by MPG Books Ltd, Bodmin

British Library Cataloguing in Publication Data
A catalogue record for this book is available from the British Library

Library of Congress Cataloging in Publication Data
Environmental problems of East Central Europe / edited by F. W. Carter and David Turnock. — 2nd ed.
p. cm.
Originally published: Environmental problems in Eastern Europe / edited by F. W. Carter and David Turnock. 1993.
Includes bibliographical references and index.
1. Pollution—Europe, Eastern—History—20th century. 2. Transboundary pollution—Europe, Eastern—History—20th century. 3. Environmental health—Europe, Eastern—History—20th century. 4. Pollution—Europe, Eastern—Case studies. I. Carter, Francis W. II. Turnock, David. III. Environmental problems in Eastern Europe.

TD186.E583 2001
363.73′2′0947—dc21

2001031755

ISBN 0–415–17403–1

Contents

Illustrations

Figures

Tables

Editors and contributors

Dr Susan Baker, Reader, School of Social Sciences, Cardiff University, Glamorgan Building, King Edward VII Avenue, Cardiff CF10 3WA, Wales, UK, e-mail: BakerSCM@cardiff.ac.uk or drsusanbaker@hotmail.com

Specialising in the theme of environmental politics she has published *Dilemmas of transition* (Cass, 1998), edited with P. Jehlicka; also – with three co-editors – *Sustainable Development: Theory, Policy and Practice Within the European Union* (Routledge, 1997); and 'Environmental policy in East and Central Europe: the role of the EU' in the book edited by P. Glasbergen titled *Environmental Policy in an International Context* (Arnold, 1995). She was previously Associate Professor in Public Administration at the Erasmus University, Rotterdam.

Valentina Brečko MA, Department of Geography, Faculty of Arts, University of Ljubljana, Askerceva 2, 1000 Ljubljana, Slovenia, e-mail: valentina.brecko@ff.uni-lj.si

A native of Celje (Slovenia), she graduated in 1991 at the Geography Department in Ljubljana and undertook postgraduate studies on landscape ecology as a junior researcher during 1993 to 1996. Since 1996 she has been assistant at the chairs for physical geography and environment, researching on the water supply of Ljubljana. She is now engaged in doctoral research on environmental quality and sustainable development for the Kamniška bistrica catchment.

Dr Stefan Bužarovski, School of Geography and the Environment, University of Oxford, Mansfield Road, Oxford OX1 3TB, UK, e-mail: Stefan.Buzarovski@Geography.Oxford.ac.uk

A doctoral student and also Larkinson Scholar at St Hugh's College, he holds an M.Sc degree in environmental change and management (with distinction) from Oxford University. A Macedonian native, he has co-ordinated numerous environmental organisations (including the Federation of the Young European Greens) as well as projects and initiatives, such as the Mavrovo National Park Campaign within Friends of the Earth Macedonia.

† **Dr F. W. Carter**, Reader in Geography, Department of Social Sciences, School of Slavonic and East European Studies (University College London), University of London, Senate House, Malet Street, London WC1E 7HU, UK.

He wrote extensively on East Central Europe over four decades both in contemporary and historical perspective, with later interest focused particularly on energy conservation, environmental degradation, urban development and present-day issues concerning ethnicity and foreign direct investment.

Dr Richard Clarke, Senior Lecturer in Environmental Conservation, Centre for Extra-Mural Studies, Birkbeck, University of London, 26 Russell Square, London WC1B 5DQ, UK, e-mail: r.clarke@cems.bbk.ac.uk

He heads the School of Natural and Social Sciences within the Faculty of Continuing Education at the University of London's Birkbeck College. At the university's Centre for European Protected Area Research, his teaching and research interests focus on protected area management both within the UK and at a European level. He has a particular interest in the relation between environmental policy and sociopolitical transition in the former Yugoslavia. He also teaches on the London University M.Sc in protected area management which is twinned with the M.Sc Natural Heritage protection programme at the University of Ljubljana.

Dr Alan Dingsdale, Faculty of Humanities, Nottingham Trent University, Clifton Lane, Nottingham NG11 8NS, UK, e-mail: alan.dingsdale@ntu.ac.uk

Educated at the Universities of Durham, Wales and Leeds, he completed his Ph.D. in 1975 and began researching on East Central Europe in 1980, with particular interest in urban and environmental issues in Hungary. From 1993 to 1998 he was Secretary of the Royal Geographical Society/Institute of British Geographers' Working Party on Post Socialist Areas. He has recently completed a book entitled *Mapping Modernities: Geographies of Central and Eastern Europe* (Routledge).

Associate Professor Vladimir Drgona, Geography Department, Faculty of Natural Sciences, Constantine the Philosopher University, Tr. A. Hlinku 1, 949 74 Nitra, Slovakia, e-mail: vdrgona@fpv.ukf.sk

His teaching repertoire includes environmental geography, theoretical geography and regional development in East Central Europe. Research is focused on the human impact on agricultural landscape, with particular reference to regional development in border regions; also European integration and transformation of the economy and society in East Central Europe. For ten years he has edited *Geographical Studies, Nitra*. He is President of the Slovak Association of Geography Teachers which is a member of EUROGEO.

Dr Adam Fagin, Lecturer in Politics, Queen Mary and Westfield College, University of London, Mile End Road, London E1 4NS, UK, e-mail: Adam.Fagin@port.ac.uk

He has published several articles on the environmental movement and the development of environmental capacity in the Czech Republic. Recent contributions include 'Sustainable development in the Czech Republic: a doomed concept?' (with Petr Jehlička) – which appeared in the book edited by S. Baker and P. Jehlička titled *Dilemmas of Transition: The Environment, Economic*

Development and Democracy in East and Central Europe (Cass, 1998) – and 'Environmental capacity building in the Czech Republic' in the book edited by M. Janicke and H. Weidner titled *National Environmental Policies: A Comparative Study of Capacity Building* (Springer, 1999).

Professor Derek R. Hall, Department of Leisure and Tourism Management, Scottish Agricultural College, Auchincruive, Ayr KA6 5HW, UK, e-mail: D.Hall@au.sac.ac.uk
He has had a research interest in Albania for thirty years. His publications include *Albania and the Albanians* (Pinter, 1994), *Reconstructing the Balkans* (Wiley, 1996) and *Europe Goes East: European Union Enlargement, Diversity and Uncertainty* (Stationery Office, 2000) – the last two co-edited with Darrick Danta.

Dr Peter Jordan, Osterreichisches Ost- und Sudosteuropa-Institut, Josefsplatz 6, 1010 Vienna, Austria, e-mail: peter.jordan@rs6000.univie.ac.at
A native of Hermagor (Carinthia) he was educated in geography and ethnology at the University of Vienna (D.Phil.) with habilitation at the University of Klagenfurt. Since 1977 he has worked at the Austrian Institute of East and Southeast European Studies in Vienna where he now heads the Geography Department and is also Deputy Director and Editor-in-Chief of the *Atlas of Eastern and Southeastern Europe*, published by the Institute.

Professor Ewelina Kantowicz, International Studies of Environmental Protection, Faculty of Geography and Regional Studies, ul.Krakówskie Przedmieście 30, 00-927 Warsaw, Poland, email: kanto@wgsr.uw.edu.pl
She has extensive research interests in human–environment relations: anthropogenic changes and threats to the environment, protection of the environment (especially in rural areas) and methodological problems. Important publications include 'Trends in regional geography' (in *Miscellanea Geographica*, 1984) and *Arid Zone Resources and Threats to Agriculture*, published in Warsaw in Polish by Wyd. Akademickie Dialog.

Zoya Mateeva, Research Associate, Institute of Geography, Bulgarian Academy of Sciences, Acad. Georgi Bonchev Str. 3, 1113 Sofia, Bulgaria, e-mail: zm@bgcict.acad.bg
As a researcher in physical geography and climatology, her research looks into regional climate change and power generation systems.

Dr Imre Nagy, Senior Researcher and Fellow, Alfold Institute, Szabo Dezso ut. 5601 Bekescsaba, Hungary, e-mail: nagyi@rkk.hu
He researches on the urban environment and environmental problems in border regions. He also lectures on urban ecology in the Department of Environment and Water Management at Tessedik Sandor College.

Mariana Nikolova, Institute of Geography, Bulgarian Academy of Sciences, Acad. Georgi Bonchev Str. 3, 1113 Sofia, Bulgaria, e-mail: mn@bgcict.acad.bg
As a researcher in physical geography and climatology, her current research examines issues of environmental risk management in Bulgaria.

Professor Petr Pavlínek, Department of Geography and Geology, College of Arts and Sciences, University of Nebraska, Omaha, NE 68182 0199, USA, e-mail: petr_pavlinek@unomaha.edu
He has recently published *Environmental Transitions: Transformation and Ecological Defence in Central and Eastern Europe*, with John Pickles (Routledge, 2000) and *Economic Restructuring and Local Environmental Management in the Czech Republic* (Edwin Mellen Press, 1997).

Professor Gyorgy Perczel, Department of Social and Economic Geography, Eötvös Lorand University, Ludovika ter. 2, 1083 Budapest, Hungary, e-mail: perczel@ludens.elte.hu
He holds a doctorate in geographical sciences and has been President of the Hungarian Society of Physical Sciences since 1990. His research interests include social geography and environmental protection. His publications include a series of maps relating to the Hungarian environment which were included in the *National Atlas of Hungary* (Geographical Research Institute, Hungarian Academy of Sciences, 1995)

Professor John Pickles, Geography Department, University of Kentucky, Lexington, KY 40506 0027, USA, e-mail: GEG198@pop.uky.edu
He has conducted research into industrial change, environmental conditions and ethnicity in Bulgaria since the late 1980s. He has edited – with A. Smith – a book titled *Theorising Transition: The Political Economy of Post-communist Transition* (1998) and written (with P. Pavlinek) *Environmental Transitions: Transformation and Ecological Defence in Central and Eastern Europe* (2000) – both published by Routledge.

Professor Dušan Plut, Department of Geography, Faculty of Arts, University of Ljubljana, Askerčeva 2, 61000 Ljubljana, Slovenia, e-mail: dusan.plut@ff.uni-lj.si
A native of Novo Mesto (Slovenia), he gained degrees in geography and history at Ljubljana University. During the 1980s, he cooperated with the Association for Environment Protection and became the Founding President of the Green Party of Slovenia in 1989. The following year he was elected a member of the Presidency of Slovenia and the National Council in 1992. Lecturing in geography at the School of Humanities in Ljubljana, he has participated in more than sixty Slovenian and international symposia and has contributed many papers on landscape ecology and degradation issues. He has also written ten books on environmental matters.

Dr Anton Popov, Faculty of Geology and Geography, 'St Kliment Ohridsky' University, 15 Tsar Osvoditel Blvd, 1504 Sofia, Bulgaria, e-mail: popov@geo.uni.sofia.bg
His current research focuses on the use of GIS for modelling regional environments and environmental management.

Professor Metka Špes, Department of Geography, Faculty of Arts, University of Ljubljana, Askerceva 2, 61000 Ljubljana, Slovenia, e-mail: Metka.Spes@Uni-Lj.si

A native of Celje (Slovenia), she was awarded an M.Sc. degree in 1985 and a Ph.D. ten years later. From 1974 to 1997 she worked at the University of Ljubljana's Geography Institute and since then has been lecturing in the Geography Department. Her research is concerned with the causes and effects of pollution in the most degraded Slovenian landscapes and between 1975 and 1990 she represented the former Yugoslavia on the Comecon Commission on Environmental Protection. She has participated in numerous international conferences and for the past fifteen years has been Editor-in-Chief of the annual publication *Geographica Slovenica*.

Dr Caedmon Staddon, Faculty of the Built Environment, University of the West of England, Frenchay Campus, Coldharbour Lane, Bristol BS16 1QY, UK, e-mail: Caedmon.Staddon@uwe.ac.uk

He has been conducting research into local-level environmental politics focusing on post-communist change in Eastern Europe since the early 1990s. He has just completed a major research project on the ecological problems of East Central Europe in the light of restructuring and privatisation.

Professor Aleksandar Stojmilov, Institute of Geography, Faculty of Natural Sciences and Mathematics, University of Ss. Cyril and Methodius, PO Box 162, 91 000 Skopje, FYR Macedonia

Dean of the Faculty and Vice Rector of the University of Skopje, he holds a Master's degree from the University of Belgrade and a Ph.D. from the University of Skopje. He has published more than 170 scientific papers, seven books and numerous maps. His areas of specialisation include the regional geography of Macedonia, tourism, spatial planning and environmental protection.

Dr Andrew Tickle, Lecturer in Environmental Policy and Management, Centre for Extra-Mural Studies, Birkbeck, University of London, 26 Russell Square, London WC1B 5DQ, UK, e-mail: a.tickle@cems.bbk.ac.uk

He has written extensively on environmental issues in the region (including *Environment and Society in Eastern Europe* published by Longman in 1998, with Ian Welsh as co-editor) and is researching transition impacts on nature and landscape conservation in the Czech Republic and Slovenia. In 1988 to 1989 he was a British Council visiting researcher in the Czechoslovak Academy of Sciences and went on to co-found Greenpeace Czechoslovakia in 1992.

Dr David Turnock, Geography Department, The University, Leicester LE1 7RH, UK, e-mail: dt8@le.ac.uk

He has been teaching and researching on the rural and regional development of East Central Europe for the past thirty-five years. He has written extensively, particularly on Romania, and set up the project on 'Environmental Problems' with Frank Carter in the late 1980s. He is currently cooperating with the World Wide Fund for Nature over its initiative for a Carpathian Ecoregion.

Dr Stefan Velev, Senior Researcher, Institute of Geography, Bulgarian Academy of Sciences, Acad. Georgi Bonchev Str. 3, 1113 Sofia, Bulgaria, e-mail: sv@bgcict.acad.bg

His research on physical geography focuses on issues related to regional climatology and the hydrology of karst landscapes in Bulgaria.

Dr Geoff A. Wilson, Department of Geography, King's College, Strand, London WC2R 2LS, UK, e-mail: geoff.wilson@kcl.ac.uk

He lectures on environmental management and agri-environmental policy. As an expert on the environmental attitudes of farmers and agri-environmental policy in Europe and New Zealand, he has published widely on these topics. He has also published books on environmental management and German agricultural policy.

Dr Olivia J. Wilson, Geography Department, De Montfort University, 37 Lansdowne Road, Bedford MK40 2BZ, UK, e mail: ojwilson@dmu.ac.uk

She lectures on rural policy and the geography of the European Union. She has published a number of articles on rural policy and rural restructuring in Britain, New Zealand and the former East Germany. She has recently published a book on German agricultural policy.

Acknowledgement

Thanks to Phil Taylor of the Faculty of the Built Environment, University of the West of England, Bristol for producing the set of basic location maps used in this volume.

Abbreviations

Measurements

bln	billion
cu.m	cubic metre
dBA	decibel
DM	Deutschmark
E	Euro
Ft	Forint (Hungary)
ha	hectare
km	kilometre
kwh	kilowatt hour
l	litre
m	metre
mg	milligram
min	minute
mld	milliard
mln	million
MW	megawatt
na	not available
pc	per capita
ptp	per thousand of population
sec	second
sq.m	square metre
sq.km	square kilometre
t	tonne
th	thousand
US$	United States Dollar
Zl	Ztoty (Poland)

Other abbreviations

AC–IMPEL	Associated Countries for Implementation of Environmental Legislation
ASEA	Asea-Brown Boveri

BiH	Bosnia and Hercegovina
BSP	Bulgarian Socialist Party
CAP	Common Agricultural Policy
CAPE	Coordinated Action for Pan-European Transport and Environmental Telematics Implementation Support
CBD	Central Business District
CCI	Chamber of Commerce and Industry
CEFTA	Central European Free Trade Association
CEP	Committee of Environmental Protection
CEPF	Central Environmental Protection Fund
CEZ	Czech Energy Works company
CFCs	Chlorofluorocarbons
CFL	Compact fluorescent lamps
C&GMOs	Chemicals and genetically modified organisms
CIS	Confederation of Independent States
CITES	Convention on International Trade in Endangered Species (1973)
CoAP	Concentration of air pollutants
CoE	Council of Europe
Comecon	Council for Mutual Economic Assistance
CPEs	Centrally planned economies
CSN	Community Support Network
CUAEPP	Committee on the Use of Atomic Energy for Peaceful Purposes
DDBR	Danube Delta Biosphere Reserve
DRPC	Danube River Protection Convention
EAP	Environmental Action Programme
EAS	Environmental accession strategy
EBRD	European Bank for Reconstruction and Development
ECE	East Central Europe
ECECs	East Central European countries
EEA	European Environment Agency
EEB	European Environmental Bureau
EIA	Environmental impact assessment
EIB	European Investment Bank
EIU	Economist Intelligence Unit
EMBO	Employee–management buy-out
ENGO	Environmental non-governmental organisation
EPA	Environmental Protection Act (and Agencies)
EPCE	Environmental Partnership for Central Europe
ERBC	Ecoregion-based conservation
ESC	Environmental Steering Committee
ESA	Environmentally sensitive area
EU	European Union
EUROGEO	European Networks of Geography Teachers' Associations
FAO	Food and Agriculture Organisation

FBiH	Federation of Bosnia and Hercegovina (within the confederation of Bosnia and Hercegovina)
FCS	Former Czechoslovakia
FDI	Foreign direct investment
FFRG	Former Federal Republic of Germany (West Germany)
FFRY	Former Federative Republic of Yugoslavia
FGDR	Former German Democratic Republic (East Germany)
FoEE	Friends of the Earth Europe
FoE-I	Friends of the Earth International
FSU	Former Soviet Union
GARC	General Assembly of Rightful Claimants
GDP	Gross domestic product
GEF	Global environmental facility
GIS	Geographical Information Systems
GNP	Gross national product
GOP	Gornoslaskie Okreg Pzemyslowy (Upper Silesian Industrial Region)
GRW	Generated radioactive waste
HCSD	Hungarian Commission on Sustainable Development
IAEA	International Atomic Energy Agency
ICPDR	International Commission for the Protection of the Danube River
IMF	International Monetary Fund
IPPC	Integrated pollution prevention and control
ISPA	Instrument for Structural Policies for Pre-Accession Aid
ISUAE	Inspectorate on the Safe Use of Atomic Energy
IUCN	World Conservation Union
JICA	Japan's International Cooperation Agency
LEAP	Local Environmental Action Plan
LUD	Lithuanian Women Farmers Association
MAP	Mediterranean Actional Plan
MERP	Ministry of Environment and Regional Planning
Minatom	Ministry for Atomic Energy
MLDBs	multilateral development banks
MoE	Ministry of Environment
MUPCEP	Ministry for Urban Planning, Construction and Environmental Protection
MWF&EP	Ministry of Waters, Forests and Environmental Protection
NATO	North Atlantic Treaty Organisation
NEAP	National Environmental Action Plan
NEM	New economic mechanism (Hungary)
NEPP	National Environmental Protection Programme
NEPPL	National Environmental Policy Plan for Latvia
NGO	Non-governmental organisation
NNR	National nature reserves

NO_x	Nitrogen oxide
NPP	Nuclear power plant
NUTS	Nomenclature of territorial units for statistics
ODS	Ozone-depleting substances
OECD	Organisation for Economic Cooperation and Development
PAN	Protected area network
PHARE	Poland and Hungary: Action for the Restructuring of the Economy
PLA	Protected landscape area
PLP	Peipsi Lake Project
PPP	Polluter pays principle
PRA	Participatory rural appraisal
PRDU	Post-War Reconstruction and Development Unit
PWR	Pressurised water reactor
RAW	Radioactive waste
REA	Regional Environmental Agency
REC	Regional Environment Centre (Szentendre)
RERP	Regional Environmental Reconstruction Programme
RHMZ	Macedonian Hydro-Meteorological Institute
RS	Republika Srpska (i.e. Serb Republic within the confederation of Bosnia and Hercegovina)
SAPARD	Special Action for Pre-Accession Measures for Agriculture and Rural Development
SFRJ	Socialist Federal Republic of Yugoslavia
SME	Small- and medium-sized enterprise
SOE	State-owned enterprise
SOF	State Ownership Fund
SPM	Suspended particulate matter
SPP	Forestry Economic Area Administration
SPSEE	Stability Pact for Southeastern Europe
TEM	Trans-Europe motorway
TETN	Trans-European Transport Network
TINA	Transport Infrastructure Needs Assessment
TNC	Trans-national corporation
TP	Thermal plants
TPS	Thermal power-stations
UN	United Nations
UN-ECE	United Nations Economic Communications for Europe
UNEP	UN Environment Programme
UNHCS	UN Centre for Human Settlements
USAID	US Agency for International Development
VOC	Volatile organic compounds
VVER	Vodo Vodyanoy Energeticheskiy Reaktor (type of Soviet reactor)
WHSs	World Heritage Sites
WWF	Worldwide Fund for Nature

Part I

Context

1 Introduction

David Turnock and F. W. Carter

The region of East Central Europe (ECE) – like all others – is prone to natural hazards including droughts, floods and landslides. There were devastating floods in the Oder valley in 1998, while the severe drought during 2000 caused heavy agricultural losses in the southern countries as well as forest fires that were particularly severe in Croatia. However, East Central Europe has also experienced major environmental dislocations through the process of economic development, and during the 1980s these amounted to a crisis which had a significant bearing on the chain of revolutions paving the way for the current transition. Although the literature gave considerable coverage to these matters at the time (especially the Chernobyl nuclear accident in the Soviet Union), no volume dealt systematically with environmental issues in each of the countries subject to a central planning system in the hands of totalitarian government.

With the encouragement of Routledge, we set about finding authors to write on the eight countries lying to the west of the USSR in order to highlight the pollution problems and consider the actions being taken to improve the situation. In the meantime the communist world broke up, and when the first edition of the book appeared in 1993 the original terms of reference had been modified to cover the prospects for the transition period – and with the unification of Germany invoked to justify the omission of the former GDR for which we had been unable to secure coverage. However, the first edition was heavily preoccupied with communist pollution blackspots arising from emissions (particularly sulphur dioxide), and discharges of raw sewage and untreated process water which were injurious to public health and biodiversity and also counter-productive through damage to the stock of physical resources (e.g. agricultural production and forest growth) and the health of the workforce.

These lessons have been slowly learned and the democratic political process, driven by civil society situated on a high learning curve, has combined effective action in reducing emissions and discharges while seeking to limit damage to employment beyond the inevitable consequences of restructuring. Meanwhile, the agenda has moved on to consider the impact of pollution on human health and biodiversity loss. We now acknowledge the need to situate the search for solutions to environmental problems more centrally in the political life of each state where policies are prioritised and resourced alongside the competing demands for job creation and economic growth. Hence the justification for a new edition which will give more attention to the ways in which environmental issues are impacting on society in the context of the post-communist reform agenda. We hope that this new edition will contribute to this broader view of what one of the contributors has referred to as the 'environment–society–economy nexus' whereby the environmental dimension is integrated into the political process which must determine priorities in line with economic and social objectives (Tickle and Welsh 1998, p.163). Since we are no longer hidebound by a monolithic communist system, the environmental issues must be discussed in the context of domestic politics, with each society attempting its own trade-offs between employment and environmental improvement, albeit in the context of global pressures (Connelly and Smith 1999).

We must also consider where environmentalism stands in the transition process now that the worst of the inherited pollution problems have been addressed. Instead of the 'hot spots' of the communist era, the focus of interest should now rest with the level of compliance with European environmental norms dedicated to sustainability and the protection of biodiversity. The EU is a major source of funding while substantially influencing the agenda because of the near-universal interest in eastern enlargement and the consequent obligation for the transition states to assimilate the whole of the '*acquis communautaire*'. This throws up a wide range of issues concerned with noise, nuclear safety and waste management which greatly widens the earlier focus of the book on air and water pollution, but while highlighting the importance of the EU we also include a contextual chapter on the former Soviet Union (FSU). Although the focus remains firmly on the former communist states outside the FSU – comprising a region that we now label 'East Central Europe' – we feel that there should be some coverage of events in a state which wielded the greatest influence over the region during the communist period and remains of great importance for the conservation of the planet. Indeed, three FSU states (Estonia, Latvia and Lithuania) are now widely regarded as integral parts of ECE with highly realistic ambitions of EU membership. We therefore acknowledge that there is a case for additional country studies instead of the compromise whereby the FSU chapter is biased quite strongly towards these Baltic States.

Despite delays arising from instability in the former Yugoslavia, we have now been able to achieve coverage of all the states in the region. We have catered for the breakup of the federations of Czechoslovakia and Yugoslavia – an adjustment which unfortunately was not possible when a partly revised edition was published in 1996 – and we have also rectified the omission of East Germany from the first edition. However, the development of the book has created problems which we cannot easily solve. The breakup of the above-mentioned two federations leaves us with thirteen countries (Figure 1.1). If we are to profile each of these states in depth and include thematic chapters, the book becomes unmanageably lengthy. Therefore the individual country studies are dealt with rather more concisely than was the case in the first edition. This may be justified on the grounds that the latter still provide the 'baseline' treatment of the communist period

Figure 1.1 The states of East Central Europe

whereas this new edition concentrates on aspects of the transition (Figure 1.1). We remain highly appreciative of the support from Routledge, and acknowledge the help of Ann Michael and Andrew Mould in handling this greatly enlarged work, following the arrangement of the earlier editions with Tristan Palmer and Sarah Lloyd.

A question for the country chapters is whether we go for a single list of thirteen territories, or use northern and southern groupings. We prefer the latter strategy on the grounds that, despite significant contrasts within each of the groups, there are factors which make for a degree of coherence. The northern group (particularly the Czech Republic, East Germany and Poland) is excessively scarred by major energy sectors, particularly evident through large coal-mines (both shaft mines and open pit workings) and generating capacities.

Although there is considerable dependence on imports from the Middle East, North Africa, Russia and Scandinavia (which constitutes a heavy economic burden), consumption of excessive quantities of fuel in power-stations carries an additional environmental cost (Green 1998). At the same time, these countries also have a superior 'capacity' for handling their problems, reflecting their ability to identify and resolve environmental problems on the basis of technology and financial resources but also legislation and enforcement regimes plus a political culture with institutions to ensure that issues are raised, alternative solutions discussed and with cooperation forthcoming in respect of agreed strategies (OECD 1994). They have significant capacities for the production of their own pollution control equipment and they are becoming highly integrated into the EU: East Germany is already within the Union, while all the others (apart from Slovakia) are negotiating for early entry, nothwithstanding some ambivalence on the part of public opinion.

The southern countries have inherited some serious pollution problems but these are relatively localised. More significantly perhaps, they have suffered from war in former Yugoslavia, and have substantial biodiversity resources which they may be ill equipped to protect using their own resources. The scale of deindustrialisation has been such as to force farmers to take more production from the land than may be sustainable, while future development pressures call for regimes of protection which are not yet in place. The Carpathians are very valuable for biodiversity, while major responsibilities in lowland areas include the wetlands of Albania which are precious assets for the future development of tourism. Civil society has not developed to the same extent as in the north where the urban population is relatively high. Awareness of the biodiversity resources is more limited and there are relatively few environmental non-governmental organisations (ENGOs) to act as pressure groups in political life and provide the voluntary labour for local conservation efforts. We hope these broad contrasts will emerge from the national profiles, although individual authors have been left to strike a balance between environmental detail and the subtleties of political debate. We also hope that the book will continue to be a useful work of reference while retaining value as a text for conventional reading.

Environmental damage under communism

The damage inflicted during the communist period should not be exaggerated, for pollution was highly localised in many areas and attempts were made to protect important landscapes and buildings. This was encouraged to some extent by the vested interests of the elite in providing high amenity areas for their own enjoyment, including select hunting grounds and parks, such as the private reserve, stocked with elephants and other exotic species, on the northern Croatian island of Brijuni for the use of the Yugoslav leader Tito and his entourage. However, substantial environmental problems have been inherited from the communist era when 'smokestack' technology, used mainly in the interest of the FSU's military-industrial establishment, gave insufficient attention to each country's own resources and circumstances. At a time when the West was moving into services and high-tech manufacturing, the planned economies of ECE stuck tightly to classical heavy industry and condoned the inefficient use of both raw materials and energy. Production plan targets were set in a 'war economy' context with investment deliberately unbalanced to realise key strategic objectives. Assets were nationalised and state-owned enterprises had free use of natural resources. The system rewarded value added rather than sustainable management of the environment and diverted attention away from the negative consequences of such an approach by insisting that environmental damage could occur only in capitalist countries (Kramer 1983).

Despite some significant gestures, insufficient attention was paid to pollution control and conservation (Singleton 1987). Oil and gas extraction rates were low and there was little recycling of waste. Energy consumption was high in relation to the volume of industrial production and, given the inadequacy of filtering systems, air pollution levels were very high in the vicinity of thermal power-stations, especially those burning lignite. Meanwhile, hydro-electric projects involved heavy resettlement costs and losses of agricultural land (some 21 million ha in the FSU alone) as well as reduced fish catches. Major environmental dislocations had serious consequences in terms of damaged ecosystems, economic inefficiency and a reduced quality of life, yet monopoly party systems combined with central planning meant 'little sensitivity to, or tactical positioning on, environmental policy issues' (O'Toole and Hanf 1998, p.97). Informed public debate and development of effective ENGOs was not encouraged and in most countries would hardly have been tolerated anyway, given the weakness of civil society. Even in Hungary there were clear limitations on political expression. However, 'the party was over' by the late 1970s and environmental problems were given greater consideration during the following decade as the production losses were calculated in terms of lost working time and wasted natural resources.

Writing about the former Czechoslovakia (FCS), A. Tickle and J. Vavroušek (1998) remind us of the long history of international relations on environmental matters, with bilateral agreements with Austria (1967) and Germany (Bavaria) in 1970, before the final accord of the Helsinki Conference on Security and Cooperation in Europe in 1975 provided a channel for environmental diplomacy. There was also some scope for action through the United Nations Economic Commission for Europe (UN-ECE) which was responsible for a Convention on Long-Range Transboundary Pollution. Yet the former socialist states were heavily constrained. Despite a rational energy policy (including waste recycling; small, slow cars; a rationally organised electricity supply system), East Germany could not square its commitment to deliver Western living standards with a clean environment once the second oil crisis (1978) made the 1980s a struggle for survival. When the FSU reduced oil exports in the early 1980s greater dependence on brown coal was necessary: the annual output of lignite reached 330mln.t and was destined to reach 450mln (West German figures say 660mln). 'Brown coal had to be mined for survival without attention to environmental impact' (Böhmer-Christiansen 1998, p.93), while fuels derived from brown coal (briquettes and coal powder for pulverised bed combustion) were exported to the West in growing amounts to earn hard currency and buy more political stability. Despite nuclear expansion, energy remained 'dirty', and the East was one of the worst transfrontier polluters at a time when West Germany was cleaning up in response to the problem of 'Waldsterben' or forest death.

Although protest was generally restrained, there were some signs of a developing opposition in Poland in the 1970s when the Nova Huta steel plant antagonised the local intelligentsia of Kraków into the spontaneous creation of a Polish Ecological Club ('Polski Klub Ekologiczny') during the first Solidarity era (1980), when a successful campaign was launched to close the Skawina metallurgical plant notorious for its fluoride emissions. Even more momentous was the undermining effect of the Chernobyl nuclear accident. When this incident blew the lid off decades of complacency, a dumbfounded public awoke to confront an apocalypse of quite staggering proportions. Tough anti-pollution laws existed only on paper because enforcement was ineffective as long as 'regulator and regulated tended to be one and the same thing' (Welsh 1997, p.20). The scale of protest was such that Western ENGOs were able to move in and add momentum to demands for more information which served to undermine the legitimacy of communist regimes (Jancar-Webster 1991; Waller 1989). Complaints about various forms of transboundary pollution made particularly good progress and helped gain acceptance for some ENGOs because national governments were not being criticised. Some governments even embraced environmental issues in order to head off more damaging critiques of 'quality of life' issues. Thus in 1989 when communism collapsed there were some organisations ready to step in: 'submerged environmental movements became highly visible political movements in incredibly short time spans, precipitating trusted environmental activists into positions of political leadership' (Welsh and Tickle 1998, p.12).

The challenge of transition

In theory a market economy should safeguard the environment through the value placed on resources. Industrial restructuring (with a switch from heavy to light manufactures) should have a beneficial effect with market forces discouraging waste in all branches. The record is by no means insignificant, as the individual country chapters will show; and, if a summary view is taken with regard to carbon emissions (Table 1.1), the absolute and per capita reductions are quite impressive. Indeed Eastern Europe's per capita figure of 7.0t in 1995 (down from 10.3 in 1980) compares favourably with 8.1 in the EU; while 8.2 in the CIS is a substantial improvement on 12.2 in 1980. The transition states therefore made a significant contribution at the Kyoto Summit in 1997 by agreeing to reduce their emissions to levels significantly below those of 1990 (a theme we return to in Chapter 19, the Conclusion).

Table 1.1 Carbon dioxide emissions 1980–95

	Total mln.t				*t/pc*				*carbon dioxide (th.t) per EmlnGDP*			
	1980	*1985*	*1990*	*1995*	*1980*	*1985*	*1990*	*1995*	*1980*	*1985*	*1990*	*1995*
Eastern Europe	978	980	831	693	10.3	10.0	8.4	7.0	6.9	6.3	5.1	4.7
Bulgaria	82	80	66	53	9.2	8.9	7.6	6.3	n.a.	n.a.	n.a.	n.a.
Czech Republic	172	177	146	111	16.8	17.1	14.1	10.8	n.a.	n.a.	n.a.	n.a.
Hungary	79	78	69	57	7.4	7.4	6.6	5.5	n.a.	n.a.	n.a.	n.a.
Poland	410	413	332	322	11.5	11.1	8.7	8.3	n.a.	n.a.	n.a.	n.a.
Romania	178	176	162	111	8.0	7.7	7.0	4.9	n.a.	n.a.	n.a.	n.a.
Slovakia	50	49	50	37	10.1	9.4	9.5	6.9	n.a.	n.a.	n.a.	n.a.
Slovenia	12	13	12	12	6.8	6.6	6.1	6.1	n.a.	n.a.	n.a.	n.a.
Baltics	n.a.	n.a.	76	43	n.a.	n.a.	9.5	5.5	n.a.	n.a.	n.a.	n.a.
CIS	3,253	3,382	3,539	2,408	12.2	12.2	12.2	8.2	5.7	5.1	4.9	5.7
Russia	n.a.	n.a.	2,416	1,497	n.a.	n.a.	16.3	10.1	n.a.	n.a.	n.a.	n.a.
Ukraine	n.a.	n.a.	721	420	n.a.	n.a.	13.9	8.1	n.a.	n.a.	n.a.	n.a.
European Union	3,003	3,046	3,086	3,045	8.4	8.4	8.5	8.2	n.a.	0.7	0.6	0.5
World	17,778	18,444	20,332	21,209	4.0	3.8	3.9	3.8	1.4	1.3	1.2	1.1

Source: European Commission 1999.

However, these improvements arise from deindustrialisation, and when emissions are related to GDP then 4.7th.t of carbon dioxide per Emln of GDP in 1995, while an improvement on 6.9 in 1980 must be compared with 0.5 in the EU. This shows that there is still a long way to go, as the discussion on energy efficiency will also emphasise. The CIS is in a worse position because, while there was an absolute decline in emissions between 1990 and 1995, this has not kept up with the fall in GDP so that there were 5.7th.t of carbon dioxide emissions in 1995 per Emln of GDP compared with 4.9 in 1990. Table 1.2 presents Eurostat data to show trends with regard to other emissions (sulphur dioxide and nitrogen oxides) and water supply between 1993 and 1997. The trend is clearly downwards, though there are substantial variations between countries – regarding both absolute per capita levels and the rate of decline – and the figures are not related to GDP. The low percentages registered by most countries for connection to waste water treatment plant are a matter of great concern because of the quantities of polluted water sinking into the ground (since septic tanks are not yet in widespread use). This problem will remain as long as sewage facilities lag behind the installation of piped water. However, there has been some benefit from reduced intensification in agriculture with less use of artificial fertilisers and, given the smaller average farm size, greater applications of pesticides by tractor rather than by helicopter.

Of course there are good reasons for the slow rate of progress under the market economy. There are high costs involved in initial clean-ups and governments have invested quite heavily to achieve the required standards in such cases as the Volkswagen purchase of Škoda in the Czech Republic and General Electric's takeover of the Hungarian company Tungsram. Moreover, the regulation system is not sophisticated enough to prevent damage through short-term considerations linked with immediate survival. Incentives may well be on offer to keep marginal enterprises in business (O'Toole and Hanf 1998, p.110) and foreign investments carrying environmental risks may be accepted in the interest of job creation. Meanwhile, peasants thrown back on the produce of their restituted small farms may wish to cut down trees and overgraze pastures for short-term gain. Thus, deindustrialisation and high unemployment 'have tended to overwhelm environmental signals in the short term' (Welch 1997, p.21). Baker and Jehlička (1998, p.5) point out how a variety of tensions and conflicts between political and economic change may have the effect of selling the environment in return for economic survival (Nefedova 1994). There tends to be resistance to radical environmental policies (based on greater energy efficiency) once the worst blackspots have been dealt with, especially in the southeast of the region where reform has been particularly burdensome in terms of poverty and unemployment.

There is certainly the potential under a democratic system for better environmental law (including judicial review procedures), with civil society in the driving seat and former state-owned industries privatised to produce a clearer division between regulators and the regulated. But strengthened institutional capacity for the framing of legislation and its enforcement still comes up against

Table 1.2 Emissions and water supply 1993–97: Absolute (th.t) and Per capita (kg)[a]

	Sulphur dioxide				*Nitrogen oxides*				*Fresh ground and surface water abstracted*				
	Absolute		*Per cap*		*Absolute*		*Per cap*		*Absolute*		*Per cap*		*WWT*[b]
	1993	*1997*	*1993*	*1997*	*1993*	*1997*	*1993*	*1997*	*1993*	*1997*	*1993*	*1997*	*1997*
Bulgaria	1,426	1,420[d]	168	170[d]	242	259[d]	29	31[d]	3,764	3,089	444	372	36
Czech Republic	1,419	700	137	68	574	410	56	40	n.a.	2,373	n.a.	230	59
Estonia	154	119	101	82	38	45	25	31	2,006	1,628	1,323	1,017	72
Hungary	757	671[d]	74	66[d]	184	189[d]	18	19[d]	6,638	6,011[d]	645	590[d]	22[d]
Latvia	73	59[d]	28	24[d]	46	35[d]	18	14[d]	726	530	281	215	n.a.
Lithuania	125	77	34	21	78	57	21	15	4,388	4,786	1,176	1,291	52
Poland	2,725	2,368[d]	71	61[d]	1,120	1,154[d]	29	30[d]	12,277	11,799	319	305	50
Romania	928	898	41	40	348	330	15	15	10,180	9,260	447	411	51[c]
Slovakia	321	202	..	37	185	123	35	23	n.a.	1,311	n.a.	244	54
Slovenia	183	120	92	60	61	71	31	36	385	457[d]	194	220[d]	30

Source: Eurostat 1999, pp.161–4.

Notes

a mln.cu.m and cu.m for water.

b Percentage of the population connected to waste water treatment 1997.

c 1994.

d 1996.

a shortage of expertise and experience when it comes to coordinating one piece of legislation with another. Further, given the high costs involved, it is not easy to find sponsors, and donors make only a limited contribution (Kolk and van der Weij 1998), despite Western aid and technology 'substituting itself to some extent for grass-roots activism and weakening environmental groups in relation to their governments' (Waller 1998, p.51). The enhanced role of emerging local governments provides a community dimension – less centralisation and more local negotiation – which should produce a more effective system of regulation, but it involves a culture change at a time when local governments tend to be overburdened and under-financed. Dilemmas arise between supporting new business and controlling pollution, raising the possibility of friction between local authorities and regional environmental inspectorates.

The new democracies have to deliver environmental gains along with economic recovery (after the initial slump of the early 1990s), while enhancing democratic freedoms and citizenship rights for more differentiated electorates including substantial underclass elements (Gelb and Gray 1991; Jancar-Webster 1995). Thus one school of thought may respond to lack of resources by going for growth, with the intention of investing the proceeds to improve technology and achieve sustainability in the future, while another might insist on a more immediate environmental focus – especially in such central areas as agriculture, energy and transport (Russell 1990; Tellegen 1996). For the region has a relatively intact natural environment and biodiversity, and 'while accession will contribute to solving technical problems of the environment as a whole, the marginalisation of nature conservation can be predicted' (Gyulai 1998, p.53). With many of the recent environmental gains secured through falling output, it is questionable how the present regulation system can deal with the pressure when growth resumes, with prevention being better than 'end-of-the-pipe' solutions. Thus 'the present time is ideal since the economic transformations now create opportunities to undertake the conversion before major new investments are sunk in place' (O'Toole and Hanf 1998, p.101).

Governments and stakeholders must appreciate 'the barriers to economic development which will result from continuing poor environmental management practices' (Waller 1998, p.51). It requires education and awareness to induce more sustainable actions and acceptance of the possibility of higher costs for environmentally friendly choices on the basis of self-interest among individuals and communities. However, the dilemma will be moderated if 'win-win' strategies can be found that are good for both the environment and the economy. This pragmatic approach, accepting economic constraints to environmental protection, was advocated as the basis of an Environmental Action Programme for the region presented at the Environment Ministers' Conference in Lucerne in 1993. It is to be expected that there will be local apologists in the face of global movements, a situation exemplified by the swift reversal in stance over nuclear power-stations in the FSU from rejection as 'imperial domination' to acceptance in the interest of 'self-sufficiency'. While some national ENGOs may still support broader 'global' perspectives, it is less easy to criticise the actions of national governments than those of the Soviet predecessors. However, it is still necessary to heighten awareness which is what the Szentendre-based Regional Environment Centre (REC) for Central and Eastern Europe is doing in recognising local authorities who achieve good results in stimulating public involvement in environmental decision-making as a basis for the dissemination of 'best practice'.

The Czech Republic has pushed harder than most transition states to safeguard the environment. Before 1989, FCS had no central body responsible for environmental problems but a Commission was established in both parts of the country in 1990. The government then took an initiative in 1991 by convening a Pan-European Conference of Environmental Ministers held at Dobřiš, near Prague (followed up at Lucerne in 1993, Sofia in 1995 and Århus in 1998) to discuss environmental ethics for sustainable development (Baumgartl 1993). Whereas key positions in environmental affairs are often held by technocrats with little political standing, substantial clout was eventually mustered in the Czech Republic with prominent figures in charge at both the Commission and the Ministry. There has been a complete redrafting of legal codes, and environmental impact assessment is now very advanced. It seems that a desire for integration within Europe has made for harmony over environmental policy. Since industrialists supported integration, EU emission standards were accepted even though they were burdensome in terms of the 'polluter pays principle' (PPP). Indeed, the Environment Ministry has continued to be proactive. The last Czech uranium mine has closed despite pleas that it should be retained (and another reopened) to guarantee fuel for the country's two nuclear power-stations. In 1999 the gold prospecting licence of TVX Bohemia permitting operations in the Kašperské Hory (South Bohemia) was cancelled because of violations near the Šumava National Park. The Ministry also wants to tax traditional energy suppliers in order to build up a fund to develop alternative energy sources – and it has criticised the government's recently approved (but controversial) plan for the future of Prague's infrastructure which could threaten the unique atmosphere of the city centre.

However, throughout the region progress can be seen in the beginnings of an effective system of central organisations (ministries and agencies), responsible for legislation and the establishment of monitoring and enforcement systems; and local responsibilities over water, sewage and waste management supported by income from taxes to finance environmental projects (Klarer 1994). Fiscal measures are being adopted to stimulate good practice: green taxes according to levels of pollution and trading in pollution rights with tax reliefs and subsidies to encourage further anti-pollution investments. However, the PPP is fundamental, and in various ways this generates some (though by no means all) the resources for national and regional ecological funds (though municipalities do not always reserve payments received under national environmental fund disbursements for ecological projects). Considerable progress has been made in cleaning up the blackspots, but it takes time and the situation varies between countries: emissions from the lead-zinc smelter and fourteen other industries contribute to a high death rate in Veles – Macedonia's most polluted city – where ENGOs are now promoting a Local Environmental Action Plan with Japanese support. Sewage capacity is being increased in the capital cities of Budapest and Warsaw which lacked proper facilities before 1989 and discharged effluent into the Danube and Vistula rivers respectively: up to one-fifth in raw form and the rest only partially treated. Rivers are now cleaner, for salmon have been reintroduced into three tributaries of the Elbe in the northwestern part of the Czech Republic after a gap of nearly fifty years. Yet in Poland, progress on the Vistula is slow, and both salt and sulphur cause problems in the Oswiecim/ Kraków immediately downstream of Upper Silesia.

A national water shortage in Bulgaria provides an incentive for tougher pollution controls (although the linkage of drainage basins by aqueducts is opposed by some conservationists). Yet contaminated water has caused deaths (through cholera) in Albania, while Latvia has recorded cases of hepatitis A and bacterial dysentery: problems which are being addressed by UN-ECE and the World Health Organisation Regional Office for Europe. Water tainted with radium and uranium has leaked from Byalata Voda uranium mine at Dolna Banya (in the Sofia region of Bulgaria), threatening the Aegean Sea because pollution is carried through the Ochusnitsa to the Maritsa. The Turnu Măgurele chemical plant was forced to shut down for repairs in late 2000 after ammonia levels reached twenty times the permitted level in Nikopol. In addition, what is being regarded as the worst environmental crisis since Chernobyl has affected fish in the Tisza on account of cyanide leaks from the 'Aurul' plant in Baia Mare (a joint Australian–Romanian gold-mining company) when the wall of a lagoon was breached during heavy rain. Recorded cyanide levels in the Tisza reached 12.4mg/l compared with the maximum permitted which is 0.1. There is much concern in Baia Mare where locals feel that their water supplies have been polluted for some time. It mirrors sensitivity over the hazards posed by other industrial installations, such as the leather processing plant started in Durrës in 1990 which was closed by health authorities and, after reopening during the 1993–8 period was accused of poisoning the local population who threatened violence.

There is increasing synergy between national efforts, given the Pan-European Conferences of Environment Ministers already referred to (Baker 1996). There is also ongoing research by NATO into such matters as catalytic and electrochemical processes for sulphur dioxide and NO_x emission abatement (including Romania and Russia); Black Sea ecosystem processes and forecasting database management systems (with Bulgaria, Romania and several former Soviet states); and reclamation of former military lands at Amar (Estonia), Borne-Sulinowo (Poland), Klaipeda (Lithuania), Liepaja (Latvia) and Ralsko (Czech Republic) which furnish material for a reference book. In 1999 the World Bank approved a loan to upgrade two waste-water plants and extend the sewer system in Budapest and to build a secondary waste-water treatment plant in Dunaújváros. The projects have been funded by a US$31.6mln loan and will help Hungary's efforts towards EU accession by supporting a least-cost compliance strategy phasing in environmental investments by local governments. Another World Bank loan of DM65mln for Croatia secured water and sewerage systems for Split, Solin, Kastela and Trogir (1998) for the benefit of water in the Adriatic. Cooperation between neighbouring states can be seen in the case of Albania and Macedonia with respect to a US$4.1mln grant for a conservation project to protect Lake Ohrid's unique plants and animals. The scheme will also involve local ENGOs coordinated by REC. In addition, the two confederal entities within Bosnia and Hercegovina are working together on environmental projects.

Given that forests suffer contamination both from domestic and transboundary pollution, it is very appropriate that there should be cooperation within Euroregions (e.g. a German–Polish–Czech Cooperation Programme for the Black Triangle was initiated in 1996 with funding by Deutsche Bundesstiftung Umwelt and the Czech and Polish offices of the Environmental Partnership for Central Europe (EPCE)). Projects in 1997 covered energy (energy savings in schools and a feasibility study for renewable energy in the Žlutice area), waste management and recycling (including a waste management study and the collection of specific wastes in Liberec) and the development of ecotourism (the Moldava–Fojtovice footpath and cycle way to link with other paths in Germany); and the foundation for ecotourism in the Krkonoše to complement the existing organisation in Poland. Wider regional programmes include the Danube River Protection Convention which came into force in 1998 after discussions dating back to 1991: this will clean and preserve the river through a regionally focused environmental strategy. There is a Danube Environmental Forum coordinating thirty-seven ENGOs to improve/preserve Danube ecology. There will also be ENGO participation in the International Commission for the Protection of the Danube River; the Danube Programme Management Task Force; and the Danube Pollution Reduction Programme (1998) which is based on activities by some fifty ENGOs and stakeholders coordinated by the REC. The Baltic Sea Comprehensive Environmental Action Programme (1995) is achieving considerable improvement in sewage treatment and there are similar arrangements for the Black Sea.

The region is certainly not alone. As early as April 1990, EU environment ministers met in Dublin to put together a comprehensive aid package to restore the ravaged environment of East Central Europe. Western expertise has played a major role over the past decade, given the shortage of trained environmental scientists in the region. Finally, disputes are being resolved, notably through the International Court ruling that the 1977 treaty between Hungary and FCS regarding hydroprojects for Gabčikovo and Nagymaros is still valid despite Hungary's withdrawal in the face of public opinion at the end of the communist era (Szirmai 1993; Waller 1992). However, while Slovakia awaits renewed co-operation – Dutch consultants having advised that completion would be beneficial for water level, navigation and underground water reserves – Hungary's Fidesz–Hungarian Civic Party government considers that the judgement is only advisory and not binding. Its priority is to 'recover' the diverted Danube, engineered unilaterally by Slovakia at Gabčikovo, rather than assist with further works downstream at Nagymaros.

References

Baker, S. 1996, 'The scope for East–West cooperation' in A. Blowers and P. Glasbergen (eds), *Environmental Policy in an International Context* (London: Arnold), pp.135–65.

Baker, S. and Jehlička, P. 1998, 'Introduction' in S. Baker and P. Jehlicka (eds), *Dilemmas of Transition: The Environment Democracy and Economic Reform in East Central Europe* (London: Frank Cass), pp.1–28.

Baumgartl, B. 1993, '"Burden-sharing" as a euphemism: delusion at the Lucerne Conference', *Environmental Politics* 2, 249–52.

Böhmer-Christiansen, S. 1998, 'Environment-friendly deindustrialization: impacts of unification on East Germany' in A. Tickle and I. Welsh (eds), *Environment and Society in Eastern Europe* (London: Longman), pp.67–96.

Connelly, J. and Smith, G. 1999, *Politics and the Environment: From Theory to Practice* (London: Routledge).

European Commission 1999, *Energy in Europe: 1998 Annual Energy Review* (Brussels: European Commission Directorate General for Energy).

Eurostat 1999, *Statistical Yearbook for Central European Countries 1998 Edition – Data 1993–1997* (Luxemburg: Office for Official Publications of the European Communities).

Gelb, A. and Gray, C. 1991, *The Transformation of Economies in Central and Eastern Europe: Issues, Progress and Prospects* (Washington, DC: World Bank Policy and Research Series 17).

Green, J. 1998, *Energy and the Environment in the European Union: Training Documents for NGOs in Accession Countries* (Brussels: Friends of the Earth Europe).

Gyulai, I. 1998, *The European Union and Biodiversity* (Brussels: FoEE/EEB).

Jancar-Webster, B. 1991, 'Environmental politics in Eastern Europe in the 1980s' in J. DeBardeleben (ed.), *To Breathe Free* (Washington, DC: Woodrow Wilson Center Press/Baltimore, MD: Johns Hopkins University Press), pp.25–55.

Jancar-Webster, B. 1995, 'Environmental degradation and regional instability in Central Europe' in J. DeBardeleben and J. Hannigan (eds), *Environmental Security and Quality after Communism: Eastern Europe and the Soviet Successor States* (Boulder, CO: Westview Press), pp.43–68.

Klarer, J. (ed.) 1994, *Use of Economic Instruments in Environmental Policy in Central and Eastern Europe* (Budapest: Regional Environment Centre for Central and Eastern Europe).

Kolk, F. and van der Weij, E. 1998, 'Financing environmental policy in East Central Europe' in S. Baker and P. Jehlicka (eds), *Dilemmas of Transition: the Environment Democracy and Economic Reform in East Central Europe* (London: Frank Cass), pp.29–52.

Kramer, J. M. 1983, 'The environmental crisis in Eastern Europe', *Slavic Review* 42, 212–16.

Nefedova, T. 1994, 'Industrial development and the environment in Central and Eastern Europe', *European Urban and Regional Studies* 1, 168–71.

OECD 1994, *Capacity Development in Environment* (Paris: OECD).

O'Toole, L. and Hanf, K. 1998, 'Hungary: political transformation and environmental challenge' in S. Baker and P. Jehlicka (eds), *Dilemmas of Transition: The Environment Democracy and Economic Reform in East Central Europe* (London: Frank Cass), pp.93–112.

Russell, J. 1990, *Environmental Issues in Eastern Europe: Setting an Agenda* (London: RIIA).

Singleton, F. 1987, 'National parks and the conservation of nature in Yugoslavia' in F. Singleton (ed.), *Environmental Problems in the Soviet Union and Eastern Europe* (London: Lynne Rienner), pp.183–98.

Szirmai, V. 1993, 'The structural mechanisms of the organization of ecological-social movements in Hungary' in A. Vari and P. Tamas (eds), *Environment and Democratic Transition: Policy and Politics in Central and Eastern Europe* (Dordrecht: Kluwer), pp.146–56.

Tellegen, E. 1996, 'Environmental conflicts in transforming economies: Central and Eastern Europe' in P. Sloep and A. Blowers (eds), *Environmental Policy in an International Context* (London: Arnold), pp.67–96.

Tickle, A. and Vavroušek, J. 1998, 'Environmental politics in the former Czechoslovakia' in A. Tickle and I. Welsh (eds), *Environment and Society in Eastern Europe* (London: Longman), pp.114–45.

Tickle, A. and Welsh, I. (eds) 1998, *Environment and Society in Eastern Europe* (London: Longman).

Waller, M. 1989, 'The ecology movement in Eastern Europe', *Journal of Communist Studies* 5, 303–28.

Waller, M. 1992, 'The dams on the Danube', *Environmental Politics* 1, 121–7.

Waller, M. 1998, 'Geopolitics and the environment of Eastern Europe' in S. Baker and P. Jehlicka 1998, 'Introduction' in S. Baker and P. Jehlicka (eds), *Dilemmas of Transition: The Environment Democracy and Economic Reform in East Central Europe* (London: Frank Cass), pp.29–52.

Welsh, R. V. 1997, 'Local government and sustainable environment: European perspectives', *European Environment* 7(1), 16–24.

Welsh, R. V. and Tickle, A. 1998, 'The 1989 revolutions and environmental politics in Central and Eastern Europe' in A. Tickle and I. Welsh (eds), *Environment and Society in Eastern Europe* (London: Longman), pp.1–29.

Internet sources

Conservation organisations and NGOs

Bund fur Umwelt und Naturschutz (De)
<http://www.bund.net>

Daphne (Sk)
<http://www.fns.uniba.sk/zp/daphne>

Environmental Partnership (Cz)
<http://www.ecn.cz/epce>

European Commission Environmental Directorate (Enlargement Page)
<http://www.europa.eu.int/comm/environment/enlarg/home.htmEuropean>

Environmental Agency
<http.//www.eea.eu.int>

European Mountain Forum
<http.//www2.mtnforum.org/mtnforum/regions/europe.htm>

European Parliament Environmental Committee
<http://www.europarl.eu.int/dg2/envi/en/default.htm>

Friends of the Earth (Hu)
<http://www.zpok.hu/mtvsz>

Friends of the Earth (Cz)
<http://www.duhafoe.cz>

Friends of the Earth Europe
<http.//www.foeeurope.org>

Greenpeace (Cz)
<http://www.greenpeace.cz>

Pan Parks: Nature Conservation and Tourism in Europe's Protected Areas
<http://www.panparks.com>

Powerful Information
<http://www.gn.apc.org/powerful-information>

Regional Environmental Centre
<http.//www.rec.org>

Society for Sustainable Living (Sk)
<http://www.fns.uniba.sk>

Sumava National Park (Cz)
<http://www.jiznicechy.cz/asc/uk/prizaj/sumava/html>

UNEP
<http.//www.grid.unep.ch/df>

Worldwide Fund for Nature
<http.//panda.org/crisis>

Media and general information

Central Europe Online
<http://www.centraleurope.com>

Open Media Research Institute Prague
<http://www.omri.cz/index.html>

Radio Free Europe/Radio Liberty Newsline
<http://www.rferl.org/newsline>

Romanian Business Journal
<http://www.starnets.ro/RBJ/>

Transitions Online Magazine
<http.//www.transitions-online.org>

United Kingdom Foreign and Commonwealth Office
<http://www.fco.gov.uk>

Warsaw Voice
<http://www.warsawvoice.com.pl/pl-iso/index/html>

Sources for individual countries

Albania
<http://www.albanian.com>

Bosnia and Hercegovina
<http://www.bosnianembassy.org>
<http://www.taponline.com/tap/life/newz/bosnia/fieldtrip.html>

Bulgaria
<http://www.bulgaria.com/index.html>

Croatia
<http://www.croatiaemb.org>

Czech Republic
<http://www.muselik.com/czech/frame.html>
<http://www.czech.cz/washington>

Hungary
<http://www.fsz.bme.hu/hungary> <http:www.hungary.com/tourinform>
<http://www.hungaryemb.org>

Poland
<http://www.polishworld.com> <http://www.polishworld.com/polemb>

Romania
<http://bucharest.com/index.html> <http://www.embassy.org/romania>

Slovakia
<http://www.eunet.sk/slovakia/slovakia/html>

Slovenia
<http://www.ijs.si/slo.html>

Yugoslavia
<http://www.yugoslavia.com>
<http://www.rworld.compuserve.com/homepage/yuembassy>

Other organisations

European Union
<http://europa.eu.int>

EU Agenda 2000
<http://www.europa.eu.int/comm/agenda2000/index.htm>

EU Baia Mare Task Force
<http://www.eu.int/comm/environment/enlarg/home.htm>

EU Environmental Acquis
<http://www.eu.int/comm/environment/enlargement/index.htm>

Organization for Security and Cooperation in Europe
<http://www.osceprag.cz>

Slavic Review
<http://ragnar.econ.uiuc,edu/-slavrev>

Soros Foundations Network
<http://www.soros.org>

United Nations
<http://www.un.org>

United Nations Development Programme: Regional Bureau for Europe and the CIS
<http://www.undo.org/undp/rbec>

University of Pittsburgh Russian and East European Studies
<http://www.pitt.edu/-cjp/rees.html>

University of Texas Russian and East European Network Information Center
<http://reenic.utexas.edu/reenic.html>

University of Toronto Petro Jacyk Central and East European Resource Center
<http://library.utoronto.ca/www/pjrc/pjrc.htm>

Woodrow Wilson Center, East European Studies
<http://www.wilsoncenter.org/ees>

World Bank
<http://www.worldbank.org>

WHO World Health Organisation
<http://www.who.dk>

World Trade Organisation
<http://www.wto.org>

2 Environmental politics and transition

Susan Baker

Introduction: transition and the politics of the environment

This chapter examines how environmental politics is being shaped by the transition process currently underway across the region. Because of the close connection between political transformation and economic restructuring, this chapter also looks at the impact of economic reform on environmental governance in the region. Since 1989, the region has undergone a complex process of transition, involving both democratisation and marketisation. There has been a tendency by academic writers, western multilateral agencies and advisers to governments to assume that this 'transition', while complex, involves a relatively unproblematic implementation of a number of economic and political reforms (see Baker (1996) for further details). This in turn would allow progress towards a set of pre-defined end-points. However, it has become clear that 'transition', far from being a linear process, is both evolutionary and path dependent (Smith and Pickles, 1998, pp.15–16). It is evolutionary in that it is based upon institutionalised forms of learning and is path dependent in that it involves complex reworking of old social relations in the light of the attempt to construct a form of capitalism on and with the ruins of the old communist system.

Studies looking at the environmental dimensions of transition typically explore the environmental legacy of communist rule, the nature of post-communist environmental problems, the environmental management strategies evolving at the state and international levels to deal with these problems and the role played by the environmental movement in the collapse of the old regimes and, subsequently, in environmental management in post-communist states. However, understanding transition in terms of a dynamic unfolding between evolutionary and path-dependent change, what we will call the 'transition dynamic', enables us to cast new light upon the impact of transition on these environmental issues and, more generally, upon the changing nature of environmental governance. It allows us to view environmental governance within the context of the complex interface that is evolving between the legacies of the old regime and the new systems of environmental management that are being introduced across the region. It helps explain the seemingly contradictory tendencies that research to date on the environmental dimensions of transition is uncovering. On the one hand, research is finding that there is a great deal of continuity with the past. Examples of this continuity include the low priority assigned to environmental protection over economic development; the limited role played by environmental NGOs in public policy formation; problems with the ongoing weaknesses in civil society; and the continuation of centralised administrative structures alongside closed and highly politicised bureaucratic cultures (Smith and Pickles, 1998). On the other hand, research is also pointing to new features, processes and influences that are arising from the democratisation of political life, from marketisation and the growth of consumerism, from the decentralisation of public administration and from the increased levels of involvement that countries in the region now have with international agencies, in particular the EU.

Acknowledging that transition is an evolving, learning process and yet one that is embedded within the socio-political and historic contexts of the particular nation-states of the region also allows us to take account of the diversity of forms of transition that are emerging across the region. There is a great deal of variation between countries in ECE in terms of how they are managing their transition processes. This variation is the result of several factors. First, it arises from diversity in the cultural, religious and ethnic make-up of the states (and at times, the sub-state levels) of the region. Second, there are differences in the historical, socio-political experiences of states under communism. Differences existed in relation to levels of industrialisation and degree of centralisation of the economy (Hungary and Bulgaria providing contrasting examples) and in relation to the legitimacy of communist rule. The use of the single term 'Eastern Europe' to refer to the entire region during the period of communist rule masked the political and economic diversity that existed within the region. Third, the countries of the region underwent varied 'revolutions' in 1989. The post-communist situation in the Balkans is less stable than those in the Visegrad countries in terms, for example, of the Balkans region experiencing high turnover rates of governments, low public acceptance of change, continuous strength of the communist successor parties and poor commitment to economic and political reform. The wars that followed the breakup of the former Yugoslavia also contributed to this instability in the Balkans region. Since 1989 these factors have contributed to the emergence of a highly differentiated transition process between the countries in the north and south of the region.

Democratisation and environmental governance

The democratisation process

By viewing transition in terms of an unfolding dynamic we are able to guard against the mistaken notion that the countries of ECE are merely 'copying' the political structures, institutions, practices and procedures of the West. Rather, we acknowledge that the transformation of the political institutions and processes is being undertaken against a background of inherited legacies (political, social, economic, administrative and cultural) from the periods prior to communism and from the communist era itself. This contextual embeddedness means that these countries are not so much copying but adapting and adjusting western-style models of democracy, at times selecting certain elements of these models while ignoring others. This is resulting in political transformation whose end-point (either for individual states or for the region as a whole) is far from clear. We also have to acknowledge that practice in actual existing democracies across the globe fits along a continuum ranging from more to less democratic and that there is no 'pure' practice of democracy which can be used as a yardstick to measure the political developments that are occurring in the region. Furthermore, while the EU is acting as a model of 'democracy', in particular for those countries that have formed Association Agreements with the EU, 'democracy' has taken several different forms in EU member states (Loughlin *et al.*, 1999) and, further, modes of democratic practice at the EU level itself have yet to evolve.

While acknowledging the variety and differences between countries, political transformation is nevertheless beginning to show certain discernible features across the region: the shift from one-party rule to pluralist, multi-party parliamentary systems; consolidation and maturation of the party system; the formation of elected and accountable government; decentralisation of political power, exercised under the rule of law; the adoption of new constitutions and in some countries a separate Bill of Rights (for example, Hungary). Democratic control over state administration has also been strengthened, as has its accountability to democratically elected bodies. This latter reform is helping to bring public administration strictly under the rule of law as well as strengthening the legality of administrative acts. There have also been moves to decentralise the bureaucratic apparatus of the state by introducing reforms in the structures and practices of regional and local administration and government. Reform at the sub-national (regional and local) levels is important for the political transformation of the region, given the highly centralised nature of the state under communist rule. There has already been considerable advancement in the formal devolution of powers and responsibilities to sub-national levels of government in Hungary and Poland. However, a number of critical problems remain, particularly in relation to substantive capacity: in most cases regional and local levels of government still lack the administrative capacity, organisational structure, expertise and resources to carry out their new duties effectively. Furthermore, decentralisation is taking place at different rates across the region. Bulgaria provides a good example of a country where the pace of regional reform is relatively slow. It is for this reason that Bulgaria is a recipient of aid from the SIGMA programme, a joint EU/OECD initiative aimed at helping the reform of public administration in the countries in transition (http://www.oecd./org/puma/sigmaweb/aboutsgm.htm, accessed 10 October 2000).

Democratisation is also contributing to the development of civil society; that is, a sphere of autonomous social action, independent of the state. A dominance of the state and a concomitant weakness of civil society marked the region during the period of communist rule. The development of civil society involves far-reaching and hence slower processes of change in political culture, including a reduction in the role of the state in social affairs. For many countries, the development of civil society also requires a radical change away from passive to active, participatory citizenship. In contrast to western democratic traditions, countries in the region developed a typical 'subject' political culture under communism; that is, the citizens often became apathetic subjects of the state (Crawford, 1996, p.24). In places, particularly in the Balkans, this passive political culture was prone to a 'messianic concept of political change', with faith usually placed in strong leaders and in finding easy solutions to social and economic problems that would not entail public debate. In some countries, however, most noticeably in Poland and Czechoslovakia, the political culture was less passive. Here some limited form of pluralism existed, which was able to catalyse a 'bottom-up' political movement to contribute to the collapse of their totalitarian regimes (Tickle and Vavroušek, 1998, p.136).

It comes as no surprise to find that there are marked differences in how the process of democratisation is progressing across the region as a whole. Hungary provides an example of a relatively smooth transition to democracy, where, for example, the existence of some form of pluralism before the collapse of the old regime allowed stable democratic political parties to develop rather quickly after 1989. In addition, this also facilitated the development of ENGOs in the realm of civil society (Rootes, 1996, p.50). Bulgaria, in contrast, provides an example of a country seeking to manage its environment in an unstable transition, with the new political system commanding only limited political legitimacy and with continuing blockages in the reforms both of the system of public administration and of local government (Baker and Baumgartl, 1998). Similar problems exist in Romania and Slovakia, where, particularly in Romania, political parties and the election process continue to show characteristics of underdeveloped party systems (Crawford, 1996, pp.250–1) or, as is the case in Bulgaria, have developed into systems characterised by 'polarised pluralism' (Gallagher, 1998, p.47). The Balkans region has experienced frequent changes of government and while free, fair, competitive elections are now held at regular intervals throughout the region, there has been some government manipulation (for example, in Romania and Albania) and in Albania, Bulgaria and Romania rural-based support for former communists (the communist successor parties) remains high (Crawford, 1996, p.167). There are also problems with the adherence to the rule of law and with uncoupling public administration and the civil service from party bureaucracy and membership. Further, many Balkan states remain extremely centralised with weak regional and local government and underdeveloped interest and pressure groups. In short, while most countries have experienced rapid change in the nature of governance since 1989, the procedural characteristics of political reform are more robust than the substantive dimensions of change, and there are noticeable differences between the depths of change in the Visegrad countries and in other states, most especially in the Balkans.

In pointing to the gap between formal and substantive change, it is important to acknowledge that democratisation requires time. Within a very short space of time, countries in the region have, by and large, managed to set in motion far-reaching procedural changes in their political structures and in their institutional and administrative procedures. The substantive changes that are part of the democratisation process are, by necessity, slower. They involve changes in the political culture, which cannot merely be imposed through 'top-down' reform, but need to stem from 'bottom-up', attitudinal changes. This has to reflect the socio-political culture of the individual states of the region. Furthermore, the full effects of institutional reforms, such as the establishment of new administrative machinery, also take time to manifest themselves, particularly given that existing systems of public administration are locked into bureaucratic ways of thinking and behaving that are a legacy of the traditions of the Austro-Hungarian or the Ottoman empires. In this context, the new procedures that many countries have established (for example, in relation to environmental monitoring and Environmental Impact Assessment) require not only new managerial skills and retraining of administrative and legal personnel, but a change in the bureaucratic political culture and policy-making style. To date this political culture, particularly that stemming from the Ottoman tradition, has resulted in a reluctance on behalf of policy-makers to question bureaucratic regulations, or to accept any degree of personal responsibility in decision-making, and has led to a plethora of minor regulations bedevilling many transactions (Crawford, 1996). Changes at this level take time.

The impact of democratisation on environmental governance

Despite the problems with democratisation, there have nevertheless been significant changes in what we may call the organisation of environmental governance. First, there have been developments within the formal institutions and structures of the state. Hand in hand with democratisation has gone a process of capacity-building through the enhancement of state (national and sub-national) institutional and legislative capacity. Among the most noticeable areas of development has been improvements in environmental legislation, with most countries passing new, framework environmental legislation within a short time after the collapse of the old regimes. Examples include the Hungarian 1992 national environmental framework law and the Bulgarian Environmental Protection Act of 1991. Capacity-building has also taken place within government, helped in many cases by the appointment of many Green activists to positions of power, including ministerial positions within the first post-communist governments, such as in Bulgaria. Furthermore, there has been capacity enhancement within the system of public administration, such as the formation of new environmental units within ministries and new monitoring bodies and agencies.

These developments can be attributed, at least in part, to increased responsiveness of governments to the continuing high level of public concern about the environment. It also acknowledges the role played by environmental groups in the collapse of the old regimes. But the changes are also due to international pressures and commitments, including those entered into under the EU Association Agreements. These Agreements help in particular to explain the focus of attention on improvements in legislation. Of central importance is participation in EU accession talks and efforts to meet the requirements of EU membership.

Second, democratisation has forced greater openness and accountability within the policy-making process. Greater openness, in particular interest group participation in new policy-making fora, is helping to change the nature of environmental governance, by supplementing the formal, traditional approach to conducting the business of government with something closer to a network style of governance, discussed below. Greater accountability is also facilitating steps towards environmental policy evaluation, critique, and ultimately may enhance the flexibility of the system to respond to changing public concerns and ecological conditions. Third, democratisation, in particular the consolidation of party politics, has opened up a new arena for environmental expression. The consolidation of multi-party systems, and the holding of free elections, has allowed expressions of environmental concern to be routed through the electoral process, through engaging in party politics, including Green parties. Democratisation, insofar as it is encouraging new attempts at decentralisation, can also help bridge the gap between environmental legislation and policy implementation, a gap that widened during the latter years of the old regime. This can lead to more effective regulatory processes (Pickles, 1998, p.188). In particular, decentralisation has the potential to allow local and regional authorities a greater say in monitoring, regulatory enforcement and the imposition of sanctions, fines, etc. for regulatory non-compliance. However, local and regional authorities are not always good at adhering to environmental regulations, especially if this is seen to act as a barrier to economic development (see Pavlínek, 1989). Still, in the long term, it is fair to say that the reform of the public sphere will lead to greater co-operation between the levels of government and improvements in environmental management.

However, the gap between formal and substantive change, discussed above, is also of importance as far as environmental management is concerned. Many of the first post-revolutionary parliaments suffered from 'legislative hyperactivity'; that is, they passed an enormous amount of legislation in a very short period of time, including new environmental framework legislation (Crawford, 1996, p.262). As a result it comes as little surprise to find that the processes of moving from formal enactment of new environmental legislation to actual implementation of that legislation suffered and continues to suffer from bottlenecks. Environmental policy management in Western Europe also experienced a similar increased intensity of legislative activity, especially at the EU level during the 1980s to 1990s, which was then followed by an implementation deficit. The problems encountered with environmental management in transition countries also bear some resemblance to those encountered in the West, where effective implementation makes financial demands upon the system and calls for resource allocation. Because governments operate with resource limitation, this requires political negotiation and bargaining, which environmental departments typically enter into from a disadvantageous and weak position. Thus, like the experience of their counterparts elsewhere in Europe, many Green activists who achieved positions of power in post-communist governments found that their hands were tied by other, pressing economic and political considerations and by the policy limitations that stem from being in coalition governments. But there are features of environmental management in ECE countries that are specific to the process of transition. Many of the new bills passed in the immediate aftermath of the collapse of communism were poorly prepared, involved little consultation with organised interest groups and have subsequently been subject to numerous amendments. The numerous amendments to the land restitution bill in Bulgaria provide an example, a bill that is important for environmental management because it is expected to have a significant impact upon nature conservation efforts, as land now designated as national parks is returned to private ownership. In addition, many environmental problems, especially those relating to pollution, occur at a local level and local governments are typically charged with the task of pollution monitoring, waste and refuse collection and ensuring policy compliance at the local level. The lack of progress in some countries in carrying out reform in regional and local governments impedes the process of effective policy implementation and more effective environmental governance.

Democratisation and the legitimisation of environmental concerns

Since 1989 mass mobilisation around environmental issues has declined throughout the region. Several explanations have been put forward to account for this decline. First, it is argued that non-environmental interests, in an effort to mobilise public opposition against the old regimes, often used the environmental 'cause' in an opportunistic manner (Waller, 1998). Once the old regimes collapsed, activists were free to focus on their underlying concerns. Second, the harsh reality of economic life in post-communist societies displaced concern about the environment. Third, account has to be taken of the fact that the environmental movement was composed of very different interests and as such began to suffer internal fragmentation once the collapse of the old regime removed the focal point that bound the movement together. Fourth, it is also fair to say that the process of democratisation has in itself reduced the perceived need for mobilisation, because it provided environmental interests with new forms of, and opportunities for, political expression. The period since 1989 has witnessed a shift in the expression of environmental concern from that of protest to more modest, and moderate, forms such as lobbying, negotiation and discussion. Environmental concern is now increasingly expressed through participation in professional, goal-orientated environment NGOs, not through forms of mass protest (Jancar-Webster, 1998). Typically, NGOs have formal organisational structures, formally enrolled members and often employ professionals on a full- or part-time basis. Unfortunately, not all the groups in the movement were able to make this change and to grasp the new opportunities presented by the transition process, and consequently some fell into decline.

However, to focus exclusively on the decline in environmental mobilisation is to run the risk of ignoring a more subtle and, in many ways, more important change in the way in which environmental concerns find expression in post-communist societies. Hand in hand with the professionalisation of the movement has gone a new engagement with institutional forms of political activity, particularly within the public policy-making process. This is because democratisation presented a new opportunity for, perhaps even imperative to, environmentalists to engage in more institutional forms of political activity. Of particular importance here has been the increased institutionalisation of consultation between environmental interest groups and public policy-makers, most noticeably at the level of central government. This is leading to new network forms of environmental governance, more typical of recent western styles of environmental management. Since 1989 there have been varying efforts by governments to open the policy-making process both to interest group participation and public debate. Examples include the involvement of Bulgarian NGOs in the new established National Centre for Environment and Sustainable Development (Baker, 2000). Similarly, in Poland, there has been an institutionalisation of consultation between the Environmental Ministry and ENGOs (O'Toole and Hanf, 1998). This contrasts sharply with typical practice during the period of communist rule, with its centralised, policy-making process that lacked both a spirit of compromise and a culture of participation and rational public debate (Toonen, 1993, pp.154–5). By presenting environmentalists with such new opportunities, democratisation has enabled the environment to pass from being a mobilising concern of often marginalised social actors to being a standard part of the business of government. Thus we can see participation in and involvement with the environmental policy-making process by ENGOs both as a product of, and as helping to consolidate, the process of democratisation in the region.

Participation is also important for consolidation of democracy in another way, in that it aids the development of civil society, which, as we have seen, remains weak. This is because participation facilitates learning about the practice of responsible and effective interest group politics. It also contributes to a self-reinforcing spiral of civic virtue, which is necessary for the consolidation of liberal democratic practices (see Cox, 1998). In this way, participation strengthens the substantive component of the democratisation process, a component that, as we have argued above, remains in many respects the more difficult aspect of democratisation to achieve. In turn, the opening up of the policy-making process has exposed governments and policy-makers to the arguments of environmental groups. This has contributed to what we may call the 'acceptance' of environmental issues as legitimate concerns of politics and of public sector policy-makers – in short it has contributed to the legitimisation of environmental concerns.

However, in the region as a whole the process of becoming insiders, or participants, in the policy-making process has been an uneven one, with Hungary, the Czech Republic and Poland leading the field. This difference is in part a product of past experience, in part a result of variations in present opportunities. The pattern of environmental mobilisation under communism differed across the region, including in relation to the strength of academic ecology, the presence of official conservation cadres and the extent of synergy between official groups and other social movements, including the independent opposition (Tickle and Welsh, 1998, p.156). This resulted in groups from different countries having varying abilities to engage in political participation in the post-communist period. The role played by the environmental movement in the collapse of communism was also important in shaping post-1989 strengths. This role varied across the region, resulting in the movement commanding different levels of popular respect in the post-1989 period. While significant in the region as a whole, only in Bulgaria did the principal challenge to the regime take the form of an environmental protest, and only there did the environmental movement stimulate the development of a mass movement capable of challenging the communist monopoly of power (Waller and Millard, 1992, pp.165–6). Following the introduction of free elections, environmentalists in Bulgaria were thus able to enter into governments and were instrumental in bringing in new environmental laws and in shaping environmental policy, at least in the immediate post-communist period (Baker and Baumgartl, 1998). Present opportunities also differ, especially given the differences in the substantive nature of the democratic reform process between countries in the region. The EIA procedures in Hungary provide an example of deepening opportunities for NGO participation, while, in contrast, the 1995 amendment to the Bulgarian Environmental Protection Act restricted the participatory opportunities available to environmental NGOs through to the EIA process (Baker, 2000).

Despite enabling environmental interests to continue to find forms of expression in post-communist, transition society, the professionalisation of the environmental movement has nevertheless been criticised. Professionalisation is largely a consequence of the incorporation of western models of political practice into the political systems, in particular the policy-making process. Within the West, professionalisation has been a means by which environmentalists have been able to become 'insiders' in the policy-making process. However, 'insider' status typically results in the moderation of environmentalists' demands and is largely confined to the consultation stage of the policy-making process. Such limited outcomes are a far cry from the hopes and aspirations that mobilised the publics of Eastern European countries in the last days of the old regime. Professionalisation has also been criticised for leading to a growing dependence of the region's ENGOs on foreign aid and support. It has been claimed that this has resulted in a shift in priorities away from domestic concerns to those most likely to receive foreign aid and support (Jancar-Webster, 1998). It has also been argued that it has loosened the ties that have anchored environmental NGOs to their own local support base. We can look here to the example of Bulgaria, where the environmental movement has now consolidated around Sofia-based academic ecologists, many of whom have weak ties to the local population (Baker, 2000), although the 1995 mobilisation of rural-based groups against state water management plans provides a notable exception (Staddon, 1998). Paradoxically, professionalisation has also the potential to contribute to, rather than undermine, the tendency towards passivity within the political culture of the region, a tendency inherited from the period under communist rule. This lent credence to the belief that change can come from 'top-down' agents who are the bearers of 'solutions' to complex social and economic problems, such as those posed by environmental deterioration. In this context, professionalisation can weaken the capacity of environmentalists to stimulate substantive change at the grass-roots level.

Democratisation of environmental governance in the region is intimately linked with the incorporation of environmental policy into a wider international, and in particular EU, regulatory and management framework. Participation in international environmental management regimes, including at the UN level, has been, for example, a key factor contributing to the institutionalisation of consultation between policy-makers and environmentalists. Bulgaria's participation in the post-Rio UNCED process, for example, was highly instrumental in the formation of the National Centre for Environment and Sustainable Development (mentioned above), and in shaping its subsequent composition (Baker, 2000).

Democratisation and the incorporation of the region into an international framework of environmental management

Since 1989, the region has seen the development of structures of regional environmental co-operation. In 1991, for example, the seventeen countries in the catchment basin of the Danube River, along with interested international institutions, launched an Environmental Programme for the Danube River Basin. Subsequently, these countries, along with the EU, signed the 1994 Convention on Co-operation for the Protection and Sustainable Use of the River Danube. This led to the creation of an international commission, which has taken over the water-related work of the Environmental Programme, mentioned below. Since then, a strategic action plan for the Danube Basin has been launched, which relies on a combination of national plans and international activities, such as international monitoring and emergency response systems. Similar structures of regional co-operation have emerged in relation to the management of the Black Sea. Much of this work is framed within the 1992 Convention for the Prevention of Pollution of the Black Sea and its three protocols (OECD, 1996). These regional co-operation structures are potential sources of improvement in ambient quality because they help enhance environmental management capacity, particularly at the transboundary level (see Hyde-Price, 1998). However, while these structures of regional co-operation are significant, Association Agreements with the EU and the closely related 'Environment for Europe' process are by far the most important external factors shaping environmental policy in the region.

By the mid-1990s, ten countries in the region had signed European Association Agreements with the EU. Under the so-called 'Copenhagen Criteria', Association countries can apply for membership of the EU, so long as they meet certain economic and political criteria. Through the Phare programme the EU has made available a limited range of funds and expertise to help countries meet these criteria. At present, Hungary, Poland, the Czech Republic, Slovenia and Estonia are judged the best candidates for the 'next round' of EU membership (Mayhew, 1998). Association Agreements are significant for environmental governance in ECE in a number of ways. First, they have laid down a number of conditions governing the management of the environment in transition countries. These include that countries approximate all their environmental laws to those of the EU; that is, that Association countries adopt the *acquis communautaire* of the EU. It is also expected that policy, including across the sectoral levels, be guided by the principle of sustainable development and that countries meet certain EU norms and standards regarding the conduct of policy-making and its implementation. This includes the establishment of appropriate monitoring and reporting mechanisms. The Agreements mean in effect that environmental management in Association countries is now anchored into a transnational, EU-level legislative and monitoring framework.

Second, Association Agreements are also leading to the release of some funding, through the Phare programme, to help with environmental clean-up. More recently, funds have been targeted at the strengthening of environmental management capacity, particularly in relation to monitoring and information gathering. However this is not without its difficulties: the environmental component of the Phare programme has been weak and under-funded, and strongly criticised for channelling much of the benefits to western consultants. Furthermore, on balance, the EU PHARE prioritisation of economic reform (privatisation and marketisation) over environmental considerations has helped legitimise the long-standing belief in ECE that economic matters should take precedence over social and other environmental goals (Baker and Jehlička, 1998). Furthermore, assistance has tended to reflect donor country priorities, such as transboundary pollution, providing further pressures to displace domestic environmental concerns.

Third, the Association Agreements are resulting in significant policy transfer, a process whereby the countries of the region import policy approaches adopted elsewhere. This is a common occurrence between existing EU member states and between them and the USA, particularly with respect to environmental policy. However, in ECE it is taking place within the context of a deepening asymmetrical relationship between the countries of the region and international agencies, including the EU (Caddy, 1997). The 'Environment for Europe' process is the most important conduit through which this policy transfer takes place. At the 1993 Lucerne Conference on 'Environment for Europe', Environment Ministers from OECD countries and from Central and Eastern Europe endorsed an Environmental Action Programme for Central and Eastern Europe (EAP). The EAP is a framework for national action and international assistance programmes which emphasises that environmental improvements depend on combined efforts at policy reform, institutional strengthening and investment. In addition, it encourages countries in the region to set priorities for tackling environmental problems. This led to the launch of national environmental action programmes, which develop country-specific priorities based on the EAP. A further 'Environment for Europe' conference was held in Sofia, Bulgaria, in 1995. Since 1991 the 'Environment for Europe' programme has provided the main strategic framework for environmental policy development in the region. As a powerful interlocking coalition of international and European institutions, the 'Environment for Europe' process has become, in effect, the key transnational regime for management of the region's environmental problems (Slocock, 1999, p.154). While the process places significant emphasis upon institutional building and policy reforms, it has been severely criticised for failing to provide countries with the necessary financial resources to effectively manage the environmental dimensions of transition (Baker and Baumgartl, 1998).

Fourth, as we have already argued, the incorporation of environmental policy in the region into a wider international framework is also of significance in shaping how expressions of environmental concern are managed within the political system. We have already pointed to the role that international influence has played in the transformation of the environmental movement into professional, goal-orientated NGOs and their increased involvement in state-level environmental policy fora. Finally, as Slocock has argued, the transnationalisation of the environmental policy process is significant in that it has reinforced the position of national environmental policy-makers in several states. This has allowed them to keep environmental *policy* firmly on the agenda even when environmental *politics* has sunk beneath a tide of other issues of popular concern (Slocock, 1999, p.157).

Economic restructuring, marketisation and environmental governance

The process of economic restructuring

The processes of democratisation and economic restructuring are intimately linked, not least because the legitimacy of the transformation process is in turn linked to the maintenance of and even the improvements in economic well-being. Paradoxically, however, the transformation process is also about the rejection of the Marxist principle of unity between politics and economy, allowing distinct spheres of political and economic life to emerge. In economic terms this means both privatisation of the industrial and services sectors and marketisation of the economy. Privatisation includes not just the denationalisation of a large share of the previously state-controlled productive capital, but also the strengthening of private enterprise, substantial deregulation and liberalisation of the national economy and strengthening the links that bind these economies to the global market. Marketisation involves an end to government subsidies, both in relation to production and consumption, liberalisation of prices and the opening up of the economy to global competition. It also involves the establishment of institutions, such as a central bank, enacting new laws, such as bankruptcy laws, and the introduction of new managerial, financial and business practices into the economic sphere.

Governments approach the task of economic reform with varying degrees of commitment, party approval and public support. As a result, developments vary significantly. In addition, countries began the task from very different starting points, especially in relation to the degree of centralisation of their economies that occurred under the period of communist rule. Hungary, as a country that was able to develop quasi-market relations under communism, provides an example of speedy reform. In contrast, the privatisation of state-owned companies has been very limited in Bulgaria, not least because of weak commitment on behalf of the country's largest political party, the Bulgarian Socialist Party (BSP) (Dimitrov, 1996). Furthermore, the Balkans region suffered economically as a result of wars in the former Yugoslavia, resulting in severe trade losses, which further delayed economic reform.

The impact of economic restructuring on the environment and its management

The impact of economic restructuring on environmental management has been complex. On the one hand, there have been some very positive developments. The purchase of state-owned companies by foreign buyers has, for example, forced many countries to clarify their laws and practices on environmental liability, such as happened in Bulgaria. The purchase of state-owned firms by foreign companies has also led at times to the penetration of ecologically modernising foreign firms into the economies of the region. Ecologically modernised firms treat the environment not as a free resource but rather as a factor of production, which leads to new efficiencies in resource use. Foreign direct investment has also brought similar positive improvements in environmental management capacity-building, because it helps the development of 'green proponents' within industry. In turn these are contributing to the acceptance of environmental norms within domestic business practices. Foreign purchase of firms has also led to environmental clean-up, an example of which can be found in the policy of the Belgian company Union Minière, which has a controlling stake in the Bulgarian Pirdop metallurgical plant. Union Minière is currently implementing a four-year environmental programme financed through international loans to the Bulgarian government (*Financial Times*, 8 March 1999). Furthermore, economic restructuring through marketisation is forcing those domestic firms that wish to stay in business to reduce production costs by minimising inefficient energy and other resource use. Under communist rule, energy-wasteful production was possible because individual production units were protected from the full vigour of the market, including bankruptcy and factory closure, by the state-run economic system.

Marketisation is also important as far as environmental policy-making is concerned because many of the policy tools currently favoured by the EU are reliant upon an effective market process. Such tools include energy and pollution taxes and ecologically sensitive pricing policies. In this way, marketisation helps reduce reliance upon a largely legislative approach towards environmental management, an approach that tends to lead to implementation deficits in both Eastern and Western Europe. Economic restructuring has also led to a reduction in environmental pollution through the collapse of industrial production, especially from old smokestack industries. The positive benefit from this can be seen in the reduction in both air and water pollution levels (OECD, 1996). While this is making a significant contribution to the quality of life, it is important, however, to point to the dangers inherent in achieving improvements in ambient quality in this way. Without lesson-learning, without matching such improvements with enhanced environmental management capacity and without taking steps towards the ecological modernisation of remaining industry, there is a danger that future upturns in the economy will result in a return to the old environmentally damaging ways.

However, economic reform is also proving problematic for the environment. By and large, foreign investors are only interested in investment in and purchase of the more modern, most profitable state businesses. As a consequence, privatisation policies are resulting in a rump of unprofitable industry often being left in the hands of local authorities or other government holding companies, which lack the capital for environmental modernisation. Furthermore, in many countries, such as Bulgaria, privatisation has often been left in the hands of the old *nomenclature*, which have a history of eschewing environmental regulations. This has enabled them to rework their old political power into a new economic hegemony. By doing little to remedy the problem of unclear and overlapping environmental policy jurisdictions, this method of privatisation has also allowed the political 'embeddedness' of Bulgaria's institutions of governance, production and environmental regulation to continue in the post-1989 period. It has also been argued that the continuity of the power of such elites in key state sectors such as energy is an obstacle to ecological modernisation, as it helps to further entrench demand-led paradigms of energy supply (Tickle and Welsh, 1998, p.172).

Thus, despite the aspirations of the movements that brought an end to the old regime, events since 1989 have shown that there is no simple one-to-one relationship between the end of communism and improvements in environmental quality, particularly at the local level. On the contrary, since 1989 many countries have seen a shift in the nature of environmental stress, with a reduction in some, traditional, sources of environmental pollution unfortunately combining with the arrival of new environmental pressures. This includes rising environmental problems associated with new consumerism, such as growth in private car ownership. Marketisation has also led to a substantial growth in the number of small and medium-sized enterprises. This is giving rise to problems of environmental monitoring and the safe disposal of waste. The proliferation of small firms in post-communist Bulgaria, for example, has not only put pressure on the limited number of waste treatment facilities but has also created a monitoring nightmare for the Regional Environmental Inspectorates, who represent the backbone of the Bulgarian environmental management system but who have little experience in dealing with small and medium-sized private enterprises (Koulov, 1998, p.224).

Economic reform is leading to a growing disparity between the very visible sector of society which is benefiting economically from the transition process and those who are becoming increasingly marginalised by it. The emergence of a *nouveau-riche* class, organised crime and the concentration of wealth into the hands of the old *nomenclature* through privatisation programmes, which together constituted the least public-spirited segments of society, has the potential to threaten the environmental gains won by transition. Such groups have already demonstrated their preference for private gain over the public good. Unless substantive changes can take place across the region, their influence will continue to retard the formation of strong environmental norms. Furthermore, the growing gap between rising consumer expectations and the harsh reality of economic life has led many to wish to see a return to the old order which, while producing a servile polity, at least guaranteed basic levels of economic well-being. Economic restructuring in this sense has the potential to undermine political change, including the move towards more open forms of political participation.

Conclusion

The transition process in the region is having both positive and negative influences on the environment. There have been positive developments as far as environmental management is concerned, including at the legislative level. In addition, the process of environmental governance has been pushed towards more western-style norms of environmental management. These norms pay greater attention not just to the results of policy (successful implementation) but also to the process through which policy is made (via more open participation), and to widening the set of the actors involved in and contributing to environmental protection (economic actors, the public and sub-national, regional and local government) than had traditionally been the case under communist regimes. Primarily as a result of the Association Agreements with the EU, environmental policy is now firmly locked into an international legislative and policy framework that supports these norms. Over time, this framework will demand that the structures and practices of environmental governance become more accountable, open and pay greater attention to what is achieved in practice.

Throughout the region, many environmental groups were able to grasp the participatory and developmental opportunities that political transition brought. Their transformation into professional ENGOs and their participation in the public policy-making process has subsequently contributed to the acceptance of environmental protection as a legitimate goal of politics. By contributing to the development of a more open political culture and policy-making style, they have contributed to the consolidation of democracy. However, the extent of participation is uneven across the region and groups have had at times to pay a price for this new role: marginalisation from their local roots and support base, and the loss of the ability to act as mobilising agents for social change.

The paradox, however, is that transition is also making environmental management more difficult. Environmental protection has to take place in the face of the hardship of transition, which has resulted in a displacement of environmental considerations by economic concerns. There are also rising levels of crime and, in some countries, rising levels of political corruption, where economic restructuring has helped to consolidate the post-communist power base of the least public-spirited sections of society. The growth of consumerism and of aspirations to adopt western-inspired life-styles also places new pressures on the environment. The problem of limited resources, combined with pressure to prioritise those environmental issues that are of greatest concern to western agencies, is also problematic. It diverts attention and resources away from environmental issues that are of domestic concern and reinforces the culture of political passivity that was the hallmark of the old regime, particularly in the Balkans region.

The kinds of substantive changes that are needed, both in civil society and the political system, to help overcome the problems of environmental management, cannot be achieved in the short term. In the meantime, we can expect to see that transition will result in the traditional business of government increasingly being combined with new network styles of environmental governance. The transition process will undoubtedly continue to involve periods of progress on environmental issues and seemingly backward slides. The marked differences in the nature of the transition process across the region as a whole will also continue to be manifest. In this context, the achievement of more sustainable management of the environment is thus far from guaranteed. Here it becomes important to acknowledge that there is no single solution, no authoritative 'right' answer that can be imposed on the complex task of environmental management in the transition countries. Rather, multiple, risky, complex and innovative solutions are required, which have to arise not only from 'top-down' reform but also from 'bottom-up' innovation and change. To the extent to which the transition process facilitates a shift in the style of environmental management – from that based on traditional modes of government to more participatory forms of governance – then such innovative responses may take hold.

References

Baker, S., 1996, 'The Scope for East–West Co-operation', in A. Blowers and P. Glasbergen (eds), *Environmental Policy in an International Context*, Vol. 3 (London: Arnold), pp.135–66.

Baker, S., 2001, 'Environmental Capacity Building in Bulgaria', in H. Weidner, M. Jänicke and H. Jeorgens (eds), *Capacity Building in Environmental Policy: A Comparative Study of 17 Countries* (Berlin and New York: Springer, forthcoming).

Baker, S. and Baumgartl, B., 1998, 'Bulgaria: Managing the Environment in an Unstable Transition', in S. Baker and P. Jehlička (eds), *Dilemmas of Transition: The Environment, Democracy and Economic Reform in East Central Europe* (London: Frank Cass), pp.183–206.

Baker, S. and Jehlička, P. (eds), 1998, *Dilemmas of Transition: The Environment, Democracy and Economic Reform in East Central Europe* (London: Frank Cass).

Caddy, J., 1997, 'Harmonisation and Asymmetry: Environmental Policy Co-Ordination between the European Union and Central Europe', *Journal of European Public Policy*, 4(3), pp.318–36.

Cox, T., 1998, 'The Politics of Social Change', in S. White, J. Batt and P. G. Lewis (eds), *Developments in Central and East European Politics* (Basingstoke: Macmillan, 2nd edn), pp.216–33.

Crawford, K., 1996, *East Central European Politics Today* (Manchester: Manchester University Press).

Dimitrov, M., 1996, 'Privatization: Its Goals, Progress to Date and Prospects', in I. Zloch-Christy (ed.), *Bulgaria in a Time of Change* (Aldershot: Avebury), pp.107–18.

Gallagher, T., 1998, 'The Balkans: Bulgaria, Romania, Albania and the Former Yugoslavia', in S. White, J. Batt and P. G. Lewis (eds), *Developments in Central and East European Politics* (Basingstoke: Macmillan, 2nd edn), pp.43–59.

Hyde-Price, A., 1998, 'Patterns of International Politics', in S. White, J. Batt and P. G. Lewis (eds), *Developments in Central and East European Politics* (Basingstoke: Macmillan, 2nd edn), pp.225–75.

Jancar-Webster, B., 1998, 'Environmental Movement and Social Change in the Transition Countries', in S. Baker and P. Jehlička (eds), *Dilemmas of Transition: The Environment, Democracy and Economic Reform in East Central Europe* (London: Frank Cass), pp.69–92.

Koulov, B., 1998, 'Post-Communist Change in Geopolitics and Environmental Management: A Case Study of Bulgaria's Burgas Region', in K. Paskaleva, P. Shapira, J. Pickles and B. Koulov (eds), *Bulgaria in Transition: Environmental Consequences of Political and Economic Transformation* (Aldershot: Ashgate), pp.201–28.

Loughlin, J., Aga, E., Bullmann, U., Hendriks, A., Lidström, A. and Seiler, D. (eds), 1999, *Regional and Local Democracy in the European Union* (Brussels: Official Publications Office of the European Communities).

Mayhew, A., 1998, *Recreating Europe: The European Union's Policy towards Central and Eastern Europe* (Cambridge: Cambridge University Press).

OECD, 1996, *Environmental Performance Reviews* (Paris: Centre for Co-operation with the Economies in Transition, OECD).

O'Toole, L., and Hanf, K., 1998, 'Hungary: Political Transformation and Environmental Change', in S. Baker and P. Jehlička (eds), *Dilemmas of Transition: The Environment, Democracy and Economic Reform in East Central Europe* (London: Frank Cass), pp.93–112.

Pavlínek, P., 1989, *Economic Restructuring and Local Environmental Management in the Czech Republic* (Lewiston, NY: The Edwin Mellen Press).

Pickles, J., 1998, 'Restructuring State Enterprises: Industrial Geography and Eastern European Transitions', in J. Pickles and A. Smith (eds), *Theorising Transition: The Political Economy of Post-Communist Transformations* (London: Routledge), pp.172–96.

Rootes, C., 1996, 'Environmental Movements and Green Parties in Western and Eastern Europe', in M. Redclift and G. Woodgate (eds), *International Handbook of Environmental Sociology* (Cheltenham: Edward Elgar), pp.319–48.

Slocock, B., 1999, '"Whatever Happened to the Environment": Environmental Issues in the Eastern Enlargement of the European Union', in K. Henderson (ed.), *Back to Europe: Central and Eastern Europe and the European Union* (London: University College London Press), pp.151–68.

Smith, A. and Pickles, J., 1998, 'Introduction: Theorising Transition and the Political Economy of Transformation', in J. Pickles and A. Smith (eds), *Theorising Transition: The Political Economy of Post-Communist Transformations* (London: Routledge), pp.1–25.

Staddon, C., 1998, 'Democratisation and the Politics of Water in Bulgaria', in J. Pickles and A. Smith (eds), *Theorising Transition: The Political Economy of Post-Communist Transformations* (London: Routledge), pp.347–72.

Tickle, A. and Vavroušek, J., 1998, 'Environmental Politics in the Former Czechoslovakia', in A. Tickle and I. Welsh (eds), *Environment and Society in Eastern Europe* (Harlow: Longman), pp.114–45.

Tickle, A. and Welsh, I., 1998, 'Environmental Politics, Civil Society and Post-Communism', in A. Tickle and I. Welsh (eds), *Environment and Society in Eastern Europe* (Harlow: Longman), pp.156–185.

Toonen, T., 1993, 'Analysing Institutional Change and Administrative Transformation; A Comparative View', *Public Administration*, 71(1), pp.151–68.

Waller, M., 1998, 'Geopolitics and the Environment in Eastern Europe', in S. Baker and P. Jehlička (eds), *Dilemmas of Transition: The Environment, Democracy and Economic Reform in East Central Europe* (London: Frank Cass), pp.29–52.

Waller, M. and Millard, F., 1992, 'Environmental Politics in Eastern Europe', *Environmental Politics*, 1(2), pp.159–85.

3 Environmental movements, nation states and globalisation

Adam Fagin and Andrew Tickle

Introduction

This chapter seeks to explore questions surrounding one aspect of a common response to environmental problems in East Central Europe (ECE): social and political mobilisation in the form of environmental movements. Recent work by a number of authors has described the socialist and post-socialist history, politics and sociology of such mobilisations both in ECE and Russia (Hicks 1996; K. Pickvance 1998; Tickle and Welsh 1998; Weiner 1999) but usually from only a national or regional perspective. Other country-based chapters within this volume also note the role such environmental movements play in national environmental politics.

However, in this contribution, we attempt to place this role within a broader context by examining the specificities of environmental movement activity in relation to agents and institutions operating outside and above the level of the nation state. In particular we wish to overview claims made regarding the role of environmental organisations within a seemingly globalised world and then examine in particular the relation of East Central European environmental groups to such external networks and circuits of power. In this analysis we seek to recognise the diversity of environmental movements in ECE along a spectrum from small, non-institutionalised groupings who favour less formalised repertoires of action at a range of spatial levels (local to global) to more organised collectivities (usually registered with state bodies as non-governmental organisations) who work within formal political opportunity structures – principally domestic but also international. In the context of globalisation, a consideration and differentiation of these often separate – yet linked – modes of engagement with power is important, though not unproblematic. This raises the question, forming a core part of this chapter, of the extent to which ECE environmental movements are part of a 'world civic politics' that has already been proposed for 'northern' (predominantly western) environmental organisations (Wapner 1995, 1996). This will also involve a reworking of the interplay of national and international government and governance issues (raised by Baker in Chapter 2) and the means by which environmental activists and movements seek to engage (or not) with such institutionalised power.

Although its extent, impacts and timescale are contested at many levels, globalisation is seen by most social theorists as one of the key contemporary processes operating socially, culturally, economically and politically across the world. In particular, the environment (and the sub-theme of environmental degradation) is commonly seen as an analytical domain where the processes and effects of globalisation are most evident (Held *et al.* 1999). However, the recognition of the global nature of some environmental impacts is a relatively recent phenomenon. In pre- and early-modern periods, local resource management issues were regulated predominantly by social customs and tradition (Bridge 1998: 221). As larger scale problems and conflicts became apparent in the nineteenth and twentieth centuries, this required increasing regulatory intervention by the state. Early examples include Britain's Alkali Act of 1863 – aimed at combating industrial air pollution – or, in the context of Central Europe, laws and decrees on nature conservation, air and water pollution control promulgated in countries such as Czechoslovakia, Poland and Hungary from the 1950s onwards (see e.g. Cole 1998). Finally, in the latter part of the twentieth century, transboundary pollution affecting 'global commons' such as the atmosphere and oceans became recognised and regulated – mostly through supranational environmental regimes, such as United Nations (UN) conventions. Such globalised environmental degradation is linked predominantly with the post-war intensification and extensification (regional and international diffusion) of mass consumption/production-driven industrialisation, in both its capitalist and socialist variants.

Notwithstanding the trend towards global environmental impact and interconnectedness, it should be noted that many forms of pollution remain 'national' in their impact as well as in terms of the most appropriate regulatory level: state boundaries are sufficiently delineated to identify culpability for the pollution of the seas and atmosphere (Held *et al.* 1999: 376–7). Moreover, the fact that exporting economic activity from Europe to the developing world is resulting in the occurrence of similar forms of environmental degradation (also requiring transnational responses) is by no means unprecedented. The environmental consequences of European involvement in Africa during the nineteenth century – in terms of transportation, exploitation of natural resources and the disruption to natural habitats – though never quantified, somewhat repudiates the suggestion that the environmental impact of globalisation is a novel and unique by-product of twentieth-century capitalism.

To some, globalisation is inextricably linked with the spread of dominant 'western' economic and political forms – in short, liberal democracy and market capitalism. In this sense 'Eastern' Europe has an inherently ambivalent relationship with this model of globalisation given the differing trajectory of socialist (communist) power in the twentieth century. Although many still seek to characterise capitalist and socialist polities as having been separate and non-interactive systems, this belies the degree of linkage that has existed historically – ranging from the involvement of American venture capitalism in the early industrialisation of the Soviet Union (Kotkin 1995: 43ff.) to the large-scale western investment which helped stave off economic crisis in Eastern Europe from 1970 onwards (Schopflin 1993: 165). Indeed it may be suggested that the failure of Comecon (Council for Mutual Economic Assistance (CMEA)) states either to compete or survive autarkically in an increasingly global market-place was a crucial factor in their economic collapse.

In relation to other linkages between the two systems, East–West environmental dialogue from the mid-1970s onwards (originating in the CSCE detente process and then embedding itself within the Geneva-based UN Economic Commission for Europe – see Füllenbach 1981) may be interpreted as a factor which led to the environment being recognised officially in ECE as a social problem, setting off expert and public discourses that reduced the legitimacy of the party state in countries such as Czechoslovakia (Tickle and Vavroušek 1998) and Poland (Hicks 1996). Similarly, evidence also exists showing that ECE environmental movements (both state-sanctioned and independent groups) were able to maintain formal and informal relationships with like-minded organisations outside the Soviet bloc (e.g. international nature conservation bodies such as the World Conservation Union (IUCN): see Weiner 1999: 255), latterly encouraging the diffusion of new policy directions in state environmental agendas (Tickle 2000).

However, in this chapter – although we recognise the path-dependent influence of the socialist period – we now wish to look in more detail at how recent, post-socialist environmental mobilisations have related to the much faster unfolding of globalising forces in the region since 1989. This is first approached by a consideration of globalisation and the environment and a description of its idealised, paradigmatic form in early transition models proposed for ECE. We then proceed to a critical analysis of this paradigm, drawing on a number of empirical studies of environmental movements in the region and a consideration of contemporary theoretical strands within social movement theory. We conclude by offering some thoughts on the future potential direction of ECE environmental movements.

Globalisation and the environment

Though the concept of globalisation is widely applied to nearly all aspects of contemporary political activity, its relevance to the environment is particularly notable. As inferred above, the transnational response required by certain forms of global environmental degradation contradicts notions of state-centric political sovereignty (Held *et al.* 1999: 377). The apparent incapacity of nation states to deal with increasingly complex environmental problems seemingly endorses the need for the transfer of decision-making and regulation away from the state to global institutions and agencies. Indeed, during the past two decades concern about such global issues as climate change, ozone depletion and biodiversity loss has led multilateral institutions such as the European Community, the Organisation for Economic Co-operation and Development (OECD) and the UN to develop environmental agendas and programmes that seek to effect change at a supranational level (Haas *et al.* 1993).

It is also claimed that the globalisation of environmental politics is occurring not just within policy elites, but concurrently at the level of civil society with the emergence of a 'global civic politics' of which the environmental lobby forms a key part (Clark *et al.* 1998; Falk 1999; Lipschutz 1996; Wapner 1995). In some instances, local environmental activism is increasingly structured by large internationally organised environmental movements who seek to link local protest in one region with similar activities elsewhere but in the context of a global agenda. Thus in an era of global connectivity where environmental issues require macro-level solutions, it does appear that 'there are ever fewer arenas of contention whose parameters are purely national, regional or local' (Rootes 1999: 299). While not all environmental problems are global and the nation state remains the focus of much environmental activism, contemporary environmental movement activity is often depicted on a continuum tending towards increasing global interaction and co-operation, linking local concerns to wider global issues and uniting activists from different regions and localities (Held 1998).

However, recent work on new social movements and globalisation (Welsh 2000) draws attention to the limitations of measuring influence solely in relation to the role of formalised environmental organisations and their involvement in institutionalised politics (e.g. transnational environmental regimes). Instead Welsh (ibid.) shows how novel mobilisations of power by dispersed networks of activists (including some in ECE) can, through internet linkages, successfully focus attention and protest on globally significant fora, such as the World Trade Organisation's (WTO) Seattle meeting in November 1999. Thus, in analysing globalisation in relation to environmental movements, we would emphasise that traditional analytical approaches (e.g. Dalton and Rohrschneider 1999; Rootes 1999) – focusing predominantly on formal collectivities and political opportunity structures (usually at an elite level) – are an inadequate basis for a properly nuanced understanding of the impact of environmental movement in a globalising world. In ECE and many other regions, power is also displayed in more dispersed and less organised settings and, though this renders meaningful analysis more difficult, this facet of environmental mobilisation should not be overlooked.

Globalisation and the environmental transition in ECE: the normative narrative

Echoing the bipolar view of world political systems, the collapse of Soviet-style socialism was widely acclaimed as evidence of the triumph of liberal democracy and the free market (Falk 1999: 4–5; Fukuyama 1992). Having spent the preceding decades theorising the contradictions of 'existing' western-style capitalism under the banner of 'transition', political analysts swiftly turned their attention to eulogising the trinity of democratisation, marketisation and privatisation as the optimum model for East Central Europe. With the frontier for western capital now extended eastwards, and with the apparent embracing of western political values by the new elites in the East, the theoretical assumptions of the hyperglobalist strand of the globalisation model seemed validated: the free market, capitalism and liberal democracy were indeed becoming globally hegemonic (Fukuyama 1992; Greider 1997; Ohmae 1995). As with nearly all aspects of the reform process in East Central Europe, this version of the globalisation narrative provided a normative conceptual framework for considering post-socialist environmental policies in the early 1990s.

It was widely argued that isolation (however questionable) from the capitalist world had accentuated environmental degradation and weakened the capacity of ECE states to deal with pollution. The 'actually-existing' socialist states had been denied (or had denied themselves) the benefits of western co-operation, expertise and technology. The absence of the market and the mechanism of central planning had led to greater inefficiency and pollution, worsened by the dominance of heavy industry in socialist economies. The political model of democratic centralism and the communist parties' monopoly of power severely impeded policy reflexivity and learning and also repressed societal pressure for environmental amelioration – both factors contributing to the state's inability to protect the environment. The geo-politics of the Cold War also accentuated the degree of pollution, not least because of the skewing of the economy towards military expenditure (thus limiting social welfare investment) and, on a more localised scale, the highly polluting impact of Soviet bases in ECE.

Foreign direct investment (FDI) and the general opening up of these countries to global political and socio-economic forces was therefore mooted as the 'solution' to environmental problems, such as outdated modes of energy generation (see Russell 1991). Competition and market forces would also reduce the emphasis on heavy industrial production. It was therefore envisaged that the incorporation of these states within the global political and economic order would – in the longer term – wipe away disparities in terms of environmental degradation and strengthen the capacity of the new regimes to mitigate the effects of pollution. In addition, both bilateral and multilateral aid and loans for short-term clean-up and environmental 'know-how' transfers were made available, though both the rationale and effectiveness of such packages have been questioned (Baker 1996; Gowan 1995).

At the same time the fledgling environmental movements in many of the countries of the region, who had helped delegitimise socialist rule, now sought international links in their attempt to establish themselves within the new political order. Various factors informed this. In line with general national and civic aspirations, groups wanted to be part of a common European 'house' or 'home', in which environmental resources (e.g. transnational ecosystems) were seen as the 'ecological bricks' (Langer 1990). Like their counterparts elsewhere, environmental movements in ECE also recognised that the framework for environmental contestation was distinctively altered by globalisation (Held 1998) – power both in the public and policy sphere now lay partly outside the nation state. Thus, establishing links to networks of European and global environmental movements, which had been successful in influencing a range of international environmental regimes (see e.g. Mazey and Richardson 1992; Princen and Finger 1994), clearly offered potential benefits plus enhanced political legitimacy both with the public and politicians. In addition, part of the western (global) transition paradigm also included the inflow of bilateral (US and European philanthropic and governmental initiatives) aid for the environment which provided many environmental movements with desperately needed funds.

The normative narrative: a critique

A decade since the collapse of 'actually-existing' socialism and the dawn of the so-called global era of liberal capitalism, the impact of the integration of ECE within the global political economy and the effects of globalisation on environmental politics look far more complex than initially envisaged. While the globalisation hypothesis still offers a viable conceptual framework for considering environmental movement activity, aid and assistance, and the dynamics of the environmental policy process, many of the more optimistic and idealistic assertions of the early 1990s need revisiting, particularly in light of the broader critique of hyperglobalist accounts of 'denationalisation' and notions of symmetrical interaction and co-operation (Held *et al.* 1999).

The experience of environmental politics in ECE during the past decade endorses many of the general criticisms levelled at the globalisation hypothesis. Thus, instead, we prioritise the significance of situative context and domestic political structures as much stronger determinants of the organisational structure, repertoires of action and efficacy of environmental movements. Indeed, the case of environmental politics in ECE provides a valuable commentary on the concept of globalisation and its application to the environment in three ways:

1 by exposing the power imbalance between indigenous and large western environmental groups and networks;
2 by stressing the role of the newly established nation states as key determinants of environmental movement efficacy and power;
3 by emphasising path dependency and structural context in shaping the interaction between local, regional and global actors.

The contention here is not whether the forces of globalisation have impacted on environmental politics in ECE, but rather the degree and nature of that impact. A decade since the collapse of communism and the onset of political reform, it is now possible to assess the power relationship between indigenous environmental movements and larger international organisations (such as Greenpeace or Friends of the Earth), to evaluate the extent to which resources remain a constraint on environmental capacity, and to consider the extent to which local action is affected by a global agenda.

Building the 'national' before the 'global'

While eschewing simplistic development models or 'exceptionalist' arguments, a first and – in many ways – fundamental point is that the environmental movements of ECE are in a very different position from their western counterparts, partly as a result of the legacy of political autocracy, but also reflecting the hiatus of social, political and economic restructuring since 1990. While the extent to which western environmental movements have become globalised is debatable (Rootes 1999), the relative underdevelopment of the environmental movements in ECE casts serious doubt on their ability to operate in formal global environmental politics.

Environmental protest during the dying days of socialist rule in ECE did not represent the emergence of a submerged western-style mobilisation capable of rapidly fulfilling the functions and role of western movements once the old regime had collapsed. In most ECE states environmental activists were in disarray in 1990; those that had not been catapulted into formal politics lacked experience of policy processes, coherent ideological platforms, internal cohesion and, most importantly, resources. What followed was a difficult period of transformation and rebuilding during the early 1990s which is by no means over (Fagin 2000). Environmental groups still lack adequate resources and remain largely dependent on external funding for their survival. There has been little investment in the movement infrastructure and in fund-raising expertise. Thus to assume that the environmental groups in ECE are capable of operating in a similar way to western movements ignores their recent turbulent history and their relative underdevelopment.

The global capacity of environmental movements in ECE is also constrained by the absence of a developed and stable political and institutional framework within such countries. By comparison, western environmental movements operate against a background of established political opportunity structures, and, as a result of the longevity of their political presence and better resources, have access to political and economic elites. This has undoubtedly helped propel western environmental movements on to the global stage. Thus, having emerged as a radical social movement within the broader climate of political protest of the late 1960s, western environmental movements have gradually become institutionalised (see Diani and Donati 1999). One key factor promoting this has been mass memberships (both active and passive), which have provided the necessary resources to make use of the established and entrenched political opportunity structures of western democracies (van der Heijden 1999). Institutionalisation has also promoted the convergence of policy discourses – notably in relation to 'ecological modernisation' – between some of the larger western environmental movements and structuring political and economic elites. This process of ecological modernisation, in which the pursuit of environmental reform measures is balanced against maintaining the long-term success of the industrial and post-industrial economy (ibid.: 200), is now a central narrative in much of the formal global interaction between environmental movements, transnational corporations (TNCs), national governments and supranational institutions.

In this sense the ability of some environmental movements to wield power in global political fora is based on their established position within domestic political opportunity structures. Put another way, what gives the large effective and 'global' environmental groups their gravitas and prominence is their attachment to powerful nation states. After all, it is the states making up entities such as the Group of Seven industrialised nations (the 'G7') and the OECD plus TNCs with their headquarters in western states that are the key political and economic actors on the global stage. Foreign investment flows are concentrated among advanced capitalist states and most TNCs remain primarily creatures of their home states or regions (Allen and Thompson 1997). Thus, not only do environmental movements in ECE generally lack a developed domestic base for global activity, they are also affected by the peripheral status of their governments and national economies in global politics.

The capacity of environmental movements in ECE to gain formalised access to the global arena is essentially constrained by two interconnected factors: their low level of institutionalisation, and the relatively undeveloped nature of the mechanisms and linkages that normally characterise the state–economy–civil society nexus in more mature polities. Even in ECE states where the political reform process is most advanced (such as Hungary, Poland and the Czech Republic), notions of partnership between government, environmental movements and other actors within the environmental policy process remain largely undeveloped. This may not be surprising given that, since the collapse of the Soviet-style systems, a tumultuous process of rebuilding the nation state and the reformulation of constitutional and political norms has been in progress (Elster *et al.* 1998). Thus, rather than eliminating the nation state as a political entity – as implied both in hyperglobalist and Marxist accounts of the process – globalisation alters the operation of the state, imposes a supranational agenda on its operations, weakens its institutions and generally challenges the boundaries of its sovereignty (Beck 2000: 2–3). But whatever the reduction or alteration in its sovereignty, we believe that the state still plays a critical role in engaging with the globalised world. That said, the role of the state as 'gatekeeper' to the forces of globalisation is thwarted in ECE by the fact that the foundations of the modern state – its sovereignty, legitimacy and the parameters of its jurisdiction – are only just being established (or in some cases, re-established). To assume that the re-emerging states of ECE, still in their infancy, will be able to adjust to the forces of globalisation in the same way and at the same pace as their western counterparts is unrealistic.

An additional complexity in the process of state-building in ECE is the ongoing decentralisation of power to new local and regional political structures, a process that is clearly elided with the notion of globalisation 'hollowing-out' the nation state. In relation to sub-national environmental activism, this is important in the sense that it provides discrete regional sites of contestation, potentially assisting policy and social learning at a local level. In many countries, however, even where local and regional tiers of government have been established, the legacy of the democratic centralism of the socialist period can serve to limit the effectiveness of local or regional administrative reforms. In the short term at least, such remnant bureaucratic cultures are unlikely to encourage responsive forms of civic consultation or participation, either directly with citizens or by proxy with NGOs. As C. G. Pickvance (1998) has suggested in relation to Hungary and Russia, decentralisation and democracy do not necessarily go hand in hand.

Furthermore, it is at the regional and local levels that the global dynamic of foreign direct investment (FDI) ultimately manifests itself, especially in relation to its environmental impact, illustrating the cross-scalar flows implicit in the term 'glocalisation' (Swyngedouw 1992). The case of FDI can also be used to illustrate how the efficacy of environmental movements in ECE is affected by a major underlying dynamic of the transition process: the region's integration within the political and economic structures of international capitalism. Here, neo-Marxist depictions of the bipolar (north–south) disequilibrium of globalisation often fail to fully appreciate the peripheral location of ECE within the political and economic hierarchy of global capitalism (Gowan 1999).

While most post-socialist states have established legislative frameworks to regulate pollution in line with EU and OECD norms (Moldan and Klarer 1997), the imperative of enticing FDI can lead to the uneven application of national regulatory processes such as EIA (see e.g. Fagin and Jehlička 1998: 119–21). While the interests of capital undoubtedly predominate in the West, the political strength of developed environmental movements, economic prosperity and the strength of regulatory structures in general afford environmental considerations an enhanced role in investment decisions. Such ecologically modern practices are hindered in ECE by the fact that governments, whether centre–left or centre–right, are concerned primarily with attracting levels of FDI necessary to sustain even modest economic growth, and by the relative weakness of ECE states to control and regulate the impact of global economic activity on their territory.

This has led to situations where environmental campaigns to prevent locally damaging projects have, of necessity, had to bypass national politics and focus, through international activist networks, on supranational institutions such as multilateral development banks (MLDBs). Successful examples of this approach include the rescinding of loans for nuclear power projects both in the Czech Republic (the US-based Exim Bank and Temelín – see Axelrod 1999) and Slovakia (the European Bank for Reconstruction and Development and Mochovce – see Williams 1995). This suggests that, although ECE environmental movements may in general lack the capacity to operate outside their national borders, in specific key instances the interests of domestic and international protest agendas can coincide to exert power.

Globalism from below: a radical critique

The other contentious element of the globalisation hypothesis as applied to ECE environmental politics is the claim about the emergence of a 'global civic politics' (Falk 1999: 3; Wapner 1995). In response to the transboundary nature of the environmental problems and to the increasing global location of political and economic power (including regional and global environmental regimes), environmental activism is portrayed as shifting along a continuum away from the national and towards the global. This 'globalisation-from-below' is assisted by new communication technologies such as the internet, which is entirely global in its spatial reach (though far from accessible to all) and largely impervious to territorial boundaries (Falk 1999: 6).

While the extent to which such global interaction routinely exists even among western environmental movements is challenged by some (e.g. Rootes 1999), the assumption that global co-operation, campaign styles and interdependence have embraced environmental movements in ECE is not borne out by the (albeit) limited empirical research that currently exists (Dalton and Rohrschneider 1999). This has led some to suggest that international civic politics is a predominantly western-based phenomenon which is limited and entirely dependent on the region's supremacy within the global arena (Allen and Thompson 1997). However, although other studies show that NGO participation at some of the largest UN-led conferences in the 1990s has become much more representative in terms of North–South balance, many southern groups felt 'restricted' by northern NGO interests (see Clark *et al.* 1998: 28, fn.78). Both of these accounts are also consonant with recent claims that such supranational fora represent emergent 'green regimes of accumulation' which pursue 'new production strategies and regulatory regimes to minimise environmental risks whilst simultaneously ensuring Northern economic advantage via the assertion of superior knowledge claims' (Welsh 1999: 48). In this interpretation, ecological modernisation unfolds as a narrative of institutionalised environmental racism and exclusion, where global asymmetry is persistently exposed rather than reduced. That western environmental movements have expanded their activities beyond the national political context to the international and in some cases the global stage can therefore not only be viewed normatively as a reflection of the conducive domestic political opportunity structures and resource availability which strengthen civil society in established democracies, but also such movements' tacit structuring within ecomodernist, globalising political regimes. Thus western civil societies' capacity to enable global environmental movement activity also reflects the position of developed 'northern' states as globalisers rather than globalised.

In contrast, environmental movements in ECE (as in other transition regions – see Pickvance 1999) operate against a background of an emerging and resource-constrained civil society; the activities, strategies and organisational structures of such movements heavily reflect this. At a very basic level, many groups lack internet access and, even where it is available, information about global campaigns does not inform their day-to-day activities or campaigns (Dalton and Rohrschneider 1999). They lack fee-paying mass memberships able or prepared to provide the time and resources necessary to mount effective national campaigns, let alone participate in global actions. A major exception to this characterisation are the newly established (post-1989) national offices of Greenpeace such as in Russia and Czechoslovakia (K. Pickvance 1998: 121ff.; Welsh and Tickle 1998: 22–3). Here the large financial and organisational resources made available by Greenpeace International have created, *de novo*, relatively successful campaign offices which have been able to withstand the vicissitudes of transitional political opportunity structures much better than indigenous EMOs. In contrast, some studies suggest that it tends to be the new, more radical, less-institutionalised elements of western environmental movements which operate globally and fuse links across states and regions (Doherty 1999). However, we would note that such radicalism tends to be largely absent in ECE. What this means in effect is that the focus of environmental movement activity in ECE tends to be on the national political context.

The assumption of symmetrical globalisation that EMO activity in ECE will rapidly take on a global dimension thus ignores the importance of the national political context. It also underestimates the complicated task of establishing such a context in post-communist states beset by post-industrial economic crises and the associated social dislocation of neo-liberal economic reforms (see Elster *et al.* 1998). This is clearly illustrated by environmental movements in ECE, which, despite their high-profile role in the events of 1989, remain in the early stages of development (Baker and Jehlička 1998). In countries such as the Czech Republic, the restructuring and adaptation to the new political environment of the early 1990s resulted in movement paralysis and in many cases decimation (Fagin 2000). Socialised under communism, lacking democratic political experience and also lacking sufficient resources, the fledgling environmental movements were unable to benefit from the overtly pro-environmental stances of the new dissident-led administrations (Fagin and Jehlička 1998). In short, the early 1990s did not see the strengthening of environmental movement in ECE and development along a continuum of professionalism and political inclusion; rather it was a period of turmoil, fragmentation and in many cases political exclusion. After a brief interlude of popularity and involvement in progressive policy-making, environmental movements were excluded from the formal policy process and their public support went into rapid decline, coinciding with the start of economic restructuring (Fagin 2000). A similar pattern of loss of environmental movement influence is confirmed for many other countries in the region including Bulgaria (Baker and Baumgartl 1998), Hungary (O'Toole and Hanf 1998; K. Pickvance 1998; Pickvance 1999), Poland (Hicks 1996; Millard 1998), Romania (Dragomirescu *et al.* 1998), Russia (K. Pickvance 1998; Pickvance 1999) and Slovakia (Podoba 1998).

Insofar as the importance of resources in enabling environmental movement activity is acknowledged, the specific complexities of realising such resources (especially financial) in post-communist ECE are particularly critical in terms of explaining asymmetries of movement efficacy. In the absence of sufficient donations from private individuals – who, as a consequence of economic restructuring, continue to lack levels of disposable income necessary to provide environmental groups with the funds on which their western counterparts commonly rely – groups often depend on small amounts of state funding and donations from foreign agencies and organisations. While such foreign assistance has been critical during the past decade in keeping many groups in existence and maintaining projects and campaigns, there has been limited investment in the infrastructure of the environmental movements. For example, both EU funds (e.g. PHARE) and grants from the UK Know How Fund cannot be used for salaries or the general development of groups, but must be targeted towards specific projects. Indigenous organisations, such as Zelený kruh (Green Circle), set up to co-ordinate the activities of Czech groups and to strengthen the movement as a whole, have been denied external funding despite their critical role. While groups have been offered help with auditing, fund-raising and general management skills, the lack of resources to employ such professionals dedicated to such activities limits the effectiveness of this assistance (see Fagin 1999).

Thus, in general, the nature of such foreign assistance has fostered dependency rather than autonomy. The fact that such financial support is largely short term – and has in many cases already been re-targeted to parts of Southeastern Europe – does not provide a basis for building a strong and sustainable environmental movement in the region. Where it exists at all, state funding for environmental groups in ECE tends also to favour capital projects delivered by largely apolitical NGOs rather than capacity-building in more politically prominent 'advocacy' groups. This funding scenario has undoubtedly encouraged a noticeable ideological shift among environmental movements towards conventional and non-radical policy stances, as well as inhibiting a long-term developmental perspective. The fierce competition for scarce state resources as well as for foreign aid has also fostered a competitive rather than co-operative culture among groups as well as harnessing them, through financial dependency, to the national political context.

In the early 1990s in the Czech Republic, nascent environmental organisations initially received funding from the new dissident-led administration (Moldan 1990), campaigned on global issues, and sought to widen their activities through membership of large international groups (e.g. Friends of the Earth International (FoE-I)). After 1992 however, in response to the virtual curtailment of state support, they quickly retreated into the limiting and often depoliticised zone of domestic environmental politics. During this period such groups not only feared the withdrawal of their funding but also felt constrained by the (generally) accurate perception that post-socialist society was deeply conservative and hostile to radical critiques (especially those focused on re-emerging national polities) or novel repertoires of protest such as non-violent direct action.

Towards global 'civic politics'

An analysis of environmental movement activity in ECE therefore provides a valuable insight into the asymmetric nature of so-called 'global civic politics'. Considering the ascribed role of environmentalists and other global grass-roots activists within this model – co-operating across national and regional boundaries to protect the interests of sections of the community that global capitalism has abandoned – it is somewhat ironic that the unequal power relationship between states and regions (an endemic feature of monopoly capitalism) should also configure the relationship between environmental movements at societal level. However, interpreted through the more radical perspective outlined above, this may not be as surprising, as groups entering into the dominant discourse (ecological modernisation) in institutional arenas may serve only to legitimise dominance in the process (see McKechnie and Welsh 2000).

While large western-based environmental movements enjoy close co-operation with similar organisations in neighbouring states, the relationship between such movements and their East European counterparts is instead power-structured where the former can dominate the conditions and nature of co-operation. For example, the relationship between Greenpeace International and one of its newly established offices in ECE provides an illustration of such relationships and their limits. In 1998, Greenpeace International undertook a geo-political reassessment of the organisation's global capacity and its potential to maximise campaigning goals into the next century. In some regions, this led to the difficult decision to rationalise resources by merging some small national offices into regional campaign entities, a process that in ECE would have subsumed the Czech office within the governance structures of Greenpeace Austria. This was deemed unacceptable by the Czech staff, given the likely negative impact of Austrian linkages both on campaign effectiveness and fund-raising potential. Although the rationalisation of the Czech office was later rescinded, virtually all the staff would have preferred to have resigned rather than stay in what they perceived as an organisation that lacked sufficient national autonomy and identity. Thus power asymmetry can be seen to have certain limits. The Greenpeace example is by no means unique: FoE-I refused the Czech environmental movement Děti Země (Children of the Earth) member status in the early 1990s on the basis that they lacked sufficient resources and support. The fact that the organisation had played, and continues to play, a key role in the post-1989 environmental policy process was apparently overlooked. In the end, FoE national partner status was awarded to the larger group Hnutí Duha (Rainbow Movement).

To some extent, the absence of sufficient resources among environmental movements in ECE and the relative inexperience of their activists facilitates an unequal power relationship. Yet there is little evidence to suggest that such an imbalance is diminishing: much as 'developing' states never become 'developed' and thus 'catch up' with the 'developed' countries, so it seems unlikely that, given the continued lack of resources and slow development of civil society in ECE, eastern environmental movements will 'catch up' and co-operate globally on equal terms with their western counterparts. In this sense, 'global civic politics' and 'grass-roots globalisation' are little more than idealistic western perspectives that are – in the short term – of limited relevance to environmental movement activity in East Central Europe.

References

Allen, J. and Thompson, G. (1997) 'Think global, then think again: economic globalization in context'. *Area* 29(3), pp.213–27.

Axelrod, R. (1999) 'Democracy and nuclear power in the Czech Republic'. In *The Global Environment. Institutions, Law and Policy* (eds N. J. Vig and R. S. Axelrod). London: Earthscan, pp.279–99.

Baker, S. (1996) 'The scope for east–west co-operation'. In *Environmental Policy in Context.* Vol. 3: *Prospects* (eds A. Blowers and P. Glasbergen). London: Arnold, pp.135–65.

Baker, S. and Baumgartl, B. (1998) 'Bulgaria: managing the environment in an unstable transition'. *Environmental Politics* 7(1), pp.183–206.

Baker, S. and Jehlička, P. (eds) (1998) *Dilemmas of Transition: The Environment, Democracy and Economic Reform in East Central Europe.* London: Frank Cass.

Beck, U. (2000) *What is Globalization?* Cambridge: Polity Press.

Bridge, G. (1998) 'Excavating nature: environmental narratives and discursive regulation in the mining industry'. In *An Unruly World?* (eds A. Herod *et al.*). London: Routledge, pp.219–44.

Clark, A. M., Friedman, E. J. and Hochstetler, K. (1998) 'The sovereign limits of global civil society: a comparison of NGO participation in UN world conferences on the environment, human rights and women'. *World Politics* 51, pp.1–35.

Cole, D. H. (1998) *Instituting Environmental Protection: From Red to Green in Poland.* London: Macmillan.

Dalton, R. J. and Rohrschneider, R. (1999) 'Transnational environmentalism: do environmental groups co-operate globally?' Paper presented at the ECPR workshop, Environmental Protest in Comparative Perspective, Mannheim, Germany, 26–31 March.

Diani, M. and Donati, P. R. (1999) 'Organisational change in western European environmental groups: a framework for analysis'. *Environmental Politics* 8(1), pp.13–34.

Doherty, B. (1999) 'Manufactured vulnerability: eco-activist tactics in Britain'. *Mobilization* 4(1), pp.75–89.

Dragomirescu, S., Muica, C. and Turnock, D. (1998) 'Environmental action during Romania's early transition years'. *Environmental Politics* 7(1), pp.162–82.

Elster, J., Offe, C. and Preuss, U. K. (1998) *Institutional Design in Post-Communist Societies: Rebuilding the Ship at Sea.* Cambridge: Cambridge University Press.

Fagin, A. (1999) 'The development of civil society in the Czech Republic: the environmental sector as a measure of associational activity'. *Journal of European Area Studies* 7(1), pp.91–108.

Fagin, A. (2000) Environmental protest in the Czech Republic: three stages of post-communist development. *Czech Sociological Review* 8(2), pp.1–18.

Fagin, A. and Jehlička, P. (1998) 'Sustainable development in the Czech Republic: a doomed process?' *Environmental Politics* 7(1), pp.113–28.

Falk, R. (1999) *Predatory Globalization: A Critique.* Cambridge: Polity Press.

Fukuyama, F. (1992) *The End of History and the Last Man.* London: Hamish Hamilton.

Füllenbach, J. (1981) *European Environmental Policy: East and West.* London: Butterworths.

Gowan, P. (1995) 'Neo-liberal theory and practice in eastern Europe'. *New Left Review* 213, pp.3–60.

Gowan, P. (1999) *The Global Gamble: Washington's Faustian Bid for World Dominance.* London: Verso.

Greider, W. (1997) *One World, Ready or Not: The Manic Logic of Global Capitalism.* New York: Simon and Schuster.

Haas, P. M., Keohane, R. O. and Levy, M. A. (eds) (1993) *Institutions for the Earth: Sources of Effective Environmental Protection.* Cambridge, MA: MIT Press.

Held, D. (1998) 'Democracy and globalisation'. In *Re-imagining Political Community* (eds D. Archibugi *et al.*). Stanford, CA: Stanford University Press, pp.96–120.

Held, D., McGrew, A., Goldblatt, D. and Perraton, J. (1999) *Global Transformations. Politics, Economics and Culture.* Cambridge: Polity Press.

Hicks, B. (1996) *Environmental Politics in Poland: A Social Movement Between Regime and Opposition*. New York: Columbia University Press.

Kotkin, S. (1995) *Magnetic Mountain: Stalinism as a Civilization*. Berkeley, CA: University of California Press.

Langer, H. (1990) 'Ecological bricks for a common Europe that can survive'. *Politische Okologie* 8(2), pp.3–5.

Lipschutz, R. (1996) *Global Civil Society and Global Environmental Governance*. Albany, NY: SUNY Press.

McKechnie, R. and Welsh, I. (2000) 'When the global meets the local: critical reflections on reflexive modernisation'. In *Sociological Theory and the Environment* (eds F. Buttel, P. Dickens, R. Dunlap, R. and A. Gijswijt). Wichita, NE: Rowman and Littlefield, pp.193–212.

Mazey, S. and Richardson, J. (1992) 'Environmental groups and the EC: challenges and opportunities'. *Environmental Politics* 1(4), pp.109–28.

Millard, F. (1998) 'Environmental policy in Poland'. *Environmental Politics* 7(1), pp.145–61.

Moldan, B. (1990) *Zivotni prostredi Ceske republiky* (Environment of the Czech Republic). Prague: Academia.

Moldan, B. and Klarer, J. (eds) (1997) *The Environmental Challenge for Central European Economies in Transition*. Chichester: Wiley.

Ohmae, K. (1995) *The End of the Nation State*. New York: Free Press.

O'Toole, L. J. Jr and Hanf, K. (1998) 'Hungary: political transformations and environmental challenge'. *Environmental Politics* 7(1), pp.93–112.

Pickvance, C. G. (1998) 'Decentralisation and democracy in eastern Europe: a sceptical approach'. *Environment and Planning C: Government and Policy* 15, pp.129–42.

Pickvance, C. G. (1999) 'Democratisation and the decline of social movements: the effects of regime change on collective action in eastern Europe, southern Europe and Latin America'. *Sociology* 33(2), pp.353–72.

Pickvance, K. (1998) *Democracy and Environmental Movements in Eastern Europe*. Boulder, CO: Westview Press.

Podoba, J. (1998) 'Rejecting green velvet: transition, environment and nationalism in Slovakia'. *Environmental Politics* 7(1), pp.129–44.

Princen, T. and Finger, M. (1994) *Environmental NGOs in World Politics: Linking the Local and Global*. London: Routledge.

Rootes, C. A. (1999) 'Acting globally, thinking locally? Prospects for a global environmental movement'. *Environmental Politics* 8(1), pp.290–310.

Russell, J. (1991) *Energy and Environmental Conflicts in East-Central Europe: The Case of Power Generation*. London: Royal Institute of International Affairs.

Schopflin, G. (1993) *Politics in Eastern Europe*. Oxford: Blackwell.

Swyngedouw, E. (1992) 'The mammon quest: "glocalisation", interspatial competition and the monetary order: the construction of new scales'. In *Cities and Regions in the New Europe* (eds M. Dunford and G. Kafkalas). London: Belhaven Press, pp.39–67.

Tickle, A. (2000) 'Regulating environmental space in socialist and post-socialist systems: nature and landscape conservation in the Czech Republic'. *Journal of European Area Studies* 8(1), pp.55–76.

Tickle, A. and Vavroušek, J. (1998) Environmental politics in the former Czechoslovakia. In *Environment and Society in Eastern Europe* (eds A. Tickle and I. Welsh). Harlow: Longman, pp.114–45.

Tickle, A. and Welsh, I. (eds) (1998) *Environment and Society in Eastern Europe*. Harlow: Longman.

van der Heijden, H-A. (1999) 'Environmental movements, ecological modernisation and political opportunity structures'. *Environmental Politics* 8(1), pp.199–221.

Wapner, P. (1995) 'Politics beyond the state: environmental activism and world civic politics'. *World Politics* 47(3), pp.311–40.

Wapner, P. (1996) *Environmental Activism and World Civic Politics.* New York: State University of New York Press.

Weiner, D. R. (1999) *A Little Corner of Freedom. Russian Nature Protection from Stalin to Gorbachev.* Berkeley, CA: University of California Press.

Welsh, I. (1999) 'Risk, "race" and global environmental regulation'. In *Global Futures: Migration, Environment and Globalisation* (eds A. Brah, J. Hickman and M. Ghail). London: Macmillan, pp.47–69.

Welsh, I. (2000) 'New social movements and power'. In *Sociology: An Annual Review* (ed. M. Haralambos). Ormskirk: Causeway Press, pp.21–46.

Welsh, I. and Tickle, A. (1998) 'The 1989 revolutions and environmental politics in central and eastern Europe'. In *Environment and Society in Eastern Europe* (eds A. Tickle and I. Welsh). Harlow: Longman, pp.1–29.

Williams, K. (1995) 'The foreign relations of independent Slovakia'. *Slovo* 8(2), pp.87–108.

4 The central importance of the European Union

David Turnock

At the moment the EU is very much in the driving seat in much of the region, having provided substantial aid since 1989 and attracted strong support for eventual membership in the interests of security and trade (Hudson 1994; Jachtenfuchs 1992). The CIS lies outside this framework but the only other group of non-candidate countries is located in the West Balkans where there is a widespread desire to overcome the recent history of war which has effectively blocked progress towards integration. Of course there are uncertainties over the enlargement process and the weaker candidates may find the financial burden of approximation unsustainable. In each country serious doubts are expressed and public opinion is sceptical. However, no alternative to early integration with the rest of Europe is seriously considered outside the CIS where Russia's links with the West are balanced by its massive stake in Asian affairs. Meanwhile there is strong support for enlargement among the present membership and a realisation in the European Commission that the promise of membership must be backed by political will to expedite entry as quickly as possible through a readiness to compromise.

On this basis it is necessary to consider European norms as a key factor influencing the shape of current environmental policy which the EU believes will be a force for stability, with a better environment reducing transboundary problems and demonstrating the benefits of reform and restructuring (Baker and Jehlička 1998, p.18). PHARE (Poland-Hungary: Assistance for Restructuring their Economies) has provided investments leading to the adoption and application of EU law and reinforcement of administrative and judicial capacity (Marwaha 1993). Such assistance has become more effective in setting realistic targets for accession, commensurate with limited absorptive capacity for investment. On the other hand, East Central European countries have accepted 'asymmetry' in the interim association agreements so as to meet the requirements of the present member states on trade and migration (Caddy 1997), for there is particular concern over competition from low-priced manufactures and excessive diversion of structural funding. However, good prospects for full membership for associated countries were confirmed at the European Council meeting in Copenhagen in 1993 which laid down key criteria; followed by a pre-accession strategy which emerged from the Essen Summit (1994) and a

White Paper on key measures to be adopted (1995). It is now clear that the Czech Republic, Estonia, Hungary, Poland and Slovenia are on the 'fast track' for accession around 2002, while negotiations began in 2000 to bring in Bulgaria, Latvia, Lithuania, Romania and Slovakia in a second wave later in the decade.

Sustainable development

There is some concern over the EU's conservation credentials and in particular the growth–conservation balance. Conservationists fear that too much growth is being anticipated – for the EU is fundamentally an economic association (Anon. 1998a, p.174), while there is talk of a 'democratic deficit' and a lack of transparency in the work of EU institutions. Scepticism arises from experience with the Common Agricultural Policy (CAP) which has promoted unhealthy intensive farming. Thus while there may well be technical solutions to problems of air and water pollution, the long-term impact of accession may be biodiversity loss if, for example, headage payments increase grazing pressure, perhaps with irreversible results, through resource depletion in sensitive areas. PHARE is perceived as having a history of preference for large projects, whereas clusters of small projects might be more effective; this suggests that conservation groups need a presence in Brussels to engage with the relevant EU organisations. However, the focus is now very much on capacity-building for sustainable development to meet the needs of the present without compromising future generations. EU Environmental Action Programmes regard environmental protection as an essential task.

During 1973 to 1976 and 1977 to 1981 the priority was to avoid damaging policy impact; and in 1982 to 1986 and 1987 to 1992 the emphasis was on risks and benefits of environmental policies for the internal market. The European Environmental Agency stated in its 'Dobříš Assessment' that 'the decline of Europe's biodiversity in many regions derives mainly from highly intensive partially industrial forms of agricultural and silvicultural landuse; from an increased fragmentation of remaining natural habitats by infrastructure and urbanisation and the exposure to mass tourism as well as pollution of water and air' (Gyulai 1998, p.45). However, during 1992 to 1999 there was a stronger accent on 'paradigmatic change' to reflect global environmental risks, with proposed structural change in favour of public transport, energy efficiency and waste prevention (including a crucial role for ENGOs and local/regional authorities). For the 1996 Cork Declaration stated that rural policies inspired by CAP 'encourage productivity and competitiveness to the detriment of environment and social concerns' (Anon. 1998b, p.73). In future, there would be integrated programmes of sustainable rural development for each region. Policy was also to be decentralised, according to the principle of subsidiarity: based on partnership and participation at all levels to harness 'the creativity and solidarity of rural communities' (ibid., p.109).

Environmental policy was bolstered by the Amsterdam Treaty of 1997 which stated that sustainable development and integration of environmental considerations in all EU-policy areas were key objectives. Environmental considerations must be integrated into all EU-policy areas. There is now pressure for a shift from 'material intensive growth' – involving higher resource inputs – to 'relative delinking' and some 'absolute delinking' (e.g. sulphur dioxide emissions). In future there will be the challenge of 'dematerialisation' involving reduction in raw material input by a factor of 10 which will bring Europe in sight of a notional 'fair share of environmental space' and address the principle of equity and social justice included in Rio Agenda 21. Of course, member states are reacting rather slowly to this pressure (e.g. regarding transposition of the 'Birds and Habitats Directives' and site designations for a European 'Natura 2000' network of protected areas). Aspects of Agenda 2000 (concerning enlargement, finance and agricultural policy and structural funding) have not been placed explicitly in a sustainability framework. Further, according to Kouvelis (1998, p.21), especially where the environment is concerned, 'the structural funds have had an impact ranging from beneficial to devastating'. However, during the first funding round (1988–93), environmental concerns were not built into the funding methodology; whereas the second round (1994–99) did incorporate environmental clauses with positive results and the 2000 to 2006 period will provide opportunity for new regulations coincident with enlargement. The sustainability agenda is being pressed more and more strongly with projects financed by ISPA (Instrument for Structural Policies for Pre-Accession Aid) and SAPARD (Special Action for Pre-Accession Measures for Agriculture and Rural Development) as well as PHARE – together worth Euro500mln annually during 1999 to 2106 – assessed on this basis. Thus ISPA finance for the Bulgarian motorway through the Kresna Gorge has brought the EU into a growing ecological controversy.

Meanwhile, the World Bank's new environmental strategy for East Central Europe recognises that air pollution has come down due to lower industrial output and that there are threats arising from higher levels of vehicle ownership and increased dependency on natural resource exploitation as a major source of export revenues and government income. While the Bank is committed to a central mission of poverty alleviation, this is to be approached through environmental interventions intended to directly benefit the poor (improved air and water quality) and changes in governance and public policies that will foster sustainable management of natural resources. Thus there is now less likelihood that East Central European countries will absorb funds through large-scale cohesion projects which could be environmentally damaging. While 'it is an interesting contradiction that native environment and cultural diversity could be best preserved in underdeveloped rural or boundary, politically isolated regions' (Gyulai 1998, p.51), there is some basis for optimism that these resources can be safeguarded through sustainable regimes in the future. The Regional Environment Centre (REC) for the area was set up at Szentendre in 1990 by the European Commission, the USA and Hungary to help fifteen transition states solve their environmental problems. Since then other nations have endorsed the charter and provided help, with the greatest national contributions coming from Denmark, Japan, the Netherlands and the USA.

Other forms of assistance are available, such as debt burden relief against environmental spending. During 2000, the European Parliament passed a resolution doubling environmental aid over the period to 2006 in respect of countries seeking EU accession, though it is unclear whether the EU will respond. Member states are doing a great deal individually to help ECECs; for example, with regard to waste water treatment to clean up waterways in which they have an interest. Thus Germany supports the Most and Teplice sewage stations which will help to clean up the Elbe River. However, this leaves heavy burdens for the candidate countries. Poland will spend Euro30.5bln on water purification, drainage, waste management and the curbing of industrial pollution over the next twelve years in order to comply with EU standards, and the bureaucratic implications of the present approximation task are very great. Agenda 2000 states that countries have an obligation to adopt the whole *acquis communautaire* (some 300 laws) including environmental provisions for air and water quality, chemicals, energy and transport – plus environmental impact assessment (EIA) which is now needed for all loans from the European Bank for Reconstruction and Development (EBRD) and the World Bank; also geology and mining, nature conservation, noise, nuclear safety; sewage treatment and waste management. Poland has found that while existing environment law often coincides with EU law, there are discrepancies in most fields (Anon. 1998a, p.164). Approximation will involve many different central bodies, including some yet to be identified (e.g. an organisation to handle noise in the Czech Republic). Slovenia has adopted a comprehensive environmental accession strategy (EAS) as an annex to the 'National Programme for the Adoption of the EU Acquis Communautaire': thirty-one working groups were appointed in 1998 to cover all areas of approximation regarding environmental legislation. However, late in 1998 officials from the ten accession countries were involved in discussions on the environment investment necessary (Amman *et al.* 1995). Furthermore, projects have to be managed by national and regional authorities of member states through monitoring committees with representatives of these authorities plus the European Commission and European Investment Bank (EIB). To encourage local participation in the enlargement process the EC Environment Directorate offers awards to Eastern European cities for their compliance with EU standards. Recent winners were Bourgas (Bulgaria), Elbląg (Poland), Kecskemét (Hungary), Maribor (Slovenia), Parnu (Estonia), Râmnicu Vâlcea (Romania), Spišska Nova Ves (Slovakia) and Svitavy (Czech Republic).

In early 1999 officials from the first group of accession countries had intensive talks with sixteen units of the Commission's environmental directorate (DGXI) to review the main obstacles. It then became clear that not enough headway was being made as a prelude to political negotiations, and that accession might be delayed for a decade in order to meet all environmental standards. Waste water treatment plants for all communities with over 2000 inhabitants will be very difficult to realise. In Slovenia, EU-compliant sewage facilities cannot be complete before 2015 to 2017, given past neglect, and nature protection is also deficient. A World Bank study has found that US$2bln will be needed to bring Czech water hygiene up to EU standards by 2005 and environmental protection legislation is still inadequate for waste management (Hempen 1998). Hungary has problems with waste management and integrated pollution prevention; she also wants to delay compliance over water quality rules since improving water quality will take up to half the E8.6bln that Hungary expects to invest to comply with environmental *acquis* (by 1994 only 42 per cent of Hungary's households were connected to secondary waste water treatment facilities compared with 90 per cent in Austria). However, during 1999 the Commission President R. Prodi signalled reduced pressure over environmental standards to enable the first states to accede in 2002. Even so, the ten states need to spend E7.5bln p.a. to bring sulphur dioxide and nitrogen oxide emissions to acceptable levels. And in the process of 'Compliance Costing for Approximation of EU Legislation in the CEEC', the EU has quoted E100–230bln as the cost of bringing the environmental protection programme in these same countries to the EU level. Even so, it seems doubtful whether this takes into account all costs needed to achieve the desired level of environmental quality rather than just compliance.

The role of non-governmental organisations

The desire for a better quality of life gave considerable authority to environmentalists in the early 1990s. However, with growing unemployment it became clear that the public would have to accept pollution in order to retain jobs (Fisher 1992; Waller and Millard 1992) and a sense of malaise in the mid-1990s gave rise to more ambivalent assessments of the non-governmental sector. But NGOs in general have no tradition in the region and so they lack both experience and expertise. At a time when there is a general lack of institutional capacity and resources, even a modest level of activity is to be welcomed as a means of carrying out environmental projects and helping to boost environmental awareness generally (Rootes 1997). Effective networking is evident in the case of Bulgarian ENGOs opposing deforestation in the Pirin National Park where new ski trails may threaten its world heritage rating; while Polish ENGOs are attracting EU scrutiny of a government decision to proceed with dams on the Vistula at Niewzawa and Odra because the objective of preventing the erosion of existing dams will create greater problems downstream. ENGOs could interact better both with the public and governments if there was less reserve over sharing influence with 'outsiders': the public tends to be informed rather than consulted and awareness of the environmental implications of EU accession is low (Toth Nagy 1993). Even in Germany, where ENGOs are consulted widely, activists feel that the authorities are cautious and do not take volunteers seriously. However, for their part ENGOs need to be objective, accurate and dispassionate to gain credibility: aiming at high-quality participation without unnecessarily negative or stalling tactics. It would therefore be unreasonable to 'give up' on ENGOs and local environmental action through small projects, for the record is by no means insignificant (Jancar-Webster 1998). In response to greater public awareness and EU pressure, most political parties have absorbed elements of the Green agenda and politicians can gain influence by championing local issues, such as the Kővágószöllös waste repository proposal in Hungary (Szirmai 1993, p.155).

Some ENGOs are particularly enterprising: for example, the Foundation for Applied Environmental Research set up in Slovakia in 1993 by young professionals dedicated to environmental research and conservation. In 1997 the organisation became known as 'Daphne: Centre for Applied Ecology' with its own publication, *Životne Prostredie*. It is much concerned with sustainable watersheds and specific projects such as the 1200ha Morava floodplain where 200ha of meadowland have been restored, while all the arable land in the Ramsar area will eventually be converted. Daphne has produced management plans for meadow biodiversity in Kroipanec-Javorina national park; also for sustainable use of meadows pastures and wetlands in Velke Borove village. A pathway through the area has been opened up and various aspects of revitalisation, sustainable agriculture and tourism have been handled through triangular cooperation including organisations in Austria ('Distelverein' and the Worldwide Fund for Nature (WWF)) and the Czech Republic ('Veronica'). Annual conferences are held and information campaigns focus on 'Wetlands for Life' and 'Meadows for Life'. The group is also involved in the WWF Green Danube project and analysis of ecological agriculture in the lowland Záhorie region. Meanwhile, the Czech Environmental Partnership (Nadace Partnerstvi) is working with sister organisations in Hungary, Poland and Slovakia – which together form the Environmental Partnership for Central Europe (EPCE) – to promote environmentally sustainable civil society by assisting non-profit community groups and ENGOs through grants, fellowships, technical assistance and networking. Many of the grants go to local government–ENGO partnerships and there is much spending on environmental issues in underdeveloped regions. A major achievement is the Prague–Vienna Greenway (opened for walkers in 1995 and for cyclists in 1997) which ties in with landscape detail, local culture and monuments. Initiated by the civic association Zelene stezky ('Greenway'), management was taken over by EPCE in 1998 and further corridors are now being provided for the Elbe–Morava Watershed, the Krkonoše and the White Carpathians. On a smaller scale, sustainable development is the role of the 'Biebrza Connoisseurs Society of Ecotourism', a local organisation established in 1999 to develop sustainable tourism in the Polish Biebrza wetland region. The aim is also to raise the level of environmental awareness among local people by demonstrating how conservation can be profitable when responsible use is made of the natural and cultural heritage through the package tour office Biebrza Eco-Travel.

To be sure, more stimulation is needed to expand the ENGO network. There are some national organisations like Latvia's Centre for the Support of NGOs, while Slovakia offers small grants to help organisations which are active in conservation. There is also assistance from European organisations. Friends of the Earth Europe (FoEE) is a network of national environmental organisations in twenty-nine countries aiming at a sustainable society with help through coordination of campaigns of member groups. The European Environmental

Bureau (EEB), representing some 125 environmental organisations in twenty-four countries, focuses on the EU role in the environment within and beyond present EU borders and monitors European institutions regarding their respect for environmental protection. There is also much effective action being undertaken by the REC because relations between government and ENGOs in Yugoslavia are particularly poor. A five-year South Eastern European Development of Environmental Society Initiative will seek to stimulate ENGOs and rebuild civil society in general. REC has also been encouraging transboundary arrangements, and in 1999 ENGOs from Albania, Bulgaria, Bosnia and Hercegovina, Croatia, Macedonia, Slovenia, Romania and Yugoslavia were able to benefit when the Swiss Agency for Development and Dutch Ministry of Agriculture, Nature Management and Fisheries asked the organisation to distribute Euro125,000 to encourage environmental projects giving people in the region better contact with and understanding of nature around them. ENGOs have been encouraged to work with local communities, farmers and park administrations to come up with proposals. Meanwhile, USAID (United States Agency for International Development) has offered US$22mln in partnership grants to promote market-oriented solutions to environmental problems in the region. Powerful Information based in Milton Keynes (UK) provides information and training for ENGOs in the region. Reference should also be made to a database for local environmental action provided by Coordinated Action for Pan-European Transport and Environment Telematics Implementation Support (CAPE).

However, there is a risk of 'donor addiction syndrome', as in Bosnia and Hercegovina where short-term technical fixes may not have a positive long-term effect because of the failure to develop viable indigenous institutions to ensure an adequate level of participation. Now is a crucial time when ENGOs should be playing a fuller role. The total money involved is considerable: E235bln during the third (2000–06) funding round, combining cohesion and structural funding and pre-accession aid, which can be doubled or tripled through loans from development banks (EBRD, EIB and the World Bank) with additional funding from national, regional and private sources (Kouvelis 1998). Better information and public participation should follow implementation of the EU Directive on Environmental Information and Directives of the Århus (1998) Convention on Public Access to Information Environmental Decision-Making and Justice. Local authorities will then have to release specific environmental information, whereas previously (for example) Polish authorities were obliged to inform only recognised ENGOs about investments that might threaten the environment. Lack of transparency through ineffective participation means that too many programmes are being run by elite consultants (transport specialists for ISPA and agriculture/regional development for SAPARD) without proper evaluation by local authorities and ENGOs which need to be involved not only in a consultative capacity but more directly through the practical implementation of funds. Indeed, the Amsterdam Treaty emphasised partnership in structural funding, requiring each member state to designate representative partners including bodies active in the environment field. A joint publication has been undertaken by Bund: Freunde der Erde (Germany) and FoEE as a resource to stimulate ENGO awareness and participation regarding regional policy (now grounded in sustainability) in both member states and candidate countries (Maier *et al.* n.d.). It is good that there is a project for increasing awareness and NGO consciousness of the EU to facilitate an active role by ENGOs in the accession process. Implementation of Poland's 1998 Cooperation Declaration on ENGOs has been slow, but the public is now gaining in self-confidence, and lack of information or consultation can result in blockades and demonstrations, while failures to follow rigorous EIA procedures (including public hearings) are now more likely to be challenged.

Transport, energy and rural issues

A consequence of EU membership is that incomes and car ownership are likely to rise, thereby increasing road congestion and raising important questions over sustainability. Will East Central Europe learn from the West and constrain traffic growth through taxation, planning and investment policies to ensure that people pay the full social cost of travel; with investment in alternatives to cars and planes? Lead-free petrol has become widely available during the 1990s. Mannesmann AG is supplying and assembling equipment for a planned second catalytic reformer unit (cost US$39.3mln) at Bulgaria's largest refinery (Neftochim) which will produce lead-free petrol after 2002 conforming with the Convention for Clean Air in Europe: excise taxes on unleaded petrol have already been halved, which will help to compensate drivers for the higher cost of lower sulphur fuel from the upgraded refinery. Meanwhile, leaded petrol went off the market in Hungary in 1999 and Poland's largest fuel retailer (CPN) also stopped selling leaded fuel in 1999, ahead of the EU ban on leaded sales which came into effect in 2000. There will be problems for owners of old vehicles, but additives are becoming available and in Poland U95 petrol with potassium-based anti-knock agents is available. The motor industry should be able to cope easily with more stringent emissions standards for new vehicles (given the industry's experience in producing for European markets), but inspection and maintenance procedures could make it very expensive to keep old vehicles on the road. Currently they are widely imported into the region and the business offers a good deal to the sellers because high disposal costs are avoided, while greater toxic pollution in ECE countries has 'turned them into the EU's dumping ground' (Higman 1998, p.48). It is almost a form of waste export, and the solution lies in lower disposal costs in the EU.

Much of EU transport policy is concerned with fair and efficient pricing for a 'citizens' network', but there is a more controversial aspect through the Trans-European Transport Network (TETN), coordinated by TINA in Vienna, which will cost E400–500bln for completion by 2015. Most of the money will come from member states, but one-tenth of funds for any one scheme can come from EU through ISPA (e.g. for the Sentjakob–Blagovica road in Slovenia).

The TETN seeks growth and employment, but new links with Eastern Europe could release a flood of cheap imports from the European centre. The European Commission has never carried out a rigorous EIA of the TETN to assess the impact of cheaper travel in terms of more journeys and increased pollution. However, given the support from the Cohesion Fund, governments will be 'sorely tempted to take the money and complete the TETN with appalling environmental consequences' (Higman 1998, p.25). Airports pose threats since increased air traffic means pressure for airport development which local environmentalists may contest. For example, a proposed freight airport at Cochstedt in the rich Magdeburger *borde* will develop a former Soviet base where an airport and business park could create up to 2000 jobs with a total investment exceeding E50mln. But it seems doubtful if full capacity will be reached and EIA has been limited: there is a danger of noise and air pollution and road congestion (since all the freight would have to be moved by road). ENGOs in the Baltic States are backing sustainable transport through a multi-model rail-based transport corridor between Klaipeda, Minsk and Moscow which might attract ISPA funding and stand as a sound alternative to road links with Western Europe. However, the Eesti Express (Tallinn–Warsaw) has been withdrawn due to lack of passengers.

Although the road transport corridors are not crucial for accession, East European governments have been investing heavily and the results in some cases could be excess capacity combined with pollution and crime associated with a growing transit traffic which does not bring substantial economic gain to the countries concerned. For example, Hungary has a ten-year US$2.6bln highway programme including a four-lane motorway to the depressed northeast, with two-thirds of the funding from ISPA. Montenegro is supporting a Greek proposal for an Adriatic highway following the coast through Albania, Montenegro, Croatia, Bosnia and Hercegovina, and Slovenia as far as Italy.

There is particular controversy over the Prague–Dresden motorway which is set to parallel the upgraded railway corridor through the České Středohoří Protected Landscape Area. Conservationists believe there is a threat to biodiversity following an inadequate EIA. Experts have recommended that the motorway should end at Lovosice where traffic could be transferred to the rail corridor: a combined system could accommodate 20,000 lorries each year. Another approach would involve a modified route including a tunnel through parts of the sensitive area. On the other hand, if the road goes ahead, subsidies for combined transport on the railway will be threatened. Meanwhile there are concerns over the northern section of the Budapest orbital road (MO). Hungarian and EU law has allegedly been breached (through an EIA flawed by inadequate information about public enquiry hearings) and residential areas near the road have been left open to pollution. In Poland the Poznań bypass, backed by PHARE and the EIB, is closely following the alignment of a 1940s German plan which threatens water resources and recreational sites: despite protests and blockades, the 'Poland Needs Roads' lobby is winning.

The most recent controversy concerns the Bulgarian General Roads Authority's desire to route the Strouma motorway (part of the Trans-European Transport

Corridor Four) through the Kresna Gorge. This follows the feasibility study by SPEA Ingegneria Europea of Italy, which seems to have considered only the biodiversity value of the Tissata reserve and not the importance of the gorge as a whole, which comprises a rocky 17km section of the Struma where juniper, pine and plane woodlands are represented. Indeed, there is no other site in Bulgaria with such a high concentration of rare and vulnerable vertebrate and invertebrate species. The biodiversity of the gorge – a bottleneck site on the Via Aristotelis bird migration route – is of international importance and will feature in the European Ecological Network 'Europa 2000': 123 species of nesting birds, 17 species of bats, 31 species of reptiles and amphibians and 14 species of fish. There are also tortoises and snakes, but the bats are of particular note: in no other part of the Balkans is there such diversity, given the climate and landscape features and the rich insect fauna; yet some of the roosts and shelters are highly vulnerable to disturbance. In addition to the upheaval of construction damage – from which many species will not recover due to their slow rate of reproduction – permanent damage will arise from the barrier posed by the road which causes the death of many animals that try to migrate across it. NGOs consider that the government commission underestimated the damage. There will also be damage to the economy of Kresna town which is a potential growth area, where nature-based tourism provides a source of income. The new road will cross the town and undermine its local trading role. Bulgarian ENGOs – Balkan Wildlife Society, the Environmental and Education Centre and Eco-Club 2000 – have spoken out against the plan from an early stage and have presented their views to the president, prime minister and the ministers both of Regional Development and Public Works and Environment and Waters, as well as the GRA. They want the road to take a different route which will involve viaducts and tunnels, with the possibility of a zero alternative based on combined road–rail transport. Since the road is to attract ISPA funding, the EU can take a legitimate interest in the controversy and endorse the views of Bulgarian ENGOs that the project should not violate EU environmental directives.

Energy issues

Energy consumption must be reduced as a matter of efficiency, given extremely high energy intensity levels (Table 4.1) and the challenge to achieve a better quality of life commensurate with 'fair shares' of energy consumption. Overall, East Central Europe is relatively more sustainable than Western Europe but may lose the advantage – given the growth in consumption expected over the next few years – if the region cannot leap-frog towards a better model of dematerialised sustainable development. There has been reduced energy demand following the power 'supply' ethos of the communist era. In addition, air quality legislation is helping to get emissions down through taxes motivated by the PPP concept: thus from 2000 Estonia is imposing an additional carbon dioxide tax on power companies. Pollution by power-stations has been reduced by better control systems (notwithstanding the much-quoted anomaly of solving air pollution

Table 4.1 Energy intensity 1980–95

	th.t oil equivalent per Emln GDP			
	1980	*1985*	*1990*	*1995*
East Central Europe	2.26	2.12	1.79	1.70
Bulgaria	2.35	2.04	1.69	1.79
Czech Republic	2.70	2.65	1.61	1.96
Hungary	1.19	1.13	1.02	1.00
Poland	2.71	2.73	2.20	1.92
Romania	2.16	1.85	2.05	1.67
Slovakia	2.01	1.96	2.07	1.97
Baltics	n.a.	n.a.	1.25	1.34
Confederation of Independent States	2.00	1.91	1.87	2.27
European Union	0.29	0.27	0.25	0.24
World	0.54	0.52	0.50	0.43

Source: European Commission 1999.

from a Czech power-station by using limestone quarried within a protected area (Baker and Jehlička 1998, p.21). 'Investment into efficient and clean technologies is desperately needed, but funding for such infrastructural needs are not being given sufficient priority in the move towards a market economy' (Green 1998, p.99). There is scope for a switch from coal given the large gas reserves of Eastern Europe and the FSU. The Black Triangle (the Czech–German–Polish border region) has already achieved reduced sulphur dioxide emissions through extension of the natural gas distribution system in Lusatia, and further change in Poland may be linked with a gas pipeline project for the Baltic which will take 3–4bln.cu.m p.a. from Norway to Poland.

All this will reduce pollution, especially in the context of revamped district heating systems (a characteristic of former socialist cities) which could be linked with refuse collection and incineration. In any case, efficient electricity generating (comparable with Western Europe) is required to improve the competitiveness of East European industry. A referendum in Slovenia in 1999 decided against the new 200MW Trbovlje power-station, proposed by Termoelektrana to maintain support of the Hrastnik and Trbovlje mines when the present plant is closed by 2004. Alternative energy options will require the economic and social restructuring of the Zasavje region. Meanwhile, Croatia plans three thermal power-stations for North/Central Dalmatia, Vukovar-Sirmium county and Zagreb to function on 'ecologically safe fuel' (not coal or nuclear energy). The Czech Republic's Kyjov district heating system has benefited from a Global Environment Facility (GEF) grant of US$5.8mln to reduce emissions of greenhouse gases and increase efficiency in the power/heat supply to Vetropak Moravia glass factory. Lignite will still be used, but with a reduced share of the total fuel mix. Finally, going against the grain (despite opposition) is Macedonia's plan to import high sulphur lignite from Greece, where environmental standards will not permit its use, in return for electricity from Bitola power-station which is not subject to the same constraints.

Some progress is being made with renewable energy, although the potential for hydropower in the Balkans is limited by risk of drought. Albania envisages eight small hydro-stations for the Vjosa valley in the south of the country: a high demand area which is inadequately served due to problems with the grid carrying power from the principal hydro-stations in the north. In 1999 a wind turbine was set up in Hungary on a 30m tower adjacent to the Bakony Power Plant's Inota thermal-station at a cost of HUF60mln. It should pay for itself in twenty years, but there will be an evaluation after two years to see if further investment is justified. Meanwhile in the Czech Republic, the Děčin geothermal energy heating project will improve the environmental health of the town by reducing consumption of non-renewable energy. However, greater efficiency in energy use is also necessary: Ukraine's Institute for Energy Saving considers that municipal and domestic consumption could be reduced by one-fifth. Greater efficiency is being encouraged by the Alliance to Save Energy and by organisations such as REC and USAID working in liaison with the Polish Network of Energy Cities (covering around forty small towns) and the Romanian Energy Policy Association. Slovakia has produced a practical guide to energy and buildings in Slovakia to be distributed among the building sector. Meanwhile, Ukrainian calculations point to scope for a staggering 80 per cent reduction in the industrial sector. Many industries cannot afford energy-efficient measures, 'but without them they will find it increasingly difficult to survive and compete in the European market particularly as energy costs increase' (Green 1998, p.99).

Nuclear power poses a dilemma: it allows the reduction in sulphur dioxide emissions and consumption of non-renewable resources, but there is great concern over safety at a time when some countries have a high nuclear dependence, especially for generation (as opposed to capacity); often exceeding the level for the EU as a whole in 1995 (Table 4.2). However, 'greater energy savings could save 30 percent of electricity generation in the applicant countries on average and as much as 80 percent in some' (Green 1998, p.85). Upgrading and modernisation over seven to ten years is recommended under Agenda 2000 for Dukovany and Temelin in the Czech Republic, Paks in Hungary and Mochovce in Slovakia, along with part of the capacity at Kozloduy in Bulgaria and Bohunice in Slovakia. However, some ENGOs doubt that western technology can make any of the Soviet reactors safe, and there is a consensus that some stations in EU candidate countries are definitely non-upgradeable and must therefore be closed. This will affect part of the Bohunice and Kozloduy complexes along with Ignalina in Lithuania. Nuclear politics however remain complex, as the countries responsible for suspect power-stations look for maximum western financial support for modernisation or decommissioning. Further, despite being a non-candidate country, Ukraine is pressurising the West to fund two new reactors in return for implementation of the decision to close Chernobyl entirely (though there are also domestic pressures for reduced nuclear dependence).

Table 4.2 Nuclear power 1980–95

	Capacity (GW)						*Generation (TWh)*					
	1980			*1995*			*1980*			*1995*		
	a	*b*	*c*	*a*	*b*	*c*	*a*	*b*	*c*	*a*	*b*	*c*
East Central Europe	70.7	1.7	2.4	93.7	8.8	9.4	323.3	10.7	3.3	361.7	55.0	15.2
Bulgaria	8.2	0.9	11.0	12.1	3.5	28.9	34.8	6.2	17.8	40.7	17.3	42.5
Czech Rep	n.a.	n.a.	0.0	13.9	1.8	12.9	52.7	0.0	0.0	60.6	12.2	20.1
Hungary	4.8	0.0	0.0	7.0	1.8	25.7	23.9	0.0	0.0	34.0	14.0	41.8
Poland	24.7	0.0	0.0	29.5	0.0	0.0	120.8	0.0	0.0	137.0	0.0	0.0
Romania	16.1	0.0	0.0	22.3	0.0	0.0	67.5	0.0	0.0	59.3	0.0	0.0
Slovakia	n.a.	n.a.	n.a.	7.1	1.6	22.5	20.0	4.5	22.5	25.6	11.4	44.5
Slovenia	n.a.	n.a.	n.a.	2.5	0.6	24.0	8.0	0.0	0.0	12.6	4.8	38.1
Baltics	n.a.	n.a.	n.a.	11.1	2.4	21.6	n.a.	n.a.	n.a.	26.2	11.8	45.0
CIS	266.8	14.0	5.2	342.6	37.9	11.1	1294.0	73.0	5.6	1292.7	182.2	14.1
Russia	n.a.	n.a.	n.a.	210.8	21.2	10.1	n.a.	n.a.	n.a.	880.6	99.5	11.3
Ukraine	n.a.	n.a.	n.a.	54.3	12.8	23.6	n.a.	n.a.	n.a.	194.0	70.5	36.3
EU	n.a.	n.a.	n.a.	538.8	119.6	22.2	n.a.	n.a.	n.a.	2328.1	810.1	34.8

Source: European Commission 1999.

Notes
a Electricity.
b Nuclear.
c Nuclear share (per cent).

The Czech Republic's Temelín project has been particularly controversial. Work began in 1986 with a view to constructing four 1000MW Soviet reactors but, although the first dissident-led post-communist government considered the design to be flawed, the centre–right Klaus administration awarded the American Westinghouse company a contract to modernise the plant in 1993. Despite studies revealing safety risks as well as inefficiency, the company was able to secure substantial loan guarantees for the Czech government from the American Export–Import Bank at a time when ENGOs were effectively marginalised by political pressure. Despite criticism of nuclear power within the ranks of the Social Democrats – who gained power in 1998 – raising the possibility of abandonment, further resources were provided to meet a predicted cost of over US$4bln for two reactors (half the number initially contemplated). ENGOs argue that the extra power is not needed and resourcing should go towards conservation measures to reduce energy consumption, but the government turned down Parliament's proposal for a national referendum and the first reactor was commissioned in 2000. There has been strong political pressure from Austria throughout the 1990s – and especially the provincial government of Upper Austria – since Temelin (near České Budějovice) is only 30km from the Austrian border. Because environmental impact studies were considered insufficient, the Czech border with Austria was blocked on several occasions by protesters during September 2000, but the power station has since opened, following attention to the malfunctioning steam generator safety-valves. Upper Austria has campaigned for the abandonment of Temelín as a condition of continued EU membership negotiations and has launched a US$8mln campaign based on the collection of environmental and safety information, though the EU does not agree with this stance. The Czechs insist that the project is economically viable but it remains to be seen if action is taken to block the country's energy exports in future. Temelín was reported to be working at 5 per cent capacity in November 2000: protests were continuing, with the Czechs refusing to talk while they remained under pressure.

Meanwhile, Bulgaria must agree to acceptable deadlines for closure of its 440MW reactors at Kozloduy before accession talks can begin. Under its 1993 agreement, Bulgaria was to close the four old reactors by 1998, but she now wants to keep them running until the end of their operational life after upgrades by Siemens and Westinghouse: hence the parliamentary strategy of closure during 2006 for the first two reactors and 2010 for others (allowing a thirty-year life span). After all, closure is costly – US$100mln for two reactors – and Bulgaria needs the power – but in order to show flexibility, revised closure dates of 2004 to 2005 and 2008 to 2010 were put forward in 1999. It is now agreed that the two oldest reactors at Kozloduy will close by 2003. Meanwhile, the two 1000MW units will be modernised during 2000 to 2005 by French, German and Russian companies at a cost of US$300mln (partly covered by a Russian loan of up to US$150mln). This will help with the decommissioning of the four unsafe 440MW units already referred to. Although the European Commission has signalled reduced pressure over compliance with the environmental *acquis* in advance of accession, nuclear safety is of paramount importance. Lithuania and Slovakia are also involved in the decommissioning of unsafe nuclear plant at Ignalina and Jaslovské Bohunice respectively. Ignalina 1 will shut by 2005 and the second probably by 2009 (with a final decision in 2004), though Lithuania wants to build a third reactor to compensate. Meanwhile, Romania has completed its first nuclear power-station at Cernavodă (440MW) and wishes to invest US$750mln in finishing a second reactor (currently 40 per cent complete), but this does not generate opposition because the Canadian Candu system (also used for a 600MW station in former Yugoslavia) is considered safe.

Uranium mining has been contracting rapidly since the better reserves have been exhausted and the activity is not profitable. One remaining mine in the Czech-Moravian Highlands has been a bone of contention between the Czech environment and industry ministries, though the EU has now been told that the mine will close by the end of 2001. Greater controversies arise over the dumping of waste which affects all nuclear countries, including Romania and Slovenia which operate the reliable Candu system (Kabala 1991). Arrangements initially negotiated with the FSU will not be satisfactory for all time since Russia will require increasingly large payments. Slovakia's arrangements to send fuel rods of Mayak have recently been put on hold. Further, while Bulgaria has been sending spent fuel to Russia (at costs 25–30 per cent below quotations from Western Europe) through barter deals involving Bulgarian food and medicines, Moscow passed a law in 1992 against disposing of nuclear waste from foreign countries on its territory. Fuel reprocessing in Russia generates a lot of liquid waste and in 1998 Minatom contemplated costs as high as US$1000 per kg. Bulgaria has also had difficulties with transit countries and Moldova's demand for higher transit fees prompted consideration of the Romanian–Ukrainian route avoiding Moldova altogether. Meanwhile the Bulgarian Atomic Committee raised the question in 1998 of a regional storage facility for high-level waste – at a cost of some US$100 million. And Yugoslavia has considered a nuclear dump drawing waste from all over Europe in the former Gabrovnica uranium mine at Kalna on the Bulgarian frontier, even though it is situated in the Stara Planina National Park and has attracted opposition from the NGO Green Table ('Zeleni Sto'). Hungary is unlikely to offer storage for waste from foreign countries since accommodating its own waste is an extremely sensitive issue, but Croatia plans a low to medium radioactive waste site in Moslavacka and Trgovska mountains in Sisak-Moslavina county.

Rural and regional policy

Here again, Eastern Europe is poised on a knife-edge in the sense that its relative advantage in terms of sustainability may easily be lost. Accession countries are tending to absorb funds through large-scale intervention which could be environmentally damaging, whereas they should be seeking maximum environmental protection with the help of ENGOs. 'In Hungary (for example) the regional plans of the counties have been prepared and in some cases are being prepared completely separately from the environmental plans whereas the links between the two should be obvious' (Maier *et al.* n.d., p.17). Thus there is scope for a partnership between ENGOs and a new generation of officials to place regional policy on a solid footing. Meanwhile, following the 1996 Cork Declaration, EU policy is that rural policy must be transparent with sustainable development programmes in each region: 'reversing out-migration, combating poverty, stimulating employment and equality of opportunity and responding to growing requests for more quality, health, safety, personal development and leisure and improving rural well-being' (Anon. 1998b, p.107). Hence there is a two-fold challenge for SAPARD geared to farm efficiency, quality production and vocational training according to development plans prepared by member states which constitute the Community Support Framework (CSF) adopted by the Commission. The CSF includes an obligatory environmental profile and WWF has included this in a STEPs test (sustainability, transparency, enforcement and partnership) which it intends to use as a methodology for monitoring SAPARD as part of a shift in cohesion policy towards sustainable regional development. Thus there is now a clear prospect of new locally and regionally based sustainability indicators requiring new skills and competencies (Kouvelis 1998, p.59).

Some reduction in pressure will be desirable, for agriculture has been a generator as well as a recipient of pollution, as evidenced by ammonia emissions which can affect microclimate (Kic 1996). There have been alarming losses of wetland as 'natural areas' are lost to uniform landscapes of cereals, dairy pastures and forest. Substantial habitat loss has followed the 'denaturalisation' of rivers in Eastern Europe's regions of intensive agriculture. With the protection of floodplains by diking, underground water resources have declined and aridisation has been accentuated (Saratpatka and Sterba 1998). Extensification is needed to overcome overproduction and reduce harmful environmental impacts: ending grassland–arable conversions with the help of a grassland premium, reducing inputs in general and livestock density in particular (through payment per forage hectare rather than premiums per head) and conversion to organic (Anon. 1998b, p.93). 'Plant production restructuring' should include zoning to re-establish forest and permanent grass in floodplains. Meanwhile, although fertilisers and pesticides undermine flora and fauna (ibid., p.27), nitrogen fertiliser dressings have gone up in the EU by 50 per cent from the start of the 1980s even though maximum guaranteed production quantities were laid down for dairy products in 1972 and for cereals in 1982 (ibid., p.13). It is important that Eastern Europe does not follow this route.

There is growing pressure in several countries for environmentally friendly farming. TERRAS in Yugoslavia advocates organic produce and has been involved in such work at Subotica since 1990. Animal rights are becoming an issue, as was evident in 1999 when there were coordinated Czech and German protests over a broiler house on the Czech–German border at Všeruby. Progress is being made in forestry, particularly regarding the pine forests. Estonia is discussing the optimum cutting level for forests: the country exports wood products, paper and furniture (17.5 per cent of the country's industrial capacity) but it may not make good business sense to rely heavily on timber-based exports when other parts of the world can produce wood more cheaply. There is also opposition to felling trees in the Czech Republic. Šumava National Park has been cutting as a response to the problem of bark beetle, although some experts say extreme measures are unnecessary. Meanwhile, the WOLF forest protection movement (Tulčik, Slovakia) aims to preserve forests with kinder logging methods. It has opposed plans to cut broad swathes through the Polana Biosphere Reserve (1998).

Yet, while the region has shifted its trade towards the EU after Comecon collapsed in January 1990, agreements have so favoured EU member countries that their agrofood balance with associated countries in the region shifted from deficit in 1992 to surplus in 1993 and 1994, with only Hungary retaining a positive balance (Anon. 1998b, p.71). Eastern European countries need to be assured of market opportunity if agriculture is guided towards sustainability, given they have traditional agricultural know-how, skilled and abundant rural labour and 0.40ha per inhabitant compared with 0.21 in the EU. Taking into account the halving of agriculture's GDP share during 1989 to 1994 and successive droughts, these countries 'have undeniably an important margin of productivity' (ibid., p.67). Yet abandonment of extensive grazing in some regions may constitute a biodiversity threat, especially where mountain grazings are already neglected (ibid., p.73). An interesting experiment is currently being undertaken in East Germany at Babbe (Sachsen-Anhalt) which is safeguarded as an ecological domain. The historic house and farm complex has been renovated and will be protected in the context of an integrated rural economy: an organic farm with meat processing and a farm shop; family accommodation and a farm restaurant; an environmental education centre; a vegetable compost plant; and nature conservation areas. The project will generate twenty-five full-time and 100 part-time jobs, with the project increasingly being financed from profits. At Wulkow (Brandenburg), sustainable agriculture is being promoted through a weekly market and annual fair, with restoration of buildings and the opening of community and conference centres. There will also be a CFC-free refrigeration unit for local production using solar energy while biomass (wood-processing residues) will provide heating.

Protected areas: the Danube–Carpathian zone

A substantial network of protected areas has been inherited from communism and this is being both extended and strengthened through more effective management systems. In the Czech Republic's Šumava region, which consitutes a national park, protected landscape area and (since 1990) a biosphere reserve with a diverse range of habitats and mountain ecosystems, the inner Zone 1 is strictly natural, while Zone 2 is available for managed activities with protection from the surrounding buffer comprising Zone 3. Sustainable tourism is achieved through the emphasis on walking/cycling routes, and canoeing/rafting with access through an ecobus service. However, sustainability should not be restricted to protected areas but should apply to all areas with the support of stronger legislation and enforcement, backed by environmental education and community/stakeholder involvement in conservation. Local conservationists need to raise community awareness and help negotiate best practice with environment managers, including economic incentives to maintain good management practice (e.g. in restituted woodlands). Czech activists from the Vsetín Environmentalists' Association in eastern Moravia are working to preserve twenty-nine rare species of orchids growing in the wild on some eight hectares of meadows. The work, supported by subsidies of 10,000 crowns per hectare, involves regularly eliminating 'aggressive' weeds and shrubs by hand and making sure that the orchid seeds fall on fertile soil.

Ecotourism could play a more central role in generating money for conservation (e.g. safeguarding forests and wild animals). Synergy between nature conservation and tourism in Europe's protected areas is being promoted through the support of the WWF and the Molecaten Group (a Dutch leisure company) for a protected area network (PAN) comprising European parks which offer outstanding wildlife and excellent tourist facilities combined with effective nature protection involving no commercial logging or hunting. Currently, the network includes Triglav in Slovenia (83.8th.ha); Retezat Mountains in Romania (54.4); Duna-Drava in Hungary (49.5); Slovak Paradise in Slovakia (32.8); and Bieszczady in Poland (29.0). Extensive areas – in excess of 25,000ha – are needed to safeguard large carnivores and herbivores. There is evident public interest in these projects: 400,000 people are now visiting Plitvice National Park in Croatia every year.

The cross-border dimension is important. In 1990 attention was drawn to the scope for environmental action in border areas through the publicising of sites presented as 'Ecological Bricks for our Common House in Europe' (Langer 1990). No overarching network has developed but a series of separate initiatives has gathered momentum (Figure 4.1). During 1995 to 1999 the PHARE programme included integrated rural development in Hungary's border areas (especially with Austria and Slovenia) involved the Naturpark concept, needing the cooperation of all local agents in sustainable development. Cross-border biosphere reserves include trilateral Eastern Carpathians Biosphere Reserve (Poland, Slovakia and Ukraine). Further, an increasing number of formal Euroregions include environmental functions in their activities (Figure 4.1); for example, the Carpathian Euroregion Working Commission on Tourism and Ecology. Reference should also be made to the German–Polish–Czech Cooperation Programme for the Black Triangle and the Carpathian Transfrontier Environmental Network; while Euromontana can help coordinate the expanding national eco-networks. Major ENGOs like IUCN and WWF are also involved in the conservation of both wetlands and mountains. A green corridor is being promoted along the Danube while the Carpathians are now being approached in a coordinated fashion by IUCN's European Mountain Forum and WWF's Carpathian Initiative for Eco-Regional Conservation.

These initiatives may be placed in the context of EU readiness to provide a stronger lead following Rio. Birds and Habitats Directives require that 'conservation measures of natural habitats and their species should have priority

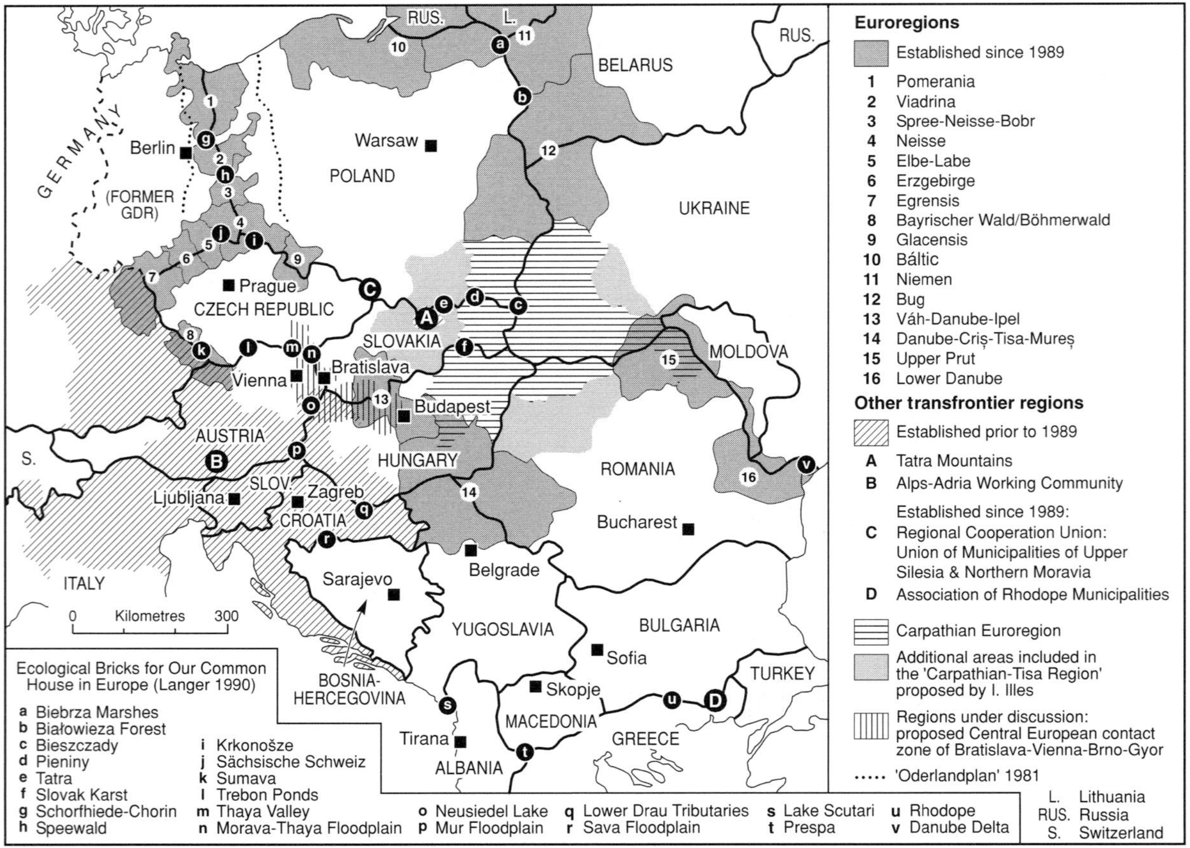

Figure 4.1 Euroregions

Source: Langer 1990 and EU briefings.

to the aspects of the exploitation they are confronted by' (Gyulai 1998, p.10). Hence the need for Natura 2000, a regional network of protected areas where member states will have to prevent deterioration. Thus East European states are now drawing up lists of habitats to harmonise with the EU classification system. But, more generally, states must 'consider impacts on the conservation and sustainable use of biodiversity across the rest of the territory beyond protected areas' and 'support recovery plans for the most threatened species' (ibid.). However, in general the East can offer nature resources to a common Europe (e.g. Carpathians). Yet there is a threat since these mountains comprise 'a typical transit area: a centre of permanently increasing transport, but also a meeting point for the dispersal of flora and fauna components which stream into the basin' (ibid., p.56). As noted above, a 'denaturalisation' of the valleys has occurred and although there is currently reduced agricultural activity this is balanced by greater tourist pressure and there is a recent history of grazing pressure on the high ground involving removal of scrub woodland. But more significant risks now arise through the cutting of privately owned woodlands and increased hunting which threatens large carnivores. In addition, there are tourist pressures through proposals that the Tatra might host the Winter Olympics, not to mention plans for new local, national and international roads. With accession, development pressures will increase and conservation will be marginalised. Unless sustainability is absolutely central to regional development, there is a distinct prospect that 'decision makers and local communities will not give enough attention to sustainable development and every development investment will contribute to the deterioration of rural areas where nature is best preserved' (ibid., p.55).

Case study: The Danube

Conservation of the Danube has been driven in the first instance by concern over pollution. The Danube River Protection Convention (DRPC) was adopted in 1994 and came into force in 1998. It built on the Helsinki Convention on Transboundary Watercourses (1992) and Convention on the Protection of the Black Sea Against Pollution (1992). In 1995 the main river was found to be nearly free of heavy to excessive pollution but it had lost much of its self-purification capacity and drinking-water supplies were at risk. Moreover, emergencies remained a strong possibility. In 1999 an International Commission for the Protection of the Danube River (ICPDR) was set up under the DRPC with priority for water protection through monitoring laboratories and an accident emergency warning system. Pollution at Novi Sad and Pančevo linked with the Kosovo War of 2000 has been a cause for concern, but this new infrastructure was particularly useful following the pollution of the Tisza River by cyanide, and the release of heavy metals in Romania during the year 2000 led to the establishment of a Baia Mare Task Force which brings together all the countries with an interest in the Tisza basin. All hazard sites are to be evaluated and placed under regular surveillance but the high-risk sites are frequently

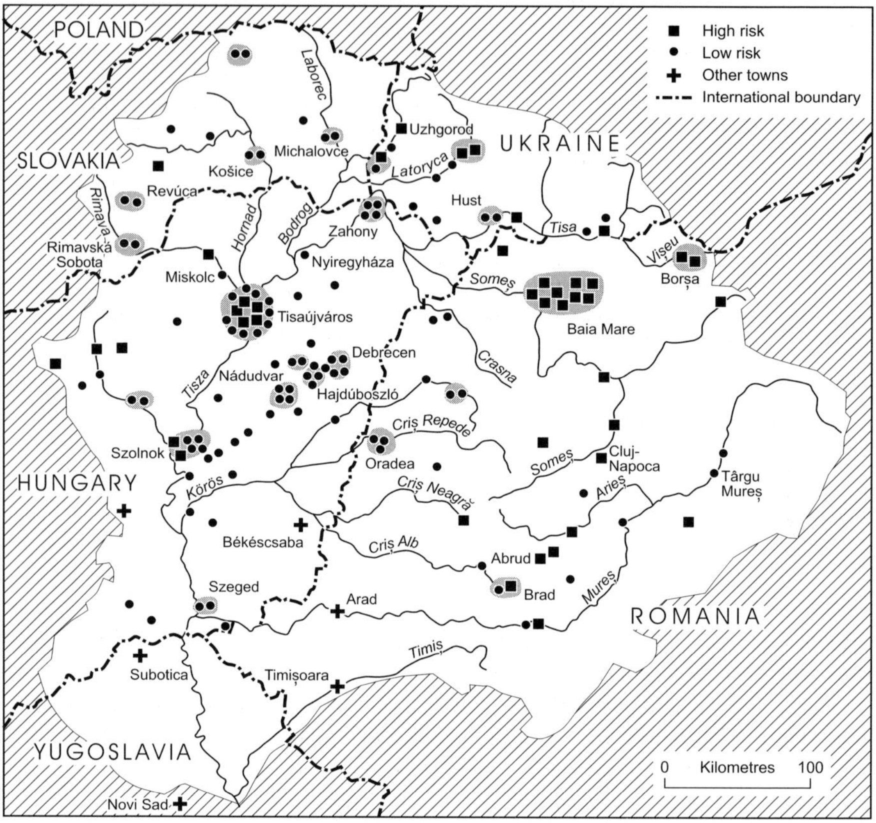

Figure 4.2 Pollution hazards in the Tisza Basin

Source: Topographical maps and statistical yearbooks.

dams and 'tailings ponds' for the storage of wastes and sludges containing heavy metals and toxic substances. Design may be poor, while lack of maintenance at abandoned sites creates dangers of leakage and collapse.

A study carried out by ICPDR has revealed 144 risk sites: forty-two are high risk and 102 are relatively low risk (Figure 4.2). Predictably, in view of the way the catchment area is divided up, Hungary and Romania account for the great majority of the sites (77.1 per cent) but while Hungary has 55.9 per cent of the low-risk sites and Romania 18.7 per cent, Romania has 57.1 per cent of the high risk 'hot spots' and Hungary 26.2. Most of the high-risk sites are concerned with mining (50 per cent) and energy (26.2), while the low-risk sites are much more widely distributed: chemicals and pharmaceuticals 21.6; mining and energy each 18.6; municipal 16.7; agriculture 7.8; metallurgy and food each 8.8 (Table 4.3). There is a pronounced clustering, since nine of the high-risk sites are grouped together around Baia Mare; while five high-risk sites

Table 4.3 Tisza Basin: inventory of pollution risks for the Baia Mare Task Force

	High-risk sites									*Medium–low-risk sites*									*Total*
	a	*b*	*c*	*d*	*e*	*f*	*g*	*h*	*All*	*a*	*b*	*c*	*d*	*e*	*f*	*g*	*h*	*All*	
Hungary	1	.	.	.	.	8	1	1	11	8	6	15	.	4	12	6	6	57	68
Romania	19	1	2	1	1	.	.	.	24	6	.	3	.	4	4	1	1	19	43
Slovakia	1	.	.	.	.	.	.	.	1	5	2	4	1	.	1	.	.	13	14
Ukraine	.	.	.	3	.	3	.	.	6	.	1	.	.	.	2	.	10	13	19

Source: International Commission for the Protection of the Danube River (ICPDR) 2000, Regional inventory of potential accident risk spots in the Tisa catchment area (Vienna: ICPDR) accessed through http://www.eu.int/comm/environment/enlarg/home.htm

Notes
a Mining and processing.
b Engineering and metallurgy.
c Chemicals and pharmaceuticals.
d Pulp and paper.
e Agriculture.
f Energy.
g Food.
h Municipal.

and ten low-risk sites are clustered around Tiszaújváros. If looser clustering is considered then six high-risk and seven low-risk sites fall to Romania's West Carpathian (Apuseni) mountains; while two high-risk and ten low-risk sites lie in and around Szolnok (including the towns and villages of Kenderes, Kisújszállás, Martfű, Mezötúr, Szajol and Törökszentmiklós). Finally, eighteen low-risk sites are to be found in eastern Hungary around Debrecen, Hajdúboszló and Nádudvar (with Balmaszújváros, Ebes, Hajdúböszörmény, Kaba, Püspökladany and Nagyhegyes).

The pollution issue must be seen in the context of a deepening tension between development and conservation. The principal valleys of the region are obvious transport axes, and in this connection the European Commission's Regional Policy Directorate has commissioned a Danube Space Study from the Austrian Institute for Space Planning. However, the rivers themselves have come under scrutiny in connection with the EU's Transport Infrastructure Needs Assessment (TINA) which has been under review at a base in Vienna since 1996. The Danube itself is recognised as a transport corridor, and this implies a further phase of development following the regulation of the past two centuries which has transformed what was once a wide, branching river into a navigation canal with a high degree of flood control linked with hydropower generation. However, TINA has raised the possibility of a Danube–Oder–Elbe Canal, intersecting with an east–west Oder–Warta–Vistula–Bug Canal (with the latter extendible through Ukraine to the Black Sea over the longer term). These waterways will have to be quite massive to accommodate boats of 185 × 11.8m with 2.8m draught, and the rivers themselves – including sections of the Danube – will require some deepening. But the justification lies in the scope for 'combined transport' which will help move freight from the roads. However, projects to develop the major river valleys are considered highly controversial in view of the threat to biodiversity. The conservationists have an alternative vision based on floodplain protection and restoration, after four-fifths of such lands were destroyed during the twentieth century and many species have become endangered as a result, such as the Dalmatian pelican and white-tailed eagle.

The alternative vision builds on work during the past decade to safeguard floodplain wetlands not only in Germany and Austria but also in the Czech Republic and Slovakia (Morava-Dyje), Hungary and Croatia (Gemenc/Kopacki Rit), Bulgaria's Danube Islands (Kitka, Milka and Vardim) and the delta lands in Romania and Ukraine (Marin and Schneider 1997). These initial projects are to be integrated into a 'green corridor' through a chain of other conservation projects (Avis *et al.* 2000; Weller and Gunther-Diringer 1999). The aim is to supplement 400,000ha of existing protected areas with 100,000ha of newly protected areas with 200,000ha of priority sites restored (some of which are included in the previous two categories) to create a continuous corridor. In 2000, Bulgaria, Moldova, Romania and Ukraine agreed to establish a green corridor for the Lower Danube with GEF/UNDP funding. ENGOs have played an important role throughout, particularly IUCN and WWF. In 1998, thirty-seven Danube NGOs agreed to form the Danube Environmental Forum to raise public awareness of the need for sustainable development and undertake small-scale wetland restoration – with secretariat provided by Slovakia's Daphne Institute for Applied Ecology. Thus there is a clash between development and conservation, with similar conflicts emerging on other major rivers of the region, for example, the Vistula, whose middle and lower sections have exceptional natural and landscape value which would be threatened by a 'Lower Vistula Cascade' – starting with Niesawa – to help stabilise the one already existing at Wloctawek (1970) and enhance the navigation. Conservationists wish to protect what is the region's only large river still retaining strong natural dynamics, in line with the Ramsar Convention to which Poland is committed. They suggest removing the unstable Wloctawek complex, which is undermined by downstream erosion, and this solution would also allow for the disposal of accumulated toxic sediment.

Case study: The Carpathian ecoregion

Conventionally, threatened species have been protected in nature reserves, with clusters of such reserves safeguarded through national parks. However, experience shows that 'islands' may not survive and that fragments need to be integrated through corridors. In addition, some species, such as large carnivores, require extensive protected areas. Thus thinking is now veering towards caring for the environment in all areas, with particular emphasis on networks of protected areas where biodiversity is a particular concern. The EU is currently developing such a system, and within these networks there is scope for co-ordination across whole ecoregions. In this connection the Carpathian mountain system has surfaced as a critical area. Not only does this region comprise attractive mountain country rising to over 2000m (the highest peak is Gerlach in Slovakia: 2665m) and exhibiting glacial features in the Retezat and Tatra, but the Carpathians constitute the most valuable area still remaining within Europe's temperate conifer and broadleaf forests, with fragments of natural mountain beech forest amounting to some 20,000ha. About half is forested and one-third consists of open, semi-natural habitats (predominantly grassland). In addition, the area supports a remarkable species richness, including many endemic species and medicinal plants. The region has not been seriously degraded and remains one of Europe's most important biocentres. The region also provides the largest refuge for megafauna species, both predators and herbivores, and species that have become extinct in many other parts of Europe, such as the brown bear (Ursus arctos), the wolf (Canis lupus), the lynx (Lynx lynx), the wild cat (Felis silvestris), and many threatened bird species. The region also functions as an important corridor for a range of animals: the wolf makes particular use of the Carpathian 'bridge' between the coniferous forests (*taiga*) of northwest Europe and dispersal areas to the south and west. Finally, the Carpathians are a major source of fresh water: still relatively clean, flowing overwhelmingly to the Black Sea (only one-tenth to the Baltic) (Nowicki 1998).

However, the Carpathians are divided between six countries and in the context of the closed frontiers of communism (excepting the Czech Republic and Slovakia which were unified as Czechoslovakia) there was little overall planning. Over recent decades, attempts to conserve the region's biodiversity have focused mainly on ensuring that the most valuable sites are formally protected, and about 6 per cent of the region's total area is now under protection (Table 4.4). However, better management is needed and there is a wider problem in that these protected areas do not safeguard all threatened species – and, even if they did, retention within small islands would not rule out the possibility of extinction. Extensive continuous areas need to be safeguarded where large

Table 4.4 The Carpathians[a] (figures in hectares except where otherwise stated) (i) Carpathians as a percentage of national territory (ii) Percentage distribution of Carpathian territory

Country	*Area*	*Carpathians*	*(i)*	*(ii)*	*Protected Cats 1–2*	
Czech Republic	78,864	6,708	8.5	3.6	204,810	
Hungary	93,030	7,735	8.3	4.1	244,162	67,201
Poland	312,685	19,716	6.3	10.4	886,575	81,508
Romania	238,391	95,566[a]	40.1	50.4	423,184	396,761
Slovakia	49,035	38,150	77.8	20.1	848,875	243,219
Ukraine	603,700	21,700	3.6	11.4	239,964	206,860
Total	1,375,705	189,575	13.8	100.0	2,847,570	995,549

Note

a Some definitions of the Carpathians include the Transylvanian Plateau but this area is excluded from these calculations which cover only the main Carpathian ranges, the Subcarpathians and contact areas where communities make use of mountain grazings.

carnivores are concerned and there is also a case for biodiversity conservation on a coordinated basis throughout whole ecoregions: hence the concept of 'ecoregion-based conservation' (ERBC) for the sustainable development of large land units that are biologically coherent in terms of species, communities and environmental conditions. Such strategic biodiversity conservation projects can realistically cope with immediate threats in the context of the socio-economic conditions and safeguard the future through a clear vision of the conservation goals and the participation required by stakeholders, working in partnership at all levels from international agencies to local communities – and with input from all relevant disciplines, for an adaptive programme of coordinated management interventions (Figure 4.3).

The Carpathians have seen substantial ecological changes in the past (Pietrzak 1998), especially in terms of deforestation and the creation of mountain grasslands with their own biodiversity value. But they are now confronted by a mix of threats which, while not overwhelming, are cumulatively significant enough to require coordinated management in the interest of biodiversity (Augustyn and Kozak 1997). There is evidently a northwest–southeast contrast. In the former region there is a 'virtuous circle' where civil society and political stability attracts investment to reinforce attractiveness; whereas in the south there is a 'vicious circle' through lack of investment (arising from lower spending power and greater distance from Western Europe) which progressively inhibits civil society and compromises political stability. The differences are reflected in variations in investment (state budgets and foreign direct investment), unemployment, poverty and some ethnic tensions. Development threats are therefore greater in the northwest, including hunting. Forests are expanding (partly due to lower farming pressure), yet air pollution damage to forests is most evident in the Beskidy of Poland, and pollution from Poland affects Ukraine. Mass tourism exists in some areas, road projects are very likely to be implemented, and the demand for electricity will increase the need to proceed with the potentially damaging Czorsztyn-Sromowce series of hydro-plants on the Dunajec in Poland. Meanwhile, development pressures in the south are weaker; yet the poorer countries with high unemployment generate threats to environment through illegal cutting of restituted forests and heavy grazing pressure by peasant farmers, whose main source of income arises from the sale of livestock. Thus, while the output of basic commodities has fallen, implying reduced pollution damage (including transboundary pollution), rural economies may be less sustainable than before. This combines with the reduced effectiveness of law enforcement in the aftermath of communism and a low level of public awareness over environmental issues. New legislation is important but cannot succeed without better law enforcement and environmental education; along with relief of poverty and measures to engage the interest of mountain communities – and their visitors – in conservation.

Biodiversity is clearly threatened by a range of factors responsible for shrinking habitats. On the physical side there are landslides and avalanches, along with heavy rainfall and frost damage, but agriculture is also a threat through conversion to farmland, drainage schemes, overgrazing and trampling; while

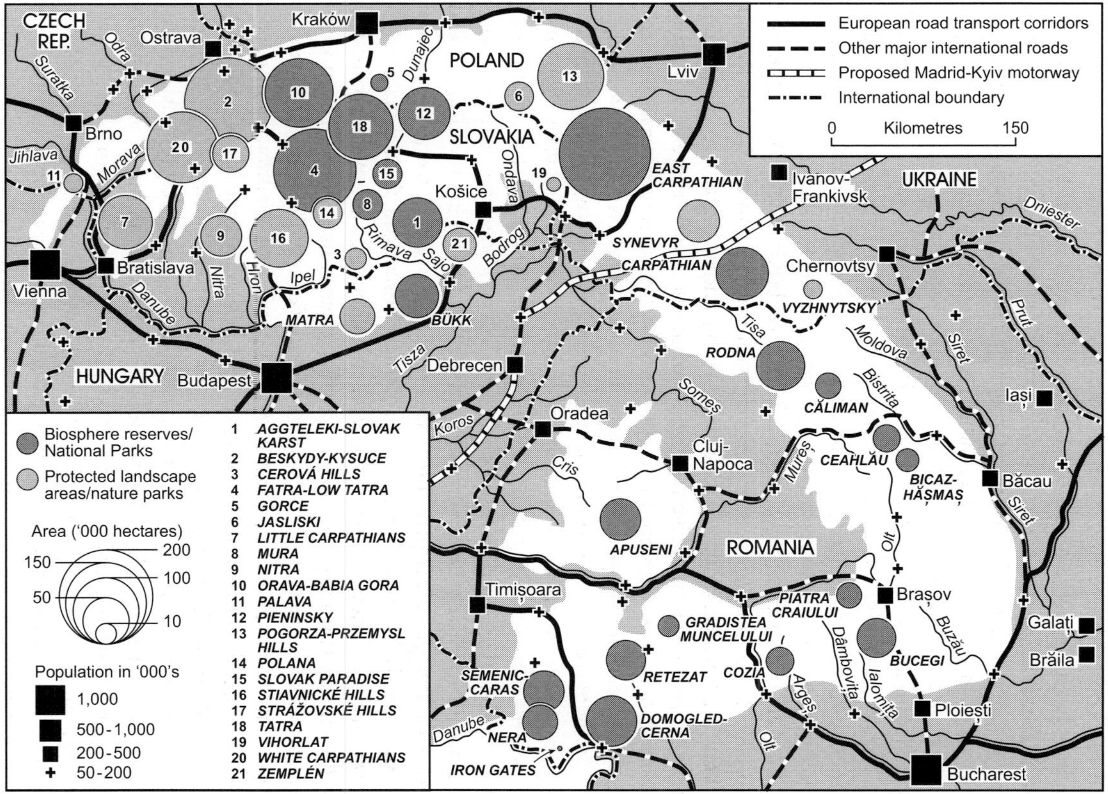

Figure 4.3 The Carpathian Ecoregion

Source: International Commission for the Protection of the Danube River.

forestry is damaged through fire, clear-felling and woodland monocultures. In addition, there is industrial pollution, peat extraction, tourism and the harvesting of medicinal plants. However, while the current transition poses a threat to the integrity of the Carpathian ecoregion, it also provides an opportunity for sustainable development to fulfil both environmental and societal needs. Now that many of the worst problems inherited from the communist era have been addressed, the agenda is quickly moving ahead to anticipate threats from future development linked with EU membership and increased spending power, suggesting that the full range of EU environmental protection systems should be brought to bear as quickly as possible. At the same time it is recognised that legislation must be backed by adequate enforcement, and ways must be found to gain the support of mountain communities and visitors for conservation measures. Delays in implementing the Natura 2000 integrated network of protected areas could be particularly damaging in Carpathian countries in terms of 'units of biodiversity lost'.

At the same time, the Carpathians on the whole are still well populated, despite some areas being emptied by resettlement (Perzanowski 1999), and any approach to conservation must recognise the universal demand for growth and higher living standards in all the countries involved; also the inevitability of investment pressure following EU accession which is now being negotiated in all the countries except Ukraine. Poverty is greatest in the southeast: Ukraine is plagued by macroeconomic and political instability, reflected by low foreign direct investment (FDI), low domestic savings, limited state funds for investment (partly due to support of state-owned enterprises) and a low absorptive capacity for investment due to institutional deficiencies and the legacy of a workforce more experienced in heavy industry than in a market-oriented service economy with a large rural tourism sector which is the key to the future (Slee 1999). However, the situation is complicated by the fact that Carpathian areas tend to be relatively backward in relation to their respective national economies (Suli-Zakar 1999). While Hungary's northeastern periphery conserves natural and cultural values, the villages in border areas have suffered depopulation since the Treaty of Trianon imposed a new border and disrupted existing patterns of production and trade. The transition has added to their backwardness, since the share of industrial production has fallen and agricultural yields are relatively low despite the reputation of the Bükk, Eger, Mátra and Tokaj wine districts. Unemployment is relatively high while per capita GDP is relatively low. Mountain conditions themselves impose constraints: Poland's Carpathian region suffers from high agricultural production costs arising from the harsh climate, poor soil and steep slopes; also poor access to local markets, linked with the quality of the infrastructure (energy, transport and business services) and poor labour markets (Gorz and Kurek 1999).

Various initiatives are being taken. The 1994 Slovak Act on Nature and Landscape Protection now provides for five levels of territorial protection and specific flora and fauna, minerals and fossils: working from the national level, through protected landscape areas (PLAs) with characteristic landscape or historical settlement forms; national parks where natural heritage protection carries overriding priority; small protected sites comprising biocorridors or biocentres of local or regional importance; and finally, nature reserves and monuments of nature. Furthermore, there is now scope for tax concessions to private landowners who set aside land for conservation purposes. Other countries are working along roughly similar lines, although there are large variations in the proportion of land protected and, despite recent designations, areas zoned for protection are not fully inclusive. Natural monuments and environmentally sensitive areas are often threatened by pollution, and (especially in Hungary) ownership of protected areas by the conservation authorities is limited. In Romania the management of national parks is inadequate to ensure the maintenance of natural values and legislation is not adequately enforced. An assessment of Rodna National Park by Romsilva and the Bucovina Centre for Ecological Information and Education (Suceava) points out that neglect of environmental education programmes is contributing to illegal tree cutting and aggravated flooding (especially in high unemployment areas like Borşa and Fălticeni), while eagles are affected by hunting and habitat loss, as is the black mountain goat. Municipalities in Ukraine often lack the financial and technological resources to monitor pollution; and operating the polluter pays principle (PPP) through fines imposed on profits gives immunity to the unprofitable enterprise: fines could be taken from assets, but government wishes to avoid creating further unemployment.

On the other hand, biosphere reserves, integrated with national parks and landscape parks, are becoming increasingly favoured over strictly protected nature reserves because of their flexibility in communication with the local communities and the scope for zoning to avoid socio-economic conflict. It provides a formula whereby half the Carpathian area could be brought under protective regimes compared with barely 15 per cent at present. There is political will to extend protected areas. Ukraine already has a Programme for Developing a National Econetwork in Ukraine 2000–2015 (following the 1998 Concept for the Preservation of Biological Diversity of Ukraine) which will contribute to a future European Ecological Network (Hamor 1998). Indeed, all the countries are taking significant initiatives although they are not yet fully coordinated. Meanwhile, through its 1997 Nature Conservation Act, Hungary has created a legal basis for effective nature conservation. Regional plans will include inventories of unique landscape features, and convergence with the EU means that land use practices are now being reconciled with a system of environmentally sensitive areas and a National Ecological Network which is currently being established. Meanwhile, a methodology to identify critical areas exists in the Czech Territorial System of Ecological Stability which was researched in the 1980s before the concept was formalised in 1992. With comprehensive coverage across the Czech Republic, it is now possible to identify biocentres and biocorridors which can relate to the National Biodiversity Conservation Strategy. Ukraine's new multi-functional forestry code seeks to preserve the habitats of vulnerable 'red-listed' species, while Poland's Directives Concerning Improvement of Forest Management on Ecological Grounds provide for nature inventories as a basis for reconstruction. Romania is considering the management of some designated national parks as 'natural parks', which will attempt to compromise between conservation and the pressure for development by the local population (Abrudan and Turnock 1999), and is playing a key role with regard to a Carpathian Large Carnivore Project. Reference should also be made to special programmes for heavily polluted areas highlighted for priority in Poland, Romania and Slovakia (Turnock 1996).

Finally, cross-border cooperation has been greatly encouraged over the past decade for environmental reasons as well as socio-economic concerns related to permeable frontiers and the reinvigoration of peripheral areas. Transborder protected areas are considered valuable for the proper management of water resources, forest and agro-ecosystems, and (especially) tourism. They also have importance for scientific and educational cooperation. Several cases arise along the Polish–Slovak border area alone, including Pieniny and Tatra which now comprise the Tatra Euroregion set up in 1994. At the beginning of the transition phase Langer (1990) drew attention to a series of potential cross-border conservation projects informally designated 'Ecological Bricks for our Common House in Europe' and several of these are now being funded. Outstanding is the Eastern Carpathians trilateral initiative aided by World Bank Global Environmental Facility (GEF) in 1997 (Stoiko 1996). The focus is on legal, institutional and administrative measures to connect the habitat fragments of isolated reserves to ensure the long-term presence of viable populations of flora and fauna that are sensitive to habitat loss. Work has involved an inventory of plant and animal wildlife, considering community, species, population and genetic levels of diversity. Knowledge of the biodiversity of all the massifs covered by the Transcarpathian Biodiversity Protection Project has increased and there are proposals for establishing a network of protected areas interconnected by ecological corridors, while conflicts have been overcome through negotiation with local people. Meanwhile, the wider Carpathian Euroregion is interested in sustainable economic development and the Carpathian Bridge (Priashev) is operating as an International Association of Public Ecological Organisations.

However, these various initiatives have in some way to be combined into a strategy which can apply across the Carpathians as a whole. This requires better international coordination and a greatly improved supply of information to educate stakeholders and the public in general over the unique resources of the Carpathians and to research potential threats like woodland restitution. Given strong commitment from the six countries, the Worldwide Fund for Nature (WWF) has drafted a mission statement 'to initiate and facilitate a process through which key stakeholders collaborate to secure conservation and sustainable development in the Carpathian region'. Reconnaissance in 1999 took account of all relevant expertise, and involved contact with selected stakeholders and key actors in order to assess the biodiversity of the Carpathians in the light of current threats and conservation efforts. This work has established that, while there are challenges which require urgent attention, there is also a high level of commitment to achieve worthwhile objectives. During 2000 to 2002, detailed biodiversity/socio-economic assessment will be undertaken, leading to a long-term biodiversity 'vision' and the initiation of specific projects. There will also be work on preparing a ten- to fifteen-year ecoregional conservation plan and an action programme for the next five years. While biodiversity lies at the core of the initiative, a socio-economic perspective is needed to identify threats and to assess the motivation, power and influence of stakeholders. In addition, while appropriate machinery is needed at the national level, the vision must also be scaled down to the grass roots through local community input, informed by expert guidance, and through negotiation with environment managers.

This Carpathian initiative can draw on the commitment of environment ministries and NGOs, while the Euromontana organisation is dedicated to the sustainable development of mountain areas and is actively extending its interest eastwards both to the Carpathians and the Caucasus (Backmeroff *et al.* 1996). It provides an appropriate feature for this chapter since it is grounded in the biodiversity resources of Eastern Europe and the threats posed by EU enlargement, while gaining momentum from Western Europe's expertise in conservation and the environmental components of the EU *acquis*. It also highlights the points which are now central to environmental issues in the region, these being somewhat more diverse – and challenging both in ecological and socio-political terms – than the pollution agenda a decade ago when this book was first conceived.

Approximating the *acquis*: the regular reports for 2000

Progress in approximation of the environmental *acquis* may be considered in the context of the regular reports which have been drawn up annually by the European Commission since 1998 to monitor each country's preparations for EU enlargement. This review is based on the reports for 2000 which record a high level of activity (European Commission 2000) (Table 4.5). Some countries have coordinated national programmes for the introduction and implementation of EU legislation and Bulgaria has national sectoral programmes for biodiversity, waste management, waste water treatment (also the phasing out of leaded petrol and ozone-depleting sustances), while Latvia has a national programme for biodiversity linked with a new department at the Ministry of Environment and Regional Development. However, other countries are proceeding in a more *ad hoc* fashion; Lithuania's transposition of the *acquis* is scattered through many pieces of legislation and not all procedures for identifying protected areas comply with those of the EU. There is a diversity of approach, as in the case of the nuclear issue. The Slovak Nuclear Regulatory Authority has planned for breakdowns and accidents and has regulations covering radioactive waste shipment; while Bulgaria has a Permanent Commission for the Protection of the Population to handle accidents and catastrophes in general, in addition to a regime for the safe management of spent nuclear fuel and radioactive waste.

Most countries were dealing with the great majority of relevant criteria during 1999 to 2000 but with much variation in emphasis. Environmental impact assessment (EIA) is a crucial test of compliance and most countries have taken appropriate action. Where no progress is recorded action is widely seen as imminent, as in the Czech Republic. Bulgaria has made a start, although further

Table 4.5 Candidate countries: environmental legislative actions 1999–2000

	NP	*SP*	*EIA*	*Air*	*C&GMOs*[a]	*IPPC*[b]	*Natr*[c]	*Noise*	*Radtn*[d]	*Waste*	*Water*
Bulgaria	+	+	+	+	+	+	+	.	+	+	+
Czech Repubic	.	.	.	+	+	+	.	+	+	.	+
Estonia	.	.	+	+	+	+	+	+	+	+	+
Hungary	+	.	+	+	+	.	+	+	+	+	+
Latvia	.	+	+	+	+	.	+	.	.	+	.
Lithuania	.	.	+	+	+	+	+	.	+	+	+
Poland	.	.	+	.	+	+	+	.	+	.	+
Romania	+	.	.	.	+	+	.	.	.	+	.
Slovakia	+	.	+	+	.	.	+	+	+	.	.
Slovenia	.	.	+	+	+	+	+	.	+	+	+

Source: European Commission Enlargement Reports http://europa.eu.int/comm/enlargement/index.htm

Notes

a Chemicals and genetically modified organisms.

b Industrial pollution prevention control and risk management.

c Nature protection.

d Radiation protection and nuclear safety.

legal changes are needed. And while Bulgaria's Ambient Clean Air Act was amended in 2000 – with further regulations limiting the content of harmful substances in fuel and requiring regular motor vehicle tests – there has been little progress over waste management and measures for the handling of hazardous wastes are not yet satisfactory. In the Czech Republic where progress over air and water is recorded, this does not yet include framework legislation: only a fraction of legislation has been transposed whereas half the EU provision has been covered in the cases of chemicals and GMOs, noise and radiation; and one-third for IPPC. Of course, the fact that no legislative progress is shown in the table for 1999 to 2000 does not mean that nothing has been done so far: the Czech Republic has approximated one-fifth of the EU provision for nature protection and waste management while in Latvia the water supply in four towns now complies with EU standards. The table is not comprehensive. Over the year 1999 to 2000 the Convention on Trans-Boundary Environmental Impact Assessment came into force in Bulgaria while the three Baltic States and Poland have started to implement the Agenda 21 sustainable development action programme for the Baltic Sea countries. International cooperation is particularly important for Hungary in view of the extent of the Tisza drainage basin. In April 2000, the environment ministers of Hungary and three neighbouring countries (Romania, Slovakia and Ukraine) agreed to cooperate over a regional inventory of 'hot spots' and implementation of the PPP.

Administrative capacity is generally inadequate: Bulgaria has a national air monitoring system but lacks capacity for implementation both at national and regional levels. However, there may also be problems of coordination when a range of ministries become involved in environmental matters – agriculture, forests and land reform, health and regional development and public works – in addition to environment and water in the case of Bulgaria. In the Czech Republic there is some confusion over water because the Ministry of Environment is responsible for the protection of waters and the Ministry of Health for drinking-water, while a new section for water management has been established at the Ministry of Agriculture. But also in Hungary, the Ministry of Environmental Protection needs more staff, and coordination between ministries concerned with environmental issues needs to be improved. Local authorities often have a role complementing that of the central bodies, but considerable effort is required to strengthen the capacity of local authorities by planning investments, as Lithuania is finding in the case of waste management and water protection. In Poland environmental functions apply at all three local government levels: voivodships (regions), powiats (counties) and gminas (municipalities). The last have the main responsibility for carrying out decisions and are directly responsible for waste management while the monitoring of harmful activities and the issue of environmental permits are concerns at the higher level where strengthening is needed – and the Inspectorate for Environmental Protection operates at the voivodship level. However, in Estonia the environmental departments of county governments have become regional departments of the Ministry of Environment in order to enhance administrative capacity.

However, an adequate system includes not only ministries and local government but also regional inspectorates, national park directorates and other special authorities (sometimes with a National Environmental Agency umbrella). Slovenia has a full range of institutions already in place – Ministry of Environment and Spatial Planning, Administration for the Protection of Nature, Chemical Institute, Health Institute, Hydrometeorological Institute, Inspectorate for the Environment and Spatial Planning, Nature Protection Institute, and a Nuclear Safety Administration – but sometimes the bureaucracy may become too complex for efficient coordination. In Latvia, where the Ministry of Environmental Protection and Regional Development shares some responsibility with the Ministries of Agriculture, Interior, Transport and Welfare, there are also some thirty subordinate bodies (including regional boards and the environmental inspectorate). Approximation therefore involves inter-ministerial groups with representatives from ministries and institutions. Some new institutions are being set up, such as the Environmental Impact Assessment Bureau and a Radiation Centre, but some restructuring is considered necessary for effective application and enforcement of the *acquis*. Lithuania also has a large number of authorities and institutions concerned with the implementation of legislation. Some rationalistion is needed to avoid duplication while some new institutions are needed to cope with all the requirements of the *acquis*.

It is difficult to work out a league table from the data available. Bulgaria shows up well in terms of the breadth of its activity during 1999 to 2000 although the EU points out that more has still to be done over IPPC, nature protection and waste. It is not yet clear how the PPP will be applied and the principle of sustainable development needs to be integrated into other sectoral policies. Estonia attracts a good deal of recognition for vigorous efforts, but the financing of environmental investments is seen as a weakness, and in every policy area except chemicals the overall assessment highlights the need for further action.

The radiation problems linked with the decommissioning of a nuclear installation and remedial work at the Sillimae tailing pond are highlighted. The money involved is considerable, but there are sharp variations. Since 1994 the Czech Republic has been investing Euro1bln each year (or 2.4 per cent of GDP): a rate that should be maintained if not increased. However, Poland's annual investments of Euro2bln amount to only 1.7 per cent of GNP and, while Latvia's spending on environmental issues rose to Euro46mln in 1999, this was 0.7 per cent of GDP. Complex problems of approximation have led to the setting up of AC-IMPEL (Associated Countries for Implementation of Environmental Legislation) in 1998 as a complement to EU-IMPEL. Clearly there is a great deal to be done, but for the moment the European project is a priority throughout the ECECs and Baltic (non-CIS) sector of the former Soviet Union.

Conclusion: a wider unity for post-communist countries

The general thrust of this chapter is that 'things are moving' and the escalator which is propelling much of East Central Europe towards EU accession will ensure that the environment remains in a central position with regard to policy, even though many of the worst problems inherited from communism have been addressed. However, most former communist states (the CIS) are not directly concerned with the EU expansion project, and, even within the more restricted transition zone, which is the main concern of this book, there is a substantial group of non-candidate states in the western Balkans. So some machinery is needed to ensure an appropriate priority for conservation in these cases, especially in the Balkans where there is the powerful legacy of war. The Bosnian War created many problems through minefields which sought to stabilise Bosniak-Croat and Serb areas. And the Kosovo War of 1999 created problems of radioactivity through the use of armour-piercing depleted uranium weapons, posing long-term health risks, not to mention the dumping of unexploded ordnance and other waste in Macedonian lakes and the accumulation of debris in the Danube which brought shipping to a virtual standstill (with heavy losses for shipping companies in Bulgaria, Romania and Ukraine). A WWF team recorded much damage to Yugoslavia's environment through chemical contamination at Novi Sad and Pančevo which was found to be spreading into surrounding areas. Soil and water samples revealed the presence of significant quantities of mercury, polycyclic aromatic hydrocarbons, ethylene dichloride and other highly toxic substances, including dioxin, which pose a threat to the rivers and especially the Danube. The scale of the damage has revealed deficiencies in the monitoring of toxic chemicals and has shown that clean-ups are needed at Bar and Kragujevac as well as Novi Sad and Pančevo.

Norway has made a grant of US$630,000 towards a US$1mln programme for 'A Clean and Green Macedonia' (operating in fifty-one localities and generating 2000 jobs) to sort out the damage associated with refugee and troop movements. And the UK government is giving US$0.55mln for a state-of-the-art medical waste incinerator (to be installed at Drisle at a state community hygiene company) after the need arose during the Kosovo War when the army hospital in Skopje was overwhelmed by medical waste. These early initiatives have developed into a Stability Pact for Southeastern Europe (SPSEE) which comprises an official international body established to address the damage arising both from the Kosovo War and previous neglect and to bring the countries of the region into the EU enlargement exercise. This arrangement for former non-candidate countries (though the list includes Bulgaria and Romania while excluding Yugoslavia from full membership prior to the demise of the Milošević presidency) provides for a Regional Environmental Reconstruction Programme (RERP), launched in January 2000 with E28mln of funding for environmental rehabilitation following wars and previous neglect. This is an initiative taken by the countries of the region themselves and it has now been endorsed by the environment ministers. The purpose of RERP is to restore an environment damaged by war, as well as by the severe drought which also occurred during 2000. With 'quick start' projects followed by longer term programmes, the resources will be used for priority local and national projects to strengthen EIA procedures, improve environmental monitoring and project preparation and enhance regional cooperation of the kind currently evident at Lake Ohrid and Lake Prespa. The reopening of the Danube shipping lane should be achieved during 2002 now that the Pact's Economic Development Working Group has approved the RERP. The EU will pay up to 85 per cent of the cost of the project agreed by the Danube Commission in Vienna (which does not include the rebuilding of bridges). More generally there will be an emphasis on strengthening institutions and policy development and on raising awareness among civil society so that ENGO/civil society networks will exert an influence. In addition to the war damage there are non-ferrous metallurgical smelters which are major hazards. Kosovo's Zvecan lead factory near Mitrovica (within the run-down Trepča mining complex) needs a fully compliant waste management system to remove the hazards posed to local people and the peace-keeping troops deployed there. But there are other serious problems in Macedonia.

References

Abrudan, I. and Turnock, D. 1999, 'A rural development strategy for the Apuseni Mountains, Romania', *GeoJournal* 46, 319–36.

Amman, M. 1995, *The Effectiveness of Financial Instruments for Environmental Investment in CEE Countries: Recipients' Perspectives* (Laxenburg: International Institute for Applied Systems).

Anon. 1998a, *EU Accession and the Environment: An Introduction* (Brussels: FoEE/EEB).

Anon. 1998b, *Training Materials on Agriculture* (Brussels: FoEE/EEB).

Augustyn, A. and Kozak, I. 1997, 'The trends of anthropogenic pressure in the Polish and Ukrainian Carpathians', in K. Perzanowski and M. Augustyn M. (eds) *Selected Ecological Problems of Polish–Ukrainian Carpathians* (Ustrzyki Dolne: ICE PAS), pp.15–22.

Avis, C., van Wetten, J., Seffer, J., Tinchev, G. and Weller, P. 2000, 'Danube river basin: wetlands and floodplains', in A. Wood, P. Stedman-Edwards and J. Mang (eds) *The Root Causes of Biodiversity Loss* (London: Earthscan/WWF International), pp.183–212.

Backmeroff, C., Chemini, C. and La Spada, P. 1996, *European Inter-governmental Consultation on Sustainable Mountain Development* (Trento: Giunta della Provincia Autonoma di Trento).

Baker, S. and Jehlička, P. 1998, 'Introduction', in S. Baker and P. Jehlička (eds) *Dilemmas of Transition: The Environment, Democracy and Economic Reform in East Central Europe* (London: Frank Cass), pp.1–28.

Caddy, J. 1997, 'Harmonisation and asymmetry: environmental policy coordination between the EU and Central Europe', *Journal of European Public Policy* 4, 318–36.

European Commission 2000, Regular reports on EU enlargement http://europa.eu.int/comm/enlargement/index.htm

Fisher, D. 1992, 'The environmental movement in Central and Eastern Europe: its emergence and role in the political changes of 1989', in D. Fisher (ed.) *Civil Society and the Environment in Central and Eastern Europe* (London: Ecological Studies Institute), pp.182–97.

Gorz, B. and Kurek, W. 1999, 'Variations in technical infrastructure and private economic activity in the rural areas of southern Poland', *GeoJournal* 46, 231–42.

Green, J. 1998, *Energy and the Environment in the EU* (Brussels: FoEE/EEB).

Gyulai, I. (ed.) 1998, *The European Union and Biodiversity* (Brussels: FoEE).

Hamor, F. D. 1998, 'Regional aspects for the introduction of a sustainable development strategy', in F. D. Hamor (ed.) *Issues of Sustainable Development in the Carpathian Region* (Rakhiv: Open), pp.44–9.

Hempen, S. 1998, *Training Materials on Waste Management* (Brussels: FoEE/EEB).

Higman, R. 1998, *Transport Training Materials* (Brussels: FoEE/EEB).

Hudson, R. 1994, 'East meets west: the regional implications within the EU of political and economic change in Eastern Europe', *European Urban and Regional Studies* 1, 79–83.

Jachtenfuchs, M. 1992, 'EC foreign environmental policy and Eastern Europe', in M. Jachtenfuchs and M. Strubel (eds) *Environmental Policy in Europe* (Baden-Baden: Nomos Verlagsgesellschaft), pp.17–41.

Jancar-Webster, B. 1992, 'Technology and environment in Eastern Europe', in J. R. Scanlon (ed.) *Technology Culture and Development: The Experience of the Soviet Model* (Armonk, NY: M. E. Sharpe), pp.1271–98.

Jancar-Webster, B. 1998, 'Environmental movement and social change in the transition countries', in S. Baker and P. Jehlicka (eds) *Dilemmas of Transition: The Environment, Democracy and Economic Reform in East Central Europe* (London: Frank Cass), pp.69–90.

Kabala, S. 1991, 'The hazardous waste problem in Eastern Europe', *RFE/RL Report on Eastern Europe* 2(25), 27–33.

Kic, P. 1996, 'Environmental control in agricultural buildings and its relation to surrounding conditions', in J. P. Smith, D. L. Holt and A. Wheeler (eds) *Aspects of Environment in Europe* (Wolverhampton: University of Wolverhampton), pp.131–46.

Kouvelis, S. 1998, *The Structural Funds of the European Union* (Brussels: FoEE/European Environmental Bureau).

Langer, A. 1990, *Ecological Bricks for our Common House in Europe: Global Challenges Network* (Munich: Verlag fur Politische Okologie).

Maier, P., Kowalzig, J. and Rostock, S. n.d., *Billions for Sustainability?: EU Regional Policy and Accession* (Brussels: Bund/FoEE).

Marin, G. and Schneider, E. (eds) 1997, *Ecological Restoration in the Danube Delta Biosphere Reserve, Romania* (Tulcea: Danube Delta Biosphere Reserve Authority and WWF).

Marwaha, M. 1993, 'Environmental degradation in Central-Eastern Europe: response of the European Community', in C. K. Varshney and D. R. Sardesai (eds) *Environmental Challenges* (New Delhi: Wiley Eastern), pp.107–21.

Nowicki, P. (ed.) 1998, *The Green Backbone of Central and Eastern Europe* (Tilburg, Netherlands: European Centre for Nature Conservation).

Perzanowski, K. 1999, 'Long term ecological research in the Bieszczady Mountains', in P. Bijok and M. Prus (eds) 'Long term ecological research: examples, methods and perspectives for Central Europe', Proceedings of the ILTER regional workshop, Madralin (Warsaw: International Centre of Ecology, Polish Academy of Sciences), pp.131–5.

Pietrzak, M. 1998, 'Development of settlement and farming from the Neolithic period to date in the marginal zone of the Carpathian Foothills between the Raba and Uszwica rivers', in W. Chelmicki (ed.) *The Carpathian Foothills Marginal Zone* (Krakow: Jagiellonian University Geography Institute, Prace Geograficzne 103), pp.15–44.

Pinder, J. 1991, *The European Community and Eastern Europe* (London: RIIA).

Rootes, C. A. 1997, 'Environmental movements and green parties in Western and Eastern Europe', in M. Redclift and G. Woodgate (eds) *International Handbook of Environmental Sociology* (Cheltenham: Edward Elgar), pp.319–48.

Saratpatka, B. and Sterba, O. 1998, 'Optimization of agriculture in relation to the multifunctional role of the landscape', *Landscape and Urban Planning* 41, 145–8.

Slee, B. 1999, *A Tourism Development Plan for the Stuzhytsa-Ushanski Park, Ukraine* (Aberdeen: University of Aberdeen Department of Agriculture).

Stoiko, S. M. 1996, 'Problems in transboundary protected areas in Ukraine', in A. A. Breymeyer and R. Noble (eds) *Biodiversity Conservation in Transboundary Protected Areas: Proceedings of an International Workshop, Bieszczady and the Tatra National Park, Poland* (Washington: National Academic Press), pp.80–5.

Suli-Zakar, I. 1999, 'Socio-geographical transition in the rural areas of the Carpathian Euroregion', *GeoJournal* 46, pp.193–7.

Szirmai, V. 1993, 'The structural mechanisms of ecological and social movements in Hungary', in A. Vari and P. Tamas (eds) *Environmental and Democratic Transition: Policy and Politics in Central and Eastern Europe* (Dordrecht: Kluwer), pp.146–56.

Tickle, A. and Vavroušek, J. 1998, 'Environmental politics in the former Czechoslovakia', in A. Tickle and I. Welsh (eds) *Environment and Society in Eastern Europe* (London: Longman), pp.114–45.

Toth Nagy, M. 1993, 'Observations on the relationship between NGOs and governments in Central and East European countries', in P. Hardi *et al.* (eds) *New Horizons?: Possibilities for Cooperation Between Environmental NGOs and Governments in Central and East Europe* (Budapest: Regional Environmental Center for Central and Eastern Europe), pp.17–22.

Turnock, D. 1996, 'Environmental issues in the Nitra valley of Slovakia', in F. W. Carter and D. Turnock (eds) *Environmental Problems in Eastern Europe* (London: Routledge), pp.188–205.

Waller, M. and Millard, F. 1992, 'Environmental politics in Eastern Europe', *Environmental Politics* 1, 159–85.

Weller, P. and Gunther-Diringer, D. 1999, *Danube Pollution Reduction Programme: Evaluation of Wetlands and Floodplain Areas in the Danube River Basin* (Vienna: WWF Danube–Carpathian Programme).

Wódż, J. and Wódż, K. 1998, 'Environmental sociology in Poland and the ecological consciousness of Polish people', in A. Tickle and I. Welsh (eds) *Environment and Society in Eastern Europe* (London: Longman), pp.97–113.

5 The Soviet Union and the successor states

David Turnock

The Soviet era

The Soviet Union has left a legacy of severe environmental problems for the successor states (Figure 5.1) as a result of decades of economic development driven by the pressures of the arms race (Bridges and Bridges 1996; Singleton 1987; Welsh 1996). Of course there has for long been a view that resources are inexhaustible (Bassin 1993; Diment and Slezkine 1993) and that nature's challenges should not 'condemn the country to permanent backwardness' (Shaw 1999, p.128). Yet there is a traditional 'ecological' conscience among the Russian people, since a nature protection organisation emerged back in 1853 and many of the leading ecologists in the Russian Empire were trained at Tartu University. The Soviet Union initially contributed to global conservation theory and practice through an All-Russian Conservation Society in 1924. However, while the first socialist government was environmentally friendly, nothing was allowed to constrain Stalin's vision of a new economic geography with a more continental emphasis, complete with major canals and river diversions (Rostankowski 1982). Stalin used science for a war on nature to rectify its defects (Burke 1956), as the early strategy in Siberia was followed up by a nuclear programme which was central to the postwar drive for growth. But 'the extreme application of this anthropocentrism became a key characteristic of the whole Soviet nation and is one of the major factors that brought the FSU to the brink of environmental crisis' (Mirovitskaya 1998, p.36). The existing system of nature conservation was practically abandoned and the environmental organisations ceased to exist (Komarov 1980).

In the 1960s the Academy of Sciences campaigned successfully against the closure of the nature reserves (zapovedniks), though the network remained restricted (Fischer 1981; Weiner 1988a, 1988b), and lobbies were able to achieve cuts in pollution, notably through the 1987 decree over pollution abatement on Lake Baikal following pressure from academics and intellectuals (Gustafson 1981). A commission of enquiry in 1986 recommended a Lake Baikal National Park covering both eastern and western shores, as well as three nature reserves (Vorobyev and Martynov 1989). Organisations grew from the late 1950s (Kelly 1976b) – examples being the Tartu Young People's Nature Protection Corps

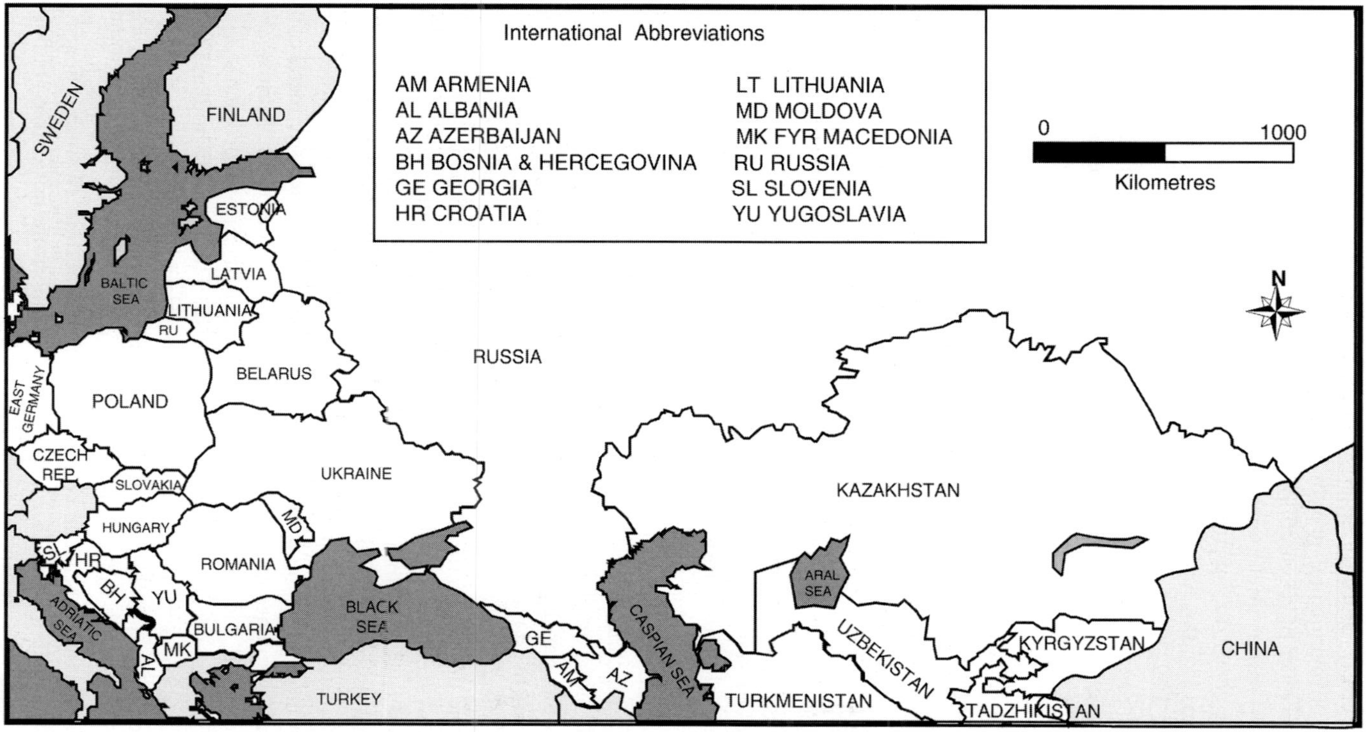

Figure 5.1 States of the former Soviet Union and East Central Europe

(which became part of a wider Students' Nature Protection Corps) and the Moscow Society of Naturalists' Student Section – to number around 1000 registered environmental groups by 1990. However, the new 1960s concept of 'ecologisation of production' did not challenge the basic goal of economic growth, and major development schemes led to serious problems of erosion and salinisation (Feshbach and Friendly 1992; Micklin 1986), especially in some specific areas like the Aral Sea (Micklin 1992). Water supply problems mounted (Rozengut 1991) and the literature drew attention to the Caspian (Rodionov 1994) and Kazakhstan (Golubtsov 1996; Kes *et al.* 1982) in particular. Nature reserves were much reduced and the conservation effort was quite inadequate (Pryde 1972, 1984, 1988, 1991). Resources were very limited (Turnbull 1991; Ziegler 1987) and legislation could not effectively control political vested interests (Gustafson 1989; Massey Stewart 1992). Environmental information remained in short supply until the glastnost era when a new agency, the USSR State Committee for Environmental Protection (Goskompriroda), was established in 1987 (Bater 1996; Jancar 1987).

There was serious waste of resources (Wolfson 1989, 1992, 1994), since the central planning system encouraged high production costs to increase output 'value', thereby boosting wages and career prospects (Csepel 1985; DeBardeleben 1985). There was no incentive for conservation until 1987 when state enterprises had to start paying for natural resources and production linked with environmental conservation became a possibility, although no elaborate structure was worked out at the time. Energy consumption remained high in relation to the volume of industrial production, while low oil/gas extraction rates and substantial pipeline leaks meant the loss of 10–15 per cent of all oil and gas produced, along with periodic explosions of the kind that wrecked trains on the Trans-Siberian Railway at Usinsk near Chelyabinsk in 1989. Some massive oil spills, such as the 60,000t released into the Pechora catchment in 1994, adversely affected wildlife and there was much damage to the tundra vegetation through the development of gas production in Western Siberia. Hydro-electric projects involved enormous costs in terms of resettlement and the flooding of agricultural land (21mln.ha); while fish catches have declined.

Much land was lost to urban development and large mining projects, despite some reclamation (Bond and Piepenburg 1990), while rural areas were affected by soil compaction arising from the use of tractors with an excessive metal content; also by indiscriminate use of pesticides, DDT and the defoliant 'Butifos' (linked with cancer and mental illness) which has created serious problems in Lake Onega and in rivers such as the Volga and Oka. The loss of woodland became a matter for concern as timber resources, estimated at 125bln.cu.m in 1935, were reduced to one-third of that level (Barr and Braden 1988). Reservoirs in the Angara/Yenisey valley flooded 5.9mln.cu.m of timber and there are continuing heavy losses because of disease and the method of timber-floating on the Siberian rivers (Figure 5.2). Acid rain became more noticeable (Hill 1997) and large areas of woodland are lost through fire because of inadequate protection systems in the taiga belt. Deforestation of the fragile tundra and taiga bionomes

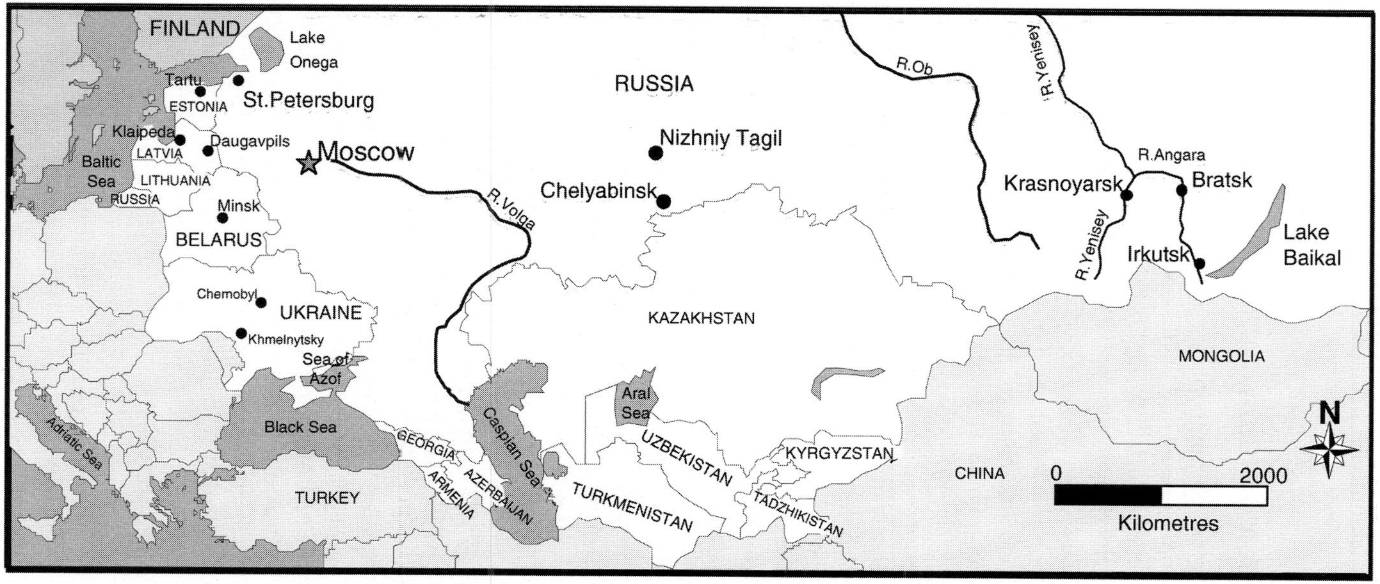

Figure 5.2 Aspects of environmental management

is seen as a major threat contributing to global warming on a scale equivalent to forest loss in the southern hemisphere (Feshbach 1995; Serebryannyy and Zamotayev 1997).

Specific pollution problems

Much pollution arose through routine economic activity (Flint *et al.* 1991) and the overall impression is one of high pollution levels in relation to GNP: indeed 2.5 times the pollution in the USA for the same output (Kelly 1976a). Water pollution (especially in the case of the Caspian and rivers feeding into it) exacerbates problems of shortage, especially in the south of European Russia. The FSU was responsible for one-fifth of global emissions of carbon dioxide and sulphur dioxide; and one-tenth of the CFCs depleting the ozone layer. Many Soviet factories had quite inadequate equipment to filter emissions and process effluent, so industrial towns commonly experienced pollution many times in excess of legally permitted levels (Gladkevich and Sumina 1982). There were many pollution blackspots suffering heavy toxic emissions (Smith 1990), like Norilsk on the Arctic Sea, with its health hazards and landscape damage arising from the smelting of nickel. The Norilsk smelter still pumps 23mln.t of sulphur dioxide into the atmosphere each year. There are large dumps of waste around the town and a concentration of toxic agents in the atmosphere higher than in any other polar region (Bond 1984).

Other blackspots include Bratsk, where carbon disulphide pollution is injurious both to public health and the local forests, and Ust-Kamenogorsk (now Oskemen) with a lead-zinc-magnesium smelter which also turned uranium into reactor fuel rods (Edwards 1993): high levels of lead and zinc occurred in the soil and chromosome damage was linked with cadmium and mercury. In the case of the iron and steel industry, the Nizhniy Tagil complex in the Urals became the source of 0.7mln.t of poisonous emissions each year. At Magnitogorsk in the same region, 12mln.t of steel were produced each year by 50,000 workers in a town of 435,000 where one-third suffered from respiratory ailments, including asthma and chronic bronchitis. In the Kola Peninsula where cobalt, copper and nickel are smelted, winds carry clouds of nephelite sand, sulphur dioxide and nitrogen oxides into Finland and Norway; a classic case of transboundary pollution which has attracted international action in recent years (Tikkanen 1995). Shaw (1999, p.130) mentions the ‘Arctic haze’ phenomenon where pollution travels great distances and destroys vegetation in highly fragile environments; also the severe health problems arising when high pressure systems trap smog – including carbon monoxide from power-stations – in valleys and depressions beneath temperature inversions.

Lake Baikal (the ‘Pearl of Siberia’) is the most voluminous lake in the world (23,000cu.km, accounting for one-fifth of the earth’s fresh water) and also the deepest (1620m); with 600 plant and 1200 animal species, many of which are unique (Galaziy 1981). These resources have been threatened by woodcutting and by the pulp mills. Over many years loggers have cleared large areas of seemingly inexhaustible taiga, but the forest has been slow to regenerate. The scenery has been affected and erosion on deforested slopes has generated landslides carrying vast quantities of silt into the rivers and subsequently into the lake. Much of the timber rafted on both the rivers has sunk (1.5mln.cu.m between 1958 and 1968) and bacteria resulting from its decay has depleted the water of oxygen (Massey Stewart 1990a, 1990b, 1991). The situation improved during the 1970s with tree-planting programmes, removal of rotting timber and the construction of new roads for timber transport; while ships replaced the rafts ('cigars') towed on the lake, and from 1988 all felling was banned along the lake shore. But problems still arise through effluent from lakeside factories (especially the Baikalsk Pulp and Paper Combine opened in 1966 at the southern end of the lake and the Selenga Pulp and Cardboard Combine of 1974 at Selnginsk on the Selenga River) and others in the Irkutsk-Kemerovo area which transfer waste into Lake Baikal via the Selenga River and also affect vegetation through air pollution (Kozhova and Silow 1998). Meanwhile, hydropower development in the Irkutsk area has raised the level of the lake by 1.2m and thereby destroyed shallow feeding and spawning grounds.

The Caspian and Aral seas

A minor disaster has arisen in connection with the mining of salt in Karabogaz Gol, a huge bay on the edge of the Caspian sealed off in 1980. Various substances (barium, bromine, lithium, strontium and potassium) are extracted but salt has been blown over fertile soil as far away as Orenburg and Stavropol (Kozyrev 1993). The barrier has now been breached and sea level has fallen. But far more serious problems have arisen in Central Asia through the drying up of the Aral Sea. Very large quantities of irrigation water are taken from rivers such as the Amu Darya and Syr Darya and transported over great distances (up to 800km in the case of the Kara Kum Canal) to irrigate the loess-covered piedmont (Micklin 1978). In the past some 45cu.km of water reached the Aral Sea each year, but the maximum now is no more than ten, and in some recent years has been zero (Lvovich and Tsigelnaya 1981). And while the Aral Sea's volume has been reduced from 106 to 60cu.km in thirty years, the salinity has tripled from 10 to 30gm/l and fishing has suffered to the extent of up to 50,000t p.a. Canneries now bring in fish from the Caspian, the Far East and from fish farms, while the fishing port of Muynak, which used to produce one-tenth of the Soviet Union's fish catch, is now a virtual ghost town. If the present rate of shrinkage continues the sea will disappear by 2020, leaving only patches of brackish water (Glazovsky 1995; Rafikov 1983).

The moderating effect of the lake on climate is no longer as evident, for the delta forest has died off and the climate has become more extreme (ironically this means that the growing season for cotton is shortened). Moreover, as the sea bed is exposed, salt is blown away over a radius of 300km: an average of fifteen storms each year distribute 75mln.t of material which threatens the agricultural land (Lipovsky 1995). The air around the Aral Sea has become so laden with toxic salts that respiratory problems abound and women are urged not to breast-feed their babies because toxins may be passed on (Martin 1994). Health problems – including very high levels of infant mortality – are aggravated by reduced water supply and water pollution arising from heavy use of fertiliser and pesticides on the cotton fields. There is also a fear that water-borne diseases – like cholera and hepatitis-B – may become endemic. A radical solution contemplated in the Soviet era involved diversion of Ob/Irtysh water along a 2200km canal to restore sea level and expand agriculture and industry. But while a feasibility study carried out in 1978 was allegedly supported by the Uzbek party hierarchy, there was opposition to the large-scale flooding of Siberian forests and a counter-proposal that existing water resources could be better managed (Darst 1988; Micklin 1983). The UN and World Bank are now trying to improve the water supply (though local people generally dislike the chlorinated water and prefer the polluted water to which they are more accustomed) and attention is being directed to irrigation water losses from unlined canals and the inadequate distribution of water where fields are uneven (so that water cannot reach all areas without considerable flooding) (Glantz 1999; O'Hara 1997; O'Hara and Hannan 1999). There is a need for agreement over division of water between the five republics, previously organised in Moscow. There will have to be economic restructuring with less emphasis on cotton, and work could be financed by the profits arising from oil and gas if there is a political will based on democratic politics reflecting local opinion (Atabaki and O'Kane 1998).

The nuclear issue

The foundations of the USSR's nuclear programme were laid in 1945 and Russia now maintains a massive Ministry for Atomic Energy (Minatom); but, despite heavy use of nuclear technology, safety has not been the highest priority. The domestic nuclear energy programme dates from 1958 when the first commercial station (semi-experimental) was set up at Beloyarsk near Sverdlovsk and subsequently capacity in nuclear power-stations increased from 950MW in 1970 to 24,100 in 1984. For many years there have been radiation hazards associated with the nuclear test site of Kurchatov near Semipalatinsk (now Semay) in Kazakhstan where 500 tests were carried out between 1949 and 1989; giving rise to many local health problems affecting half a million people. Radioactivity in river water has arisen through the direct-flow method of cooling at the plutonium factory at Krasnoyarsk, but much more serious problems have arisen at the Mayak nuclear reprocessing facility (Chelyabinsk 65) in the Techna valley between Chelyabinsk and Ekaterinburg (formerly Sverdlovsk) in the Urals: untreated waste was dumped in a lake, from where contamination spread as the lake overflowed into the Techna between 1948 and 1951; and as radioactive dust was distributed by wind-blow during periods when the lake was dry (Monroe 1992).

There was also the Kyshtym incident in 1957: an explosion linked with a failure in the cooling system which contaminated 59,000ha farmland (40,000 of which has since been put back into agriculture) and led to the evacuation of

10,000 people over an extensive area (Edwards 1994). Concern has also been expressed over the Tomsk plutonium factory opened in 1949 (geared to nuclear weapons) and still an important reprocessing plant: there was a plutonium leak in 1993 which contaminated an area of 120sq.km and there is continuing concern over the arrangements for the storage of nuclear waste at the site. There are further hazards through the abandonment of nuclear submarines and ice breakers at various points around the coast (such as Novaya Zemlya) and the dumping of nuclear waste in the seas of the Arctic and the Far East; likewise the dumping of up to 400,000t of Nazi gas canisters in the Baltic and Barents Seas after the Second World War, with the safety period for the metal barrels now running out (Pryde and Bradley 1994).

There was a spectacular accident at Chernobyl nuclear power station in Ukraine in 1986 (Marples 1986, 1988, 1992). While testing a turbine during a scheduled shutdown, power output rose to 100 times the normal, thus leading to an explosion and the burning out of the graphite core of Reactor Four. Radioactive material was released for ten days with contamination of agricultural land over much of Europe (Zeimetz 1990), according to the prevailing winds and the extent to which rain washed poisonous dust from the clouds: contaminating large areas of Belarus (2600sq.km), Ukraine (560) and Russia (460) (Medvedev 1990; Rich 1991). Chernobyl contributed to destabilisation of the USSR; for despite glastnost the West first learned of the disaster through Swedish monitoring stations (Young and Launer 1991). There were also serious internal ramifications since Ukrainians referred to ethnic genocide while environmentalism featured in arguments for independence in the Baltics given the presence of a RBMK nuclear power station (similar to Chernobyl) at Ignalina in Lithuania (Gould 1990; Hewitt 1986). So strong was the anti-nuclear protest – a significant point of access for western ENGOs – that by 1991 the Soviet authorities had cancelled some sixty projects involving 160,000MW of nuclear capacity (Dawson 1995; Marples 1993b). Chernobyl also 'heightened tensions with the Soviet allies in Eastern Europe amongst whom access to nuclear power had been a symbol of trust and status' (Tickle and Welsh 1998, p.13).

The post-Soviet era: CIS states, especially Russia

Ecological problems were at first neglected by the Russian reformists, for neither the 1969 programme of the Democratic Movement nor the 1998 programme of the Democratic Union mention environmental issues (Sedaitis and Butterfield 1991; Ziegler 1991). But nationalists made the running based on the moral message from 'rural writers' (derevenshchiki) in the 1960s that 'the destruction of natural life and of harmony between environment and human beings has resulted in the ruin of nature' (Zisserman 1998, p.677). Water diversion proposals meant Russian sacrifices for non-Russian peoples. Meanwhile, Solzhenitsyn's 'Letter to Soviet Leaders' in 1973 called for zero growth, with greater use of manual labour and renouncement of gigantism. The glastnost era projected the ideas of environmental groups like Zelenyi Mir (the equivalent of Greenpeace) and, on a wider front, professionally competent ecological groups took over from Marxism by advocating international control of national government policies. Green movements such as the Socio-Ecological Union of Russia (established in the mid-1980s) have championed sustainable development and healthy living conditions (Pickvance 1997) – highlighting the life-styles of some small minorities, such as the Nenets of the Yamal Peninsula, largely ignored in the Soviet period (Jancar-Webster 1995). Greenpeace Russia has opposed the large-scale development ethos in general, and the high-speed Moscow–St Petersburg railway has drawn opposition from conservationists aware of its threat to the Valdai National Park.

However, despite Russia's significance for global sustainability, the element of movement surrogacy (environmentalism as an approach to system change), coupled with concern over jobs, has limited the appeal of the movement given the low level of integration with EU norms (Doktorov *et al.* 1993). While recruiting protesters in the early days, green movements are now struggling, with people unused to taking initiatives, compared with civil society in the West (Yanitsky 1993a, 1993b, 1996). Although there are some Green parties they have not been able to win seats, and 'most environmental groups lack the basic financial and organisational supports necessary to run a successful social movement' (Smith 1999, p.111). However, there have been some successful urban struggles against planners (Lang-Pickvance 1997; Oldfield 1999; Yanitsky 1991a, 1991b) and attention is being drawn to public health issues (Malkhazova *et al.* 1997).

Considerable efforts have been made to overcome the Soviet legacy and new institutions have emerged (Artobolevsky 1991; Peterson 1993, 1995a, 1995b; Shaw and Oldfield 1998). Environmental zones were first designated in 1991 (Pryde 1994). These include urban and mining-industrial areas such as the Kuzbass, Kola Peninsula, Moscow and the Urals; also areas of soil and water pollution such as the Black Sea and the Caspian, the Chernobyl fallout area, the Kalmyk Republic, Lake Baykal, the Sea of Azov and the Volga. Next, sixty critical locations were listed in 1993. Meanwhile the Law on Protection of the Natural Environment (1991) is an effective 'ecological code', including the polluter pays principle (PPP) (Bond and Sagars 1992), and has been hailed by Shaw (1999, p.144) as 'a laudable attempt to place environmental protection in Russia on a new footing after the dismal record of the Soviet years'. This has been a launching pad for a series of related pieces of legislation, following the Rio Summit in 1992: the concept of ecological security (1994), the concept of specially protected areas (1995), the concept of transition to sustainability (1996) and the concept of air pollution (1999). The Ministry of Environmental Protection and Natural Resources (now the State Committee for Environmental Protection, Goskomekologiya) has been working on environmental action programmes, and the one for 1996 to 1997 is linked with a model for sustainable development including new afforestation plans (ibid., p.145; Ponomarenko 1994). There are now 38mln.ha of protected areas, but this does not cover all priority areas and additions are being suggested, such as the virgin Manchurian-type wetlands and forests in northern Ussuriland (Mikhailov and Shibnev 1998).

Some writers see ENGOs rendered largely impotent by bureaucracies inherited from the communist period which see environmental reform as a threat to economic progress through demands for unreasonably high pollution standards. Hence the likelihood in Russia that FDI, 'particularly that targeted on resource extraction and utilization, will be pursued at the expense of severe environmental degradation' (Tickell and Welsh 1998, p.175). Indeed, these two authors see environmentalism as being highly vulnerable to authoritarian modernisation which may provoke a confrontation between federal/regional elites and an immense underclass of 'counter-modernisers'. Certainly there are grounds for concern because pollution seems to have declined less than GNP, given the growth of motor traffic, and water quality continues to decline (Bond 1993), while unregulated hunting and poaching in Central Asia and Kamchatka points to the need for more money and information. Oldfield (2000) has noted that some polluting industries (e.g. Volga chemicals and petrochemicals) have comparative advantages and may create greater problems if and when the Russian economy recovers. Although a stronger emphasis on energy and raw materials (rather than on downstream processing) may reflect more realistic raw material costs, 'the apparent simplification of the Russian economy suggests that the country's substantial resource reserves will be utilised more intensively, at least in the short term, as individual regions take advantage of the opportunity to earn hard currency' (ibid., p.83).

Yet there are state instruments which activists and pressure groups can use and 'the notion of ecological security developed in Russia could, given the necessary political will, be combined with the European concept of ecological modernisation to create a strong sustainablity approach to modernisation' (Tickell and Welsh 1998, p.176). There is also a measure of international coordination through the CIS and regional agreements concerning the Arctic and the Baltic, Barents and Black Seas (Ivanov and Oguz 1998). For example, sewage is a major concern in the Baltic and a upgrade in Kaliningrad will be an important step forward. Measures are also being taken to reduce pollution damage, often through international action. Finland's Kumera Corporation is using technology from Reynolds Metals to substantially reduce pollution from Russia's Kandalaksha aluminium smelter. In the process the US$50mln project will boost capacity from 62,000t to 70,000. There is some financial support from Finland since the smelter is on the Kola Peninsula and there is concern about fluoride emissions. At the same time, Pechenganikel is being refurbished with the help of the new Scandinavian partners of the private owner (Norilsk Nickel) which is also being helped by direct budget support and export tariff exemptions from the Russian government. 'Norilsk Nickel was able to exploit the very "dirtiness" of the Pechenganikel plant in order to cheat the market forces that threatened the survival of its operations in northwestern Russia' (Darst 1997, p.61).

New environmental threats: nuclear power

New hazards are coming to the fore in mining areas that play a key export role (DeBardeleben and Hannigan 1994), with particular reference to oil and gas developments in the Caspian and Central Asia (Blum 1998; Bolukbasi 1998). However, the nuclear debate continues to attract much attention. Russia has admitted the dumping of nuclear waste at sea during the Soviet era and has attracted western financial support for alternative arrangements – even with help from Japan (despite its territorial claims in the Kurile Islands) for a reprocessing facility in the Far East in 1996. But there is concern over Russia becoming a dumping ground for nuclear waste, and Moscow passed a law in 1992 against disposing of nuclear waste from foreign countries on its territory. However, the Bulgarian Atomic Committee has now raised the question in 1998 of a regional storage facility for high-level waste – at a cost of some US$100mln. There is certainly a need for some co-operation, since Bulgaria has difficulty in sending waste to Russia through the sensitivities of transit countries like Moldova, while Slovakia has been forced to suspend arrangements made with Mayak on account of high costs. Meanwhile a new nuclear power programme is underway after pressure to halt construction in 1990 (Marples 1993c). This has attracted some protest, notably at Krasnoyarsk (a nuclear waste-processing site) and the power station locations of Kostroma and Rostov. But demands for referenda, provided for under the constitution, have drawn a bureaucratic response (demanding an impossibly large number of petitioners), while there has been some media hostility to perceived interference by western ENGOs, and Rostov protesters have been branded as foreign agents (Smith 1999, p.111).

Meanwhile, conservation groups have emerged to oppose nuclear power in Ukraine. They have been assisted in their propaganda by an embarrassing series of further mishaps (post-Chernobyl) including events in May 1992: Khmelnytsky suffered a damaged turbine and a circuit closure (related to heavy rain and high humidity levels) while two reactor generator problems occurred at Zaporizhya and defects were found in all three reactor units at South Ukraine. In addition, Chernobyl's Reactor Two suffered a turbine fire in 1991 and had to be closed down. Although not serious, the problems point to the extent of the problem of worn and defective equipment. In view of the high risk of another Chernobyl-type disaster, international opinion has suggested the closure of twenty-one reactors currently supplying three grids (Scott *et al.* 1995). There is some concern over the Russian-designed pressurised water reactors (PWRs) of the VVER type, for the designs have built-in weaknesses and do not meet internationally recognised design standards. But there is particular concern about the stability of the graphite-moderated light water reactors (RBMK) of 1000MW. These reactors were installed at Chernobyl and it has been revealed that the accident that occurred there was particularly severe because of defects in the design of the control and safety rods (deficiencies that were known about in the Soviet Union before the accident but not corrected). There were earlier failures at Leningrad (now St Petersburg) in 1975 and Chernobyl in 1982 (Ebel 1995; Egorov 1999).

However, the anti-nuclear movement has largely vanished in the harsh light of independence: 'nuclear power plants once denounced as odious manifestations of Soviet imperialism, if not outright genocide, were now transformed into valuable bulwarks of national independence' (Darst 1997, pp.61–2). It is not realistic to contemplate sweeping closures of the inherited stations, in view of the huge cost of decommissioning and of providing new capacity where demand is high. In Ukraine, nuclear stations are responsible for half the country's electricity, while Lithuania's Ignalina Power Station – discussed below – supplies almost 90 per cent of that country's power. The emphasis must rest with improving safety and reliability at stations that will continue to function. Nuclear politics are such that alarm in the West over nuclear safety has prompted help to subsidise the modernisation of existing nuclear power plants in the former Soviet bloc and to assist in completion of some of the reactors whose construction had been interrupted by the anti-nuclear surge of 1988 to 1990. Meanwhile, Ukraine is happy to retain the plant (linked with completions at Rivne and Khmelnytsky) while pressuring the West to increase donations. Following the 1986 explosion, the G7 have provided almost USD1bln for safety work, decommissioning and sarcophagus repair. Loans exceeding US$2bln have also been advanced to help develop a more reliable and stable energy sector, while the EBRD is financing two new nuclear power-stations in Ukraine linked with the closure of Chernobyl.

A Russian presidential decree in 1992 anticipated construction of over seventeen Gigawatts of new nuclear capacity by 2010. Such a programme will complete reactors under construction at the time of Chernobyl and construct others needed to replace reactors scheduled for closure and to increase nuclear capacity overall by developing the clusters of reactors already in existence (Marples 1993b, 1994). The Russian Pacific fleet survey and research ship *Urals* is being turned into a floating nuclear power station to supply military garrisons and coastal villages in maritime territory of the Russian Far East (Bradshaw and Kirton 1988; Bradshaw and Lynn 1998). Russia is helping to build a 1900MW nuclear plant near Lake Balkhash in Kazakhstan (the only former Soviet country interested in co-operating with Russia in this area) with three VVER-640 reactors. The project started in 1999 and the reactors should come into operation in 2005, 2007 and 2008. Russia is also to provide training, nuclear fuel and reprocessing. At the same time, Russia wants Japanese help in building new fast-breeder reactors.

Energy efficiency and renewable power

Pressure for more nuclear capacity could be moderated through greater energy efficiency, driven by high energy costs: heating and hot water averaged US$30–50p.c.m. for a typical Russian apartment in 1995 while the average monthly wage was around US$120. Municipal government subsidies for residential heating and hot water absorbed about one-third of all municipal budgets in 1996. Energy use is inefficient because technological change was inhibited by lack of incentive to innovate, since the output of monopoly producers was always in demand and centralised systems were assumed to have advantages *a priori* (Nekrasov 1993). There is also scope for renewable energy, especially wind power in the North Caucasus: a flat, windy area with a high population density and a severe energy deficit in the regional grid (Martinot 1992). There is also potential for solar energy in the south, as well as supplementary water heating throughout Russia. In addition, for the 20mln people not connected to the grid, hybrid wind–diesel systems and biomass-fired plants could replace or supplement diesel and oil generators. Biomass could be used for district heating in towns with boilers less than 10MW and geothermal resources have been identified in Stavropol, Krasnodar, Sakhalin and Kamchatka. Russia has the technical capabilities, while still lacking 'the associated market-oriented skills and institutions to take full advantage of these technological capabilities' (Martinot 1998, p.909). There are capital, informational, infrastructural and institutional barriers but there is scope for joint ventures and technology transfer (Schipper and Martinot 1994).

Post-Soviet era: the Baltic States

The Baltics have seen the strongest environmental movements since the Estonian Society of Nature (Looduskaitse Selts) was founded in 1966 with membership of several thousand. The Green Movement was active in Lithuania as early as 1986 (Kritkautsky 1995). Latvia's Environmental Protection Club was founded in the same year and the Estonian Heritage Society was formed in 1987. In some cases, concern was driven by a national culture which draws inspiration from the environment: for example, the 'nature culture' of Latvia, celebrated in folk arts and crafts, is the bedrock on which the current independence movement has been based (Bunkse 1992). However, the Baltics have provided some evidence to support the notion of 'movement surrogacy' where radical reformers hide behind surrogate causes that target appropriate audiences (Dawson 1996). This may result in environmental issues being exploited as a means to achieving other goals, whereupon the surrogate issues may be downgraded: nuclear power-stations initially condemned as 'imperial domination' may be accepted in the interest of 'self-sufficiency'. The argument may be vindicated to some extent by the split between the Estonian Green Movement and Estonian Green Party in 1989, arising out of the latter's political aspirations; although even here the charge that politicians used the environment as a facade is reversed by claims that the Green Movement is naive in its political inactivity (Hosking 1992). This situation is now changing with the EU's stronger environmental approach to pre-accession funding. ENGOs are included in Estonia's Special Preparatory Programme launched in 1998 to provide a framework for effective use of EU structural funds. There has also been involvement in Estonia's National Forestry Policy.

However, it is to be expected that some reassessment will arise after system change when environmental goals must be prioritised in the context of financial resources and safeguarding jobs. Surrogacy arguments can also be overstated to the point of denying genuine grievances over pollution and the waste of resources, linked with 'mindless' industrial policies requiring Russian immigration to sustain environmentally unfriendly industries linked with the Estonian oilshales. On the whole, 'the words and deeds of "glastnost" era Baltic environmentalists suggest that genuine abhorrence for Soviet industrial schemes rather than clever adoption of environmental rhetoric for non-environmental ends inspired popular opposition', though there was also the economic autonomy concern (Auer 1998, p.670). Latvian Greens were concerned about pollution from industrial centres like Olaine, with air pollution and hazardous waste problems linked to chemical and pharmaceutical industries. There was also concern over air pollution linked to transport and district heating, eutrophication resulting from untreated municipal waste water and agricultural run-off (note also phosphorous and nitrogenous pollution in the Gulf of Riga), army training areas and transboundary pollution (Vilnitis 1997). The controversial Riga Metro building also attracted much public attention, but the main target for environmentalists was the Daugavpils hydropower project (1991) (Muiznieks 1987). This did not involve contamination and it lacked a 'colonial' dimension because the power was intended to supply Latvia. But environmentalists argued that small dams would be better, especially in view of experience with the Plavinas hydro project which destroyed cultural treasures along the Daugava River. There was also opposition from Belarus which stood to lose two-thirds of land flooded by the project (eighty towns and villages) and from some Russian organisations. Also in Latvia there was a campaign in 1989 to 1990 against the antiquated Sloka pulp and paper mill at Jurmala on the Gulf of Riga because local people had to suffer the pollution so that the whole of the USSR could be supplied with certain grades of paper. In this case it seemed that Soviet planning was ignoring local interests and susceptibilities.

All the Baltic States – but especially Estonia and Latvia – face serious environmental problems over nuclear, chemical and mechanical pollution resulting from former military bases. In Estonia there was also much concern over oilshale strip-mining at Kohtla-Jarve which damaged the landscape and caused local health problems through inappropriate disposal of the radioactive upper layer. In addition, river pollution arose through phenol leaching from waste tips; while much damage was caused by dust and sulphur emissions from oilshale power plants in Narva. Problems are also associated with phosphate mining at Maardu (started as a means of fertilising some of Russia's poorer soils) and the prospect of more damage from a further venture at Rakvere on the Gulf of Finland. Protest was very strong in view of the intention to push ahead immediately with a scheme that would (incidentally) pollute a river supplying irrigation water, contribute a further increase to oilshale power station capacity and take shale-mining and processing beyond the peak of 30mln.t reached in 1980 (compared with around 1mln when the Soviets took over in 1941) (Barabner and Kaganovich 1998). The shales supplied power to Estonia but also much electricity to Russia (from Narva) and gas to Leningrad (Auer 1998). Protests in 1987 led to the concession by Bruno Saul, Chairman of the Presidium of Estonian SSR Council of Ministers, that the project would be suspended until proper scientific studies had been made. But the Green Movement continued to grow, and in 1988 the Estonian Greens called for Saul's replacement because he had failed to defend national interest against 'colonial' plans. The Greens wanted an end to phosphorite and shale-mining, and also a halt to Russian immigration and recognition of Estonia's autonomy. They were also provoked by plans for a third reactor at Ignalina, as were a number of post-Chernobyl activists from Belarus, Moldova and Ukraine.

The Green Movement is still strong, with a keen sensitivity over corruption in government (Banks 1998; Banks and Todd 1995) – though attacking the new national governments is more difficult than blaming the Soviets – and a belief that democracy must prevent the environment from being degraded for the benefit of a few. Inspired projects like the environmental film studio in Latvia have been completed and a common position is being sought among Baltic ENGOs for long-term sustainable transport. EU funding for a multi-model rail-based transport corridor from Klaipeda to Minsk and Moscow is preferred to further road construction. However, wider international cooperation is vital (Banks 1994, 1997) and considerable help is coming from the West. Thus the UK group Powerful Information, set up in 1990 to provide information and training projects for ENGOs, has made contact with the Lithuanian Women Farmers Association (LUD) reformed in 1992 and dedicated to improving the lives of rural women. It now has around 100 groups and a total membership of over 2000. The UK Lotteries Charities Board has funded a three-year joint project whereby Powerful Information will help LUD launch a newsletter and provide training in alternative activities: organic farming (including bee-keeping, herb-growing, the commercial production of mushrooms and the rearing of fur-bearing animals) along with traditional crafts and tourism. A Green Library has been set up in Riga to provide environmental information. Meanwhile, the Estonian Fund for Nature and its Latvian counterpart are undertaking a joint project to protect biodiversity in 140sq.km of border wetlands. This cross-border protected area has attracted support from the Regional Environment Centre (Szentendre) and will provide a habitat for many rare and endangered species. Estonian and Russian Greens have been brought together to secure government agreement on joint management of Peipsi-Chudskoye Lake (i.e. Peipsi Lake Project (PLP)) (O'Lear 1997). Latvian ENGOs generally operate at the local level, apart from the Environmental Protection Club, strong in the late 1980s, and the Latvian Fund for Nature referred to above, although WWF has a presence as does Coalition Clean Baltic. Nevertheless, more public participation is needed both through Green parties and NGOs. Better environmental law is needed as well as the capacity to implement conservation projects.

Environment management

In Latvia, overall responsbility for the environment – inspection, monitoring, data collection and environmental impact assessment (EIA) – has rested with the Ministry of Environmental Protection and Regional Development (previously the Environmental Protection Department) since 1993. The first National Environment Policy Plan for Latvia (NEPPL) was drawn up in 1994 (with help from The Netherlands and Sweden) and was adopted in 1995 (though subject to regular review) and a National Environmental Action Programme was derived from it in 1996. There are plans for an Integrated Environmental Management System combining policy planning, operational planning, implementation monitoring and feedback. Following the umbrella law on Environmental Protection (1991) there have been additional measures passed on hazardous waste and protected areas (1993), nuclear safety (1994), and the Natural Resources Tax (1996). There is provision for environmental education, and a trilateral agreement between the three Baltic countries (covering early warning and transboundary pollution) has – like the Baltic Environmental Forum (for the exchange of information and experience) – operated since 1995. However, while the machinery in Latvia could be improved (including education and public awareness) the fundamental problem concerns the lack of finance (regular MEPRD budget cuts) and expertise. Further, if Latvia is to progress with EU accession, 'then the whole system of environmental standards and norms has to be completely changed' (Vilnitis 1997, p.192).

Protected areas now comprise a substantial network. Latvia has a national park (gauja) administered by the State Forest Service, while nature reserves are handled by two administrative units: Slitere State Nature Reserve in the west and Teicu SNR in the east. There is also a North Vidzeme Nature Protection Complex managing a range of protected areas in the north. However, across the country in general, low-intensity land use 'allowed preservation of the natural forests, meadows and swamps where rich animal and plant populations are to be found now' (Vilnitis 1997, p.171). Soils are generally clean with a low heavy metal content. Indeed, the neglect of agriculture has resulted in an extension of woodland from 28 per cent of the land in 1939 to 42 per cent today. But with land back in private ownership there is a danger that short-run needs will lead to over-exploitation of minerals, timber and peat (all the more so since Latvia is poor in energy resources and this could intensify pressure on bogs and forests, whereas in the Soviet period extraction did not exceed natural growth) (Kavaliauskas 1994). Agriculture is already quite sustainable – especially given the reduction in pesticide use since 1991 – but since it now accounts for 40 per cent of GDP there could be problems over any anticipated intensification. And Estonia is asking the EU to amend the 1992 Habitats Directive to take account of the need to control wild populations of wolves, lynx, bears and beavers: hunting is necessary to reduce damage to agriculture and the threat to human life. Estonia is also discussing the optimum cutting level for forests: the country exports wood products, paper and furniture (17.5 per cent of capacity), but it may be poor business sense to rely heavily on wood exports when other parts of the world can produce wood more cheaply. Meanwhile, the Lithuanian Greens oppose the expansion of military training in forests.

Aspects of international cooperation

On the whole there is good co-operation over transboundary pollution and other environmental concerns (Valiunas *et al.* 1994). Water resources are a priority (Juodkazis 1994). The USA and Lithuania are leading a NATO project to study the effects of nutrient-rich river water and build models for water management. Fieldwork will be carried out in Estonia (Emajogi amd Narva rivers), Latvia (Daugava River) and Lithuania (Nemunas and Neris rivers). The chosen rivers are the largest and are important for their aquatic organisms and drinking-water supplies. Sweden is paying for a river clean-up in the Baltic States and is also assisting sewage treatment. The waste-water plant inaugurated at Klaipeda in 1998, supported by Finland, Sweden, the World Bank and the EU, will help clean up one of the biggest sources of pollution, according to the Helsinki Commission on Baltic Environmental Protection. Meanwhile, waste-water treatment is improving in other Lithuanian towns – Kaunas, Palanga, Siauliai and Vilnius – in order to reduce pollution in the Baltic. Finnish technology is helping to cope with emissions at Estonia's Narva Power Station (also subject to tax, with an additional carbon dioxide tax payable from 2000).

A PHARE project is on hand to provide a hazardous waste management system in northeastern Estonia where there is a large stockpile of such waste from the Soviet period. The Sillamae Radioactive Tailings Pond constitutes the most serious environmental problem of Estonia and one of the most environmentally hazardous objects of Europe. Situated beside a factory in Narva, closely linked with the military, the pond contains 8mln.cu.m of radioactive waste including 1200t of radioactive uranium and 600t of thorium – on a 33ha site which threatens the Baltic with massive pollution in the event of a fragile dam collapsing. An additional 7.7th.t of waste is being generated each year but current policy forbids incineration and so storage is crucial. Lack of consultation among local residents induced NIMBY attitudes which then ruled out the best site (Aseri), and also at Sillamae, where public protests forced the local authority to cancel the deal. After that, consultation produced a deal at a third site of Vaivara which is now operating with EU aid. 'Because local concerns were ignored and public participation discouraged the start of the project was delayed, time and money were wasted and the authorities' credibility tarnished' (Maier *et al.* n.d., p.38). Rehabilitation work started in 1999 and eventually the tailings pond should be stable for 1000 years after a Euro20mln spend financed mainly from Scandinavia.

Meanwhile, Latvia generates only 150–300kg pc/p.a. of waste (compared with 300–450kg in the West), but the quantity is increasing with more western imports and waste management is quite inadequate. Only half is collected and none of the 558 landfills operate to applicable health standards, though a former Soviet missile base is being used as a central store for hazardous waste. Lithuania, however, has the problem of an unsafe nuclear reactor which does not meet western nuclear safety standards (according to the West European Nuclear Regulators' Association in 1999) (Vilemas 1995). Lithuania is the most nuclear-dependent country in the world and the government would be happy if the 1998 upgrading would allow operation until 2020 or even later. Indeed, a Swedish report found that early closure could cost US$3.3 to 3.9bln in lost opportunities alone. However, officials have said that a timeline for closure should be a precondition for Lithuania to enter EU accession talks and it has been announced that the first reactor is to close during 2000 to 2005, conditional on foreign assistance, while the second will be decommissioned in 2010 (closely in line with the original 'design life' of 2009 to 2014). However, the total cost of the shutdown could be between US$325 and 375bln and some opinion in Lithuania would prefer upgrading to allow operation until 2020 or even later, though this is clearly not now possible.

Approximation of the EU environmental *acquis*

According to the EU regular reports on the enlargement process, the Baltic States made good progress with the approximation of EU environmental legislation during 1999 to 2000. This involves a range of criteria, including EIA, air quality, chemicals and GMOs, IPPC, nature protection, noise, radiation protection and nuclear safety, waste management and water protection. Latvia has a national programme for biodiversity linked with a new department at the Ministry of Environment and Regional Development. There has also been adoption of Agenda 21 sustainable development policies by the Baltic Sea region as a whole. Finance is probably not yet sufficient. Although Latvia's spending increased to E46mln in 1999, this was only 0.7 per cent of GDP which compares unfavourably with 2.4 per cent in the Czech Republic. Estonia attracts a good deal of recognition for vigorous efforts but the financing of environmental investments is seen as a weakness and in every policy area except chemicals the overall assessment highlights the need for further action: the radiation problems over the nuclear decommissioning and remedial work at the Sillimae tailing pond are highlighted. Other weaknesses have been picked up. Lithuania's procedures for identifying protected areas do not all comply with those of the EU, while for waste management the transposition of the *acquis* is scattered through many pieces of legislation.

There is a need to strengthen administrative capacity in Latvia. The Ministry of Environmental Protection and Regional Development – with some thirty subordinate bodies (including regional boards and the environmental inspectorate) – shares some responsibility with the Ministries of Agriculture, Interior, Transport and Welfare. Approximation therefore involves inter-ministerial groups with representatives from ministries and institutions. Some new institutions are being set up, such as the Environmental Impact Assessment Bureau and a radiation centre, but some restructuring is considered necessary for effective application and enforcement of the *acquis*. Lithuania also has a large number of authorities and institutions concerned with the implementation of legislation. Some rationalisation is needed to avoid duplication while some new institutions are needed to cope with all the requirements of the *acquis*. At the same time, a considerable effort is required to strengthen the capacity of local authorities by planning investments, as Lithuania is finding in the case of waste management and water protection. However, in Estonia the environmental departments of county governments have become regional departments of the Ministry of Environment in order to enhance administrative capacity.

Conclusion

This can be no more than a thumb-nail sketch of a vastly complex set of issues which are particularly significant because of the sheer scale of issues such as the nuclear risk and the threats to deforestation. The former Soviet Union is all the more important because it is the source of the model imposed on East Central Europe during most of the second half of the twentieth century. Today we are reminded that this theatre is not (with the exception of the Baltic States) subject to the sustainability agenda which is being beamed out of Brussels. As it happens, transition has benefited the environment through reduced output, but economic stress has increased with marketing difficulties and low competitiveness with western manufacturers. Municipalities have responsibilities but they frequently have insufficient resources (including trained personnel) despite some taxation revenue. Although some growth is now being resumed, only limited resources are available for the environment and some controversial projects (e.g. in the nuclear domain) abandoned after opposition in the Soviet era have been restarted under the 'economic independence' banner. Furthermore, given the tensions underlying the breakup of the Soviet Union, there have been inevitable disagreements among the successor states. Because of non-payment of bills, the Russian town of Ivangorod was cut off for a time by the Estonian company in Narva providing water and sewage, although this meant that untreated sewage was dumped into the Narva River just six kilometres from the Gulf of Finland. Latvia has objected to Lithuania's new oil terminal close to its border, although the Lithuanians believe that the installation is adequately protected and have rejected what they see as inappropriate political lobbying by firms fearing competition for Ventspils over the export of Russian crude oil. However, with an underlying readiness to co-operate, there is stiff challenge to marshall financial and institutional capacities commensurate in scale with the ongoing biodiversity threats.

References

Artobolevsky, S. 1991, 'Environmental problems in the USSR'. *Geography Review* 4(4), 12–16.

Atabaki, T. and O'Kane, J. (eds) 1998, *Post-Soviet Central Asia* (London: Tauris Academic Studies).

Auer, M. R. 1998, 'Environmentalism and Estonia's independence movement'. *Nationalities Papers* 26, 659–76.

Banks, A. J. 1994, 'Environmental cooperation in the Baltic Sea Region: a Lithuanian perspective'. *GeoJournal* 33, 37–43.

Banks, A. J. 1997, 'International assistance for environmental purposes in Lithuania 1991–1994'. *Journal of Transforming Economies and Societies* 4(3), 18–30.

Banks, A. J. 1998, 'Environmentally sustainable society in Lithuania?' *Lithuanian Papers* 12, 61–9.

Banks, A. J. and Todd, J. J. 1995, 'Ignalina nuclear power plant: problems and prospects'. *Lithuanian Papers* 9, 30–4.

Barabner, N. I. and Kaganovich, I. Z. 1998, 'Oil shale production and power generation in Estonia'. *Energy Policy* 21, 703–9.

Barr, B. M. and Braden, K. E. 1988, *The Disappearing Russian Forest: A Dilemma in Soviet Resource Management* (Totowa, NJ: Rowman and Littlefield).

Bassin, M. 1993, 'Turner, Soloev'ev and the frontier hypothesis: the nationalist signification of open spaces'. *Journal of Modern History* 65, 473–511.

Bater, J. H. 1996, 'Natural resources management in the Soviet and post-Soviet eras', in J. H. Bater, *Russia and the Post-Soviet Scene: A Geographical Perspective* (London: Arnold), pp.154–75.

Blum, D. W. 1998, 'Domestic politics and Russia's Caspian policy'. *Post-Soviet Affairs* 14, 137–64.

Bolukbasi, S. 1998, 'The controversy over Caspian Sea mineral resources: conflicting perceptions and clashing interests'. *Europe-Asia Studies* 50, 397–414.

Bond, A. R. 1984, 'Air pollution at Norilsk: a Soviet worst case?' *Soviet Geography* 25, 665–80.

Bond, A. R. 1993, 'Environmental disruption during economic downturn: White Book Report'. *Post-Soviet Geography* 34(1), 74–8.

Bond, A. R. and Piepenburg, K. 1990, 'Land reclamation after surface mining in the USSR: economic political and legal issues'. *Soviet Geography* 31, 332–65.

Bond, A. R. and Sagars, M. J. 1992, 'Some observations on the Russian Federation environmental protection law'. *Post-Soviet Geography* 33, 463–74.

Bradshaw, M. J. and Kirton, P. 1998, 'The energy crisis in the Russian Far East: origins and possible solutions'. *Europe-Asia Studies* 50, 1043–64.

Bradshaw, M. J. and Lynn, N. J. 1998, 'Resource-based development in the Russian Far East: problems and prospects'. *Geoforum* 29, 375–92.

Bridges, O. and Bridges, J. W. 1996, *The State of the Russian Environment* (Aldershot: Avebury).

Bunkse, E. V. 1992, 'God thine earth is burning: nature attitudes and the Latvian drive for independence'. *GeoJournal* 26, 203–9.

Burke, A. E. 1956, 'Influence of man upon nature: the Russian view – a case study', in W. L. Thomas (ed.), *Man's Role in Changing the Face of the Earth* (Chicago, IL: University of Chicago Press), pp.1036–51.

Csepel, A. 1985, 'Marxism and the ecological crisis'. *East European Reporter* 1(2), 41–3.

Darst, R. G. 1988, 'Environmentalism in the USSR: the opposition to the river diversion projects'. *Soviet Economy* 4, 223–74.

Darst, R. G. 1997, 'Bribery and blackmail in East–West environmental politics'. *Post-Soviet Affairs* 13(1), 42–77.

Dawson, J. I. 1992, 'Anti-nuclear activism in the USSR: the case of the environment', in J. Massey Stewart (ed.), *The Soviet Environment: Problems, Policies and Politics* (Cambridge: Cambridge University Press), pp.64–87.

Dawson, J. I. 1996, *Eco-nationalism: Anti Nuclear Activism and National Identity in Russia, Lithuania and Ukraine* (Durham, NC: Duke University Press).

DeBardeleben, J. 1985, *The Environment and Marxism-Leninism: The Soviet and East German Experience* (Boulder, CO: Westview Press).

DeBardeleben, J. and Hannigan, J. (eds) 1994, *Environmental Quality and Security after Communism: Eastern Europe and the Soviet Successor States.* Boulder, CO: Westview Press.

Diment, G. and Slezkine, Y. 1993, *Between Heaven and Hell: The Myth of Siberia in Russian Culture* (New York: St Martin's Press).

Doktorov, B. Z., Firsov, B. M. and Safronov, V. V. 1993, 'Ecological consciousness in the USSR: entering the 1990s', in A. Vari and P. Tamas (eds), *Environment and*

Democratic Transition: Policy and Politics in Central and Eastern Europe (Dordrecht: Kluwer), pp.249–67.

Ebel, E. R. 1995, *Nuclear Energy Safety Challenges in the Former Soviet Union: A Consensus Report of the CSIS Congressional Study Group and Task Force* (Washington, DC: Center for Strategic and International Studies).

Edwards, A. 1993, 'A broken empire (Kazakhstan)'. *National Geographic* 183(2), 2–53.

Edwards, M. 1994, 'Pollution in the former USSR: lethal legacy'. *National Geographic Magazine* 186(2), 70–115.

Egorov, N. N. 1999, *The Radiation Legacy of the Soviet Nuclear Complex* (London: Earthscan).

Feshbach, M. 1995, *Ecological Disaster: Cleaning up the Hidden Legacy of the Soviet regime* (New York: Twentieth Century Funds Press).

Feshbach, M. and Friendly, A. 1992, *Ecocide in the USSR: Health and Nature Under Siege* (London: Aurum).

Fischer, D. 1981, 'Nature reserves in the Soviet Union: an inventory'. *Soviet Geography* 22, 500–22.

Flint, V. E. 1991, *Environmental Status Reports 1990: USSR* (Cambridge: International Union for the Conservation of Nature).

Galaziy, G. I. 1981, 'The ecosystem of Lake Baikal and problems of environmental protection'. *Soviet Geography* 22, 217–25.

Gladkevich, G. I. and Sumina, T. I. 1982, 'Measuring the environmental impact of industrial centres in the natural-economic regions of the USSR'. *Soviet Geography* 23, 155–65.

Glantz, D. M. 1999, *Creeping Environmental Problems and Sustainable Development in the Aral Sea Basin* (Cambridge: Cambridge University Press).

Glazovsky, N. F. 1995, 'The Aral Sea Basin: a man-made ecological catastrophe'. *GeoJournal* 35, 7–70.

Golubtsov, V. V. 1996, 'Vulnerability assessment of the water resources of Kazakhstan to anthropogenic climatic change and the structure of adaptation measures'. *Water Resources Development* 12, 193–208.

Gould, P. 1990, *Fire in the Rain: The Democratic Consequences of Chernobyl* (Oxford: Polity Press).

Gustafson, T. 1981, *Reform in Soviet Politics: Lessons of Recent Policies on Land and Water* (Cambridge: Cambridge University Press).

Gustafson, T. 1989, *Crisis Amid Plenty: The Politics of Soviet Energy under Brezhnev and Gorbachev* (Princeton, NJ: Princeton University Press).

Hewitt, E. A. 1986, 'Economic and political consequences of Chernobyl'. *Soviet Economy* 2, 97–130.

Hill, M. R. 1997, *Environment and Technology in the Former USSR: The Case of Acid Rain and Power Generation* (Aldershot: Edward Elgar).

Hosking, G. A. 1992, 'Popular movements in Estonia', in G. A. Hosking, J. Ares and P. J. S. Duncan (eds), *The Road to Post-communism* (London: Pinter), pp.180–201.

Ivanov, L. I. and Oguz, T. 1998, *Ecosystem Modelling as a Management Tool for the Black Sea* (Dordrecht: Kluwer Academic Publishers).

Jancar, B. 1987, *Environmental Management in the Soviet Union and Yugoslavia* (Durham, NC: Duke University Press).

Jancar-Webster, B. 1995, 'Environment and politics in the FSU', in W. Rudig (ed.), *Green Politics*, Vol. 3 (Edinburgh: Edinburgh University Press), pp.251–9.

Juodkazis, V. 1994, 'Ground water quality and its monitoring in the Baltic States'. *GeoJournal* 33, 63–70.

Kavaliauskas, P. 1994, 'Land management in Lithuania: past and future'. *GeoJournal* 33, 97–106.

Kelly, D. R. 1976a, *The Economic Superpowers and the Environment* (San Francisco, CA: W. H. Freeman).

Kelly, D. R. 1976b, 'Environmental policy making in the USSR: the role of industrial and environmental interest groups'. *Soviet Studies* 27, 570–89.

Kes, A. S., Krenke, A. N., Minayera, Y. N. and Tsigelnaya, I. D. 1982, 'The present state and future prospects of using local water resources in Central Asia and South Kazakhstan'. *Soviet Geography* 23, 414–25.

Komarov, B. 1980, *The Destruction of Nature in the Soviet Union* (London: Pluto Press).

Kozhova, O. M. and Silow, E. A. 1998, 'The current problems of the Lake Baikal ecosystem conservation'. *Lakes and Reservoirs Research and Management* 3(1), 19–33.

Kozyrev, Y. 1993, 'Marooned in a sea of salt'. *Geographical Magazine* 65(12), 34–9.

Kritkautsky, R. 1995, 'Lithuania', in P. R. Pryde (ed.), *Environmental Resources and Constraints in the Former Soviet Republics* (Boulder, CO: Westview Press), pp.125–39.

Lang-Pickvance, K. (ed.) 1997, *Environmental and Housing Movements: Grassroots Experience in Hungary, Russia and Estonia* (Aldershot: Avebury Press).

Lipovsky, I. 1995, 'The deterioration of the ecological situation in Central Asia: causes and possible consequences'. *Europe-Asia Studies* 47, 1109–23.

Lvovich, M. I. and Tsigelnaya, I. D. 1981, 'The potential for long term regulation of run-off in the mountains of the Aral Sea drainage basin'. *Soviet Geography* 22, 471–83.

Maier, P., Kowalzig, J. and Rostock, S. n.d., *Billions for Sustainability? EU Regional Policy and Accession* (Brussels: Bund/Friends of the Earth).

Malkhazova, S. M. 1997, 'Public health and environmental pollution: a methodology for the assessment of critical ecological situations'. *International Geographical Union Bulletin* 47(1), 5–16.

Marples, D. R. 1986, *Chernobyl and Nuclear Power in the Soviet Union* (London: Macmillan).

Marples, D. R. 1988, *The Social Impact of the Chernobyl Disaster* (London: Macmillan).

Marples, D. R. 1992, 'Chernobyl and nuclear energy in post-Soviet Ukraine'. *Radio Free Europe/Radio Liberty Research* 1(35), 54–8.

Marples, D. R. 1993a, 'The post-Soviet nuclear power program'. *Post-Soviet Geography* 34, 172–84.

Marples, D. R. 1993b, 'Ukraine, Belarus and the energy dilemma'. *Radio Free Europe/Radio Liberty Research* 2(27), 39–44.

Marples, D. R. 1993c, 'The legacy of the Chernobyl disaster in Belarus'. *Radio Free Europe/Radio Liberty Research* 2(5), 46–50.

Marples, D. R. 1994, 'Nuclear power in the CIS: a reappraisal'. *Radio Free Europe/Radio Liberty Research* 3(3), 21–6.

Martin, K. 1994, 'Central Asia's forgotten tragedy'. *Radio Free Europe/Radio Liberty Research* 3(30), 35–48.

Martinot, E. 1992, 'Wind-generated electric power in the former Soviet republics: geographical prospects'. *Post-Soviet Geography* 32, 219–36.

Martinot, E. 1998, 'Energy efficiency and renewable energy in Russia: transaction barriers, market intermediation and capacity building'. *Energy Policy* 26, 905–15.

Massey Stewart, J. 1990a, 'Baikal's hidden depths'. *New Scientist* 126(1772), 42–6.

Massey Stewart, J. 1990b, 'The great lake is in great peril'. *New Scientist* 126(1773), 58–62.

Massey Stewart, J. 1991, *Lake Baikal: On the Brink?* (Cambridge: IUCN East European Programme, Environmental Research Series).

Massey Stewart, J. (ed.) 1992, *The Soviet Environment: Problems, Policies and Politics* (Cambridge: Cambridge University Press).

Medvedev, Z. A. 1990, *The Legacy of Chernobyl* (Oxford: Blackwell).

Micklin, P. P. 1978, 'Irrigation development during the 10th Five Year Plan (1976–1980)'. *Soviet Geography* 19, 1–24.

Micklin, P. P. 1983, 'Water diversion proposals for the European USSR: status and trends'. *Soviet Geography* 24, 479–502.

Micklin, P. P. 1986, 'The status of the Soviet Union's north–south water transfer projects before their abandonment in 1985–6'. *Soviet Geography* 27, 287–329.

Micklin, P. P. (ed.) 1992, 'Aral Sea crisis'. *Post-Soviet Geography* 33, 269–331.

Mikhailov, K. E. and Shibnev, Y. B. 1998, 'The threatened and near-threatened birds of northern Ussuriland, southeast Russia, and the role of the Bikin River basin in their conservation'. *Bird Conservation International* 8(2), 141–71.

Mitrovitskaya, N. 1998, 'The environmental movement in the former Soviet Union', in A. Tickle and I. Welsh (eds) *Environment and Society in Eastern Europe* (London: Longman), pp.30–66.

Monroe, S. D. 1992, 'Chelyabinsk: the evolution of disaster'. *Post-Soviet Geography* 33, 533–45.

Muiznieks, N. R. 1987, 'The Daugavpils hydro station and "glastnost" in Latvia'. *Journal of Baltic Studies* 18(1), 67.

Nekrasov, A. S. 1993, 'Russia's energy system: development alternatives'. *Studies on Russian Economic Development* 4, 477–513.

O'Hara, S. L. 1997, 'Irrigation and land degradation: implications for agriculture in Turkmenistan, Central Asia'. *Journal of Arid Environments* 37(1), 165–79.

O'Hara, S. L. and Hannan, T. 1999, 'Irrigation and water management in Turkmenistan: past systems, present problems and future scenarios'. *Europe-Asia Studies* 51, 21–42.

O'Lear, S. 1997, 'Electronic communication and environmental policy in Russia and Estonia'. *Geographical Review* 87, 275–90.

Oldfield, J. 1999, 'The environmental impact of transition: a case study of Moscow city'. *Geographical Journal* 165, 222–31.

Oldfield, J. D. 2000, 'Structural economic change and the natural environment in the Russian Federation'. *Post-Communist Economies* 12, 77–90.

Peterson, D. J. 1993, *Troubled Lands: The Legacy of Soviet Environmental Destruction* (Boulder, CO: Westview Press).

Peterson, D. J. 1995a, 'Russia's environmental and natural resources in the light of economic regionalisation'. *Post-Soviet Geography* 36(5), 291–309.

Peterson, D. J. 1995b, 'Building bureaucratic capacity in Russia: federal and regional responses to the post-Soviet environmental challenge', in J. DeBardeleben and J. Hannigan (eds), *Environmental Security and Quality After Communism* (Boulder, CO: Westview Press), pp.107–26.

Pickvance, K. 1997, 'Social movements in Hungary and Russia: the case of environmental movements'. *European Sociological Review* 13(1), 35–54.

Ponomarenko, S. 1994, 'The great green wall of Russia'. *Geographical Magazine* 66(2), 17–20.

Pryde, P. R. 1972, *Conservation in the Soviet Union* (Cambridge: Cambridge University Press).

Pryde, P. R. 1984, 'Biosphere reserves in the Soviet Union'. *Soviet Geography* 25, 398–408.

Pryde, P. R. 1988, 'The future environmental agenda of the USSR'. *Soviet Geography* 29, 555–67.

Pryde, P. R. 1991, *Environmental Management in the Soviet Union* (Cambridge: Cambridge University Press).

Pryde, P. R. 1994, 'Observations on the mapping of critical environmental zones in the former Soviet Union'. *Post-Soviet Geography* 35, 38–49.

Pryde, P. R. (ed.) 1995, *Environmental Resources and Constraints in the Former Soviet Republics* (Cambridge: Cambridge University Press).

Pryde, P. R. and Bradley, D. J. 1994, 'The geography of radioactive contamination in the former USSR', *Post-Soviet Geography* 35, 557–93.

Rafikov, A. A. 1983, 'Environmental change in the South Aral region in connection with the drop in the Aral Sea level'. *Soviet Geography* 24, 344–53.

Rich, V. 1991, 'An ill wind from Chernobyl'. *New Scientist* 130(1765), 26–9.

Rodionov, S. N. 1994, *Global and Regional Climate Interaction: the Caspian Sea Experience* (Dordrecht: Kluwer Academic Publishers).

Rostankowski, P. 1982, 'Transformation of nature in the Soviet Union: proposals, plans, reality'. *Soviet Geography* 23, 381–90.

Rozengut, M. 1991, *The Soviet Water Crisis: Exposing an Environmental Disaster* (Boulder, CO: Westview Press).

Schipper, L. and Martinot, E. 1994, 'Energy efficiency in former Soviet republics: opportunities for the West'. *International Journal of Global Energy Issues* 6(3–5), 216–27.

Scott, M. J., Dagle, J. E., Gaustad, K. L., Placet, M., Roop, J. M., Schienbein, L. A. and Ulibarri, C. A. 1995, 'Dark after Chernobyl?: closing former Soviet power reactors'. *Energy Policy* 23, 703–17.

Sedaitis, J. B. and Butterfield, J. (eds) 1991, *Perestroika from Below: Social Movements in the Soviet Union* (Boulder, CO: Westview Press).

Serebryannyy, L. R. and Zamotayev, I. V. 1997, 'Problems of forested regions in Arkhangel'sk Oblast'. *Polar Geography* 21(3), 203–12.

Shaw, D. J. B. 1999, *Russia in the Modern World: A New Geography* (Oxford: Blackwell).

Shaw, D. J. B. and Oldfield, J. 1998, 'The natural environment of the CIS in the transition from communism'. *Post-Soviet Geographies and Economies* 39, 164–72.

Singleton, F. (ed.) 1987, *Environmental Problems in the Soviet Union and Eastern Europe* (London: Lynne Rienner).

Smith, G. 1999, *The Post-Soviet States: Mapping the Politics of Transition* (London: Arnold).

Smith, M. 1990, 'Polluted Russia: a tale of death and destruction'. *Geographical Magazine* 62(3), 10–14.

Tickle, A. and Welsh, I. (eds) 1998, *Environment and Society in Eastern Europe* (London: Longman).

Tikkanen, E. (ed.) 1995, *Kola Peninsula Pollutants and Forest Ecosystems in Lapland* (Helsinki: Finnish Forest Research Institute).

Turnbull, M. 1991, *Soviet Environmental Policies and Practices: The Most Critical Investment* (Aldershot: Dartmouth Publishing).

Valiunas, J. 1994, 'Geoscience for environmental planning in Lithuania'. *GeoJournal* 33, 91–6.

Vilemas, J. 1995, 'Nuclear energy in Lithuania: present status and hopes for the future'. *Energy Policy* 23, 719–22.

Vilnitis, V. 1997, 'Latvia', in J. Klarer and B. Moldan (eds), *The Environmental Challenge for Central European Economies* (Chichester: Wiley), pp.169–92.

Vorobyev, V. V. and Martynov, A. V. 1989, 'Protected areas of the Lake Baikal basin'. *Soviet Geography* 30, 359–70.

Waller, M. 1989, 'The ecology issue in Eastern Europe: protest and movenments'. *Journal of Communist Studies* 5, 303–28.

Weiner, D. 1988, *Models of Nature: Conservation Ecology and Cultural Revolution in Soviet Russia* (Bloomington: Indiana University Press).

Weiner, D. 1988, 'The changing face of Soviet conservation', in D. E. Worster (ed.), *The Ends of the Earth* (Cambridge: Cambridge University Press), pp.252–73.

Welsh, I. 1996, 'The USSR: ecological legacy'. *Environmental Politics* 5, 147–50.

Wolfson, Z. 1989, 'Natural resources development: an economic and environmental problem in the USSR'. *Communist Economies* 1, 79–88.

Wolfson, Z. 1992, 'The environment: from stagnation to collapse', in M. Ellman and V. Kontorovich (eds), *The Disintegration of the Soviet Economic System* (London: Routledge), pp.241–50.

Wolfson, Z. 1994, *The Geography of Survival: Ecology in the Post-Soviet Era* (London: M. E. Sharpe).

Yanitsky, O. 1991a, 'Environmental movements: some conceptual issues in east–west comparisons'. *International Journal of Urban and Rural Research* 15, 524–41.

Yanitsky, O. 1991b, 'Fight for Lefortovo, Moscow: resolving the conflict between urban planners and residents', in T. Deelstra and O. Yanirsky (eds), *Cities of Europe: The Public's Role in Revitalising the Urban Environment* (Moscow: Mezdunarodnye Otnoshenya Publishing House), pp.356–71.

Yanitsky, O. 1993a, 'Environmental initiatives in Russia: east–west comparisons', in A. Vari and P. Tamas (eds), *Environment and Democratic Transition: Policy and Politics in Central and Eastern Europe* (Dordrecht: Kluwer), pp.120–45.

Yanitsky, O. 1993b, *Russian Environmentalism: Leading Figures, Facts and Opinions* (Moscow: Mezhdunarodniye Otnoshenija Publishing House).

Yanitsky, O. 1996, 'The ecological movement in post-totalitarian Russia: some conceptual issues'. *Society and Natural Resources* 9(1), 65–76.

Young, M. J. and Launer, M. K. 1991, 'Redefining "glasnost" in the Soviet media: the contextualisation of Chernobyl'. *Journal of Communication* 41, 102–24.

Zeimetz, K. 1990, 'Chernobyl and the Soviet food economy', in K.-E. Wadekin (ed.), *Communist Agriculture: Farming in the Soviet Union and Eastern Europe* (London: Routledge), pp.72–89.

Ziegler, C. E. 1987, *Environmental Policy in the USSR* (London: Pinter).

Ziegler, C. E. 1991, 'Environmental politics and policy under "perestroika"', in J. Sedaitis and J. Butterfield (eds), *Social Movements in the Soviet Union* (Boulder, CO: Westview Press), pp.113–32.

Zisserman, D. 1998, 'The politicization of the environmental issue within the Russian nationalistic movement'. *Nationalities Papers* 26, 677–86.

Part II

Country studies

Northern group

6 Czech Republic

Petr Pavlínek

Introduction

Today's Czech Republic, the western part of former Czechoslovakia, is considered to be one of the most environmentally devastated countries in the whole of Europe (Figure 6.1). In particular, environmental degradation is concentrated in the industrial and urban regions such as northern Bohemia, northern Moravia and Prague (Figure 6.2). The reasons for the rapidly deteriorating quality of the environment were similar to those in other state socialist countries as they were deeply embedded within the logic of the state socialist development model. Environmental quality rapidly deteriorated during post-Second World War socialist industrialization, and the development of an extensive regime of accumulation focused on the rapid expansion of heavy industries largely fuelled by low-quality brown coal and lignite. Although the state socialist regime had developed a comprehensive environmental legislation to deal with the pollution problems, it had not been efficient in enforcing its own strict pollution limits. The state socialist planners had always considered production to be primary and feared that too much environmental consideration would endanger the plan fulfilment (see e.g. Carter 1985, 1993; Pavlínek 1992, 1995, 1997; Vaněk 1996; Tickle and Vavroušek 1998).

The goal of this chapter is to provide a brief assessment of the environmental effects of the post communist transformation in the Czech Republic that took place after 1989. The collapse of the state socialist regime in Czechoslovakia was understood by many as a historic opportunity to end the environmental degradation associated with its state socialist model and to launch a more environmentally sustainable future. However, disagreements occurred almost immediately over how to achieve this goal. For neo-liberals, the mere introduction of a market economy would lead to rapid environmental clean-up. Others believed that a more sophisticated approach toward environmental reconstruction was needed. Two contrasting views of the post-1989 period in terms of its environmental effects have been presented. First, many politicians and governmental officials consider the transformation to be a complete success in terms of its environmental outcomes and they believe it led to a remarkable environmental recovery (e.g. Holman 1995, p.5). They point toward rapidly declining pollution levels and improving environmental indicators as the evidence of this 'success'. The second view,

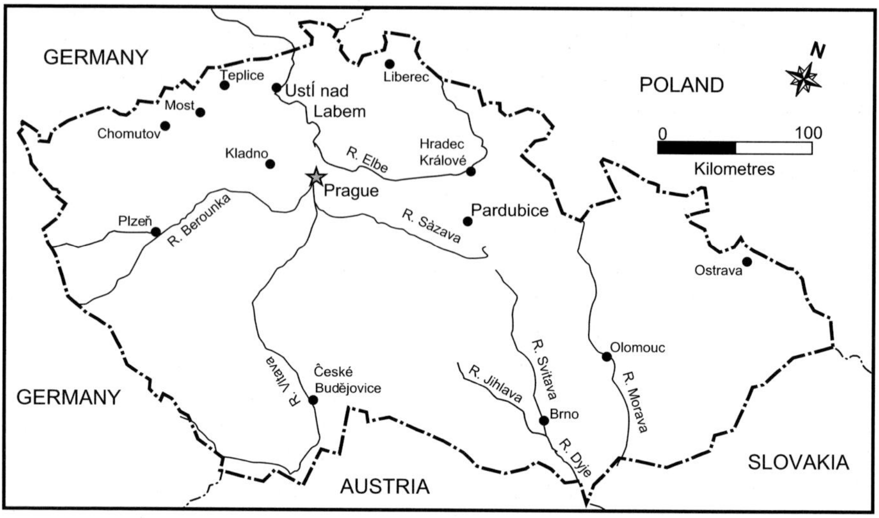

Figure 6.1 Salient features of the Czech Republic

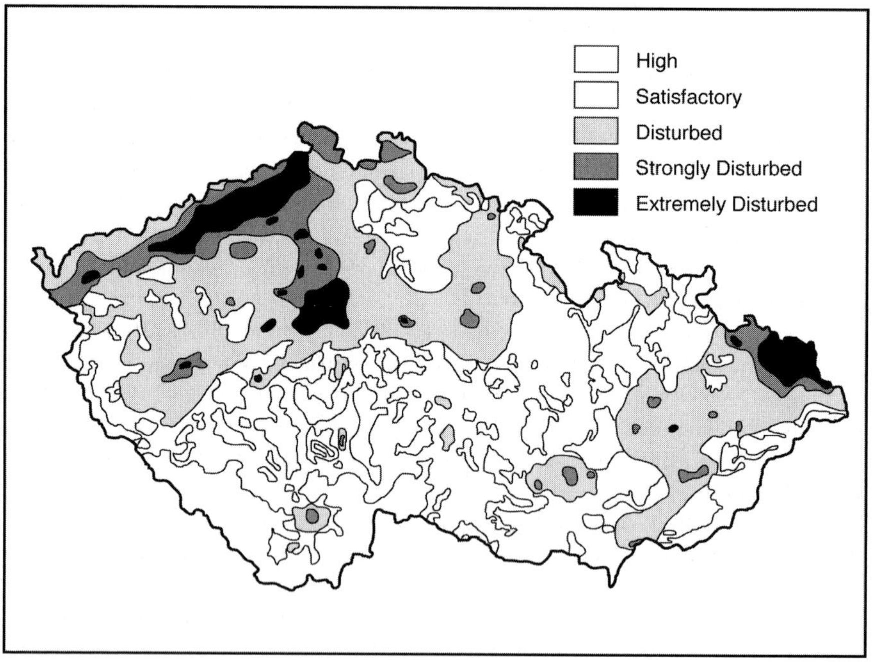

Figure 6.2 Environmental quality in the Czech Republic in the early 1990s

Source: Adapted from Geografický ústav ČSAV and Federální výbor pro životní prostředí (1992, 11-1).

expressed by many environmentalists, judges these 'successes' as overly optimistic but instead it evokes the notions of 'failed transition' introduced by Roger Manser (1993) and missed opportunity to set the Czech Republic and the rest of East Central Europe (ECE) on a pathway toward a sustainable future. Environmentalists consider improved environmental statistics to be by-products of economic collapse associated with neo-liberal economic policies, shock therapy launched in the early 1990s, and the result of the introduction of unsustainable 'end of pipe' technologies from the West to control excessive pollution.

In this chapter, I want to provide a more realistic assessment of the post-1989 environmental outcomes in the Czech Republic. The chapter shows that the levels of pollution declined considerably. At the same time, however, it stresses the contradictory nature of environmental reconstruction: strong state socialist legacies persisted in the 1990s, approaches toward the environment and environmental management were slow to change in both political and economic spheres, Western environmental practices were often adopted uncritically without considering any alternatives, and the development of civil society was slow. The chapter begins with the evaluation of the post-1989 environmental record. It then discusses the political struggles over the environment, development of environmental legislation and policy, and the role of the West in this process.

Post-1989 environmental record

The Czech Republic has experienced dramatic declines in air pollution since 1989. Between 1989 and 1997, solid emissions declined by 81 per cent, SO_2 emissions dropped by 65 per cent, NO_x emissions decreased by 54 per cent, and the emissions of hydrocarbons (C_xH_y) declined by 21 per cent. CO emissions remained unchanged after 1989, although they declined by 17 per cent between 1990 and 1997. As in the case of other CEE countries, it is unclear to what extent industrial decline after 1989 contributed to the drop in emissions and to what extent new governmental environmental policies and legislation were responsible for the decline. SO_2 and particulate emissions had begun to decline in the mid-1980s before the collapse of state socialism. These declines resulted from efforts by the government to decrease emissions of air pollutants and to meet its international commitment to cut SO_2 emissions by 30 per cent from 1980 levels by 1993.[1] In this sense, the political changes of 1989 did not represent a major turning point in pollution levels for SO_2 and solid emissions, but instead were a continuation of an earlier trend that was in turn strongly reinforced by sharp industrial decline and increased efforts to combat air pollution after 1989. In fact, industrial production declined faster than emissions between 1989 and 1993. After 1993, however, emissions kept falling, even after industrial production began to increase again. Thus, the situation in the Czech Republic suggests that emission declines are not temporary as some environmentalists have suggested and that industrial collapse is not the primary cause of emission declines. In fact, particulate and SO_2 emissions had been the target of various governmental policies since the 1980s.

The factors responsible for emission declines after 1989 included the economic decline, the installation of anti-pollution equipment in power plants and other major polluters, industrial restructuring away from energy-intensive heavy industries toward a more service-oriented economy, and the introduction of anti-pollution measures by the Czech government. The overall economic decline and the collapse of industrial production by 36 per cent between 1991 and 1993 led to reduced demand for electricity and hence declines in energy resource extraction and use. In practice, however, electricity production decreased by less than 6 per cent during this period (MoE and ČSÚ 1996, p.60). The gradual installation and use of more effective dust scrubbers at the large coal-based power plants as part of the 1985 to 1990 Clean Air Protection Programme of the Czech Energy Works Company (České energetické závody (ČEZ)) contributed to the dramatic decline in solid emissions in the early 1990s (Černá *et al.* 1995, p.387). Some industrial producers also reduced their particulate emissions substantially. The cement industry, for example, decreased solid emissions by 97 per cent after 1989 through modernization of production, largely financed by foreign capital (HN 1996, p.6). Similarly, the chemical industry lowered its air pollution considerably between 1993 and 1999. The planned 1999 solid emissions were 93 per cent lower than in 1993, SO_2 emissions were 70 per cent lower, NO_x emissions were 43 per cent lower and organic substances were 22 per cent below their 1993 levels (Maxa *et al.* 1998, p.14).

Declines in SO_2 emissions can be explained by production declines, mild winters in the early 1990s that contributed to lower demand for heat and electricity, a gradual switch of entire communities from low-grade coal and lignite to natural gas for heating, and the gradual desulphurization of all major power plants (a process completed in November 1998) (Pavlík 1996, p.62).[2] Between 1990 and 1997, heat and energy consumption patterns changed substantially, also contributing to declines in particulate and SO_2 emissions: the share of solid fuels decreased while the share of liquid and gaseous fuels, electricity and heat increased (Figure 6.3). Subsidies on electricity for businesses and industry have been removed, and prices of electricity, heat and natural gas for the public are gradually being liberalized. The effect has been price increases and corresponding reduction in household energy use. At the same time, the share of environmental investment as a percentage of GDP increased from 0.7 per cent in 1989 to 2.7 per cent in 1996 and then declined to 2.3 per cent in 1997 (Beneš and Héniková 1993, p.296; MoE 1997, p.75, 1999a, p.79) (Table 6.1).[3]

By the end of 1998, all coal-based power plants operated by the Czech Energy Works Company, accounting for about 77 per cent of electricity production in the Czech Republic, had desulphurization and denitrification equipment installed (twenty-eight units with a combined production capacity of 5930MW) and, as a result, complied with the Clean Air Act emission limits that came into effect on 1 January 1999 (Čápová 1999, p.4; Baroch 1998, p.3; ČEZ 1998). The company had closed several coal-based power plants, desulphurized 70.2 per cent of its total electricity production capacity, constructed seven fluid furnaces in four power plants, increased the efficiency of scrubbers, and decreased emissions

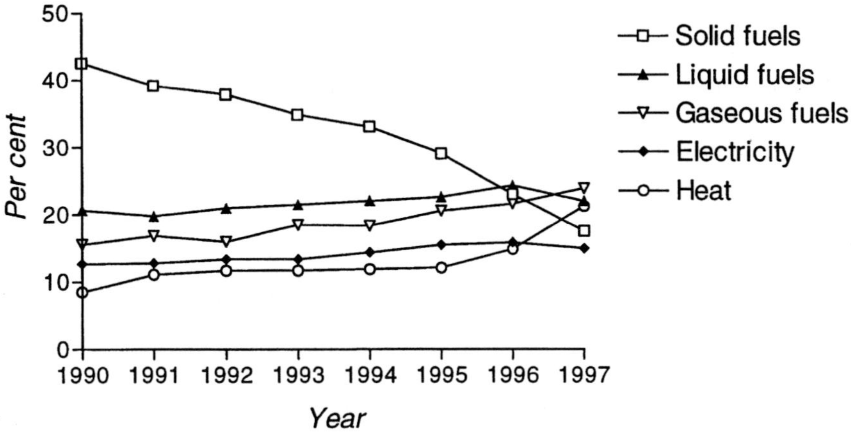

Figure 6.3 End consumption of fuels and energy in the Czech Republic, 1990–97
Source: MoE 1999, p.49.

Table 6.1 Share of environmental investment on GDP in the Czech Republic: 1989–97

Year	*Share of environmental investment on GDP*
1989	0.7
1990	1.1
1991	1.3
1992	2.1
1993	2.2
1994	2.7
1995	2.6
1996	2.7
1997	2.3

Sources: Beneš and Heníková (1993, p.296); Ministry of the Environment and ČSÚ (1996, p.267); MoE (1997, p.75, 1999, p.79).

of nitrogen oxides and carbon dioxide. As a result, emissions from power plants decreased dramatically by 1999: SO_2 emissions by 91 per cent, solid emissions by 87 per cent, and NO_x emissions by more than 50 per cent (ČEZ 1996, p.4, 1998; Baroch 1998, p.3; and see Figure 6.4).

It is unclear to what extent the citizens of the Czech Republic will benefit from this dramatic decline in emissions from large power plants. Even officials from the MoE admit that the dramatic declines in SO_2 and solid emissions will be of more benefit to neighbouring countries than to the citizens of the Czech Republic. Bohuslav Brix from the MoE argued in 1995 that better scrubbers and desulphurization of all power plants would have 'minimal impact' on the air quality in Prague and other large cities in the Czech Republic because ground-level pollution

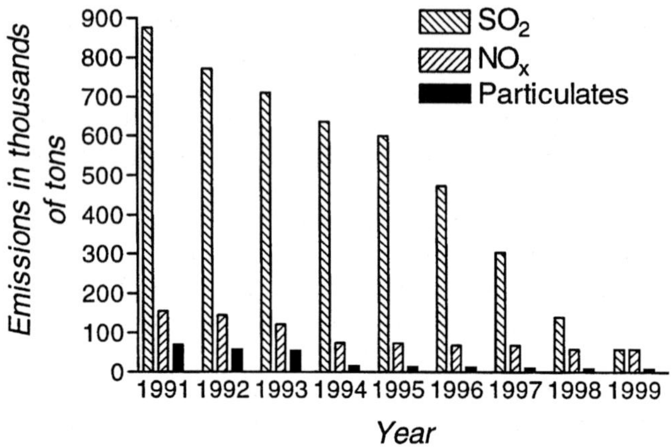

Figure 6.4 Annual emissions from the Czech energy works' power plants, 1991–99
Source: EZ 1998.

is caused mainly by automobiles and domestic heating using low-quality coal (HN 1995, p.3).[4] In fact, while SO_2 concentrations have been declining as a result of the introduction of desulphurization technology into power and heating plants and switching household heating from coal to natural gas and electricity, the concentrations of nitrogen oxides have rapidly increased because of the growing number of cars and increased car usage (Bílý 1997, p.4).

Although the 1991 Clean Air Act contributed to a rapid decline in emissions from large polluters such as power plants and heating plants, it did not deal with pollution from small stationary sources (those with thermal output smaller than 0.2MW).[5] It also did not consider mobile sources of air pollution, even though these have become a major environmental problem in urban areas (Seják 1994, p.38). Moreover, most privatized companies seem not to have changed their environmental policies since 1989, and few have any idea how they should respond to pressures to protect the environment (LN 1995, p.7).

Aggregate levels of water pollution have also decreased substantially since 1989. As with air pollution, levels of water pollution began to drop in the mid-1980s under state socialism, but this trend has been strongly reinforced during the post-1989 period. Overall withdrawals and use of water have declined considerably since 1989. Between 1990 and 1997, agriculture cut its water use by 83 per cent, the industrial sector by 36 per cent, the energy sector by 19 per cent, and drinking-water withdrawals dropped by 28 per cent (MoE 1999a, p.15). The amount of waste water released by these sectors correspondingly declined. The sharp declines in water withdrawal by agriculture and industry resulted from production slumps in these sectors following the introduction of shock therapy measures in the early 1990s. Household water consumption declined after governmental subsidies on the prices of drinking-water were removed and prices for water rose rapidly.

Water pollution from point sources also declined considerably between 1990 and 1997: biological oxygen demand (BOD5)[6] dropped by 76 per cent, pollution by undissolved substances decreased by 61 per cent, pollution by petroleum products (absorbable hydrocarbons) by 86 per cent, pollution by dissolved inorganic salts by 36 per cent, and acidity and alkalinity of surface waters declined by 87 per cent. Only 9 per cent of waste water was discharged without any treatment in 1997, compared with 27 per cent in 1989. The number of long-term monitored water quality sampling sites with heavily and very heavily polluted water (the worst two water quality categories) declined from 77 to 32 for BOD5, from 191 to 112 for chemical oxygen demand, from 133 to 69 for ammonia nitrogen content, and from 168 to 78 profiles for phosphates content (MoE 1997, pp.22–4, 1999a, pp.15–16). These improvements in surface water quality were achieved because of decreases in the overall amount of discharged waste water, construction and operation of new effluent treatment plants, industrial restructuring, and decreases in the use of industrial fertilizers and pesticides by agriculture.[7] The number of effluent treatment plants increased from 626 in 1990 to 905 in 1996 (MoE 1996b, pp.17–18, 1997, p.28). New environmental legislation requires all cities with more than 15,000 inhabitants to be equipped with basic mechanical-biological water treatment plants by the year 2000 and all settlement units with more than 2000 inhabitants by the year 2005 (MoE 1999a, p.20). The quality of ground water has also slowly improved since 1989 in most measured characteristics (see MoE and ČSÚ 1996, pp.127–9; MoE 1997, p.27, 1999a, pp.18–19 for data). Overall, the post-1989 transformation has resulted in significant environmental improvements in the Czech Republic. It is important to realize, however, that this change is uneven, highly contested and far from complete. Struggles over the direction of environmental change and over the role of the environment during the transformation have continued in national, regional and local contexts. It is to these issues that we now turn.

Struggles over the environment after 1989

Following the 1989 revolution, the environment became an important priority for the new Czechoslovak government. During this period, both the enthusiasm of the government for rapid environmental clean-up and public awareness of environmental problems were high. The Ministry of Environment (MoE) of the Czech Republic was established in 1990.[8] At the federal level, the Federal Committee for the Environment (FCE) was created after the parliamentary elections in the summer of 1990. The goal of the FCE was to rebuild and further develop the government's role in environmental management based on recent parliamentary legislation and under conditions of an emerging market economy. The FCE also saw its role as increasing environmental awareness, providing accurate information about environmental quality, sustaining and extending the new monitoring system, supporting environmental NGOs, and fostering cooperation with Western and other East Central European countries in the field of environmental management and protection. The post-1989 federal government supported these activities. Throughout its existence, however, the work of the FCE was hindered by internal divisions between Czech and Slovak commission members, and these divisions became more pronounced as tensions over the breakup of Czechoslovakia increased.[9]

Post-1989 environmental euphoria lasted only until mid-1991 (Jehlička and Kára 1993, pp.13–14). After that, economic concerns regained their priority as the country began to feel the effects of shock therapy introduced in January 1991. In the Czechoslovak government, the proponents of fast environmental clean-up were losing ground in their struggles with supporters of a neo-liberal view of the environment. The environmentalists, led by the FCE chairman Josef Vavroušek, emphasized the need to tackle environmental problems first and even at the expense of short-term economic growth. The neo-liberals, led then by the Federal Finance Minister and later by the Prime Minister of the Czech Republic, Václav Klaus (1992–7), maintained that only successful economic transformation would create favourable conditions for solving the environmental crisis. By mid-1992 the neo-liberal view of the environment had prevailed in the Czechoslovak government amid rapidly declining economic output associated with shock therapy. Environmentalists lost their positions in the government after the 1992 parliamentary elections when the FCE was abolished (Pavlínek 1997, p.102).

Václav Klaus was openly hostile toward environmental NGOs and considered the environment to be only the 'whipped cream on the [economic] cake' (Bisschop 1996, p.43). In 1995, Klaus argued that 'ecology is not a science. It has nothing in common with science. It is ideology' (MF Dnes 1996, p.4). In June 1997, he claimed in front of dozens of journalists and scientists from the Czech Academy of Sciences that the greenhouse effect was nonsense and a 'quackish theory'. He also argued that 'scientists attempt to make a fool out of the rest of society with their experiments in order to get more money' (Ekolist 1997, p.8). In successive governments, Klaus' environmental ministers were always in a weak position compared to ministers of economy and finance, and the MoE saw even these limited capacities reduced and resources shifted to other ministries such as land use planning (to the Ministry of Economy), new mining law (to the Ministry of Industry and Trade) and water management (to the Ministry of Agriculture) (Kužvart 1994, p.29; Sequens 1996, pp.5–7). In 1999, parliamentary deputies from Klaus' Civic Democratic Party initiated the process aimed at taking away additional decision-making powers from the Environment Minister for his refusal to approve a highway construction through the České středohoří nature reserve in central Bohemia (Baroch and Bartoníček 1999, p.5). Klaus and his governments were hostile toward more than one hundred independent environmental NGOs that emerged after 1989. František Benda, the Environment Minister in Klaus' government, refused any dialogue with NGOs, and several environmental NGOs (Greenpeace, Duha, Děti Země, Animal SOS) appeared on the list of extremist organizations to be monitored by the Czech Security Information Service in January 1995 (Baroch 1997a, p.4).

The Czech government has decided to complete construction of a nuclear power plant at Temelín in southern Bohemia, at a site planned and started before 1989.[10] The Temelín power plant represents an important state socialist legacy and was originally built using Soviet technology. This is currently being modified by installing Western safety features. Although the original design and technology of the plant have been substantially modified since 1989, and only two 981MW units are being built instead of the four originally planned, opposition to construction has become one of the rallying points for the environmental movement in the Czech Republic. Completion of the Temelín power plant is also strongly opposed by neighbouring Austria and since 1989 there has also been significant opposition against the power plant from within the Czech government, mainly from various ministers of the environment. The disputes within the Czech government over the Temelín power plant completion intensified in 1999 after an independent international committee of experts concluded that the Czech Republic would not need electricity from a large nuclear power plant in the next ten to fifteen years (Lipold 1999; Švehla 1999). However, in May 1999 the government upheld its decision to complete the power plant.

At the local level, some results of successful earlier environmental struggles have been gradually eroded during the second half of the 1990s. For example, in 1991 the Czech government imposed 'ecological mining limits' to protect northern Bohemian communities from further demolition because of open-cast coal-mining.[11] 'Ecological mining limits' are the territorial boundaries drawn around the communities endangered by coal-mining beyond which mining is prohibited. Government action followed popular resistance from citizens and local governments in several towns and villages slated for demolition to make way for open-cast coal-mining and their struggle with the mines (see Pavlínek 1997, pp.224–9).[12] However, in 1997 and 1998 coal-mining companies exerted strong pressure on the government to re-evaluate its 1991 decision and to allow them to mine beyond the ecological mining limits. If accepted, this would result in the demolition of five additional villages in the region (Baroch 1997b, pp.1–4; MF Dnes 1998, p.5).

Czech Greens: the politics of failure

One indicator of environmental euphoria during the 1989 revolution was the establishment of the Green Party in former Czechoslovakia. The Green Party was founded in Prague on 21 November 1989, just four days after the first student demonstration that began the 'revolution'. The speed with which the Green Party was established is surprising given the chaotic nature of late November 1989 in Prague. Another surprising fact is that the Green Party was established by people who were not previously active environmentalists. These facts led to accusations that the Green Party was orchestrated by the still communist secret police in the Czech Republic, particularly in Prague (the situation in Slovakia was different) (Jehlička and Kostelecký 1992, p.74, 1991, pp.6–9, 1995, p.228; Chorváthová 1996, p.76). Another green party called the Green Alternative

(*Zelená alternativa*) was established in Prague in December 1989. As in the case of the Green Party, the founding members of the Green Alternative did not include any recognized environmentalists, but claimed to be completely independent of political forces (Jehlička and Kostelecký 1991, p.7).

Contrary to the situation in Hungary and Poland, environmental awareness among the Czech population did not decline dramatically after 1989, and in many ways ran counter to the government's perception of the importance of the issue. At the national level, between 1992 and 1994, the environment consistently ranked high among the 'very or rather urgent problems' facing the country, together with crime, living standards, economic reform and health care (Brchaňová and Hrušková 1994, p.3). Support for environmental issues certainly did decline, but remained relatively high: in 1990, 83 per cent of citizens claimed that it was 'very important' for them 'to live in a healthy environment', by 1991 this number was 76 per cent and by 1992, 67 per cent (REC 1995, p.43). Based on opinion polls conducted in the Czech Republic in 1994, only 17 per cent of the 800 respondents prioritized economic prosperity over the environment, 40 per cent prioritized the environment over economic prosperity, and 35 per cent thought that economic prosperity and the environment were equally important for them (LN 1994, p.9). In environmentally devastated regions, such as the Most district of northern Bohemia, local citizens considered the poor quality of the environment to be by far the most important single problem they had to face in the near future (Pavlínek 1997, pp.334–6).

However, this environmental awareness did not translate into votes for the Green Party during the national elections. In the June 1990 elections, the Green Party won only 4.1 per cent of the vote for the Czech Parliament and 3.3 per cent for the Federal Parliament in the Czech Republic (Jehlička and Kostelecký 1991, p.15). As a result, the party failed to enter the Czech and Federal Parliament. Support for the party was surprisingly low even in the most environmentally devastated districts of northern Bohemia. The strongest level of support attained was only 9 per cent in the district of Chomutov (Jehlička and Kostelecký 1991, p.17). As part of a three-party coalition, the Green Party won three parliamentary seats in the Czech Parliament and three in the Federal Parliament during the 1992 parliamentary elections. At the same time, however, the Green Party lost two-thirds of its members who objected to the dubious coalition for various reasons (see Jehlička and Kosteleck 1995, pp.225–9). Plagued by internal disputes, membership losses, marginality and overall disintegration, the Czech Greens failed to become a parliamentary party during both the 1996 and 1998 parliamentary elections and any electoral support they may have had vanished.[13]

Environment legislation and policy after 1989

Czechoslovakia had quite comprehensive but largely ineffective environmental legislation during the state socialist period. Major environmental laws were enacted in the late 1960s and 1970s and by 1972, over 350 environmental regulations had been enacted into law. Among other measures, fines and fees

Table 6.2 Selected environmental legislation in the Czech Republic

Enacted before 1989	*New or revised 1990–98*
Nature Protection Law (No. 40/1956)	Clean Air Act (No. 309/1991 amended by No. 211/1994)
Public Health Act (No. 20/1966)	State Administration Obligations in Air Protection and Provisions for Air Pollution Charges Act (No. 389/1991 amended by No. 212/1994)
Air Purity Law (No. 35/1967)	Forest Protection Act (No. 229/1991 amended by No. 289/1995)
Water Act (No. 138/1973)	Agricultural Land Protection Act (No. 334/1992)
Agricultural Land Protection Act (No. 103/1976)	Waste Act (No. 125/1997, previously No. 238/1991 amended by No. 300/1995)
Land Use Planning and Building Order (No. 50/1976)	State Administration in Waste Management Act (No. 311/1991 amended by No. 300/1995)
Forestry Act (No. 61/1977)	Fees for Deposition of Waste Act (No. 62/1992 amended by No. 41/1995)
Protection and Use of Natural Resources (Mining Law) (Nos 44, 61, 62/1988)	Environmental Fund Act (No. 388/1991)
	Environmental Protection Act (No. 17/1992)
	Protection of Nature and the Landscape Act (No. 114/1992 amended by No. 289/1995)
	Environmental Impact Assessment Act (No. 244/1992)
	Environmental Inspection Act (No. 282/1992)
	Ozone Layer Act (No. 86/1995)
	Endangered Species Act (No. 16/1997)
	Environmental Information Act (No. 123/1998)

Sources: GAO (1994, p.28); MoE (1995, pp.4–5, 1996b, p.49); Hrbáček *et al.* (1989, pp.144–6); Neumann (1998, p.19).

for air and water pollution, compensation for conversion of agricultural land to other uses, and protective measures for land and forested areas were introduced (Carter 1985, pp.36–7; Hrbáček *et al.* 1989, pp.144–6; Andrews 1993, p.13) (Table 6.2). Pollution charges, fines and user fees were so low, however, that they provided no incentive for the polluters to reduce pollution discharges (World Bank 1992-I, p.17).

After 1989, the newly established Czech MoE had to set up a completely new system of environmental protection and management including basic environmental administration, information systems, environmental funds and more detailed agendas for each major environmental issue and programme. The Czech MoE published its Rainbow Programme in 1991: the Environmental recovery programme for the Czech Republic (Moldan 1991), which set the basic principles for environmental protection and management during the period of political and economic transformation. The foundations of the new system of environmental law and regulation were laid in Czechoslovakia in 1990 and 1991. Huba

(1996, p.116) has called this period the 'legislative revolution'. Between 1990 and December 1992, when Czechoslovakia broke up, several principal environmental laws were enacted and a Charter of Fundamental Rights and Freedom was incorporated into the federal and national constitutions. The Czech Republic was in the process of introducing its own more detailed provisions and regulations based on the Federal Laws. Fifty-one environmental laws and regulations were passed between 1990 and 1992 (MoE 1993, pp.52–5). The general goal of these new environmental laws was to bring Czechoslovakia's (and later the Czech Republic's) legal system as close as possible to the legal system of the European Union (Andrews 1993, p.26). After the split, all federal environmental laws and regulations were adopted by the Czech Republic, and the Czech Parliament enacted an additional eight laws and amendments and eighteen environmental regulations between 1992 and 1995 (Baltus 1993, p.12; MoE 1996a, P-12). Overall between 1990 and 1995, the Czech Republic adopted fourteen new environmental laws, numerous amendments and dozens of other legal instruments to protect the environment (MoE 1995, p.4).

The efficiency of new environmental legislation and the feasibility of many deadlines and limits set by new environmental legislation have been questioned. Although the new environmental legislation was drafted with the specific aim of bringing Czech standards closer to those of the EU, many of the laws enacted during this period, such as the Waste Management Act, did not comply with the EU standards or were not adequate in certain areas and subsequently had to be amended. Following criticism from the EU, the then Environment Minister Jiří Skalický revealed in 1997 that the Czech Republic would need 200 new or amended environmental laws and regulations in order to comply with EU environmental laws. This would account for two-thirds of the 300 existing EU's environmental norms (MF Dnes 1997, p.2; Čech 1997, p.7). There has also been much criticism of the very low levels at which fees and penalties have been set for air pollutants, which are thought to be too small for industries to actually reduce emissions (MoE 1995, p.4). The allocation of resources by the Czech Environmental Fund has also been criticized because decisions were made by only one person (Špaček 1993, p.13). Finally, new environmental legislation has been prepared and enacted so quickly that questions have been raised about its quality. For example, the Czech MoE has been criticized for blindly copying environmental legislation from Western Europe without regard to the specifics of the Czech situation (Špaček 1993, p.13, 1993 interviews). The number of laws passed during 1991 and 1992 that have since had to be amended gives credence to this claim. New environmental legislation did not give any significant new powers to municipalities and district offices to regulate large sources of pollution; instead jurisdiction was retained by the Environmental Inspectorates, creating tensions between the central state, district offices and large cities (see Pavlínek 1997, pp.307–18).

Czech environmentalists and the Ministry of the Environment were under permanent attack from liberal economists led by the then Prime Minister Václav

Klaus. Klaus consistently challenged the importance of environmental protection and management during the transformation and undermined efforts to deal with the legacy of environmental devastation. He saw environmental protection as a secondary problem that should be dealt with only after economic transformation of the country had been 'successfully' completed and then only through economic levers. He also repeatedly refused to accept the concept of a state environmental policy because it included the principle of sustainable development. For Klaus, sustainable development was a Western invention that was not relevant for the Czech Republic.[14,15] He argued, for example, that

> ecology is first and foremost an economic problem. Environmental policy is based on a country's resources, and the economy is the limiting factor whether we like it or not. Environmental policy, as understood and implemented by this government, is pragmatic and not fundamentalist. It is not separated from the overall context of societal transformation. It is not based on some concepts separated from reality, such as the concept of 'sustainable development', but it is focused on the protection and improvement of the environment based on our means. We can neither exceed these means nor change the overall direction of the transition process somewhere to [the sphere of] planning practices.
>
> (HN 1993, p.9)

A state environmental policy was eventually adopted in 1996. This policy was not to last long, however. A new state environmental policy began to be prepared after a social democratic government was formed following the 1998 parliamentary elections. The working version of the document released in 1999 argued that the new government was reconsidering the state environmental policy because it felt the necessity to include the idea of sustainable development and its related environmental principles. Another major reason for a new state environmental policy was the need to introduce and integrate environmental principles into sectoral economic policies and the fact that the existing policy was not respected by governmental ministries with the exception of the MoE (MoE 1999b, p.2; Baroch 1999, p.4). During the 1990s, development strategies prepared by governmental ministries for their respective sectors largely failed, as they did under state socialism, to consider environmental effects of their proposals. For example, the government approved a national strategy of transport development without having its environmental impact assessed (Baroch 1999, p.4). However, political parties continue to disagree about the need to pursue more effective environmental policies. Different perceptions of the environment by political elites are being translated into often contradictory environmental policies pursued by sometimes short-lived and unstable governmental coalitions with potential long-term effects. Such a situation makes the development of a long-term, consistent state environmental policy almost impossible and in its effect hinders the environmental reconstruction.

The Western role in environmental reconstruction

The two most important areas of the Western role and influence in overcoming the state socialist environmental legacy include the various forms of Western assistance and the Czech Republic's effort to approximate the EU's environmental legislation. All prospective new members of the EU have been required to meet the EU's environmental requirements early on in the process of application. These requirements include some 300 pieces of EU environmental legislation (known as environmental *acquis*) that must be included into national legislation of CEE countries. As mentioned above, despite the 'legislative revolution' largely driven by the attempts to approximate environmental legislation with the EU's standards after 1989, the Czech Republic still needed to prepare or amend about 200 environmental regulations to comply with the EU's standards in 1997.

The most important Western assistance has been associated with the PHARE programmes financed by the European Commission. After the breakup of Czechoslovakia in 1992, PHARE spent about ECU 20 million to support three types of projects: environmental studies, supplies of monitoring and analytical equipment, and credits to environmentally friendly investments such as switching from coal to natural gas and electricity for heating in the most polluted areas (MoE 1996a, p.68). In the second half of the 1990s, PHARE support focused on projects designed to improve the environmental information systems, environmental education in the area of environmental impact assessment, and the introduction of an Eco-Management and Audit Scheme into the country. PHARE-GTAF (General Technical Assistance Facility) supports the process of approximation of Czech environmental legislation with the EU's legislation. PHARE-CBC (Cross-Border Cooperation) finances environmental and infrastructural investment projects in the Czech–German and Czech–Austrian border regions. The Black Triangle Programme launched in 1991 as a part of PHARE II is an example of transnational environmental activities funded by PHARE. The programme was aimed at environmental reconstruction of heavily polluted border regions of northern Bohemia in the Czech Republic, Saxony in Germany and the adjoining area of Poland. PHARE financed the completion of a joint air-quality monitoring network in the Black Triangle and it supports small-scale investment projects that benefit the environment in the region. The US Agency for International Development financed twelve environmental projects in the Czech Republic between 1992 and 1997. The projects focused on the investigation of health impacts of the environment in the most polluted regions such as regions of northern Bohemia and northern Moravia (Teplice and Slezsko Projects) and on the development of economic instruments of environmental protection in the Czech Republic (MoE 1996b, 1997, 1999).

Conclusion

The post-1989 period has been associated with important improvements in the quality of the environment in the Czech Republic. Air quality improved as solid, SO_2 and other emissions plummeted, and water quality also improved as water consumption and corresponding discharge of effluents declined considerably. It is questionable whether these improvements are permanent, however. The improvements in environmental quality resulted from a combination of factors such as the sharp economic decline after 1989, effects of anti-pollution measures launched in the 1980s, new environmental policies launched in the 1990s, gradual economic restructuring away from heavy industries, and desulphurization of all major power plants. These improvements were not achieved automatically by a mere introduction of the market economy but were won in many struggles that began in the early 1980s. Despite the improvements, the levels of air pollution are still well above Western European standards.

At the same time, however, new, potentially dangerous environmental developments associated with the introduction of capitalism in the country emerged. Examples include rapidly growing and unregulated car traffic in urban areas, a dramatic increase in plastic packaging, and a danger of Western 'dirty investments'. Moreover, the state socialist legacies remain surprisingly strong. They are represented by the continuation of perceptions and approaches toward the environment inherited from the previous period. One consequence of this situation is that no consensus exists among the political parties and members of the same parties about the need to develop an efficient system of environmental management and integrate economic and environmental policies. As a result, the 1990s were typified by inconsistency in approaches toward the environment: the early 1990s was a period of environmental enthusiasm and rapid progress; the period between 1992 and 1998 was dominated by technocentric approaches toward the environment and reassertion of economics over the environment; the period after the 1998 elections was associated with attempts to strengthen the environmental regulation despite the strong opposition from within the government itself and from the opposition Civic Democratic Party. Environmental reconstruction and the legislative changes favouring the environment seem to be driven mostly by the EU's requirements for new potential members and not by anything which points toward the important geopolitical dimension of the post-1989 environmental reform. The influence of independent ENGOs on state environmental policies continues to be weak as the government ignores these groups. One reason for this weakness is a lack of public support for ENGOs. The public seems to be over-occupied with economic difficulties associated with the transition to a market economy. No strong green party emerged in the 1990s. Thus the environmental results of the first post-communist decade are mixed. It remains to be seen whether the 1990s was a decade during which a long-awaited environmental change toward sustainable development really began.

Notes

1 Protocol to the 1979 Convention on Long-range Transboundary Air Pollution on the reduction of sulphur emissions or their transboundary fluxes by at least 30 per cent. Not all the pre-1989 governmental efforts have been successful, however. For example, Soviet technology was bought to desulphurize the power plant Tušimice II

in northern Bohemia in the Czech Republic in the 1980s. The construction and equipment cost about 2 billion crowns (today about US$70 million) but the entire unit has never been initiated because of enormous technological problems. All the invested money appears to have been wasted (interview with René Pisinger, adviser to the Minister of the Environment, Teplice, 4 August 1993).

2 The 1991 Clean Air Act was designed to desulphurize not only large power plants, but also an additional 1600 pollution sources with more than 5MW of thermal output and 20,000 sources with 0.2–5MW of thermal output (Baltus 1993, p.12).

3 In former Czechoslovakia public investment in environmental protection averaged about 0.5 per cent of GDP between 1971 and 1975, 0.4 per cent between 1975 and 1985, 0.3 per cent between 1980 and 1985, and 0.6 per cent between 1986 and 1990 (World Bank 1992, p.18).

4 It has been estimated that household furnaces account for about 45 per cent of total solid and CO emissions and more than 50 per cent of emissions of hydrocarbons, heavy metals and other toxic substances in the Czech Republic (*EKOjournal* 1994, p.3).

5 Research in Poland has shown that environmental damage from small stationary pollution sources with low smokestacks such as household furnaces is about twelve times larger with regard to particulate matter emissions and 2.5 times larger with regard to SO_2 and NO_x emissions compared to larger pollution sources with high smokestacks (Seják 1994, p.38).

6 Biological oxygen demand is the amount of oxygen consumed for aerobic biochemical decomposition of organics contained in water for five days under standard conditions.

7 Gross agricultural production per hectare declined by 27.5 per cent between 1989 and 1994. The 1996 production was at 76.1 per cent of the 1989 level (Ládr 1997, p.10). Consumption of drinking-water declined by 35 per cent in the Czech Republic between the early 1990s and 1997, from daily per capita consumption of 180 litres to 116 litres (MF Dnes 1997b, p.1).

8 The Czech Ministry of the Interior and the Environment had been in existence since 1988.

9 Interview with Jan Kára, Prague, 18 August 1993.

10 See Carter (1988, pp.269–87) on the development of the Czechoslovak nuclear power programme before 1989.

11 (1) Resolution No. 287 passed on 2 November 1990 – the set of measures to restore the environment in northern Bohemia; (2) Resolution No. 166 passed on 5 May 1991 – the review report about the implementation of tasks from Resolution No. 287 and the proposal of its actualization; (3) Resolution No. 331 passed on 11 September 1991 about the further development of the Chabařovice open-cast mine; and (4) Resolution No. 44 passed on 10 October 1991 about the report dealing with territorial limits of coal-mining and energetics in the north Bohemian brown coal basin (Pěgřímek 1992, p.152).

12 For example, the local government of the town of Chabařovice (located in the Teplice district) was elected in the 1990 free local elections to save the town (Bystrov 1993, pp.6–7). In the case of the village of Horní Jiřetín (in the Most district), the civic resistance against the plans of coal-mining enterprises to raze the village culminated in the organization of a meeting in April 1991 which was attended by the Minister of the Environment and his deputy, the first deputy of the Minister for Economic Policy, an official from the Ministry of Culture, and several Members of Parliament. At this meeting, the inhabitants of the village and the local government voiced their concerns over plans to raze their village and determination to fight such plans (interview with Miroslav Štýbr, the mayor of Horní Jiřetín, 3 August 1993).

13 The importance of the environment for the public has continuously declined. In December 1997, the environment ranked ninth among the most important problems perceived by the Czech public. It ranked fourth in December 1994 (MF Dnes 1998, p.4).

14 It is important to note that sustainable development is not considered a radical or green concept because it accepts the primacy of economic growth and human welfare over the environment (Doyle and McEachern 1998, p.35).

15 See Slocock (1996, pp.508–14) for a detailed analysis of Václav Klaus' attitude toward the environmental protection and policy. See also Pavlínek (1997, pp.357–8).

References

Andrews, R. N. L. (1993) 'Environmental policy in the Czech and Slovak Republic', in A.Vari and P. Tamas (eds) *Environment and Democratic Transition: Policy and Politics in Central and Eastern Europe*, Boston: Kluwer Academic, pp.5–48.

Baltus, J. (1993) 'Právní úprava ochrany životního prostředí v České republice po zániku federace' (Legal adjustment of environmental protection in the Czech Republic after the end of federation), *Ekomagazín*, autumn:12–13.

Baroch, P. (1997a) 'Zákon usnadní přístup k ekologickým informacím' (Law will make access to environmental information easier), *Mladá Fronta Dnes* 8:5, 14 August.

—— (1997b) 'Ekologická hnutí nemají zatím velký vliv' (Ecological movements have so far had little influence), *Mladá Fronta Dnes* 8 (71):4, 25 March.

—— (1998) 'I po odsíření hrozba smogu zůstává' (The threat of smog remains even after desulphurization), *Mladá Fronta Dnes* 9 (226):3, 13 November.

—— (1999) 'Kužvart usiluje o zřízení ekologických odborů' (Kužvart wants to establish environmental departments), *Mladá Fronta Dnes* 10 (91):4, 19 April.

Baroch, P. and Bartoníček, R. (1999) 'Skupina poslanců chce ministru Kužvartovi vzít část pravomocí' (Group of parliament members wants to take away part of minister Kužvart's authorities), *Mladá Fronta Dnes* 10 (85):5, 12 April.

BCE (1998) *Key Data 1990–97*, Business Central Europe, May, www.bcemag.com

Beneš, J. and Héniková, S. (eds) (1993) *Životní prostředí České republiky: Ročenka 1992*, Praha: MŽP ČR and ČEÚ.

Bílý, J. (1997) 'Znečištění našich měst se v poslední době mění' (Pollution of our cities has been changing recently), *Lidové noviny* 10:10, 23 October.

Bisschop, G. (1996) 'Optimism wanes for a prompt cleanup', *Transition* 2 (10):42–5.

Brchaňová, M. and Hrušková, M. (1994) *Výsledky výzkumu veřejného mínění* (Results of opinion polls), Internal material of the Czech Ministry of Environment, No. 2280/OSV/'94.

Bystrov, V. (1993) 'Chabařovice: město s nadějí' (Chabayovice: the town with hope), *Lidové noviny* 6:229.

Čápová, H. (1999) 'Zhluboka dýchat' (Take a deep breath), *Respekt* 3:4, 11–17 January.

Carter, F. W. (1985) 'Pollution problems in post-war Czechoslovakia', *Transactions of the Institute of British Geographers (New Series)* 10 (1):17–44.

—— (1988) 'Czechoslovakia: nuclear power in a socialist society', *Environment and Planning C: Government and Policy* 6:269–87.

—— (1993) 'Czechoslovakia', in F. W. Carter and D. Turnock (eds) *Environmental Problems in Eastern Europe*, London and New York: Routledge, pp.63–88.

Čech, P. (1997) 'Varování z Evropy' (Warning from Europe), *Nika* 18 (3–4):7.

Černá, A., Tošovská, E. and Cetkovský, P. (1995) 'Economic transformation and the environment', in J. Svejnar (ed.) *The Czech Republic and Economic Transition in Eastern Europe*, San Diego: Academic Press, pp.377–94.

ČEZ (1996) *Cesta k rovnováze* (Road to balance), Prague: Kuklik.

—— (1998) 'Tečka za odsiřovacím programem ČEZ, a.s.' (ČEZ finished its desulphurization programme). ČEZ Press Release, 13 November.

Chorváthová, A. (1996) 'Strany zeleného typu', in M. Huba (ed.) *Ponovembrové Slovensko IV*, Bratislava: EuroUniPress, Spoločnosť pre trvalo Udržateľný život, pp.76–7.

Doyle, T. and McEachern, D. (1998) *Environment and Politics*, London and New York: Routledge.

EKOjournal (1994) Národní program ochrany ovzduší: Malé zdroje znečištěni ovzduší – vysoká rizika (National air pollution sources – high risks), *EKOjournal* 3 (6):4–5.

Ekolist (1997) 'Václav Klaus na skleníkový efekt nevěří' (Václav Klaus does not believe in the greenhouse effect), *Ekolist* 2 (7):8.

GAO (1994) *Environmental Issues in Central and Eastern Europe: U.S. Efforts to Help Resolve Institutional and Financial Help*. Report to the Chairman, Committee on Energy and Commerce, House of Representatives, Washington, DC: United States General Accounting Office.

HN (1993) 'Nevládní organizace versus vládní politika' (Non-governmental organizations versus governmental policy), *Hospodářské noviny* 37 (241):9, 10 December.

—— (1995) 'Snížení emisí z elektráren a chemiček obyvatelům velkých měst moc nepomůže' (Emission declines from power plants and chemical factories will not benefit citizens of large cities very much), *Hospodářské noviny* 39 (2):3, 3 January.

—— (1996) 'Cementárny snížily emise prachu za pět let o celých 97 procent' (Cement producers decreased solid emissions by 97 per cent in five years), *Hospodářské noviny* 40 (61):6, 26 March.

Holman, R. (1995) 'Mají pravdu ekologičtí radikálové?' (Are ecological radicals right?), *Lidové noviny*, 8 September, p.5.

Hrbáček, J., Binek, B. and Mejstřík, V. (1989) 'Czechoslovakia: historical development of pollution and its control', in E. J. Kormondy (ed.) *International Handbook of Pollution Control*, New York: Greenwood Press, pp.137–51.

Huba, M. (1996) 'Životné prostredie' (Environment), in M. Bútora and P. Hunčík (eds) *Slovensko 1995: Súhrnná správa o stave spoločnosti*, Bratislava: Nadácia Sándora Máraiho, pp.115–28.

Jehlička, P. and Kára, J. (1993) 'Ups and downs of Czech environmental [awareness and] policy: Where is it going?', paper presented at the ECPR Joint Sessions of Workshops, University of Leiden, April.

Jehlička, P. and Kostelecký, T. (1991) 'The Greens in Czechoslovakia', paper presented at the Green challenge workshop during the ECPR conference in Colchester, April.

—— (1992) 'The development of the Czechoslovak Green Party since the 1990 elections', *Environmental Politics* 1 (1):72–94.

—— (1995) 'Czechoslovakia: Greens in a post-Communist society', in D. Richardson and C. Rootes (eds) *The Green Challenge: The Development of Green Parties in Europe*, London and New York: Routledge.

Kužvart, P. (1994) 'Legislativa v oblasti životního prostředí' (Environmental legislation), *Nika* 15 (3):29.

Ládr (1997) 'Zemědělství je pod statním paternalismem' (Agriculture is under state paternalism), *Hospodářské noviny* 41 (54):10, 18 March.

Lipold, Jan (1999) 'Komise expertů zpochybnila Temelín' (The expert committee cast doubt upon Temelín), *Mladá Fronta Dnes* 10 (57):1–2.

LN (1994) 'Ekonomická prosperita a životní prostředí v ČR' (Economic prosperity and the environment in the Czech Republic), *Lidové noviny* 7 (212):9, 9 September.

—— (1995) 'Ekologie zůstává firemní popelkou' (Ecology continues to be neglected by firms), *Lidové noviny* 8:7 13 September.

Manser, R. (1993) *Failed Transitions: The Eastern European Economy and Environment Since the Fall of Communism*, New York: New Press.

Maxa, M., Dlouhý, J. and Řeháček, V. (1998) *Udržitelný rozvoj chemického průmyslu – část 2: Ekologická charackteristika* (Sustainable development of the Czech chemical industry, Part 2: Ecological characterization), *Chemický průmysl* 4:14.

MF Dnes (1996) 'Ropákem roku se stal úředník, nejpitomější výrok řekl Klaus (A bureaucrat declared ropák of the year, Klaus said the silliest statement), *Mladá Fronta Dnes* 7 (13):4, 13 April.

—— (1997a) 'Skalický: ekologii čeká dvě stě nových zákonů' (Ecology: two hundred new pieces of legislation expected), *Mladá Fronta Dnes* 8 (94):2, 22 April.

—— (1997b) 'Obyratele velkých měst začíná čím dál více dusit doprava (Transportation increasingly chokes inhabitants of large cities), *Mladá Fronta Dnes* 8:4, 2 August.

—— (1998) 'Obce ohrožené těžbou zpochybňují pravomoc vlády (Villages endangered by mining cast doubt on government's jurisdiction), *Mladá Fronta Dnes* 9 (130):5, 5 June.

Ministry of Environment (1993) *Zpráva o stavu životního prostředí České Republiky* (Report on the state of the environment in the Czech Republic), Prague: Ministry of Environment of the Czech Republic.

—— (1995) *State Environmental Policy*, Document Approved by the Government of the Czech Republic on 23 August 1995, Prague: Ministry of Environment of the Czech Republic.

—— (1996a) *Zpráva o stavu životního prostředí České Republiky v období 1992–1995* (Report on the state of the environment in the Czech Republic in the period 1992–1995), Prague: Ministry of Environment of the Czech Republic.

—— (1996b) *Report on the Environment in the Czech Republic in 1995*, Prague: Ministry of Environment of the Czech Republic.

—— (1997) *Zpráva o stavu životního prostředí České Republiky v roce 1996* (Report on the state of the environment in the Czech Republic in 1996), Prague: Ministry of Environment of the Czech Republic.

—— (1998) *Statistická Ročenka životního prostředí České Republiky 1998* (Statistical environmental yearbook of the Czech Republic 1998), Prague: Ministry of Environment of the Czech Republic.

—— (1999a) *Zpráva o stavu životního prostředí České Republiky v roce 1997* (Report on the state of the environment in the Czech Republic in 1997), Prague: Ministry of Environment of the Czech Republic.

—— (1999b) *Návrh státní politiky životního prostředí* (Proposal of the state environmental policy), Prague: Ministry of Environment of the Czech Republic.

Ministry of Environment and ČSÚ (1996) *Statistická Ročenka životního prostředí České Republiky 1996* (Statistical environmental yearbook of the Czech Republic 1996), Prague: Ministry of Environment of the Czech Republic and the Czech Statistical Office.

—— (1998) *Statistická Ročenka životního prostředí České Republiky 1998* (Statistical environmental yearbook of the Czech Republic 1998), Prague: Ministry of Environment of the Czech Republic and the Czech Statistical Office.

Moldan, B. (ed.) (1991) *Duhový program: Program ozdravení životního prostředí České republiky* (Rainbow programme: Environmental recovery programme for the Czech Republic), Prague: Ministry of Environment of the Czech Republic and Academia.

Neumann, J. (1998) 'Legislativa v oblasti životního prostředí v roce 1997' (Environmental legislation in 1997), *Zpravodaj Ministrerstva Životního Prostředí* 8 (1):19.

Pavlík, Z. (ed.) (1996) *Human Development Report: Czech Republic 1996*, Prague: Faculty of Science, Charles University.

Pavlínek, P. (1992) 'Regional transformation in Czechoslovakia: towards a market economy', *Tijdschrift voor Economische en Sociale Geografie* 83 (5):361–71.

—— (1995) 'Regional development and the disintegration of Czechoslovakia', *Geoforum* 26 (4):351–72.

—— (1997) *Economic Restructuring and Local Environmental Management in the Czech Republic*, Lewiston, NY: Edwin Mellen Press.

Pěgřímek, R. (1992) 'Perspektiva a podmínky dalšího rozvoje těžební činnosti v SHR ve světle probíhající společenské a ekonomické transformace (Perspective and conditions of further development of mining activities in the North Bohemian Brown Coal Basin in the light of unfolding social and economic transformation), *Uhlí Rudy* 1 (5):152–5.

REC (1995) *Status of National Environmental Action Programs in Central and Eastern Europe: Case Studies of Albania, Bulgaria, the Czech Republic, Croatia, Hungary, Latvia, Lithuania, FYR Macedonia, Poland, Romania, the Slovak Republic and Slovenia.* Budapest: Regional Environmental Center for Central and Eastern Europe.

Seják, J. (1994) 'Jsou ekologická opatření v České republice efektivní?' (Are ecological measures taken in the Czech Republic effective?), *Planeta* 2 (3):38–9.

Sequens, T. (1996) 'Kompetence v pohybu' (Competences in motion), *EkoJournal* 5 (12):5–7.

Slocock, B. (1996) 'The paradoxes of environmental policy in Eastern Europe: the dynamics of policy-making in the Czech Republic', *Environmental Politics* 5 (3):501–21.

Špaček, J. (1993) 'Situace volá po reformě' (Situation calls for reform), *Hospodářské noviny*: 13, 7 October.

Švehla, M. (1999) 'Nepotřebná energie' (Energy that is not needed), *Respekt* 10:1–3, 7 March.

Tickle, A. and Vavroušek, J. (1998) 'Environmental politics in the Former Czechoslovakia', in A. Tickle and I. Welsh (eds) *Environment and Society in Eastern Europe*, Harlow: Longman, pp.114–45.

Vaněk, M. (1996) *Nedalo se tady dýchat: Ekologie v českých zemích v letech 1960 až 1989* (It was impossible to breathe here: Ecology in the Czech lands between 1960 and 1989), Prague: Maxdorf.

World Bank (1992) *Czech and Slovak Federal Republic: Joint Environmental Study*, A joint report of the Governments of Czechoslovakia, the Czech and Slovak Republics, the European Community, the United States Government, and the World Bank. November 1991.

7 East Germany

Olivia J. Wilson and Geoff A. Wilson

Introduction

The former German Democratic Republic (GDR), now referred to as the 'new Länder' of reunified Germany, occupies an anomalous position within Europe (Figure 7.1). After the Second World War, the division of Germany split the country between the 'communist east' and 'capitalist west'. The GDR was a key player in the development of the socialist economies of Central and Eastern Europe (CEE) for forty years, while West Germany was a founder member of the European Economic Community (now the European Union or EU). Since German reunification in 1990, however, the former GDR has become a member of the EU. Thus, while the environmental problems facing the former GDR have many similarities with those of other CEE states, the policies to solve them are now formulated both by the federal government of Germany and the EU. The former GDR must therefore be seen as a 'special case' in the context of this edited collection.

In order to analyse environmental problems and policies in Germany's new Länder, this chapter is structured in three parts. First, it is important to outline the environmental legacy of the GDR as a basis for a comparison between the environmental situation in the GDR and former West Germany on the eve of reunification. Second, the policy framework within which environmental problems in the new Länder are being tackled will be examined. Third, progress in solving environmental problems in the former GDR will be assessed.

The GDR legacy

Environmental standards in the GDR are widely recognised to have lagged well behind those in West Germany in terms of air and water pollution, soil contamination, and countryside and nature conservation (*Der Spiegel*, 1990; Ardagh, 1991; Jones, 1994). However, as Huber (1993) noted, this was not always the case, and is an over-simplification of the situation. Until the first oil crisis of 1973, for example, the two German states were similar in both having a relatively poor environmental record. If anything, the GDR was more advanced in some aspects of its environmental policy. For example, the GDR

Figure 7.1 Salient features of East Germany

was the first nation in the world to incorporate a commitment to environmental protection into its constitution (in 1968), and by establishing an Environment Ministry in 1971 it was one of the first countries to provide an institutional framework for tackling environmental problems.

In contrast, West Germany did not establish a federal Environment Agency until 1974. However, West Germany soon caught up with and overtook the GDR, largely due to the rapid development of the Green Movement and the politicisation of environmental issues through the West German Green Party (*Die Grünen*). Yet, it took until 1986 – largely as a response to the nuclear catastrophe at Chernobyl in the Ukraine – for the West German government to establish a federal Ministry for the Environment, Nature Protection and Nuclear Safety (Bundesumweltministerium (BMU)) with formal environmental legislative powers (Bulmer and Paterson, 1987). In the GDR meanwhile, environmental regulations promised during the 1970s were never implemented because of increasing political and economic pressures within its socialist decision-making structures (Huber, 1993). By the early 1980s, the failure of the GDR state to control environmental degradation became obvious when the state refused to continue to publish any data on the environmental situation in the country (Fulbrook, 1995). After 1982, environmental information became a state secret, thereby preventing GDR citizens and Western observers from monitoring the environmental mismanagement that was occurring at an ever-increasing pace (Smyser, 1993). The full extent of environmental pollution in the former GDR, therefore, has only come to light since reunification in 1989.

How can this rapid divergence in environmental policy priorities between the GDR and former West Germany best be explained? The 1973 oil crisis is seen by many commentators as a decisive turning point in the fortunes of the GDR (e.g. Huber, 1993). The resulting rise in oil prices prevented the GDR from diversifying its energy source away from indigenous brown coal (lignite) towards imported oil. As a result, the GDR had to continue to rely on highly polluting brown coal for power generation until reunification in 1989, while West Germany, mainly due to its increasing economic power and rapid technological development, managed to adopt a more diversified energy policy based on various different energy resources (Jones, 1994).

The GDR's continuing heavy reliance on coal became the primary contributory factor to the GDR's poor record on air pollution. While between 1980 and 1990 sulphur dioxide emissions in West Germany decreased by 67 per cent, in the GDR they increased by 25 per cent (Huber, 1993). As a result, by 1988 sulphur dioxide emissions were almost ten times higher in the GDR than in West Germany, and most of these emissions came from power-stations (Table 7.1). Air particle levels (dust, soot) were also ten times higher than those in West Germany. The most polluted areas were the industrial conurbations of the south (present-day Sachsen-Anhalt and Sachsen) around Leipzig, Halle, Bitterfeld, Chemnitz and Cottbus (Figure 7.2). This was the industrial heart of the GDR, specialising in mining, chemicals, metallurgy, textiles and machine tools (Jones, 1994). Furthermore, by the late 1980s the GDR had the highest levels of carbon dioxide emissions per capita in the world (about 21 tonnes per capita in 1986 compared to 11.7 tonnes per capita in West Germany) (BMBau, 1991).

As a result of sulphur dioxide pollution, the GDR's forests (in particular the Erzgebirge and Thüringer Wald) suffered severely from acid rain damage (*Waldsterben*). Acid rain 'exported' from the GDR also affected nearby forests in West Germany, in particular the Harz and Bayerischer Wald (BMBau, 1991). Although much blame was apportioned to the GDR at the time by the West German government, the West German Environment Ministry none the less had to acknowledge that nitrogen dioxide pollution from West German car exhaust emissions also contributed to acid rain damage. West German (and French) pollution, for example, was the primary factor behind *Waldsterben* in the Black Forest in south-west Germany (Ardagh, 1991).

Not surprisingly, the health of the GDR population living in these industrial centres suffered considerably. The combination of air particles and sulphur

Table 7.1 Air pollution levels by source, GDR and West Germany, 1988

	GDR		*West Germany*	
	Kilotonnes	*% total*	*Kilotonnes*	*% total*
Sulphur dioxide				
Power stations and industry	4704	88.3	995	77.4
Households and small users	505	9.5	147	11.4
Transport	116	2.2	144	11.2
Total	5325	100	1286	100
Tonnes per km²	*49.2*		*5.3*	
Kg per population	*319.3*		*21.3*	
Air particles				
Power stations and industry	1960	87.7	423	79.8
Households and small users	238	10.6	35	6.6
Transport	39	1.7	72	13.6
Total	2237	100	530	100
Tonnes per km²	*20.0*		*2.1*	
Kg per population	*134.2*		*8.7*	
Nitrogen dioxide				
Power stations and industry	398	72.0	890	31.1
Households and small users	11	2.2	120	4.2
Transport	144	26.0	1850	64.7
Total	553	100	2860	100
Tonnes per km²	*6.6*		*11.5*	
Kg per population	*31.3*		*46.7*	

Source: BMBau (1991, p.109).

dioxide caused winter smogs which aggravated asthma and bronchitis. In the winter of 1988 to 1989, for example, Erfurt and Leipzig both had smog alarms on twenty days (BMBau, 1991). Deaths from respiratory diseases were twice as high as the European average, while the life expectancies of men and women were 2.5 and 7 years lower respectively than in West Germany (Huber, 1993). Interestingly, some commentators have argued that the GDR had a lower rate of allergies than West Germany, despite higher air pollution levels. This phenomenon has been attributed to less pollution from car exhausts (see below) and less processed foods in people's diets (Newnham, 1997), although this may also simply indicate the fact that the system of reporting allergies in the GDR was more poorly developed compared to West German standards.

It would be wrong, however, to argue that the environmental records of both West Germany and the GDR always came out in favour of the former. For example, in contrast to its poor record on sulphur dioxide and carbon dioxide emissions, the GDR fared better than West Germany with nitrogen dioxide emissions (see Table 7.1). This can be attributed to the overall lower consumption of oil in the GDR by cars and industry. Indeed, by the late 1980s car ownership levels in the GDR were only 296 per 1000 population compared to 485 per 1000 in West Germany (Paas *et al.*, 1994). Rail transport was more

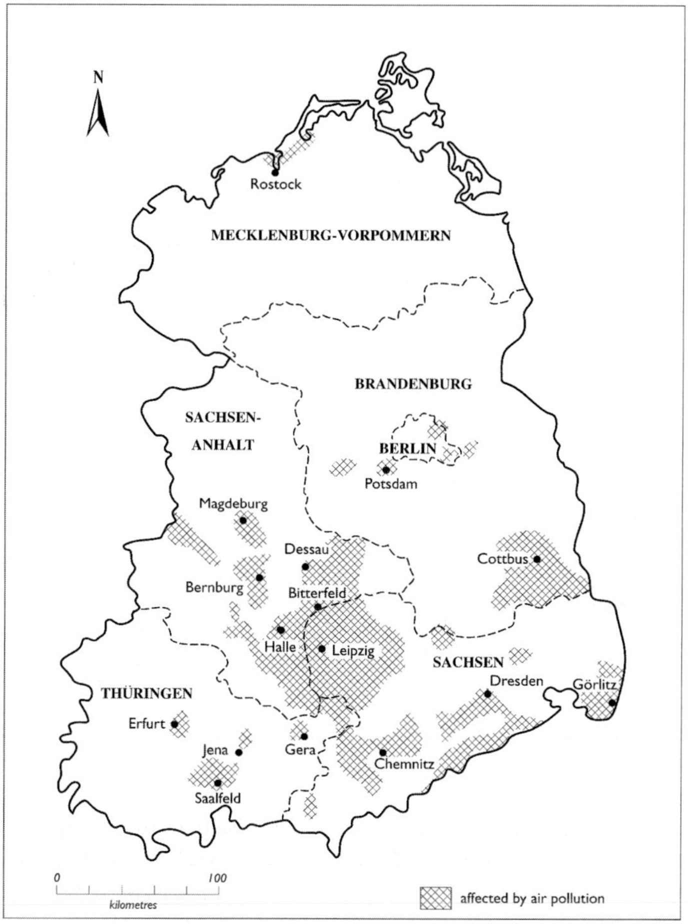

Figure 7.2 Areas of East Germany subject to middle to high levels of air pollution
Source: BMBau 1991, p.115.

important in the GDR both for freight and passenger transport than in West Germany, carrying 73 per cent of freight transport and 55 per cent of passenger transport, compared to only 45 per cent and 20 per cent respectively in West Germany (Huber, 1993). The West German emphasis on private car ownership (strongly supported by the very powerful car lobby), and its relative neglect of rail as a means for transporting goods, therefore, placed a serious strain on the environment – a factor that was much less pronounced in the GDR.

As well as air pollution, water quality in the GDR also lagged behind that in West Germany. To start with, the GDR had few surface water supplies, and many of these were heavily polluted by industrial, agricultural and domestic wastes. For example, 95 per cent of industrial waste water was piped directly into rivers with no or only minimal treatment, and only 36 per cent of households were connected to sewage treatment plants (BMU, 1995). In addition, almost 70 per cent of drainage pipes were damaged. As a result, only 5 per cent of surface water in the GDR was suitable for drinking without treatment, and only a further 55 per cent was suitable with treatment. The River Elbe, which together with its tributaries the Rivers Elster, Mulde and Saale flows through the main industrial centres of Sachsen and Sachsen-Anhalt, was as polluted in 1989 as the River Rhine in West Germany had been in the early 1970s. The River Elbe was also a major contributor to North Sea pollution, thereby directly affecting the quality of fish resources for most northern European countries (BMU, 1995). The shortage of suitable surface water supplies meant that 71 per cent of the GDR's drinking-water was extracted from ground water supplies, but much of this was also polluted, particularly by nitrates and phosphates from intensive and unregulated agricultural practices (BMBau, 1991). In 1988, 141kg/ha of nitrate fertiliser and 56kg/ha of phosphate fertiliser were being applied to arable land (Zentralverwaltung für Statistik, 1989) compared to 125kg/ha and 49kg/ha respectively in West Germany (Hagedorn, 1992). Another source of localised nitrate pollution came from the dumping of slurry waste from intensive livestock farms – a practice which was much more regulated in West Germany.

A further source of soil and water contamination in the GDR came from inadequate landfill facilities. Although household waste production per capita was lower than in West Germany (only 180kg/p.a. compared to 375kg/p.a.), facilities for disposal were much poorer. Of 11,000 landfill sites only 120 were fully regulated depots, 1000 were controlled dumping sites, and the rest were uncontrolled (BMBau, 1991, p.114). There were about 2000 industrial waste depots in the GDR of which 800 were for toxic waste (BMBau, 1991). Industrial waste disposal was a particular problem, especially in the heavy industrial areas of the south where toxic wastes were produced (e.g. toxic gases, heavy metals, oils, fats, tars, detergents). These were exacerbated by exports of toxic waste from West Germany, which at the time was happy to pass on the problem of disposal to its neighbour and exploit the fact that environmental standards in the GDR were by then much more lenient than in West Germany (Ardagh, 1991). The high military presence of the Soviet army also contributed to localised soil contamination (BMU, 1995) – a factor over which the GDR had little control due its weak power position *vis-à-vis* the former Soviet Union.

Finally, the already high pollution levels in the GDR were further exacerbated by extensive coal- and uranium-mining operations (Figure 7.3). Uranium was exported to the Soviet Union for its nuclear energy and weapons programmes (Ardagh, 1991). The uranium mines in southern Sachsen and Thüringen represented a radioactive threat to the local populations, and radioactive waste from spoil heaps was also used for aggregates in the construction industry. Brown coal was mined by open-cast methods, thereby generating large amounts of dust and contributing to localised air pollution. Altogether, 128,000 hectares were mined in two areas: the Lausitz field in eastern Sachsen and the middle German field in southern Sachsen-Anhalt and north-west Sachsen. About 260 worked-out open-cast pits were used as landfill sites, some for the dumping of toxic waste.

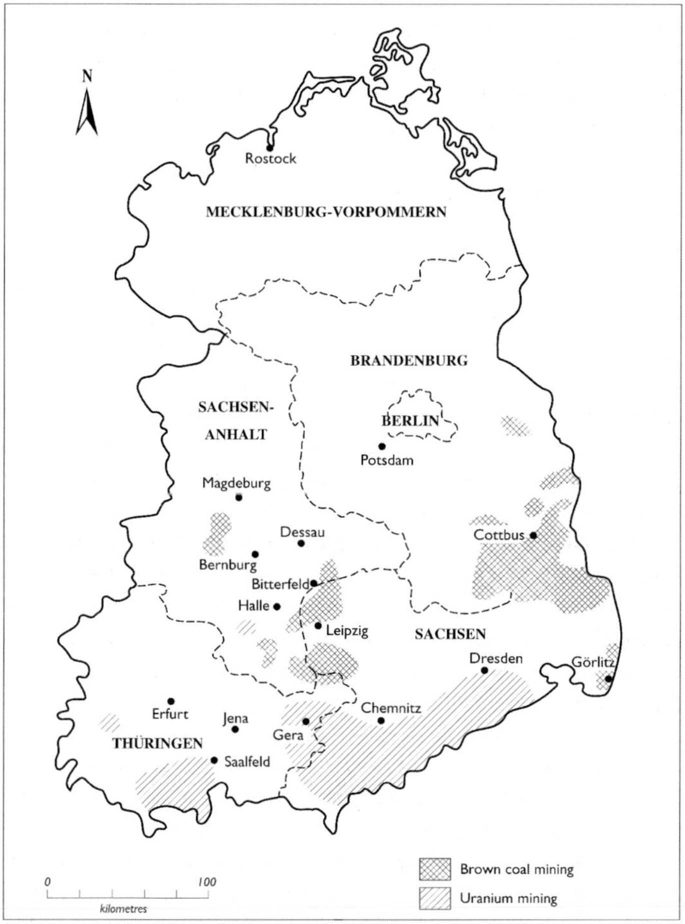

Figure 7.3 Areas of East Germany affected by brown coal and uranium mining

Source: BMBau 1991, p.115.

Inevitably, abandoned open-cast mines became eyesores in the landscape as well as posing environmental hazards from ground subsidence and air pollution. The mines needed to be pumped to prevent water-logging and in places this pumping led to a fall in the water table of 80 metres (BMU, 1994).

Yet, despite this poor environmental record, the GDR contained a high degree of biodiversity (Huber, 1993), particularly in the sparsely populated and largely rural northern part of the country. This emphasises the fact that despite high overall levels of pollution in the GDR, sources of pollution were relatively localised around the industrial centres of the south. Although the GDR state did not designate any national parks, it did have a network of nature protection areas that aimed to preserve important ecological habitats in a semi-natural state (with more or less success as long as these areas were not required for industrial or intensive agricultural production). Further, under the UNESCO 'Man and Biosphere' programme two biosphere reserves were designated, and under the RAMSAR Convention eight wetlands of international importance and thirty-eight of national importance were designated (Reichhoff and Böhnert, 1991). Paradoxically, one of the most biodiverse habitats in the GDR was created in the five kilometre-wide border strip (*Sperrgebiet*) between East and West Germany which, although not purposefully protected for conservation, provided a safe habitat for many rare plant and animal species simply because access was strictly controlled (BMBau, 1991).

Environmental policy in reunified Germany

With reunification, the environmental law and policy of West Germany (the old Länder) was transferred with few changes to the new Länder. Article 34 of the Unification Treaty stated the following objectives for the environment of the new Länder: the protection of the natural requirements of life in accordance with the principles of precaution, polluter pays and co-operation, and the achievement of environmental conditions at least as good as those in the old Länder (BMBau, 1991). Where no suitable policies existed, new policies were formulated. For instance, a policy on uranium-mining had to be drawn up, as West Germany had not had any need to establish such a policy before reunification (BMU, 1995, p.vi).

West German environmental policy is based on three basic principles:

1 *Precautionary principle*: that policies should aim to prevent possible environmental damage, even where scientific evidence for damage is as yet unproven (O'Riordan and Cameron, 1994). This principle is applied by the federal government, for instance, when it sets maximum permitted levels of air and water pollutants, or for drinking-water quality (e.g. Dieter, 1992).
2 *Polluter pays principle*: that those responsible for causing pollution should bear the costs of cleaning it up. This principle encourages businesses to adopt environmentally responsible methods to avoid pollution in the first place.
3 *Co-operation principle*: that through co-operation between the state and private actors better environmental outcomes can be achieved. In Germany, the state encompasses different layers of government: the federal, Länder and local levels, each with differing environmental responsibilities (BMU, 1997, 13).

Since the 1992 Rio Conference on Environment and Development, Germany, like other nations, has been committed to the principle of sustainable development. Responsibility for formulating and implementing environmental law and policy is divided between the federal, state and local government levels. At federal level, the Federal Ministry for the Environment, Nature Protection and Nuclear Safety has responsibility for waste disposal, air quality, noise levels and nuclear safety. Nature and landscape protection, water management and agri-environmental policy are the joint responsibilities of the federal and Länder governments. Local government is responsible for provision of drinking-water, drainage, sewage and waste disposal infrastructure (BMU, 1997, p.14).

The severity of environmental problems in the new Länder and the legal, political and social difficulties involved in reunifying the two states, however, necessitated a transitional period for the application of the new environmental policy framework and a few special measures were allowed. The major concession was an 'exemption' clause for contaminated land (*Altlastenfreistellungsklausel*). This exempts owners/managers of industrial/commercial sites in the new Länder from the normal requirement according to the polluter pays principle to bear the clean-up costs where contamination was caused before monetary union in July 1990. This clause was necessary because of the scale of contaminated land in the new Länder caused by poor environmental practices under the GDR state, and the hindrance it posed to the regeneration of the economy of the new Länder. It illustrates the primary importance placed by the federal government on economic regeneration in the new Länder in the immediate aftermath of reunification, which at times seemed to override environmental objectives. It also took a long time for the scale of land contamination to become apparent (partly due to missing environmental data during the GDR period – see above). By the end of 1993, almost 70,000 hectares of contaminated land had been identified and a further 80,000 hectares were suspected, with the highest concentration in Sachsen and Sachsen-Anhalt (BMU, 1996a).

Reunification did not only mean that the West German environmental policy framework became the overarching regulatory mechanism in the former GDR, as with political unification of the two Germanies in October 1990 the new Länder also became subject to EU environmental law and policy. Yet this has had relatively little impact on environmental policy in the new Länder (with the exception of agri-environmental policy – see below), as EU law and policy conforms largely to German standards. In some cases, German environmental policy is even stricter than the EU level (Dieter, 1992; Hoggart *et al.*, 1995). The principle of subsidiarity, that action should be taken at the appropriate level of government, is also central to EU environmental policy, and protects Germany's autonomy on environmental issues (Williams, 1996). The main benefit of EU membership for the environment of the new Länder is therefore not the new EU environmental policy framework, but the additional EU funding available for environmental clean-up projects. Between 1991 and 1999 the new Länder will have received ECU16.6 billion in EU structural fund money for economic regeneration, which includes funding for environmental clean-up projects. In addition, the EU's LIFE initiative provides funding to promote innovative environmental technologies (BMU, 1995; European Commission, 1995).

An important federal government subsidy for environmental projects in the new Länder has come from subsidised work schemes to reduce unemployment and provide training opportunities for the workforce of the new Länder. This is important from an environmental management perspective, as many of the work schemes funded under this programme have been environmental projects

(BMU, 1995, p.ix). Between 1991 and 1994, over DM32 bln was paid out by the federal government for work creation schemes under the joint federal–Land task for regenerating the economy of the new Länder (the *Aufschwung Ost* programme).

Environmental policy priorities for the new Länder

As discussed in the first part of this chapter, the main areas of pollution in the new Länder that needed to be tackled were air and water pollution from power generation, industry, agriculture and domestic waste, and soil contamination from industry, military activities and mining. In addition, more extensive nature and landscape protection measures needed to be introduced in line with West German and EU requirements. This was a formidable task for the governments of the newly created Länder, requiring skilled labour (which was not readily available given the GDR's neglect of environmental management) and substantial levels of investment in clean-up operations and new infrastructure provision.

Emergency measures

The immediate priority, even before political union of the two Germanies, was to tackle the worst environmental problems that posed a threat to the health of the population. A number of emergency measures were launched by the new Länder with funding from the federal government under the *Aufschwung Ost* programme. These included measures to improve water supplies, drainage, air quality and waste disposal. Most funding was directed to the most polluted Länder of Sachsen-Anhalt and Sachsen (BMU, 1995). Altogether, DM1.2 bln was paid out by the federal government between 1990 and 1992. These emergency measures also included the closure of the Soviet-designed Greifswald nuclear power station on the Baltic Sea coast of Mecklenburg-Vorpommern, a freeze on the construction of a nuclear power station at Stendal in Sachsen-Anhalt, and the closure of all uranium mines because of safety concerns (Ardagh, 1991; BMU, 1995).

Contaminated land

Clean-up of contaminated land has been a priority because of the hindrance it poses to economic regeneration. A deadline of 28 March 1992 was set for applications for exemption from liability under the *Altlastenfreistellungsklausel* (BMU, 1996a), and about 70,000 applications for exemptions were lodged (BMU, 1995, p.vii). Where non-liability is accepted, the clean-up costs are divided 60:40 between the federal government and the Land (Karl and Krämer-Eis, 1997). In addition, twenty-three large-scale clean-up projects were identified by a federal–Länder working party. These projects included former industrial areas with large areas of contaminated land, such as the chemical complex at Bitterfeld, the oil and gas refinery at Gommern near Magdeburg, the shipyards at Rostock, and the River Spree industrial estate in Berlin (BMU, 1995). The costs of these projects – which were ongoing in 1998 – are split 75:25 between the federal and Länder governments.

A special package was put together to clean up the brown coal-mining areas. For reasons outlined above, reunification led to a sharp drop in demand for brown coal. While in 1989 there were thirty-seven working mines in the GDR, by 1993 only fourteen survived (nine in the Lausitz coalfield and five in the middle German coalfield), and only eight mines are thought to have a secure future (BMU, 1994, p.10). Reclamation of recently closed pits and other formerly abandoned pits which were never reclaimed by the GDR state is therefore a key priority. The brown coal reclamation programme focuses on three tasks: the reclamation of abandoned mines, demolition and reclamation of processing plants, and measures to restore the water-table where it has been lowered by coal-mining. The estimated volume of water required to restore ground water levels in affected areas and to flood worked out mines comes to 21.3bln.cu.m. This may be problematic, as both the Lausitz and middle German coalfields are located in regions of low rainfall where most rainfall is required for drinking-water and to maintain river levels. As a result, it will take between fifty and ninety years for ground water supplies to be replenished in some places, highlighting the fact that most of the clean-up operations have to be planned as long-term concerns rather than as quick remedial action (BMU, 1994). Unemployed miners have been involved in reclamation work through the government's subsidised work programme mentioned above. By the end of 1993, 15,000 workers were employed on almost 200 reclamation projects, and over 5000 hectares had been reclaimed and were being prepared for new uses. A further 60,000 hectares remained to be reclaimed, a task that will not be completed before the early twenty-first century.

Altogether, it is estimated that the total clean-up costs for contaminated land will come to DM24bln, of which DM17bln will come from the federal budget. Ten billion will be spent on cleaning up individual industrial/commercial sites under the non-liability clause, 7.5bln on brown coal sites and 6.5bln on the large-scale projects (BMU, 1995, p.x). Even for a powerful economy such as that of Germany, these costs place a severe burden on the national budget. The financial burden of reunification has therefore attracted criticism from concerned parties in former West Germany and has at times created severe tensions between the two parts of Germany (Ardagh, 1991; Smyser, 1993).

Improvements in air quality

Rapid improvements were achieved in air quality, but this was largely due to the widespread closure of the GDR's industry rather than to environmental policies (parallels may be drawn with the improvement of the UK's record on air pollution due to the closure of many coal-fired power-stations in the 1980s). In March 1990, before reunification, all state-owned enterprises were handed over to a newly created agency, the *Treuhandanstalt* (THA). Following monetary

Table 7.2 Reduction in air pollution levels in the new Länder (million tonnes)

	Pre-1990	*Post-1990*	*% reduction*
Carbon dioxide	345 (1987)	177 (1993)	49
Sulphur dioxide	5.33 (1988)	3.02 (1992)	43
Carbon monoxide	3.74 (1985)	2.56 (1992)	32
Dust particles	2.24 (1988)	0.91 (1992)	59

Source: BMBau (1991); BMU (1995, p.xii).

union between the two Germanies in July 1990, the THA was given the task of preparing these enterprises for privatisation. Altogether, some 8000 enterprises employing four million people were to be privatised. Of these 8000 enterprises, 1200 were deemed to be unviable and were liquidated. The worst affected industries included those responsible for the greatest amount of pollution – particularly mining, chemicals and textiles (Jones, 1994). As a result, and as Table 7.2 highlights, carbon and sulphur dioxide emissions after 1990 were down 49 per cent and 43 per cent respectively compared to pre-1990 levels, while dust particle levels were down by 59 per cent – with continuing downward trends (BMBau, 1991).

This highlights the fact that economic restructuring and industrial closures in the former GDR following reunification were initially more important in reducing the worst environmental problems than were targeted environmental policies. The latter will none the less gain a hold in the medium to long term, as any new industrial development has to comply with West German and EU environmental regulations. Since 1996, for example, federal emission standards for large power-stations have applied in the new Länder. Power-stations must now either be modernised with new clean-air technology or closed down and rebuilt. Many of the older brown coal-fired power-stations have been closed, with concurrent dramatic reductions in air pollution from air particles and sulphur dioxide (see Table 7.2). It is planned to further diversify electricity supplies away from coal in the future, but this is dependent on new power-stations being built and electricity infrastructure being renewed and linked into the West German network (Jones, 1994). Air pollution from domestic heating is a more difficult, although smaller, problem to tackle. Sixty per cent of houses and flats in the GDR were heated with solid fuel, and it will be a long time until more environmentally friendly central heating is widely installed together with improved house insulation technologies such as double-glazing (BMBau, 1991).

The only source of air pollution that is likely to increase is car exhaust emissions due to rising car ownership levels in the new Länder. Usually the first material change in former GDR households has been the purchase of at least one new car, and as a result between 1988 and 1994 car ownership levels rose by 40 per cent from 296 to 487 per 1000 population. In comparison, car ownership levels in the old Länder increased by only 18 per cent over the same period, although car ownership levels are still higher in the west than in the east (590 per 1000 in 1994 in former West Germany) (Statistisches Bundesamt, 1996). Yet car ownership statistics do not necessarily correlate with the amount of air pollution produced. Since 1986, for example, all new cars sold in West Germany have been fitted by law with catalytic converters, which substantially cut down on car exhaust emissions of carbon monoxide and nitrogen dioxide, while the old two-stroke engine of the East German *Trabi* produced a disproportionate amount of pollution, despite much lower car ownership rates (Ardagh, 1991). Since reunification, the law related to catalytic converters has also applied to the new Länder (which was one of the reasons why the *Trabant* could no longer be built). While this regulation should keep emissions of carbon monoxide and nitrogen dioxide in check, the continually and rapidly rising numbers of cars in the whole of Germany far outweigh the potential gains of the new, cleaner technology in terms of environmental degradation.

Improvements in water quality

Water quality has also improved due to closure of industry, but for long-term improvements renewal of water infrastructure (drinking-water supplies, drains and sewage works) is required. Table 7.3 shows the example of the reduction in heavy metal pollution of the River Elbe – the most severely affected watercourse in the GDR – which has also required co-operation with the neighbouring Czech Republic (BMU, 1996b). As a result of rapid implementation of policies related to water management, the percentage of the population connected to mains drainage systems in the new Länder has increased from 60 per cent in 1991 to 70 per cent in 1995. More importantly, 95 per cent of industrial waste water and two-thirds of domestic waste water are now filtered through modern sewage works with a biological cleaning process (BMU, 1996b, p.34). Water quality in the catchment area of the Baltic Sea has also improved through a 46 per cent reduction in biological contamination and a 66 per cent reduction in phosphates from sewage (BMU, 1995). In particular, the introduction of

Table 7.3 Reduction of pollutants in the River Elbe (mg/kg)[a]

	1986	*1993*
Quicksilver	36	10
Cadmium	9	14
Nickel	103	81
Copper	336	192
Chrome	342	215
Zinc	2095	2052
Lead	191	161

Source: BMU (1996b, p.31).

Note
a Taken from the Schnackenburg measuring station just downstream from Magdeburg in Sachsen-Anhalt.

phosphate-free washing powders (not available in the former GDR) has enabled a 36 per cent reduction in phosphate pollution in the surface water of the new Länder.

Policies for the management of the countryside

Reunification has also resulted in a completely new policy framework for agriculture and countryside management in the new Länder. While the policy framework in the GDR was 'productivist' in orientation, the policy framework in the EU and West Germany in the 1990s can be termed 'post-productivist', with an emphasis on the conservation and amenity value of countryside as well as agricultural production (Whitby and Lowe, 1994; Wilson and Wilson, 2001).

The agricultural sector of the new Länder has had to be integrated with that of the old Länder and has become part of the Common Agricultural Policy (CAP). This integration has required a fundamental restructuring of agricultural structures and production patterns in the new Länder. Agriculture in the GDR was based on vast, industrialised co-operative farms, geared towards intensive production regardless of environmental costs, whereas agriculture in West Germany was based on family farming and, since the late 1970s, had moved towards less intensive and more environmentally friendly farming methods. Indeed, by the late 1980s West Germany was seen as a leader in EU agri-environmental policy and a key promoter of the EU's agri-environment regulation 2078/92/EEC which requires all EU member states to implement agri-environmental schemes (Höll and von Meyer, 1996).

Overall levels of agricultural production in the new Länder have decreased since reunification, largely because of loss of guaranteed markets and lower prices for farm products. This decrease is particularly marked for livestock farming where livestock numbers have dropped by 50 per cent or more (Agra Europe Bonn, 1995). Further, federal government agricultural grants to help new farm businesses to modernise include controls over stocking rate levels (the proportion of livestock to land) and slurry disposal (BML, 1995). Much marginal land has also been taken out of arable production and put into the CAP's set-aside scheme or has been converted to grassland or forestry. These changes, while placing financial pressures on farmers, have brought rapid benefits for the environment.

In the medium to longer term, a polarisation of farming in environmental terms is predicted (Bruckmeier and Grund, 1994). First, farming in marginal areas will become more environmentally friendly (Klinkmann and Tremel, 1995). Since reunification, agri-environmental policies (including organic farming) have been promoted in the new Länder, although at a slower pace than in the old Länder, as agricultural policy-makers in the new Länder lack the experience and expertise of former West Germany with regard to implementation of agri-environmental policy measures (Wilson, 1994). Implementation of agri-environmental policy is also a costly process, and although most funding for agri-environmental schemes in the new Länder comes from Brussels (Höll and von Meyer, 1996), there are still substantial sums that need to be paid by the individual Länder governments.

This means that moneys devoted to agri-environmental schemes have varied considerably between the new Länder, with the wealthier Länder of Sachsen and Thüringen, for example, spending on average about ECU30/ha of utilised agricultural area (UAA) on agri-environmental schemes, while poorer Länder such as Mecklenburg-Vorpommern or Sachsen-Anhalt have spent below ECU10/ha UAA (compared, for example, to ECU47/ha in Baden-Württemberg) (Schramek, 2000). None the less, recent figures indicate that agri-environmental implementation in the new Länder is quickly catching up with its West German counterparts.

Second, farming in fertile agricultural areas is likely to become more intensive (in particular, on the loess soils that stretch from Magdeburg in Sachsen-Anhalt to Dresden in Sachsen). The restructuring of farm business structures has resulted in the emergence of new corporate farms and farm partnerships with the potential to utilise economies of scale to maximise productivity (Bruckmeier and Grund, 1994; Wilson, 1996). Already, productivity levels have caught up with, or have even overtaken, those in the old Länder (Agra Europe Bonn, 1995).

A much greater emphasis has also been placed on conservation and landscape protection in the new Länder since reunification. The designation of ecologically important surviving remnants of semi-natural vegetation mentioned above has been a priority, and the first steps were taken even before political reunification. In March 1990, a national park programme was launched by the GDR government under which five national parks, six biosphere reserves and thirteen nature protection areas were temporarily designated. The national parks include two on the Baltic Sea coast (Vorpommersche Boddenlandschaft and Jasmund on the island of Rügen); one in southern Mecklenburg-Vorpommern (Müritz); one in Sachsen-Anhalt (Hochharz) and one in Sachsen (Sächsische Schweiz). The designations were confirmed in September 1990, and included in the Unification Treaty (Reichhoff and Böhnert, 1991). These initial designations covered 4.5 per cent of the land area of the new Länder, but since reunification the number of protected areas has increased through extensions of existing designated areas and new designations by the Länder governments. Thus, although the former GDR had managed to preserve some areas in their near-natural state (mainly because they were of only marginal economic use rather than for intrinsic conservation reasons), the new conservation policy structures have now formalised the conservation status of these protected areas.

Conclusions

On the eve of German reunification in 1990, environmental conditions in the two Germanies could hardly have been more contrasting: West Germany was one of the most environmentally progressive countries in Europe, while the GDR had one of the worst environmental records in Europe. However, as this chapter has pointed out, the environmental record of the two neighbours was not as 'black and white' as is popularly perceived (as the export of West German toxic waste to the GDR amply illustrated).

The Unification Treaty led to the extension both of West German and European Union environmental law and policy to the new Länder, with the goal of bringing environmental conditions in the new Länder up to those in the old Länder. Major improvements in the key problem areas of air and water pollution and soil contamination have already been achieved in the years since reunification, with air and water quality levels now approaching those in the old Länder. This has been achieved as much due to economic restructuring with the widespread closure of some of the most polluting industries, as to environmental policies. Such rapid environmental improvements would not have been possible without considerable financial support from the federal government and the European Union. Environmental infrastructure (e.g. water supplies, drainage and sewage systems, power supplies) has been renewed, giving the new Länder the benefit of the most up-to-date technology. The reclamation of contaminated land will be a longer term process, however, and it is estimated that it may take up to eighty years to restore ground water levels in former mining areas.

Although substantial progress has been made with regard to ensuring the permanent conservation of protected areas (through the creation of new national parks), implementation of agri-environmental policy aiming at enhancing countryside management has been more problematic, due mainly to the lack of experience of the former GDR with these 'post-productivist' policy mechanisms and a relative lack of finance in some of the new Länder. Agriculture is becoming polarised between environmentally friendly farming in marginal areas and intensive farming in fertile areas.

The positive environmental gains in the former GDR since reunification must, however, also be offset against the fact that with rising living standards among the population of the new Länder, general consumerism has already increased dramatically (especially car ownership levels and household waste production). The prognosis for the future is therefore a complex one. While strict environmental policy standards will inevitably lead to further improvements in the former GDR's environmental performance, these positive effects may be offset in the long term by the adoption of much more environmentally destructive consumerist life-styles related to capitalist thinking and modes of production. Environmental attitudes of the inhabitants of the former GDR will therefore be the key determinants for the successful implementation of environmental sustainability in the new Länder. Most people in the former GDR are likely to adhere to utilitarian and consumerist attitudes in the near future, partly because many have lacked the 'environmental' and 'Green' education that most people in West Germany enjoyed during the 1970s and 1980s, and partly because standards of (consumerist) living still vary considerably between the East and the West. Only when the new generation of young people growing up in the new Länder have 'caught up' with the relatively Green thinking of their West German counterparts – and will consequently implement and adhere to new Green environmental policies and politics – will long-term environmental sustainability be achieved in the new Länder.

References

Agra Europe Bonn (1995) 'Die Agrarwirtschaft in den neuen Ländern' (36), 4 September.
Ardagh, J. (1991) *Germany and the Germans*, London, Penguin.
BMBau (Bundesministerium für Raumordnung, Bauwesen und Städtebau) (1991) *Raumordnungsbericht*, Bonn.
BML (Bundesministerium für Ernährung, Landwirtschaft und Forsten) (1995) *Für unsere landwirtschaftlichen Unternehmen*, Bonn.
BMU (Bundesumweltministerium) (1994) *Ökologischer Aufbau: Braunkohlesanierung Ost*, Bonn.
BMU (1995) *Bilanz und Perspektiven der ökologischen Entwicklung in den neuen Bundesländern*, Umwelt (9), Bonn.
BMU (1996a) *Ökologischer Aufbau: Altlastensanierung*, Bonn.
BMU (1996b) *Umweltpolitik: Wasserwirtschaft in Deutschland*, Bonn.
BMU (1997) *Auf dem Weg zu einer nachhaltigen Entwicklung in Deutschland*, Bonn.
Bruckmeier, K. and Grund, H. (1994) 'Perspectives for environmentally sound agriculture in East Germany', in D. Symes and A. Jansen (eds) *Agricultural Restructuring and Rural Change in Europe*, Wageningen, Wageningen Agricultural University, pp.180–94.
Bulmer, S. and Paterson, W. (1987) *The Federal Republic of Germany and the European Community*, London, Allen and Unwin.
Der Spiegel (1990) 'Das Land der 1000 Vulkane', 1990(2): 27–32.
Dieter, H.H. (1992) 'German drinking-water regulations, pesticides, and axiom of concern', *Environmental Management* 16(1): 21–32.
European Commission (1995) *The Environment and the Regions: Towards Sustainability*, Luxembourg, Office for Official Publications of the European Communities.
Fulbrook, M. (1995) *Anatomy of a Dictatorship: Inside the GDR 1949–1989*, Oxford, Oxford University Press.
Hagedorn, H. (1992) 'Die Bauernfrage in Deutschland-Ost', *Berichte über Landwirtschaft* 70: 396–409.
Hoggart, K., Buller, H. and Black, R. (1995) *Rural Europe: Identity and Change*, London, Arnold.
Höll, A. and von Meyer, H. (1996) 'The response of the member states: Germany', in M. Whitby (ed.) *The European Environment and CAP Reform: Policies and Prospects for Conservation*, Wallingford, CAB International, pp.70–85.
Huber, J. (1993) 'Ökologische Modernisierung: Bedingungen des Umwelthandelns in den neuen und alten Bundesländern', *Kölner Zeitschrift für Soziologie und Sozialpsychologie* 45(2): 288–304.
Jones, A. (1994) *The New Germany: A Human Geography*, Chichester, Wiley.
Karl, H. and Krämer-Eis, H. (1997) 'Contaminated sites – environmental and regional problems in Germany', *European Environment* 7: 149–55.
Klinkmann, D. and Tremel, S. (1995) 'Ökologischer Landbau im Rahmen des Extensivierungsprogramms in den neuen Ländern', *AID Informationen für die Agrarberatung* 5507: 7–10.
Newnham, D. (1997) 'Domestic violence', *Guardian Weekend*, 30 August, pp.28–35.
O'Riordan, T. and Cameron, J. (eds) (1994) *Interpreting the Precautionary Principle*, London, Earthscan.
Paas, D., Timmermann, D., Hand, K., Riedel, W., Schenkel, H. and Frahm, E. (1994) *Lösungswege einer Sinnvollen Dorferneuerung in den neuen Bundesländern*, Bonn, BML.

Reichhoff, L. and Böhnert, W. (1991) 'Das Nationalparkprogramm der ehemaligen DDR', *Natur und Landschaft* 66(4): 195–203.

Schramek, J. (2000) 'Implementation of Regulation 2078/92/EEC in Germany', in H. Buller, G. A. Wilson and A. Höll (eds) *European Agri-environmental Policy*, London, Ashgate.

Smyser, W. (1993) *The German Economy: Colossus at the Crossroads* (2nd edn), Harlow, Longman.

Statistisches Bundesamt (1996) *Bevölkerungsstruktur und Wirtschaftskraft der Bundesländer*, Wiesbaden.

Whitby, M. and Lowe, P. (1994) 'The political and economic roots of environmental policy in agriculture', in M. Whitby (ed.) *Incentives for Countryside Management: The Case of Environmentally Sensitive Areas*, Wallingford, CAB International, pp.1–24.

Williams, R. (1996) *European Union Spatial Policy and Planning*, London, Paul Chapman.

Wilson, G. A. (1994) 'German agri-environmental schemes I: a preliminary review', *Journal of Rural Studies* 10(1): 27–45.

Wilson, G. A. and Wilson, O. J. (2001) *German Agriculture in Transition: Society, Politics and Environment in a Changing Europe*, Basingstoke, Palgrave.

Wilson, O. J. (1996) 'Emerging patterns of restructured farm businesses in Eastern Germany', *GeoJournal* 38(2): 157–60.

Zentralverwaltung für Statistik (1989) *Statistisches Jahrbuch der DDR*, Berlin.

8 Hungary

Alan Dingsdale, Imre Nagy, Gyorgy Perczel and David Turnock

Introduction: historical overview

Problems of environmental protection in Hungary have arisen out of economic development which accelerated after 1867. Regulation of the Tisza opened the way for the intensive cultivation of 3.4mln.ha of agricultural land, linked with the growth of the food-processing industry, which included distilleries as well as meat and sugar factories, especially in the Budapest area. There was also an expansion of mining and metallurgy, although most capacity was located in the outer regions of the Carpathian Basin which were lost after the end of the First World War. The inter-war years saw greater use of the coal resources within the limits of Trianon Hungary, following the southwest–northeast axis of the country, and the industrial centres based on coal-mining, energy production, aluminium and chemicals grew even more rapidly under communism: Miskolc, Ózd and Salgótarján in the north, along with Ajka, Almásfüzitõ, Balatonfüzfõ, Dorog, Pét, Szony and Tatabánya in Central Transdanubia. Since industry also expanded in the Budapest agglomeration and around Gyõr, the northern half of the country began to suffer from pollution (Figure 8.1).

However, only eight settlements were declared new towns between 1900 and 1945: Balassagyarmat, Békéscsaba, Csongrád, Kalocsa, Mohács, Mosonmagyaróvár, Salgótarján and Szekszárd. Altogether, Hungary had only fifty towns by the end of the Second World War, accommodating little more than one-third of the total population. As an agrarian-industrial country, with 51 per cent of the people working in agriculture, Hungary's urban-industrial growth was weaker than in Western Europe and pollution problems were not so serious. However, pollution increased under communism because of the emphasis on heavy industry. There was rapid growth in the 1950s and 1960s (though falling after the mid-1970s), with power-stations and basic industries reflecting the natural resources of central Transdanubia and the north, while development in the Budapest agglomeration was well above average in terms of processing industries. Heavy industry advanced from 50 per cent in 1950 to 70 per cent in 1990, while light industry and the food industry declined from 20 to 12 per cent and from 29 to 16 per cent respectively. At the same time the value of agricultural production doubled as large-scale and technologically advanced farming became dominant.

Figure 8.1 Salient features of Hungary

Heavy industry was often wasteful of raw materials and energy and there were other negative effects, since mining damaged karstic water reserves and reduced landscape quality through the accumulation of waste and incomplete restoration. Manufacturing worked at a mediocre technical level which limited quality while generating further waste. Atmospheric emission levels were relatively high while agriculture created further problems (most noticeable in areas of intensive cultivation) through pollution of water reserves by nitrates and gradual acidification. Although the effects on human health were severe, there was only limited spending (less than 1 per cent of GDP) by the 'paternal state' on environment management and rational use of resources.

Membership of Comecon (a relatively closed economic bloc) afforded some protection, but it also led to technological backwardness with high levels of raw material and energy consumption and poor overall efficiency compared to the developed countries. More rational use was made of local resources in the 1960s and several Hungarian industrial and food products also sold well both in developed and developing countries; but the emphasis was still on extensive sources of economic development until the mid-1970s. Meanwhile the geography of industry was modified by growth in the county centres across the Great Hungarian Plain while small subsidiary enterprises were established by the collective farms, although many did not survive the end of the communist era (Table 8.1). However, while the distribution of production became more diverse, the greatest environmental problems remained along the southwest–northeast industrial axis in the hill country with one-tenth of the area but 40 per cent of the population, including the capital. Pollution is greatest in the Borsod region (Sajó Valley), central Transdanubia (Ajka, Tatabánya and Várpalota) and along

Table 8.1 Regional distribution of industrial workers 1949 and 1991 (%)

Area	*1949*	*1991*
Budapest	54.0	20.8
Pest County	1.7	6.3
Industrialised Counties:		
North[a]	13.0	14.5
Central Transdanubia[b]	10.2	13.2
Györ-Moson-Sopron County	5.3	4.9
Baranya County	4.3	4.0
Less-Industrialised Counties:		
Trans-Danubia[c]	3.6	10.9
Great Hungarian Plain	7.9	25.3
Total	100.0	100.0

Source: Perczel 1996, p.299.

Notes
a Counties of Borsod-Abaúj-Zemplén, Heves and Nógrád.
b Counties of Komárom-Esztergom, Fejér and Veszprém.
c Counties of Somogy, Tolna, Zala and Vas.

the Danube between Komárom and Esztergom. There are also considerable emissions in the industrial centres of Dunaújváros, Pécs and Százhalombatta (near Budapest) (Perczel 1992).

Meanwhile urbanisation accelerated. During the early communist period (to 1970), Budapest developed into a world city with two million inhabitants, and nineteen new towns emerged: both 'socialist towns' (Ajka, Dunaújváros, Kazincbarcika, Komló, Oroszlány, Tiszaújváros and Várpalota) and other places where considerable organic growth occurred: Keszthely, Sárospatak, Siófok, Szarvas, Szigetvár, Tata and Törökszentmiklós. Ninety more new towns were created between 1970 and 1990; twenty during the 1970s following all-round development (Békés, Körmend, Kiskörös and Siklós) with another seventy in the 1980s through the emergence of new growth centres and a new administrative system of districts whose centres were declared towns in three stages (1984, 1986 and 1989). As a result, the total number of towns in Hungary reached 166. The process has continued into the 1990s: there were 200 towns in 1995 compared with 166 in 1989. However, the infrastructure in the new towns is generally poor: they have few institutions or amenities and only partially meet the functional requirements of their areas. Visual pollution arose through 'blockhouse' architecture (generally avoided in smaller towns): the expansion of Budapest in the 1960s and 1970s involved suburbs that were a 'horror of mediocrity' (Hinrichsen and Lang 1996, p.105), not to mention the airport noise created by Soviet planes (ibid., p.104). Meanwhile, settlement consolidation everywhere resulted in the destruction of traditional architecture after 1960. Villages came to resemble 'industrial living districts' (Ruda 1998, p.93). Modern houses typically included much wasted space (e.g. best rooms whose main function is 'passive storage'), a tendency that may be reversed in the context of individual farming and a return to historical styles and values (ibid., p.96).

In 1990, replacement of the communist government accelerated the transition to a market economy. There is now less pressure on the environment because GDP fell by one-fifth and industrial production by 30 per cent during 1990 to 1993, but an effective system of environmental protection is not yet in place, and funds are insufficient to buy all the equipment and technology necessary to meet the standards required for EU membership. Further, while foreign investment is much needed there is a danger that low environmental standards will result in 'eco-colonisation'. The situation will only improve over the longer term on the basis of international co-operation to achieve sustainable development. Nevertheless, the Hungarian environment may be considered relatively good since the forced development of polluting heavy industry was less evident than elsewhere (Hinrichsen and Enyedi 1990; World Bank 1990). Furthermore, the collapse of some industries in the early 1990s and a consequent fall in raw material and energy consumption has brought some environmental benefits (Anon. 1998). Much remains to be done in terms of water purification and sewage, while pollution has also increased due to road traffic. A quarter of all freight and three-quarters of all passenger-kms were handled by roads in 1990, compared with only 2 and 6 per cent respectively in 1950; to say nothing of the absolute growth in total demand.

Air pollution

Pollution levels increased greatly under communism with half coming from industry and some 40 per cent from transport. Carbon dioxide (three-quarters from power-stations and heating plants along with industry and domestic boilers) is easily the most important pollutant, destroying the ozone layer and aggravating the greenhouse effect. CFC compounds, emitted by the freezing industry, insulating material producers, and various chemical industries (such as cosmetics) have similar effects, although their significance in Hungary is declining. Sulphur dioxide, nitrogen oxides and volatile organic compounds (VOC) contribute to the acidification of the environment. The first two air-polluting components increased greatly with the growth in the capacity of coal-fired power-stations (with subsequent reductions through changes in the energy structure), though industry is another major source and transport contributes 45 per cent of the nitrogen oxides (Table 8.2). However, these gases spread to distant areas where they contributed to acid rain and the acidification of soil and surface waters. They also have a corrosive effect on equipment. VOC pollutants (half from transport and filling-stations and one-third from manufacturers of paints and solvents) contribute to secondary and tertiary compounds that are dangerous through the formation of tropospheric ozone. Background pollution (sulphur and nitrogen compounds) from local or foreign sources is only of moderate significance (Aujeszky 1997).

Table 8.2 Sources of sulphur dioxide and nitrogen oxides 1980–94 (th.t/yr)

	1980		*1994*	
	SO_2	*NO_x*	*SO_2*	*NO_x*
Households	290.6	18.9	6.8	13.6
Services	44.9	7.1	25.1	4.7
Transport	49.1	111.3	7.4	94.2
Power-stations	654.7	69.0	423.2	42.2
Other heat generation	33.3	4.1	13.4	4.2
Industry	522.2	53.3	162.9	25.3
Agriculture	38.1	9.9	12.2	3.3
Total	1632.8	272.9	750.0	187.5

Source: Környezetvédelmi és Területfejlesztési Minisztérium 1997, pp.1–3.

Tabel 8.3 Regional distribution of SO_2 and NO_x in 1990 (th.t)

	SO_2	*NO_x*		*SO_2*	*NO_x*
Baranya	68.6	10.5	Nógrád	10.4	2.7
Bács-Kiskun	21.3	6.9	Pest	56.0	19.6
Békés	13.2	5.0	Somogy	8.8	5.4
Borsod-Abaáj-Zemplén	185.0	21.4	Szabolcs-Szatmár-Bereg	31.0	6.3
Csongrád	8.5	4.6	Jász-Nagykun-Szolnok	19.0	5.3
Fejér	44.7	9.4	Tolna	9.7	2.9
Györ-Moson-Sopron	26.3	5.9	Vas	10.5	2.9
Hajdú-Bihar	24.0	6.7	Veszprém	56.4	17.8
Heves	133.1	12.6	Zala	4.9	4.0
Komárom-Esztergom	123.5	9.3	Budapest	37.6	27.5
			Total	892.5	186.7

Source: Környezetvédelmi és Területfejlesztési Minisztérium 1997, pp.1–6.

Hungary is high among post-socialist countries for sulphur dioxide and nitrogen oxide emissions: combining these emissions in relation to GNP gave rates nine times higher than the EU average in the early 1990s (Bochniarz 1992, p.180), reflecting gaps in technology and spending on environmental protection and restructuring. But pollution is extremely uneven (Table 8.3). Following a decree in 1971, a law was passed in 1976 to stop the increase in air pollution and limited results were achieved with reduction of dust from power-stations and cement plants, and reduction in the lead content of petrol from 0.4g/l in 1985 to 0.15 in 1992 and 0.10 in 1996. A decree in 1986 stipulated that air quality control must be taken into account when planning new industries and pollution limits were set on a plant basis. Later on, recession in certain industries also helped to reduce emissions of most pollutants even in the 'hot spots'. Heavy investments have been made to reduce the scale of the problem in Budapest and Györ, but dust levels are still high (often above the legal maximum) in Dunaújváros, Kecskemét, Vác and Várpalota. Sulphur dioxide pollution has been decreasing since 1975, mainly due to the changes in the structure of primary energy resources and modernisation of the heating systems in flats and institutions. Oil (since the 1960s and 1970s) and now gas are replacing coal in power-stations, although some subsidies for coal remain through a special government-brokered arrangement between mines and generators. There are also changed arrangements for domestic heating where solid fuel formerly comprised 40 per cent of domestic energy consumption.

There are fifty settlements where the legal maximum for sulphur dioxide was exceeded on five to forty-five days a year between 1987 and 1992, the worst cases being Tatabánya, Dorog and Miskolc. Nitrogen dioxide pollution has also started to decrease, but it is very much a function of road traffic: the legal limit was exceeded in sixty-six settlements in the early 1990s, led by Pécs, Györ and Dorog with the main road intersections in town centres the worst affected areas. Only 4 per cent of Hungary can really be considered polluted, with another 9 per cent moderately polluted, but 29 per cent and 24 per cent of the inhabitants respectively live in these areas which include the main components of the northern industrial axis: Borsod, the Zagyva River belt (including Ajka, Budapest, Gyöngyös, Várpalota and Veszprém), along with Györ, Komárom-Dorog, Mosonmagyaróvár and Tatabánya. There are also some isolated, badly polluted areas outside the main industrial zone such as Dunaújváros, Pécs and Szolnok. As energy has become relatively more expensive and the economy has slowed down, investment in the environment has become more difficult. Further efficiency gains at power-stations are needed.

Water pollution

Demand for water grew from 3000mln.cu.m in 1970 to 8000mln in the early 1990s (almost 80 per cent being used by power-stations and chemical and metallurgical industries), but as sewage treatment has not kept pace water quality has declined through increased pollution. Hungary has traditionally suffered from having too much water or too little. Aridity in summer is a major risk factor: allowing for irrigation there is a surplus in the west and a deficit in the east (Retvari 1994, p.340). The extremes have been ironed out to some extent through dams, dikes and irrigation canals – and additional measures were taken following the floods of 1974 when emergency reservoirs were provided (Goda 1997). However, national water quality is generally poor, especially in the northern hill region (MERP 1996). With growing acidification nitrate pollution is on the increase: many wells are contaminated and people have to rely on bottled water or supplies piped from neighbouring communities. Two-thirds of drinking-water has no protection from sewage contamination, such is the 'utility gap' between water supply and sewerage capacity. In 1980, 2.4mln flats received piped water but only 1.2mln had sewerage, while in 1993 the figures rose to 3.5mln and 1.7mln respectively, i.e. 77 per cent of all flats had piped water but only 43 per cent had sewerage. This problem is now being addressed as a priority (World Bank 1993), partly through provision of septic tanks with a subsidy programme for municipalities in respect of the collection and treatment of effluent.

Only 27 per cent of the polluted water is fully purified, while one-tenth is not purified at all before it is discharged into rivers. On the other hand, 63 per cent is only partly purified and it is especially the biological and chemical purification that must be significantly improved rather than the first-stage processing. In 1965, the country generated 350mln.cu.m of sewage water; rising to more than 900mln in 1992. The problem is of course greatest in the main centres, with one-third originating in Budapest alone. Even in 2000, half the waste water from Budapest will still have been released into the Danube after only biological treatment (rather than mechanical, biological and chemical which should become the rule). Sludge from Budapest sewage treatment plant is discharged into the Danube at Csepel Island. A modern sludge condensing plant is needed because the heavy metal content precludes incineration.

Water pollution does not arise solely through the inadequate treatment of sewage. Damage from waste dumping, chemicals and pesticides greatly increased from the 1950s until water management improved and better quality standards were introduced from the mid-1970s. Industrial farming generates liquid manure: some 48mln.t p.a. of which 12mln goes into rivers and streams. Seventy per cent of all polluted water is carried by streams in the Danube basin compared with 28 per cent for the Tisza and the rest for the Dráva. In the case of the Danube basin two-thirds is discharged directly into the Danube itself while 10 per cent goes into the Mosoni–Danube and Rába in the northwest and 7 per cent goes into the Séd–Nádor Canal near Veszprém. In the case of the Tisza basin the Tisza itself takes 27 per cent while the Körös receives 26 per cent and the Sajó 21 per cent. However, three-quarters of the water pollution arises from rivers flowing into Hungary from abroad and this high level of interdependence makes international cooperation all the more important, as is also the case with air pollution and acid rain (Hanf and Roijen 1995). The nitrate and phosphate content of the Danube is a problem, as is eutrophication in the Kapos, Sala and Tisza. The Danube has been improved considerably since the late 1980s, especially with regard to traditional pollutants, but biological pollution is still a matter for concern (Varga *et al.* 1990).

The deterioration of Lake Balaton was reduced by the measures taken in the early 1980s to divert some of the polluted water to other areas, control discharges from large animal farms, regulate the use of fertilisers and build purification plants. There is now a sewer network in place and waste water treatment is undertaken (Lang 1978), but high erosion in the hills increases pollution through nitrogen and phosphorous. Fish have died when rapid temperature increase reduces oxygen and triggers algal bloom (with Balaton susceptible due to already high nutrient content). The open water of the lake is now clean apart from the bays at Szigliget and Keszthely, but the quality of the water is not yet stable and it is still possible to observe large numbers of dead fish and the sudden spread of algae. Regarding pollution by bacteria, the limit is exceeded at some places at the lake shore and sometimes in the bay at Keszthely.

Meanwhile it is important that karstic and artesian subsurface waters are protected in the national interest because only one-third remain first-class in quality:

65 per cent have no reliable natural defence against pollution threats from nitrates, bacteria, toxic metals and organic micropollutants. It was assumed that underground waters would be naturally protected from pollution, but most of these reserves are liable (partly through the geological structure) to pollution sinking into the ground from the surface. Since the quality of about one-third of all water drawn from wells has deteriorated since the start of use (because reserves were not properly protected by nature), improvement is an urgent task, especially where the water is so polluted that it cannot be used. Ground water is polluted along the Danube by ammonium, iron, manganese and lead. Untreated sewage discharged into rivers in general helps to pollute ground water aquifers. Up to 2mln.cu.m of sewage leaks into the soil each year: a major danger is the deterioration of farmland through sewage contamination. Only 8 per cent of the water supplied needs no treatment, whereas 65 per cent requires limited treatment (reduction of gas or manganese) and 27 per cent needs further treatment to extract arsenic, nitrate and salt. This does not contradict the statement that one-third of the underground reserves are of first-class quality because wells are not necessarily operating where these waters are to be found.

However, major calamities still arise. Early in 1999 Hungary suffered major damage as the result of a cyanide leak into the Tisza originating at the Aurul plant of the Australian-Romanian gold-mining company in Baia Mare. Exceptionally heavy rain caused the wall of a lagoon to collapse and heavily polluted water overflowed into the Tisza via the Lapos and Szamos. It was three days before matters were brought under control by which time 100,000cu.m of cyanide-polluted water had been released. Maximum permitted cyanide levels are 0.1mg/l but levels of up to 12.40 were recorded; these decreased downstream to 2.8 in the Szolnok area and 1.0 on the Yugoslav border. Up to half the fish were killed and fishermen's livelihoods were thereby threatened. Water supplies for Szolnok were disrupted when abstraction had to stop: 80,000 people were left without water. Aurul lost US$350,000 a week during the shut-down but Hungary is now seeking US$100mln in compensation, although there is no valid international compensation accord between the two countries and The International Court of Justice may need to be used. The Regional Environment Centre (Szentendre) is calling for a decisive co-operative effort through a Tisza River Response Project. Euro20,000 has been donated by the Dutch government to help NGOs cope with urgent work in connection with what is being described as the worst environmental crisis in Europe since Chernobyl. Money is needed for the clean-up, as well as for legal assistance over claims, multi-stakeholder dialogue and general information.

In April 2000, the environment ministers of Hungary and three neighbouring countries (Romania, Slovakia and Ukraine) agreed to co-operate over a regional inventory of 'hot spots' and implementation of the PPP. Five high-risk sites and ten low-risk sites are clustered around Tiszaújváros. These are concerned with oil refining and storage (MOL and the Columbian Chemical Company); also chemicals and paints, power generation and waste incineration. Two high-risk and ten low-risk sites lie in and around Szolnok (including Kenderes, Kisújszállás, Martfü, Mezötür, Szajol and Törökszentmiklos). The main hazards relate to oil storage and waste water discharge – with organic nitrogen content – from a sugar factory. The lesser hazards involve fuel storage, wastes and sludges linked with engineering, fertiliser, leather and paint industries and tetrachloroethylene from the machine works in Kisújszállás. Nineteen low-risk sites are to be found in eastern Hungary around Debrecen, Hajdúboszló and Nádudvar (with Balmaszújváros, Ebes, Hajdúböszörmény, Hortobágy, Kaba, Püspökladány and Nagyhegyes). These relate to oil storage and processing and the manufacture of drugs, fertiliser and machinery where hazards exist in the form of acids, oil and oily waste, sludge and solvents. But there are also food industries (abattoirs, canneries and milk and sugar factories) with high discharges of waste water, large pig farms generating large amounts of effluent and some storage of hazardous waste.

Pollution of land

Mining operations still affect many areas, even though the number of mines fell from 2500 in the mid-1980s to around 2000 in 1992. Open-cast workings are particularly damaging since they affect the landscape in terms of biodiversity, drainage and scenery, as can be seen along the main industrial axis. Moreover, since production is of relatively low economic value, the large amounts of waste have generated some 3000 tips: these are unsightly and they produce much dust and gas, while reclamation costs are very high. Coal- and bauxite-mining has lowered the water-table in the karstic area of the Transdanubian Hills and mineral water springs have been affected: the healing lake at Hévíz and the thermal springs of Budapest were being threatened in the late 1980s when mining activity was at last restricted. This leaves the problem of spoil heaps, though many in Baranya are now partly reclaimed. Meanwhile, the search for hydrocarbons produces a large amount of sludge (a hazardous waste) as well as air, soil and water pollution: note the air pollution from Százhalombatta oil refinery and leakages from the 'Friendship' oil pipeline in Szabolcs-Szatmár-Bereg county where poor-quality material was used in construction.

Although there are some benefits from intensive cultivation arising from roots and stubble which return nutrients to the soil, damage arises from irrational use of fertiliser and inappropriate irrigation, while air pollution causes problems as sulphur and nitrogen compounds settle on the land, especially close to main roads. The result is salinisation and acidification: 13 per cent of the total agricultural area is highly acid, while 19 per cent is moderately or highly sensitive to acidification (meaning that pH may increase within a short time); evident through poor nutritive return, the growing mobility of toxic materials and reduced fertility. Aggravated through soil compaction due to the use of heavy machinery, there has also been degradation of soil structure: the air and water content changes, fertility declines, the biological potential is reduced and the surface becomes vulnerable to the forces of erosion.

Erosion by wind and water affects about 3.5mln.ha in Hungary, particularly on the Great Hungarian Plain (between the Danube and Tisza) and in Nyírség in the northeast, while the water erosion is evident over some 1.5mln.ha in the hills where 80–100mln.cu.m of topsoil is washed or blown away each year, leading to the loss of 1.5mln.t of organic material. The erosion is mainly a result of the inappropriate cultivation methods and crop combinations on sloping land and insufficient attention to conservation measures. Some farmland is also polluted through seepage from waste tips (some of them illegal), and from heavy metals contained in sewage water and sewage mud. Leaking fuel tanks and pipes in former military installations, along with inappropriate storage of chemicals, have also caused pollution damage costed at US$250mln.

Forest dieback (*Waldsterben*) affects 22 per cent of Hungarian forests. Linked with acid rain from the north, oak woods along the Slovak border are badly affected and low pH values mean that soils are poor in buffering agents (Hinrichsen and Lang 1996, pp.102–4). Nearly half the land has nutrient limitations, being too acid, alkaline or sandy, and hence heavy fertiliser dressings are needed to maintain agricultural output (over 300kg/ha active substance). 'Hungarian farmers often seem to be waging a chemical war on their land' (ibid., p.102), with resulting problems of rehabilitation for degraded land as lime and humus declines. Intensive fertilisation of unstable soils with few reserves of nutrients reduces pH and increases salt content. Forest health has improved through reduced pollution (sulphur dioxide by half since 1980 and NO_x by 30 per cent since 1987) but there is still a tendency to damage given the lack of water. This is a consequence of river regulation since the eighteenth century, aggravated by the incidence of dry and hot summers in the Carpathian Basin during the late 1980s and 1990s (Szepesi 1997).

Former Soviet armed forces were responsible for substantial damage over some 50,000ha of land involving 171 garrisons and 340 separate sites (Perczel and Reiniger 1995). About 40 per cent of the damage concerned water pollution, 30 per cent vegetation and landscape and 20 per cent inadequate waste disposal. The main polluting substances were sewage sludges and stored wastes, along with hydrocarbons on the airfields and heavy metals at shooting ranges. Hydrocarbons alone contaminated some 3.0mln.cu.m of soil and 1.0mln.cu.m of ground water. Extensive clean-up operations were mounted, beginning in 1992. Action began at twenty sites where work was completed (Dombóvár, Esztergom, Györ, Lovasberény, Székesfehérvár Szombathely, Tab and Veszprém) or contamination was substantially reduced by 1995. Since then around 140 sites have been tackled through short- or long-term intervention, but the programme is not scheduled for completion until 2010 at a total cost of ft50bln (1991 prices).

Waste management

Waste levels have been increasing, notably through the introduction of Western-style packaging in the 1980s without accompanying recycling systems. In the mid-1990s, 90mln.t of solid waste were produced annually, of which 86mln.t came from industry and agriculture and the rest from private households. Waste material pollutes water, air and soil and is therefore a threat to man and the biosphere via the food chain, not to mention the risk of epidemic diseases. The Hungarian strategy of waste management is based on prevention, through long-term modernisation of production systems and technology generally, recycling and neutralisation treatment. Recycling now affects half the production waste and constitutes 4–5 per cent of manufacturers' total consumption of materials. But Hungary's record is still only half as good as the performance of the industrially advanced countries.

In the past waste has usually been deposited in landfills but in 1995 more than two-thirds of some 2600 dumps operating around the country did not comply with hygiene standards and environmental protection regulations (Perczel and Reiniger 1995). Some waste is dumped in rural areas by small industries set up by the former co-operatives. For example, chrome-tanned leather sent from Budapest to a village in Bekes county for further processing led to dangerous waste being dumped on agricultural land without proper care. After paying a fair price for land in a state of 'complex melioration', the new owner found drainage ditches clogged with waste and, since the Budapest company has closed, the local environmental supervising committee decided that the restructured co-operative should move the waste to an incinerator at an estimated cost of Ft100mln (US$0.5mln). The co-operative (which in the communist period also had trouble with smallholders whose bees were killed by aerial spraying) wants the state to help because the material was not originally considered dangerous and the problem has arisen because of tighter regulations.

Still further progress is needed in dealing with domestic waste because much valuable material is still being lost. Much waste is not treated, or is dealt with inadequately, thus leading to environmental damage. Until recently garbage from Budapest was simply dumped in clay and sand pits in various parts of the city. An incinerator was opened in 1982, though slag and ash have still to be disposed of and there is some air pollution, and across the country many facilities do not comply with regulations (Lehoczki and Balogh 1997, p.141). Some 4.3mln.t of solid household and communal waste is collected (about 200kg per capita compared with the 200–320kg range for developed countries in Europe): about a quarter from Budapest, half from the other towns and the remaining quarter from villages (where a considerable amount of waste is used by the local people following rural tradition). The amount varies according to the heating system in use and the local climatic conditions, but there is an upward tendency because per capita waste in Budapest was 0.6kg per day in the 1950s but 1.1 in 1990 (only half the level of New York but double the quantity in Calcutta). The composition of rubbish varies according to living standards and life-style: in Budapest there is more ash, kitchen waste and plastic, but less paper and glass compared with the European average. Only 8 per cent is incinerated and 2 per cent recycled.

Hazardous waste

Data from the 1980s showed that Hungary produced far more hazardous waste (batteries, chemicals and medicines) in relation to the total value of industrial output than advanced industrial countries, three-quarters coming from industry: largely in food processing, chemicals and metallurgy located in the main industrial axis. However, much depends on the registration system in force because Hungarian statistics usually include some 1.0mln.t of 'red mud' from aluminium plants which has to be stored at special disposal sites since there is no technology available to deal with it. Discounting this item, the amount of hazardous material declined from 2.98mln.t in 1989 to 2.24 in 1992 as a result of recession (Lehoczki and Balogh 1997, pp.143–4), though the proportion being treated is falling as more accumulates inside the plants generating the waste (0.48mln.t in 1989; 1.14 in 1992). About one-tenth of the waste is recycled (metals, paper and textiles) where sorting is possible, but most goes to Hungary's 2000 tips, most of which do not meet current health and environmental regulations. Thus as controls on tipping become tighter, incineration ('thermal neutralising') is becoming more popular among the professionals and, as already noted, a unit has operated in Budapest since 1982. But there is generally opposition from local people who dislike any kind of waste treatment site. Organic material can be separated out and processed into compost to improve soil fertility in an environmentally friendly way. About 80 per cent of the material is properly handled (and 0.2mln.t is recycled), though the remaining one-fifth constitutes a potential danger. A decree in 1996 adopted the spirit of the Basel Convention.

A dangerous situation arose in the 1970s when Budapest Chemical Works, producing tetrachloride benzol, started storing residual chlorine benzol at Hidas (a mining village in Baranya) where some 8000t of hazardous chemicals accumulated. However, after transfer to Garé (southern Baranya) in 1980 for final storage (with additional deliveries directly from Budapest until production stopped in 1987), 10–15 per cent of the chlorine benzol seeped into soil and contamination became evident by 1993 to 1994. Unease was expressed in the Villány wine-producing area and at Harkány thermal spa; while exposure by Green activists made the issue particularly sensitive in view of the proximity to the Croatian border. The problem has been solved through an ambitious programme, initiated in the 1980s, to construct a network of up-to-date landfills, interim storage capacities and incinerators. Because of lack of finance the programme was not completed by the end of the decade but, following PHARE assistance, most hazardous waste is now treated by incineration through a network of regional units which include Aszód, Balmazújváros, Dorog, Eger, Hernádkércs and Nyíregyháza, often established in the face of opposition from local residents. Aszód and Dorog neutralise dangerous waste from virtually all parts of Hungary, while the other four depots have purely regional functions. These are communal depots owned by local authorities in contrast to industrial waste material treatment plants which are in private or company hands. Further communal incinerators and depots are needed, because the existing capacity is not enough to solve the problem.

Nuclear waste

Hungary imports more than half of its energy resources and depends heavily on Russia for oil, gas and uranium. Moreover, domestic resources are declining (oil and gas), or are extracted uneconomically due to low quality (solid fuels). Solid fuels production declined between 1983 and 1994 from 25 to 14mln.t per annum and employment fell from 50,000 to 21,000 (Green 1998, p.33). Given the shortage of conventional fuels the nuclear option has been taken up. Four 440MW units were installed at Paks between 1983 and 1987, using domestic uranium with additional supplies from Russia. However, this raises the sensitive issue of waste disposal which, under the Nuclear Energy Act of 1996, provides for a Central Nuclear Financial Fund to finance interim storage and final disposal. Low- and medium-activity waste amounts to 370cu.m of solid material and 250cu.m of liquid. The solid material is compacted and stored in 200l metal barrels in a concrete-lined repository, first opened in 1977 at Püspökszilágy in Pest county in connection with medical and experimental institutions. Plans for a repository in granite at Magyaregregy or Ófalu (Baranya county) were abandoned in the early 1990s through lack of scientific consensus and public opposition (the local infrastructure was also inadequate) and studies are being made in the Paks area: Udvari and Diósberény for underground storage and Bataapati for surface disposal (Balogh *et al.* 1995).

High-activity waste amounts to 42t every five years. This is mainly spent fuel (burnt fuel rods), shipped to Russia under the agreement reached between Hungary and the FSU. Nevertheless, an interim dry storage facility (to be operational for fifty years) is under construction and a 4500cu.m facility for final disposal will have to be provided in the future. After 2020 replacement at Paks will create some 25,000cu.m of low/medium-activity waste and over 2000cu.m of high-activity waste. Research for an optimal site began in the Mecsek in 1989. Here the geological formation has favourable properties but former uranium mines (over 1000m deep) are not suitable. Other formations have been considered in Tolna county and the selection process is ongoing (Green 1998, p.34).

Noise

The first regulations on noise pollution were formulated in the early 1980s in response to public demand for a quiet environment as an important aspect of healthy living: the World Health Organisation limits are 55dBA in the daytime and 45dBA at night because eight hours a day under 85dBA or more can be seriously damaging. Noise disturbs sleep and detracts from efficiency at work. The tendency is for noise to increase, especially traffic noise which about half the population is concerned about. This means a noise level of 10–20dBA by day and 15–23dBA at night where motorways enter Budapest, but there are higher levels in the city centre. By contrast only one-tenth of the population is concerned about the railway although the level may exceed 10–15dBA close to junctions and freight yards. Air traffic can be a nuisance close to airports, and there were strong protests at Ferihegy (Budapest) in the late 1980s when limits of 65–70dBA were set in the outer zone compared with 71–5 in the central zone. The problem of noise at military airfields has still to be dealt with. Finally, industrial noise affects 2–5 per cent of the population in certain specific regions, especially at night. About 70 per cent of factories inspected in 1985 exceeded the prescribed limits, although the problem was much reduced by the early 1990s because of recession and the enforcement of regulations. There is also concern over noise linked with entertainment, restaurants and sports facilities especially when major events are held. Clearly reducing noise requires action by manufacturers to produce quieter machines and vehicles, while drivers and operators must also take due care and passive noise protection measures should be taken. There are also implications for town planning in terms of traffic management and the segregation of land use through green areas and other protection zones.

Conservation areas

Through its 1997 Nature Conservation Act, Hungary has created a legal basis for effective conservation. Regional plans will include inventories of unique landscape features and convergence with the EU means that land-use practices are now being reconciled with a system of environmentally sensitive areas and a national ecological network. Protected areas involve all major habitat types. National parks include Aggtelek, Hortobágy and Kiskunság, and two landscape protection areas – Pilis (1979) and Lake Fertö (1990) – are designated biosphere reserves, while Ramsar sites include Hortobágy, Kis-Balaton and Tisza wetlands. Dry grasslands are well represented in nature reserves including the puszta grasses of Hortobágy and Kiskunság. Moreover, the Nature Conservation Act creates the category of natural areas which will be maintained although not supported by special conservation measures. When combined with other categories of protection it should ensure the survival of all existing habitat types, exceeding the provisions of the Habitats Directive which only provides special protection for certain habitats and species (Gyulai 1998, p.52). Nature conservation is administered by the National Authority for Nature Conservation within the Ministry of the Environment. There are nine regional directorates, each with responsibility for one national park. New rules on land use, game management and hunting, forestry, water management, mining and regional development are an improvement but still do not provide sufficient opportunity for conservation. Thus, while there is a need for wetland reconstruction after all the works in straightening and diking rivers, the aim of water management is still to ensure quick run-off of water to prevent flood risk. In addition, agricultural subsidies provide funds for land reclamation and afforestation using exotic species. There is a need for more policy integration and more highly skilled staff.

After 1990 land owned by statutory conservation organisations increased to 20 per cent of all protected land – and another 200,000ha of protected land is to be acquired by the government under legislation for the Rehabilitation of Status of Protected Areas. However, the conservation organisations (statutory and NGO) do not have enough labour, capital and management skills to look after their sites and large areas are leased out for management. Most protected areas are still privately owned and this produces conflicts of interest: however, new decrees on the compensation of losses caused by nature conservation and subsidies for nature-friendly management aim to solve these problems, but NGOs play a very supportive role. MME (the Hungarian Ornithological Society), E-Misszio, Duna Kör, Holocen and Somogy all have a strong influence on nature conservation: indeed the activities of MME cannot be separated from those of the statutory nature conservation organisation. The Hungarian Society for Karst Research and Speleology also has a long tradition of co-operation with government. The NGOs are influential because they are trusted by the public to a greater extent than the statutory authorities. Hence they play a useful role in environmental education. Help has come from the Netherlands over an ESA system and from Denmark in respect of habitat restoration and ecotourism in Hortobágy National Park. ESAs ensure appropriate farming practices in such varied areas as saline plains and wetlands, lowland freshwater supply areas, the higher Danube–Tisza interfluves, karst areas and potential buffer zones for Lake Balaton. More Green corridors are needed in connection with conservation in agricultural areas as a whole, plus environmental education and encouragement of organic farming (Fesus 1992).

Reference should also be made to the PHARE cross-border co-operation programme (1995–99) since integrated rural development in border areas with Austria–Slovenia and Romania invoked the 'Naturpark' concept whereby local agents were to develop an area in a sustainable way based on local resources and co-operation. Such a centre is now under construction in Hungary with cycle routes, border crossings, waste collection (for a group of thirty villages) and an international ecological children's camp. Hungary is also involved in 'PAN Parks' based on the concept of synergy between nature conservation and tourism within a protected area network (PAN) where no hunting or commercial logging is allowed. The idea is being developed by the Worldwide Fund for Nature (WWF) and the Dutch Molecaten Group, a leisure company. The parks will offer outstanding wildlife and excellent tourist facilities throughout Europe combined with effective nature protection, comparable with Yellowstone and Yosemite. Hungary's involvement arises through the Duna–Drava park (49,400ha, established in 1996). It consists of two portions: the floodplains of the Lower Drava on the Croatian border (one of the longest stretches of unspoiled river in Europe) and the Gemenc–Beda floodplains along the Danube south of Budapest. Floodplain forests and meadows remain at risk from the continuing diking programme, geared to safer navigation and flood protection, dating back over two centuries, yet they harbour one of the most versatile fish-fauna communities in Hungary. Burbot and sterlet are still common while pike hunt in backwaters and tributaries. The Drava is an important habitat for aquatic organisms highly sensitive to water quality and thought to be the only place in the world where the Caddisfly species (Platyphylax frauenfedldi) occurs. Protected plant species include the white water lily and fringed water lily and the rich bird life includes colonies of the endangered black stork and white-tailed eagle: indeed the area became a Ramsar site in 1997 since it provides a habitat for a quarter of the whole of Hungary's black stork population. Local people hope to benefit through ecotourism based on the angling and canoeing facilities, observation towers for bird-watchers and a narrow-gauge railway in Gemenc Forest (providing a link with the Gemenc Forest School nearby). Better facilities are needed but visitor pressure must be carefully regulated and in the Drava section thirty visitors every third day is considered the maximum.

Hungary is participating in a co-ordinated programme of ecoregion-based conservation for the Carpathians which will affect the northeastern part of the country where there are large protected areas covering Bükk, Mátra and Zemplén. The main biodiversity threats are being researched to provide a basis for discussion among stakeholders as to how management can be adjusted to minimise damage. However, this territory is relatively poor and there will be strong pressure to prioritise higher living standards. The northeast was dynamic under communism but villages in border areas were neglected and suffered depopulation after the Treaty of Trianon imposed a new border and disrupted existing patterns of production and trade (Suli-Zakar 1992). The transition has added to their backwardness while the share of industrial production has fallen. Since 1989, stocking and cultivation has been reduced and the level of processing in the region is low. Agricultural yields are relatively low as is the general technical level of farming, though the wine districts of Bükkalja, Eger, Mátraalja and Tokaj are world-famous. According to the National Conception for Rural Development, eighteen of the twenty-three statistical units in the region are socio-economically backward. Hence the importance of a regional policy that can combine higher incomes with conservation goals. Tourism could be based on the cultural and environmental values, including medicinal waters, while ecoregion-based conservation could help safeguard the water supplies which include the pure underground reserves of the Bükk Mountains and Aggtelek Karst. However, waste management is inadequate since most small villages and recreation areas are lacking a regular waste collection while arrangements for landfill, hazardous waste and sewage are unsatisfactory.

Environment policy and EU accession

Hungarian environment policy effectively began in the 1960s to 1970s and, apart from the law on air purity (1973), these measures were largely concerned with conserving resources needed for production (Horváth 1986). Hungary accepted the recommendations of the UN Stockholm Conference of 1972, and the Environmental Protection Act of 1976 (still in force in 1989) 'was the first explicit establishment of principles of environmental policy, established the formal right of citizens to live in a healthy environment and proclaimed protection of the human environment a responsibility of the broad society' (O'Toole and Hanf 1998, p.100). Despite some specific pieces of legislation, difficult to enforce with inadequate penalties, the state tended to deal with technical matters rather than regulation and the ministry was relatively weak (Bandi 1993). Policies needed better 'design' in order to prevent pollution (Kerekes 1993, p.146).

There was little public protest in the 1970s, but issues in the 1980s included pollution of water supplies by hazardous waste stored by Chinoin at Vác; dust pollution at Ajka from the local power station; resistance to the planned hazardous waste incinerator at Dorog; and the halting of a proposed low-level radioactive waste disposal facility (Ófalu in Baranya county) and especially the Danube Movement which campaigned against the Bös-Nagymaros hydropower schemes and attracted far more public support than was usually forthcoming on environmental issues. This latter issue has become a long-running saga because, while Hungary pulled out of its agreement with the former Czechoslovakia, Slovakia (taking over responsibility in 1993) refused to recognise this decision and proceeded unilaterally at the upstream site where the Gabcikovo installation now operates (Fitzmaurice 1996; Golombos 1993). The International Court ruled in 1997 that the original treaty was still valid and the Horn government resumed work at the second (downstream) site of Nagymaros on the basis of this obligation, supported by Dutch consultants who saw completion as beneficial for water level, navigation and underground water reserves. But the new Hungarian government, led by the Fidesz-Hungarian Civic Party, considers the Court judgment purely advisory and, far from accepting an obligation at Nagymaros, wishes to 'recover' the diverted Danube in the Gabčikovo area. The dispute continues with Slovakia, naturally, taking the view that Hungary must co-operate (for further discussion see Chapter 10).

In 1987, MERP was reorganised to bring environmental issues and water management together (Enyedi and Zentai 1986) but in 1990 water management was transferred to the Ministry for Transport, Telecomnunications and Water Management – and some environmental duties were distributed to other units. The Ministry for Public Health now deals with the prevention and reduction of adverse health impacts of environmental pollution, while the Ministry for Internal Affairs directs municipalities over solid and liquid waste treatment. The Ministry has not enjoyed a strong reputation, given 'a persistently weak and somewhat grudging portfolio' (O'Toole and Hanf 1998, p.104). However, although there is an element of competition when environment is linked with other functions, some think the Ministry should reincorporate tasks lost in 1990 including water management since this cannot be easily separated from water protection which is a MERP responsibility. There are twelve regional environmental inspectorates with autonomy from MERP, though the Chief Inspector for Environmental Protection is appointed by the Minister. But the inspectors lack resources and, while they control emissions, ambient air quality standards are set by public health authorities while mobile sources of emission are the concern of the Ministry for Transport, Telecommunications and Water Management, and the

National Meteorological Service measures background air pollution away from settlements. So 'changes have been made in the rules for controlling environmental pollution but these changes do not get to the heart of the enforcement and coordination questions' (ibid., p.106).

Public environmental consciousness was still low in the 1990s (MERP 1991), although the supply of information was increasing: when the then Environmental Agency's first report came out in 1986 it was classified 'top secret' but by 1989 such material was being published in newspapers (Reeves 1993). But there was little support for environmental measures that might threaten the economy (Flaherty 1993, p.93) and although ready to organise against large companies like Henkel in Bekes county, people were not really sensitive over agricultural pollution. However, the business community was far-sighted enough to consider such issues and support regulatory measures (Vastag 1994); while environmental issues could be used as a lever to gain power and influence locally. There were only vague commitments in the National Renewal Programme 1990, but in 1992 MERP announced an ambitious and comprehensive strategy to overcome the environmental legacy of the communist period (Dingsdale 1996) with priority for environmentally sound technologies, including a modern transport system and advanced technology for waste and sewage treatment. But the emerging National Environmental Programme (NEP) was clearly dependent on Western aid and the government representatives attending the Rio Earth Summit (including the then Environment Minister) made it clear that economic considerations would continue to have priority. However, there was considerable spending on cleaning up former state industries affected by privatisation. One case was Tungsram in Budapest which was purchased by General Electric. Another concerned the Lehel refrigerator factory at Jászberény, a relatively modern plant bought by Electrolux with half the money spent on disposal of the accumulated hazardous waste.

A Central Environmental Protection Fund (CEPF) was set up as an extra-budgetary fund to assist all areas of protection and nature conservation and to encourage an economically sustainable economic structure. The fund operates under MERP, and the Minister is accountable for the way the fund is used. Specific fines may be allocated to specific uses: in 1991, 70 per cent of the waste water fine was spent on water quality protection measures (the first case of strict earmarking). In 1992 the environment product charge on transport fuel was geared to transport-related pollution projects. Environmental fines are a prominent part of the regulation system, though they are self-reported and it is easy to evade emissions (e.g. sewage fines for discharges into public waterways, waste water fines for damaging surface water, air pollution fines for exceeding limits and noise and vibration fines in building and reconstruction work and hazardous waste fines for breaking regulations). There are also fines for unpermitted use of ozone-depleting substances (ODS), hazardous waste importation without a permit, soil pollution and failure to maintain land productivity. Environmental product charges payable to the CEPF were also levied on transport fuel in 1992: 'the first economic instrument to aim at taxing the consumption of a product for environmental protection purposes' (Lehoczki and Balogh 1997, p.155). This appears to be a feasible substitute for a direct emission tax with the advantage of low administrative costs and high correlation with emissions, while funding schemes to reduce transport-related environmental damage.

In 1996 measures were taken to cope with the disposal of tyres and refrigerators with ODS, as well as the recycling of batteries and packaging materials. This reflected the change of government in 1994 and the start of a shift in opinion away from concentration on growth, with income to improve technology and environment, to a stronger environmental focus in agriculture, energy and transport: 'the present time is ideal since the economic transformations now create opportunities to undertake the conversion before major new investments are sunk in place' (O'Toole and Hanf 1998, p.101). However, budgetary stringency has continued and the view of the environment as antagonistic to economic and social welfare is widespread. There was no broad policy until the 1995 Environmental Protection Act sanctioned a six-year National Environmental Protection Programme inspired by the National Environmental and Nature Conservation Policy Concept which Parliament originally asked the government to undertake (i.e. MERP 1995).

Regulation is also being decentralised in the interests of greater co-ordination and regulation at county level, but there is a need for more professionals, for 'only a few counties employ special engineers qualified in environmental protection' (Benkone Lodner 1995, p.31). Professionals should be encouraged by more tasks. Individual settlements have responsibilities for environmental protection which they do not have the resources to discharge (Lodner 1991). On the other hand, the centre has a role. Umbrella management of the environmental problems of settlements 'would require the substantial involvement of county governments, but this is not possible within the present framework of legal regulations' (ibid., p.15). Greater local input is significant in the context of declining water quality (detrimental to health in one-third of all villages) and dumping at many sites which do not meet the present environmental protection and public health regulations; while 55 per cent of the population still suffer from traffic noise pollution. 'Active participation in the regional and global protection of the Carpathian Basin' is therefore recommended (ibid., p.7).

EU accession

Although Hungary is particularly interested in the economic opportunities of membership, a commitment to harmonisation in environmental matters was made in 1995 and the EU has given substantial financial, technical and expert assistance. However, the environmental standards which form part of the *acquis communautaire* have been continually upgraded, while the 1998 report on the status of candidate countries (measured against the 'Copenhagen Criteria': political institutions, economic practices and adoption of the *acquis*) noted that Hungary's environmental preparations were slowing down. The Commissioner responsible for environmental matters (Ritt Bjerregaard) called for 'real action' to improve the environment in the pre-accession period, reinforced the following year with a call for 'additional implementation efforts' to speed up the adoption of legislation, along with monitoring and enforcement, especially in waste management (Hempen 1998).

> The Hungarian environmental law and institutional system is, for the most part, appropriate for the accession requirements. Nevertheless there are some concrete tasks to be solved: sewage treatment, air pollution caused by energy generation and transport, waste disposal and transport.
>
> (Anon. 1998, p.156)

By this time the Hungarian Environment Minister had announced priority for ecology in all sectors of the economy. Hungary would spend US$15bln over twelve years to meet EU environmental protection requirements (8–10bln coming from budget). Hungary would also sign the Kyoto Protocol on cutting carbon dioxide emissions and would cut greenhouse gas emissions to a level 6 per cent below the levels of the late 1980s. It was anticipated that basic harmonisation would be complete in time for an anticipated accession in 2002 (albeit with latitude over air and water quality and waste disposal), but early in 1999 it was conceded that concessions might be needed – over a six- to eight-year transition period – in respect of twenty-three detailed requirements, including drinking-water quality and the collection and purification of sewage. However, Hungarian politicians are confident that environmental problems will not obstruct entry, given the huge costs involved in meeting even the average environmental standards of member states in the short term (up to one-third of GDP compared with 1.2 per cent currently spent) and the European Commission duly announced a softer approach in 1999 which could defer some aspects of compliance until after accession. While a World Bank loan of US$31.6mln will upgrade two waste water plants in Budapest with extended sewer systems and secure a secondary waste water treatment plant for Dunaújváros, Hungary will probably be unable to connect all settlements with a population of more than 2000 in the near future and will probably seek concessions in this area. Hungary has no system of integrated pollution prevention and control (IPPC). There are no major structural problems over the necessary legislation but putting the system into practice will be more difficult, given the need for administrative structures: the designation of competent authorities and the setting up of monitoring systems. New resources will be needed, including a possible merging of Regional Environmental Inspectorates and Regional Water Authorities (Emmott 1997).

However, Hungary now has an overall accession strategy through its National Environmental Protection Programme (NEPP). Progress was made on a wide front during 1999 and 2000, as the EU November 2000 report makes clear. There has been progress on air quality following the prohibition of leaded petrol in 1999, while a framework law on waste management is now in force. Hungary now has four modern regional landfills serving fifty-one municipalities, but there are some 2700 waste storage facilities altogether and only 30 per cent comply with domestic standards while illegal dumps constitute a potential source of environmental damage. Attention has also been given to chemicals, nature protection, noise, radiation and water although progress is still very limited over the IPPC Directive. More is needed on waste management and water, while the measurement of noise impact is still an important necessity. The Ministry of Environmental Protection needs more staff, and co-ordination between ministries concerned with environmental issues should be improved. However, implementation and enforcement of environmental legislation is the work of twelve Regional Environmental Protection Inspectorates, nine National Park Directorates and a Nature Conservation Authority. More staff training is considered necessary to increase the effectiveness of the inspectorates. In addition to monitoring they grant licences and enforce standards by imposing fines. They also play a crucial role in compiling information on pollution sources and licensing which is fed into the NEP system. Unfortunately, only the data on waste water is compatible with EU standards and in other areas the lack of a unified data register complicates the exchange of information at international level.

Prospects for sustainability

A sustainable future is still a long way off. As in other candidate countries, a new generation of civil servants in Hungary should be fully geared up to appropriate environmental standards. Yet 'regional plans . . . are being prepared completely separately from the environmental plans whereas the links between the two should be obvious' (Maier *et al.* n.d., p.17). Despite the creation of the Hungarian Commission on Sustainable Development (HCSD) in 1993 to co-ordinate national efforts in sustainable development (HCSD 1994), 'even now there are no national programmes incorporating principles of sustainable development into the routines and regular decisions of the government or the implementers of environmental efforts' (O'Toole and Hanf 1998, p.101). There should also be an effective partnership between government and ENGOs, and while the political parties talk up environmental issues to get votes, resources are very limited and there is not yet 'a solvent middle class demanding environment-friendly products' (Szirmai 1993, p.154).

Regime change has reduced the scale of pollution associated with large farms: soils and ground waters are no longer affected as severely by liquid manure, while aerial spraying and stubble burning has been stopped. But while legislation requires agriculture to proceed without causing permanent damage to the soil or to flora and fauna, there is still excessive nitrate dressing through competition between private farms and co-operatives to increase yield and improve appearance in the eyes of consumers and processors. Smallholders often misuse fertiliser, distributing it barehanded, and using excessively strong sprays because the cucumbers 'look nicer'. Farmers are not minded to adopt ecological planning unless it serves their economic goals. To moderate agricultural pressure there should be better zoning to relate cropping to characteristics of natural regions and secure the required production in the most sustainable fashion (Tirczka and Ferencsik 1998). There are different perspectives by (1) agri-businesses which see good scope for competition with low levels of chemical use and low per capita stocking, and (2) advocates of organic farming who argue that 4.5mln.ha of cultivation (as of today) will be untenable: only 3mln in the long run using traditional methods and reduced use of chemicals. Other land will need to be rehabilitated, perhaps with greater specialisation in extensive stock-rearing with greater scope for pigs and poultry. Hungary may also exploit its comparative advantages for seed production, while dry farming should also be extended to simplify water management (Gurzo and Marton 1997), and more land may be taken into conservation with EU support. Laws presently cover animal health, fodder production, foodstuffs and nature protection but not plant protection and health.

The food industry has some pollution problems through meat/poultry waste; heavy energy use in milling, canning and the manufacture of vegetable oil; water pollution in meat/poultry, dairying and canning; water needs for sugar, beers and other drinks. This is a crucial matter since one-fifth of Hungarian exports to EU countries are supplied by the food industry. In most cases there is no specific environmental legislation concerning such activities, and responsibilities have to be deduced from general rules concerning water use, treatment of sewage and waste material, air and noise pollution. However, there will be a heavy price for compliance with EU requirements by SMEs in the food industry compared with large, foreign-owned companies that can cope relatively easily; hence the case for government help (Kerekes and Kiss 1998, pp.179–80).

Another problem concerns hazardous material emissions by power plants burning fossil fuels which may be well in excess of European norms and require installation of sulphur scrubbers. Meanwhile, the additional 900MW capacity expected from the private sector should meet all international standards on technology, waste treatment and air quality. The government has abandoned energy price controls to encourage higher environmental standards although rising costs will have socio-economic consequences. Meanwhile, there should be more emphasis on alternative energy but, given the monopolistic position of electric grid operators, independent producers are not allowed to link up. There is a need for liberalisation here and 'instead of large-capacity, centralised power plants preference should be given to the construction of small "local" plants whose heat can be utilised in a number of ways' (Kerekes and Kiss 1998, pp.185–6). In addition, energy demand (as well as raw material input) must be reduced in industry through greater efficiency, stimulated by fiscal incentives, recycling and government regulation to encourage greater use of public transport.

There is much particulate pollution from household heating (high in winter) while nitrogen oxides come mainly from vehicular traffic (high in summer) as cars have increased in number from 1.18mln in 1984 to 1.94mln in 1990 and 2.26mln in 1996 (Kerekes and Kiss 1998, pp.181–2). Hence transport policy is an important issue, especially in view of the EU's 'Eurocorridors' or Trans-European Networks, since the likelihood is that pollution from vehicular traffic will increase despite the ending of leaded petrol sales in 1999. Some conservationists think that major road projects should be reconsidered, for by no means are they a condition of accession: 'easing congestion at border crossings, the replacement of transit routes cutting through major settlements with by-pass roads may be more important than insisting on large, new projects that lack adequate domestic resources or EU assistance' (ibid., p.183). The Ministry of Transport has promoted motorway construction, including the northern section of the Budapest Orbital motorway which is being pushed through Budapest after an inadequate EIA: although residential areas near the road will be polluted, there was inadequate information about public inquiry hearings and conservationists believe that Hungarian and EU laws have been breached. Both the European Bank for Reconstruction and Development and the World Bank have warned of the danger of excess road capacity. Transit traffic yields very little income to compensate for congestion and crime. 'With the improvement of the combined [rail and road] transport system the bulk of transit through the northwest–southeast corridor could be managed efficiently at substantially lower costs and with less harm to the environment' (ibid., p.185).

Conclusion

This chapter has described the main themes of environmental problems and responses as they have emerged under three successive and contrasting sociocultural regimes in Hungary. Before the Second World War little thought was given to the relationship between industrialisation, urbanisation and the environment. The environmental impact of these processes was localised, they proceeded at a relatively slow pace and they did not form a major theme of the political agenda. Following the Second World War the new communist regime enforced a strategy of rapid economic development marked by urbanisation and extensive heavy industrialisation to achieve a primary goal of economic modernisation. The wilful disregard for environmental effects in this rush for economic growth was embedded in a system that discounted environmental costs, lacked accountability and manipulated outcomes within a syndrome of secrecy. Popular movements of opposition to communist governments in the 1980s gathered around environmental issues, but since 1990 this focus has faded and environmental activities have been restricted to small, underfunded, local groups. Encouraged by the prospect of EU membership, governments have engaged more seriously with environmental issues, but this has been within a political climate that has downgraded state regulation and planning yet promoted grand proposals. This has endowed environmental issues with a priority in political rhetoric, but not in political practice. Successive governments since 1990 have been unable to find a suitable administrative ministerial niche for the environment or funds to extend gains associated with the collapse of heavy industry. The prevailing sense of individual action has restricted ideas of regulating road transport, now a principal source of pollutants. Hungary has been the most stable of the new democracies, but has failed to tackle environmental issues. There seems to be little prospect of a change in this situation in the near future.

References

Abel, I. 1993, *Environmental Implications of Economic Restructuring: The Case of Hungary* (Budapest: Budapest University of Economic Sciences).

Anon. 1998, *EU Accession and the Environment: An Introduction* (Brussels: FoFF/EEB).

Aujeszky, P. (ed.) 1997, *Környezetstatisztikai adatok* (Budapest: Központi Statisztikai Hivatal).

Balogh, J. *et al.* 1995, 'Problems of site selection for nuclear waste disposal in a loess-covered hill environment in Hungary', *GeoJournal* 36: 261–8.

Bandi, G. (ed.) 1993, *Environmental Law and Management System in Hungary: Overview Perspectives and Problems* (Budapest: Environmental Management and Law Association).

Benkone Lodner, G. 1995, *The Legal-administrative Questions of Environmental Protection in the Republic of Hungary* (Pecs: Centre for Regional Studies of the Hungarian Academy of Sciences, Discussion Papers 20).

Bochniarz, Z. 1992, Enviroment and Development in Hungary: *A Blueprint for Transition* (Budapest and Minneapolis: University of Minnesota).

Dingsdale, A. 1996, 'Environmental problems in Eastern Europe with special reference to Hungary', in J. P. Smith, D. L. Holt and A. Wheeler (eds) A*spects of Environment in Europe* (Wolverhampton: University of Wolverhampton), pp.15–33.

Emmott, N. 1997, 'The theory and practice of IPPC: case studies from the UK and Hungary and the implications for future EU environmental policy', *European Environment* 7(1): 1–6.

Enyedi, G. and Zentai, V. 1986, *Environmental Policy in Hungary* (Pecs: Centre for Regional Studies of the Hungarian Academy of Sciences, Discussion Papers 2).

Fesus, I. 1992, *Interaction Between Agriculture and Environment in Hungary* (Cambridge: World Conservation Union).

Fitzmaurice, J. 1996, *Damming the Danube: Gabčikovo and Post-communist Politics in Europe* (Boulder, CO: Westview Press).

Flaherty, M. (ed.) 1993, *An Environmental Brief: Privatization and Environment in Hungary: A Case Study* (Boston, MA: Tufts University, Center for Environmental Management).

Galombos, J. 1993, 'An international environmental conflict on the Danube: the Gabčikovo–Nagymaros dams', in A. Vari and P. Tamas (eds) *Environment and Democratic Transition: Policy and Politics in Central and Eastern Europe* (Boston: Kluwer), pp.203–5.

Goda, P. 1997, 'Sustainable development related to water management in the Körös Region', *Alföldi Tudományos Intézet (Békéscsabai Osztaly) Alföldi Tanulmányok* 16: 17–34 (in Hungarian with an English summary).

Green, J. 1998, *Energy and the Environment in the EU* (Brussels: FoEE/EEB).

Gurzo, I. and Marton, J. 1997, 'The possibilities and limits of sustainable development in the agriculture of Bakes County', *Alföldi Tudományos Intézet (Békéscsabai Osztaly) Alföldi Tanulmányok* 16: 65–76 (in Hungarian with an English summary).

Gyulai, I. 1998, *The European Union and Biodiversity* (Brussels: FoEE/EEB).

Hanf, K. and Roijen, M. 1995, 'Water management networks in Hungary: network development in a period of transition', *Environmental Politics* 3: 168–96.

Hempen, S. 1998, *Training Materials on Waste Management* (Brussels: FoEE/EEB).

Hinrichsen, D. and Lang, I. 1996, 'Hungary', in F. W. Carter and D. Turnock (eds) *Environmental Problems in Eastern Europe* (London: Routledge), pp.89–106.

Hinrichsen, D. and Enyedi, G. (eds) 1990, *State of the Hungarian Environment* (Budapest: Statistical Publishing House).

Horváth, E. (ed.) 1986, *A környezet állapota és védelme* (Budapest: Központi Statisztikai Hivatal).

Hungarian Commission on Sustainable Development (HCSD) 1994, *Hungary: Towards Strategy Planning for Sustainable Development: National Information to the UN Commission on Sustainable Development* (Budapest: HCSD).

Kerekes, S. 1993, 'Economics technology and environment in Hungary', *Technology in Society* 15: 137–47.

Kerekes, S. and Kiss, K. (eds) 1998, *Termeles piac termeszeti kornyezet* (Budapest: Hungarian Academy of Sciences).

Környezetvédelmi és Területfejlesztési Minisztérium 1997, *Adatok hazánk környezeti állapotáról* (Budapest: Környezetvédelmi és Területfejlesztési Minisztérium).

Lang, I. 1978, 'Hungary's Lake Balaton: a program to solve its problems', *Ambio* 7: 164–8.

Lehoczki, Z. and Balogh, Z. 1997, 'Hungary', in J. Klarer and B. Moldan (eds) *The Environmental Challenge for Central European Economies in Transition* (Chichester: Wiley), pp.131–68.

Lodner, D. 1991, 'Local self-governments and environmental protection', in G. Horvath (ed.) *Regional Policy and Local Governments* (Pecs: Centre for Regional Studies), pp.127–37.

Maier, P., Kowalzig, J. and Rostock, S. n.d. *Billions for Sustainability? EU Regional Policy and Accession* (Brussels: Bund/Friends of the Earth).

Ministry for Environment and Regional Policy (MERP) 1991, *National Report to the UN Conference on Environment and Development* (Budapest: MERP).

MERP 1995, *National Environmental and Nature Conservation Policy Concept* (Budapest: MERP).

MERP 1996, *The Quality of Waters in Hungary 1995* (Budapest: MERP).

O'Toole, L. and Hanf, K. 1998, 'Hungary: political transformation and environmental change', *Environmental Politics* 7: 93–112.

Perczel György, G. 1992, 'A gazdaság és a környezetszennyezés néhány összefüggése Magyarországon', *Földrajzi Közlemények* CXVI(1–2): 39–44.

Perczel, G. (ed.) 1996, *Magyarország társadalmi-gazdasági földrajza* (Budapest: Eötvös Kiadó).

Perczel, G. and Reiniger, R. 1995, 'Municipal solid waste – hazardous waste – air pollution – social conflicts relating to the environment – pollution at former Soviet military sites', in Geographical Research Institute (ed.) *National Atlas of Hungary*, Supplementary map lift-out series, Part 5 (Budapest: Geographical Research Institute, Hungarian Academy of Sciences).

Reeves, A. D. 1993, 'The present state of the Hungarian environment', *Area* 25: 141–3.

Retvari, L. 1994, 'Natural resources and the environmental problems of their utilization in Hungary', *GeoJournal* 32: 337–42.

Ruda, G. 1998, 'Rural buildings and environment', *Landscape and Urban Planning* 41: 93–7.

Santha, A. 1998, *Sustainable Agricultural Development in the Region of Lake Balaton* (Pecs: CRS, Discussion Papers 24).

Suli-Zakar, I. 1992, 'A study of state borders as factors blocking socioeconomic progress in northeastern Hungary', *Geographical Review* 40: 53–64.

Szepesi, A. 1997, 'Forest health status in Hungary', *Environmental Pollution* 98: 393–8.

Szirmai, V. 1993, 'The structural mechanisms of the organization of ecological-social movements in Hungary', in A. Vari and P. Tamas (eds) *Environment and Democratic*

Transition: Policy and Politics in Central and Eastern Europe (Dordrecht: Kluwer), pp.146–56.

Tirczka, I. and Ferencsik, I. 1998, 'Establishment of crop production database for natural regions and its role in cropping', *Landscape and Urban Planning* 41: 99–105.

Varga, P., Abraham, M. and Simor, J. 1990, 'Water quality of the Danube in Hungary and its major determining factors', *Water Science Technology* 22: 113–18.

Vastag, G. 1994, *How Corporate Executives Perceive Environmental Issues: Comparing Hungarian and Global Companies* (Chapel Hill: University of North Carolina, International Private Enterprise Development Research Center).

World Bank 1990, *Hungary: Environmental Issues* (Washington, DC: World Bank Background Paper).

World Bank 1993, *Financing Water and Sewerage Investment in Hungary* (Washington, DC: World Bank Discussion Note).

9 Poland

F. W. Carter and Ewelina Kantowicz

Introduction

The past decade has witnessed a dramatic increase in environmental concern in Poland (Figure 9.1). Since 1990 there has been greater freedom of access to previously unobtainable information and the development of an ecological movement which has influenced considerable changes in public opinion. One of Poland's early post-communist priorities during the transition to a market economy was to tackle its environmental problems. Environmental protection has been given a much higher pre-eminence, due to serious problems inherited from the communist past with air pollution, inadequate water treatment and contaminated soil. Untreated industrial discharges, unmarked toxic waste dumps and unacceptable water supplies are found throughout the country. The early efforts were focused on specific chronic environmental cases; later endeavours were concerned with the provision of more long-term solutions linked to sustainable environmental management (Clarke and Cole, 1998).

The Polish government itself proved the main stimulus in trying to overcome the ecological crisis. This involved setting into motion major conceptual changes for policy implementation at the national level, through establishing environmental goals and the creation of an institutional support system. In conjunction with these developments, subnational policies were introduced at the regional and local levels. The lower administrative levels have become increasingly important as greater emphasis in Polish society has been placed on environmental decision-making (Warner, 1999). The aim of this chapter is to discover how much progress has been made in solving Poland's environmental problems during the 1990s. This will involve some appreciation of state environmental policy, the impact of major polluting factors, the role of biodiversity and nature conservation and character of environmental movements in the country.

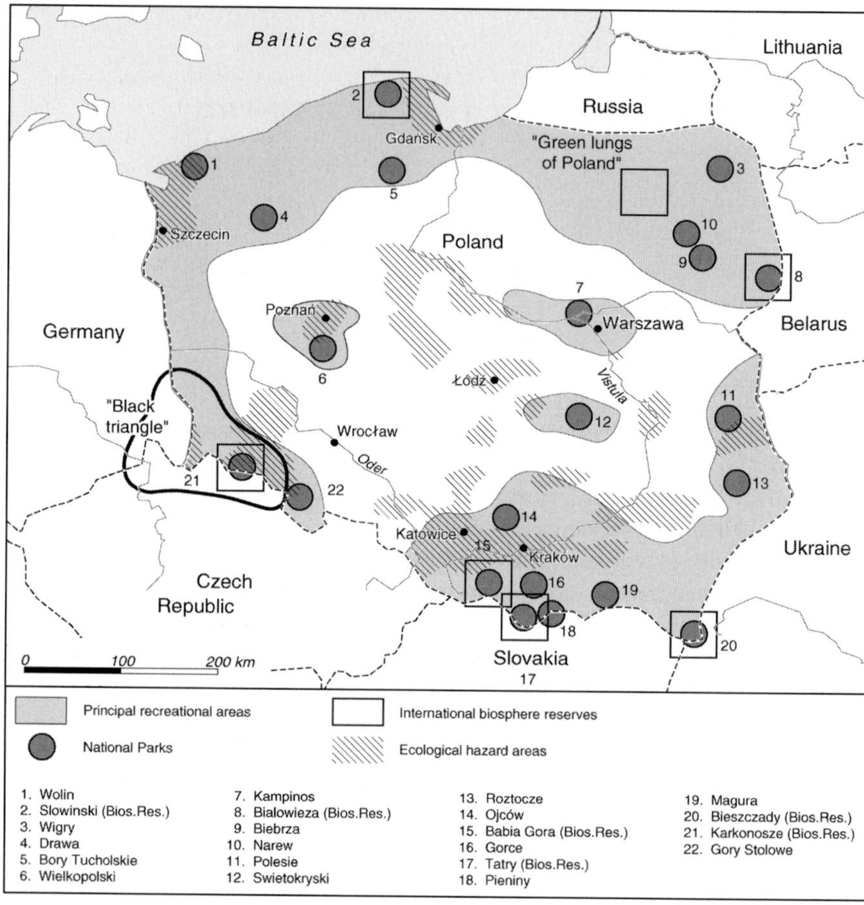

Figure 9.1 Salient features of Poland

Source: Nowicki 1992; Biosphere Reserves in Poland 1994; Stan srodowiska w Polsce 1993.

State pollution policy

Changing priorities in environmental protection

The National Environmental Policy (1991) obliged the state to solve the problems of environmental protection within the principles of 'sustainable development'. This gave Polish citizens the right to proper information about the quality and protection of its environment. At the beginning of the 1990s, Poland faced the very difficult task of reshaping its environmental policy and adapting it to radical changes in the political and economic system. To a large degree, degradation of Poland's environment was related to inefficiency in the former central planning system: wasteful use of resources, a bias towards heavy industry and an ageing, inefficient, highly polluting capital stock. Economic and structural reform was thus essential to encourage a more friendly environment. There was also a need for the new policy to depart from what was once a narrow understanding of environmental protection, to a broader goal of sustainable development. In 1990, Poland's Parliament adopted sustainable development as an underlying principle for socio-economic policies. Several tasks were set; first, there were short-term priorities envisaged for completion by 1994 to 1995, due to danger to human health. Second, medium-term priorities, for completion by 2000, which would enable Poland to move closer to EU environmental standards. Third, long-term priorities aimed for 2020 and relating to the sustainable development principle for the whole economy. These included improving management efficiency in the energy, water resources and mining/mineral resource sectors, industrial restructuring and protection of natural resources (National Environmental Policy of Poland, 1991).

In 1992 Poland agreed to implement the Rio Declaration. This would be developed through EC directives, OECD decisions and UN-ECE conventions. It required amendments to Polish laws, which had been drafted when information was more strictly controlled and democracy less developed. In particular, it meant that information on pollution emissions and inventories of toxic substances from industrial plants, whether publicly or in the process of being privately owned or being privatised, should no longer be confidential. Poland is a member of the Commission on Sustainable Development concerned with the domestic implementation of Agenda 21. In 1997, the national report to the Commission summarised the period 1992 to 1996; it was accepted together with an elaboration of the new environmental policy begun in 1998. Results from this first environmental policy may be summarised in two ways: an improved state of the environment, and new emerging problems (Agenda 21, 1998).

Improved state of the environment

Evidence shows that Poland's environment improved in the 1990s, with a decrease in air pollution, an increase in protected areas, and a decline in the excessive use of energy and natural resources. There was also an increase in the population of endangered species (e.g. beaver, aurochs), and a decline in the defoliation coefficient in forested areas. By 1992 Poland had already met its medium-term quantitative goals for reducing air pollution emissions for 2000. Energy intensity was very high in the Polish economy. Among the main causes were heavy industry, low energy prices and production inefficiencies under central planning. Between 1989 and 1992, energy consumption fell considerably, mainly caused by an economic slump. Total energy consumption decreased by over one-fifth (22 per cent), electricity consumption fell by more than 17 per cent, although domestic production dropped only about 9 per cent while electricity imports were severely curtailed. Consumption of residential district heating fell by about one-tenth. Nevertheless, Poland's energy intensity did not improve between 1989 and 1992.

Since 1996, however, Poland has adopted a new energy policy, and in 1997, Parliament passed the Energy Law Act. Energy policy up to 2010 includes creating markets, ensuring supply security, providing proper environmental protection, increasing efficiency, providing high-quality supplies to the residential sector, and freeing prices to rise towards European levels.

Poland's system of protected areas has been progressively enlarged. The system is based on the international IUCN classification and consists of nature reserves, national parks, landscape parks and areas of landscape protection. An act in 1991 gave regional and local authorities power to designate three further types of protected area; namely ecological areas, documentation sites and natural landscape zones. These new categories allow for the protection of sensitive habitats in private ownership that are too small to be designated under standard categories. The pine stands showed the greatest improvement among the coniferous trees, while beech and birch stands were the best among deciduous varieties.

New environmental problems

New attitudes have emerged in Polish society towards the existing limits of natural resource use, as well as greater consciousness/deeper awareness of environmental risks. For example, according to sociological studies, the percentage of people accepting the role of local government in solving environmental problems increased from 63 per cent to 67 per cent between 1992 and 1997. There are about 1200 organisations involved in environmental activity. Conversely, Polish society is slowly beginning to feel the impact of a market economy and this is creating new social, economic and also environmental problems. Excessive and unbalanced consumption is already present; for example, there is a growing amount of waste and vehicle emissions, especially in large cities. Currently, the poorest social strata in Poland are buying new cars at the fastest rate. This is unique worldwide, given the restricted financial capabilities of Poland's inhabitants where the new car purchase index should be two to three times lower. There is a lack of critical reflection over the consumption effects on the environment, energy and space. Education will be a key factor in the choice between building a highly developed and over-consuming society or taking advantage of underdevelopment in search of a society based on 'sustainable development'. Education is a useful tool for environmental protection. About one-third of Polish society comprises better educated/younger people and city residents and forms the potential for a group supporting sustainable development. Education takes first place among the eight main implementation methods of Polish policy towards sustainable development. The others include adoption of new legal acts with a clear division of responsibility between the state and local bodies. This involves the introduction of cost–benefit analysis on natural resources, creation of environmental charges, physical planning and rational transportation, better use of soil potentiality to augment food production, and increased afforestation from the present 28 per cent to a minimum of 30 per cent.

Pollution problems

In 1996, the fund for protection of the environment and water economy, contributed from national, regional and local sources, totalled more than Zl1.4 mld. The largest amount (93.3 per cent) was allocated to pay for the country's ecology, of which 46.5 per cent was payment to repair air pollution damage, 19.9 per cent for sewage disposal, 13.4 per cent for ameliorating water quality, and 13.5 per cent on water storage. Fines levied by the state for failure to uphold environmental protection laws comprised 4.7 per cent of the total, of which 3.6 per cent was related to above-norm untreated sewage. Clearly, the major investment was for methods of improving air quality.

Air pollution

By 1990, air pollution in Poland was exceptionally excessive, reaching one of the highest levels in Europe (Karaczun, 1994; Węcławowicz, 1996). In the early 1990s there was a considerable decrease in SO_2 emissions both in public power production and industrial processes. Between 1988 and 1994, Polish emissions of SO_2 diminished from 4.2 to 2.6mln.t – a decline of 37 per cent (Elvingson, 1997a). Several elements motivated this decline, but the major reason was the difficulties associated with transition to a market economy which checked production. However, by 1995, an improved economic situation meant that Poland was again one of the highest air polluters amongst the OECD countries, with 68 kg SO_2 pc p.a., the sixth highest in Europe (OECD, 1997; Johannesson, 1998). In that year it produced 1.3mln.t of dust, 2.3mln.t of SO_2, 1.1mln.t of NO_x and 330mln.t of CO_2 (GUS, 1997, 1998b).

The four main culprits were the energy sector, industrial processes, transport, and transboundary circulation. The energy sector was the leading offender in SO_2 and NO_x emissions (Table 9.1). Within the energy sector there are three main polluters: public power production, industrial power generation and district heating. The large thermal electric power-stations and the numerous smaller plants emitted over three-quarters (76.1 per cent) of all SO_2 into the atmosphere

Table 9.1 Poland: air pollution sources, 1996 (th tons/%)

Sources	*SO_2*	*NO_x*
Public power production	1195 (50.5%)	360 (31.2%)
Industrial power generation	406 (17.1%)	128 (11.1%)
Industrial processes	200 (8.5%)	118 (10.2%)
District heating and small stationary sources	521 (22.0%)	131 (11.4%)
Motor vehicles	46 (1.9%)	417 (36.1%)
Total	2368 (100.0%)	1154 (100.0%)

Source: Mały Rocznik Statystyczny, 1998 (Główny Urząd Statystyczny).

in 1996. The reasons for these high emissions stem from Poland's heavy reliance on coal combustion for electricity production and heating (Salay, 1996a). This is in spite of a general restructuring of the power plant sector following the more rigid restrictions on air pollution emissions after 1991 when power plant operators began using lower sulphur content coal with a higher heat value, and an improved efficiency within power plants (Bilkowski, 1995; Salay, 1996b). Polish research has also concentrated on the development of methods to meter dust particulate concentrations and liquid/noxious substances in flue gases (Anon., 1995a; Wojciechowski, 1995; Zawitaj, 1995).

Various solutions have been proposed for reducing these emission levels. One of the more popular recommendations is conversion to gas, a much cleaner energy source for modern central heating systems, and cheaper to use than coal (Pudlis, 1995a; Karaczun, 1997). Dwellings with gas conversion could also be fitted with metering systems. The consumer could then become more accountable for the amount of energy consumed and metering units would allow customers to budget properly for their energy costs. This may be further helped by the introduction of cellular concrete in the construction industry, which has a very high level of heat insulation (Anon., 1995b). Another recent occurrence has been efforts to stimulate the sale of energy-efficient lamps to Polish consumers. Since 1995, Poland's small Global Environment Facility (GEF) (Pudlis, 1995b) has been subsidising the price of compact fluorescent lamps (CFLs) by 30–50 per cent as part of the Poland Efficiency Lighting Project (Bialic, 1995). Sales of CFLs have increased by almost half, saving consumers nearly Zl240mln (US$79mln). The programme has also reduced CO_2 emissions from Polish power plants by more than 0.5mln.t (Anon., 1997a). Another project began in 1994, partly financed by the World Bank, to make greater use of geothermal energy sources in Poland. The first stage aims to provide 15,000 consumers in four towns of the Podhale region of southern Poland with enough geothermal energy for heat and hot water over the period 1995 to 2000 (Anon., 1995c).

In 1991 the National Environmental Policy introduced short- and medium-term policies to reduce air pollutants and improve the industrial environment. This appears to have been more successful for NO_x than for SO_2, thanks perhaps to the introduction of low spin-type emission burners (Anon., 1995d). By 1996, these figures had declined to 8.5 per cent for SO_2 and 10.2 per cent for NO_x. In 1989, industrial power generation and processes together reached an alarming 47.5 per cent of total particulate emissions which, in 1996, had risen even further to 49.8 per cent.

Industrial activity is largely located in a few specific areas of the country, most notably the Silesian region which covers 6650km^2 with ten urban centres, and a total of 4mln inhabitants (Mazurski, 1994; Pyka, 1996; Czaja *et al.*, 1996). The region accounts for all of Poland's hard coal and zinc production, half of its steel production and one-third of its coke (OECD, 1994). It is even more shocking when it is noted that one-third of the country's twenty-nine major industrial dust and gas producers are located in the Upper Silesian industrial region. For example, Chorzów in the Katowice region is one of the most polluted cities in Poland. In 1997, the former Katowice voivodship (province) alone produced one-fifth (21.8 per cent) of Poland's particulates and a quarter (24.7 per cent) of its gas emissions; it also contained one-tenth (10.1 per cent) of the country's population with a density of 588 persons per km^2 (Slocock and Sowiński, 1996; GUS, 1998b).

The transport sector produces about one-third of all CO_2, CO, NO_x, hydrocarbon and lead emissions in Poland (Anon., 1997b). Furthermore, accidents in the transportation of highly dangerous industrial materials by road, rail and broken pipelines affect two-thirds of Polish territory, containing 20–25mln inhabitants. One-fifth (20.3 per cent) of the total occurred in the former Gdańsk, Bydgoszcz, Bielsko, Katowice, Opole and Częstochowa voivodships, and another tenth (9.9 per cent) in Tarnów and Krosno (Forowicz, 1997). However, motorised vehicles are the main polluters, with urban areas producing half of the enveloping NO_x concentrations. Part of the explanation is that Poland remains one of Europe's liveliest car markets; registered private car ownership has risen from fourteen per 100 inhabitants in 1990 to twenty-two per 100 in 1997 (GUS, 1998b). Part of the reason for this increase is the rapid emergence of an economically viable elite capable of buying high-quality vehicles on the world market. This phenomenon has been further supported by the expansion of private banks and lending societies willing to loan money to employed customers for car purchase. This is part of the new culture in Poland, whereby car ownership is seen as a sign of status indicating wealth and mobility. However, from a pollution viewpoint, the problem is mainly exacerbated by the purchase of older cars exported to Poland from the West in the early 1990s. A considerable proportion of these vehicles had failed to comply with lead emission standards in Western Europe, but could still be bought openly in Poland. In 1996 a more regulated system emerged: cars without catalytic converters would henceforth not be registered and included reducing the lead content in petrol to 0.15 grams per litre and sulphur content to 0.2 per cent in weight. The older imported cars, lacking conversion possibilities, were subject to a higher tax. This particularly affected the heavily polluting communist-designed cars, such as the former East German-produced Trabant and Wartburg.

Transboundary air pollution is another serious problem facing Poland (Marszał and Nalewajko, 1995). Environmental hazards do not respect international boundaries and air pollution contributes greatly to ecological degradation. South-west Poland on the borders of former Czechoslovakia and East Germany lies within the notorious 'Black Triangle' industrial area, still one of the most polluted areas in Europe (Anderberg, 1999). In the early 1990s, the extent of deposition from beyond Poland's border was considerable. For example, in 1992, 52 per cent of SO_2 and 80 per cent of NO_x accumulated over Poland were considered to have arrived from outside the country. Even so, Poland herself exports large amounts of transboundary pollution especially to the FSU. Poland is also guilty of exporting more than importing some very harmful pollutants such as arsenic (As) and cadmium (Cd) to neighbouring countries.

Water pollution

Poland's adverse water balance is particularly alarming with disposable reserves totalling a mere 22 km^3 p.a. In many parts of southern and southwestern Poland it has assumed catastrophic proportions, aggravated by a high degree of surface water contamination. The water surface in Poland covers 8265km^2, or 2.6 per cent of its territory; it is dependent on large areas of wetland, marshes and peat for cleansing purposes, such as the biggest, the Biebrza Marshes, totalling some 1400km^2 (Wiśniewski, 1993). Observation of major rivers has revealed a growth in sewage discharge, toxic impurities and the release of pollutants increasingly resistant to self-purification processes. Over three-fifths of Poland's river length contains water that is too contaminated for municipal purposes and two-fifths is unsuitable for industrial use. Large rivers suffer most pollution, especially the Vistula, Oder, Warta and Bug. In 1996, only 4.8 per cent of rivers were in the Class I category, compared with 32 per cent in 1964 when river analysis first began. Over the same period rivers with excessive pollution rose from 29 to 33 per cent (GUS, 1998b). Three-fifths of the country's 500 lakes have become seriously degraded due to sewage and industrial waste contamination, as well as airborne pollutants. Of these, 130 lakes are now monitored with the aim of observing regional differences and their causes (Maciążek, 1995). The impact of all this pollution has meant increased water treatment supply costs both for the population and industry while fish stocks have declined, ecosystems have been degraded, and irrigation for agricultural and horticultural purposes has been affected. Water impurity has also led to the prohibition of bathing on certain beaches and to outbreaks of disease among animals and humans.

Sewage disposal is a major problem with about one-third discharged into surface waters without any treatment at all. In 1997, sewage cleansing facilities served less than half the population (47 per cent) compared with 90 per cent in West European countries. The urban/rural comparison in Poland reveals that while 72 per cent of urban areas have sewage treatment possibilities, the villages, with 38 per cent of the population, have only 6 per cent (GUS, 1998b). Moreover, 150 urban centres and about two-fifths of economic enterprises do not possess sewage purifying centres. Even in the capital, Warsaw, on the badly polluted Vistula River, where drinking-water meets neither Polish nor international standards, sewage treatment plants are outdated (Pudlis, 1995c), but plans are afoot to build the capital's first rubbish-burning plant (Chełstowski, 1997). In Gdańsk, 25,000 households do not have proper connections to the sewerage system (Hanneberg, 1996). Only 115 urban centres in 1997 had sewage plants equipped with the latest cleaning methods for increasing the reduction of azote and phosphorous. These plants process 216mln.cu.m of sewage, or 13 per cent of the country's total, compared with between 50 and 90 per cent in West European states.

Another source for cleaning up Poland's environment has come in the form of foreign aid. In 1994, as part of the PHARE programme, Poland obtained Euro30mln to finance eight programmes concerned with 'the development of agriculture and the rural terrain' (EUC, 1993). One of its most successful projects has resulted in a three-year pilot study for limiting water pollution in agriculture at a cost of Euro1.3mln. A Polish/Austrian joint-venture company was responsible for constructing wells, organic and liquid animal manure containers and sanitary appliances on farms located in eleven chosen rural parishes (PHARE, 1997). Knowledge obtained from this project will now be used on a much wider scale for other contaminators of the food chain such as ammonia (Johannesson, 1998).

One side-effect of the energy pollution problem has been increased research into underground geothermal water sources, which have been located in a large swathe from the western Baltic coast to the southeastern border. The largest sources have been discovered in the Polish lowlands (Szczecin–Łódż–Warsaw) located at 2800 to 3800m below ground level. The calorific value of the geothermal waters is estimated to be equivalent to 610t of crude oil or 1.2mln.t of coal. In the long term, once energy from geothermal waters is used more widely in Poland, it is expected to reduce coal consumption by some 300,000t and decrease air pollution levels by 75–80 per cent (Anon., 1995b) and eventually to provide between 2 and 5 per cent of the country's energy balance. In the Baltic, which drains 99.7 per cent of Poland's territory, sea water impurity is a serious problem. This is further complicated by the inflow of organic land matter from Poland's communal, agricultural and industrial sources with the deposition of such toxic substances as heavy metals, pesticides, phenols and detergents. Poland has a long Baltic coastline (528km) and in the early 1990s it was estimated to contribute the largest volume of pollution. It annually deposited 222,000t of nitrogen compounds and 22,000t of phosphate compounds into the Baltic Sea. In 1991, twenty-seven beaches were closed through pollution and bathers were cautioned about pollution dangers at twenty-five other beaches. Near Gdańsk, human waste bacteria in the sea water were equivalent to one-seventh of the amount of untreated sewage.

Soils and vegetation

Considerable depletion of any part of a country's natural wealth leads to future problems (Witek, 1993). In Poland over the past decade, an average of 6000ha of land a year has been legally converted to non-agricultural/forestry purposes. In 1997, 1200ha of land was transferred, of which 83 per cent (1000ha) was agricultural land; three-quarters (77.4 per cent) of this contained the best soil (i.e. classes I–III), 18.3 per cent Class IV and 4.3 per cent the lowest class. Over 30 per cent of this appropriated agricultural land was utilised for building housing estates (GUS, 1998b). One of Poland's problems is soil erosion, which is thought to have contributed to the flooding in the southwest of Poland during July 1997, when more than 4000sq.km were inundated, followed by loss of life and temporary evacuation (Jankowski, 1997). This process (through rain-wash, gullying and eduction) spoils the soil's surface composition. The main causes are eolian winds, precipitation, gravitational water mass movement and arable cultivation, all leading to surface and gully erosion. In Poland by 1997, wind erosion had affected 8843th.ha (28.2 per cent of the surface total of the country); surface water erosion had caused damage to 8744th.ha (27.9 per cent), while gullying had ruined 5685th.ha (18.2 per cent). Another specific form of this phenomenon is footpath erosion. Increasing numbers of visitors are spending time in the country's national parks, either for summer walking or winter sports. Footpath erosion is becoming serious in the Tatras, because of its small area and easy accessibility. Severe overcrowding can often occur, particularly in the higher areas, which can attract 3–4000 visitors daily thanks to cable-car facilities. Some tourists forsake marked tracks (theoretically forbidden) and produce new footpaths, leading to severe erosion as well as litter and noise (Wiszka and Hindson, 1991).

Air and water pollution damage crops and reduce soil fertility through excessive salinity. In 1997, Poland had 75.6th.ha (0.2 per cent of the country's territory) of devastated and degraded land, but the location pattern varied considerably. Over half of Poland's surface area containing such land was situated in nine of the more industrialised voivodships, namely Bydgoszcz, Chełm, Ciechanów, Jelenia Góra, Katowice, Konin, Opole, Płock and Tarnobrzeg. In the same year, land improvement was achieved on 3291ha and recultivation on a further 2206ha, mainly returned to forestry and for agricultural use (GUS, 1998b).

Poland has a smaller woodland index (28.4 per cent in 1997) than the European average (31 per cent). In forestry, part of the problem has been related to the inappropriate planting of certain tree varieties, together with an uneven age structure, which has resulted in too many young forests (Szujecki, 1995). On the basis of long-term forecasts, attempts are now being made to promote a forest cover at least equal to 30 per cent of Poland's territory. Polish forests have suffered greatly from exposure to heavily polluted air; SO_2-induced acid rain has caused serious forest damage and a disastrous woodland situation. This has been reflected in a Polish forest-damage survey of all tree species with more than a quarter defoliation (i.e. classes 2–4) (Table 9.2). Maltreatment of forests covered over 400th.ha; degradation was particularly noticeable in areas such as the Upper Silesian industrial region, while transboundary airborne pollution from the former Czechoslovakia and East Germany had resulted in the disappearance of a number of forests both in the Sudetan and Carpathian Mountains (Rytkowski, 1993). It was noticeable that in the southernmost parts of Poland, where the forest situation was worst, trees were definitely more susceptible to drought than those in the north, suggesting that air pollution increased their drought sensitivity (Elvingson, 1997b).

Table 9.2 Poland: forest damage, 1991–97 (%)

1991	*1992*	*1993*	*1994*	*1995*	*1996*	*1997*
45	49	50	55	53	40	37

Sources: Elvingson (1997b, 1998).

Environmental legislation

Detailed legal regulations are covered by more than 1000 acts, decrees and executive orders of which the Environmental Protection and Management Act passed in 1980 and amended in 1989, 1990 and 1993 is perhaps the most important. It governs environmental protection up to the present with specific legal sections on the air, against waste and non-ionising radiation, as well as the prevention of noise/vibration and conservation of rural and urban flora and fauna.

In the organisational context the administrative structure at the national level has been changed. In 1999, administrative reform reduced the number of voivodships from forty-nine to sixteen, and introduced new municipal governments run by elected councils. Several environmental responsibilities were moved from central to regional and local levels. The implementation of national policy takes place at these provincial and local levels. Each voivodship has a Department of Environmental Protection. It issues permits and licences and sets maximum allowable emission and discharge levels, collects environmental fees and certain nature conservation measures are implemented. In addition, voivodships have Inspectorates of Environmental Protection, Sanitary Inspectorates and State Forest Boards. The main environmental responsibilities at the local level (2121 municipal governments) include municipal water supply, waste water treatment and solid waste management; they also control local land-use planning and management, monitoring of environmental regulations and the issuing of location permits for local investment projects.

Poland has adopted about fifty conventions dealing with environmental issues at international level. Thus environmental policy is being increasingly integrated with OECD and EU countries. Major efforts to incorporate environmental concepts developed abroad into its own national law are also being made. Closer ties have been established with neighbouring countries and with regional international organisations. Changes over the past decade have been very significant and should have a positive influence on environmental protection. International environmental policy is strongly oriented towards solving acute air and water problems, part of which threaten the health of the Polish population. Attempts are being made to adapt Polish governmental structure and its environmental laws to European and international standards and legal norms. The conventions adopted by Poland concern air and water pollution, climatic change, the Baltic Sea, the ozone layer, biodiversity, industrial accidents, EIA and liability and compensation for international environmental damage.

The problem of waste

In the late 1990s this specific problem has helped create newly emerging environmental problems. These are linked to changes in Poland's socio-economic system. During the transition period most environmental efforts have been concentrated on air and water management. In the meantime problems regarding waste disposal have grown and progress on waste management has been relatively slow.

In the late 1990s it has become one of the state's most urgent environmental policy priorities.

1 Industrial waste accounts for over 90 per cent of Poland's total waste volume. It includes both liquid and solid substances generated during the production process which have no value and are an environmental burden. There has been a significant decline in the tonnage of industrial waste produced annually. Since 1992, it has remained stable or has slightly increased (120–5mln.t). In 1996, 124.5mln.t of industrial waste was produced, at 0.93kg per unit of GDP. This made Poland one of Europe's major industrial waste generators. The main waste sources are from mining (especially coal – 44.42 per cent) along with flotation sludges (23.7 per cent) energy production, basic metal industries (16.3 per cent) and phosphorus–gypsum (3.25). Geographically, they are concentrated in the former Katowice and Legnica voivodships. Another three voivodships – Szczecin, Piotrków and Tarnobrzeg – between them generated more than 3.0mln.t of industrial waste in 1996 (Agenda 21, 1998). Only about half of the industrial waste generated is on average used in the economy, mainly for the construction of dwellings/roads, and land devastation. While the percentage of industrial waste used in the economy increased from 50.2 per cent in 1985 to 64.4 per cent in 1997, its accumulation in landfill sites is still the main option for industrial waste management teams. Total accumulation of industrial waste dumped in landfill sites in 1996 was 54.8mln.t; land used for industrial waste disposal totalled 11,400ha. The decreasing capacity of landfill sites remains a pressing issue. Only 0.3 per cent of industrial waste is actively treated due to lack of incinerators (Agenda 21, 1998).

2 Hazardous waste was defined up to 1998, when The Waste Act came into force, in two separate ways. The first identified 162 substances divided into four groups ranging from the most (e.g. mercury) to the least hazardous (e.g. coal-processing waste). The second category distinguished three kinds of waste: infectious, radioactive and harmful (e.g. poisonous). Waste disposal in this second category requires special government approval. In 1996, several major types of hazardous waste were identified: slag from the non-ferrous metals industry (23.3 per cent); sediments from mechanical/chemical sewage (11.8 per cent); slag and fly ash from steel industry treatment facilities (11.7 per cent). There is insufficient knowledge about hazardous waste in Poland. This waste is sometimes deliberately under-reported, often being mixed with non-toxic waste by generators for reporting, fees and co-disposal purposes. There is also a fear that with decentralisation some small and medium-sized companies could practise uncontrolled dumping. Municipal waste also contains many harmful substances such as batteries. Hazardous waste from industry totals about 3.5 per cent of all industrial waste forms. Production has decreased from 8mln.t in 1991 to 5.2mln.t in 1997, due to changes in category type.

The collection, safe treatment and disposal of hazardous waste are unsatisfactory. In 1996 about 1.3mln.t (24.6 per cent) were deposited, irrespective of waste type, in landfill sites, sometimes at highly undesirable locations. Treatment for hazardous waste is limited and no central collection points exist. Information on toxic waste sites is insufficient. Another serious environmental problem is linked to exceeding the prescribed time limit for pesticides. They are deposited in so-called 'burial grounds' (mogilniki). More than 300 such places have been reported in different parts of Poland, mostly in the voivodships of Szczecin, Kielce, Piotrków, Gorzów, Poznań, Bydgoszcz and Olsztyn (PIOŚ, 1998).

3 Municipal waste is defined as objects and solid substances directly generated by human beings in sectors such as households and commercial outlets which have no use and are an environmental burden. Domestic sewage accumulated in tanks lacking an outflow is included as well as discarded cars. The composition and amount of municipal waste differ according to settlement size, dwelling house type and heating system. Food and garden waste form nearly two-fifths of municipal waste (38 per cent), with paper/paperboard in second place at one-fifth (18.6 per cent). In addition, it contains glass, metals and other materials, particularly plastic, which has steadily grown over recent years in the composition of municipal waste. The generation trends of municipal waste remained stable from the late 1980s to 1992, but since 1992 there has been a continuous increase; studies predict subsequent increases through an influx to Poland of product packaging (Table 9.3). In 1992, municipal waste totalled 368kg per person p.a. Poland was just below the OECD European average of 370kg/pc p.a. (OECD, 1994). This ratio increased dramatically in 1997 when it reached 776kg/pc p.a. (GUS, 1998a). Almost all the waste was deposited in landfill sites without separation. In 1996, 44.7mln.cu.m of solid and 15.2mln.cu.m of liquid waste went into 884 landfill sites, covering 3058ha. Since 1975, the area of landfill sites has tripled and average waste accumulation has been estimated at 143cu.m/sq.km. In many towns the amount disposed of is already beyond official capacity and one-third of those sites still working should have been abandoned; also in many cases the technical standards are unsatisfactory (PIOŚ, 1998). Owing to the limited

Table 9.3 Municipal waste generation trends, 1990–97

Year	*Million tonnes*	*Million cubic metres*		
		total	*solid waste*	*liquid waste*
1990	12.8			
1991	13.3			
1992	14.1			
1996	29.9	59.9	44.7	15.2
1997	30.0	60.2	44.4	14.8

Sources: Agenda 21, 1998; OECD, 1998.

> capacity of authorised landfill sites up to 10,000 unauthorised dumps exist. Only half of the population have access to waste collection agencies, so proper collection and disposal services are difficult to obtain. In addition, there is strong opposition to new landfill sites and incinerators, especially when waste from another local authority is involved (Szczygielski, 1997).

Waste management objectives were outlined in the 1991 National Environmental Policy document. Of the ten short-term priorities to be completed by 1994 to 1995, only one referred to the waste problem. It aimed at rapid reduction of solid waste through the use of an adequate management system for industrial/municipal solid waste and disposal of toxic wastes. This has been partly achieved, especially regarding industrial waste. Since 1990, there has been a significant decline in industrial waste tonnage produced annually due partly to lower production, but also to waste reduction by some industries and the imposition of disposal fees, which discourage waste generation. Implementation of an adequate management system is more difficult. The amount of industrial waste used in the economy has grown, but it is still much lower than the West European average. However, the share of safe treatment for industrial, hazardous and municipal waste is unsatisfactory. Relatively slow progress reflects the small investment for earth surface protection compared with expenditure for the other environmental components. Earth surface protection expenditure formed only 6–7 per cent of total environmental protection expenses.

Medium-term priorities (completed by 2000) aim at the proper processing and safe storage of all hazardous waste; a 20 per cent reduction of solid industrial waste requiring disposal together with an increase in the waste use rate; the creation of a system of separation and recycling of municipal waste and the introduction of techniques of composting, incineration and biogas production. Long-term priorities for 2020 aim at the introduction of environmentally friendly, modernised manufacturing techniques throughout all production processes, with support being given to clean technology rather than expensive 'end of pipe' cleaning equipment; the repair of environmental damage and creation of a system preventing its repeated occurrence.

By 1994, the Executive Project for the National Environmental Policy had been completed, and waste management investment outlined up to 2000. This included the building of two municipal waste incinerators, each with a daily capacity of 600 tonnes; construction of ten compost plants, each with a daily capacity of 100 tonnes; waste segregation and management systems in twenty cities; building ten new large regional landfill sites, which would meet environmental protection requirements, and a local municipal landfill site system. With reference to hazardous waste, there is a priority need for the reutilisation of fly ash and slag from coal combustion and pyrites tailing generated by the desulphurisation of hard coal. Also needed are the recovery and reutilisation of metals from waste generated by steel mills and metal-refining plants, reuse of phosphorus-gypsum derived from fertiliser production, and the construction of toxic waste incineration plants. Excellent progress has been made in controlling the import/export of hazardous waste using the available authorities and rapid co-ordination. After the discovery of illegal imports of 46,000t of toxic waste in 1989, the Polish authorities tightened controls and strengthened law enforcement. Co-operation between ministries resulted in a virtual clamp-down at borders on all waste. Further, bilateral agreements were signed with adjacent countries providing notification and regulation of waste shipments (OECD, 1994).

In 1997, the Waste Act was passed and provided a comprehensive and consistent regulatory framework for waste management at national, regional and local authority levels and the private sector. It provided definitions for waste/hazardous waste in particular, and the criteria for substance classification. It also included waste disposal and transport requirements, routine measurements and maintenance records by managers, and economic and administrative methods for waste management, and requirements for the international transport and circulation of waste. Since 1992, Poland has approved the Basel Convention on the Control of Transboundary Movement of Hazardous Wastes and their Disposal, and has co-operated with Interpol in controlling the unauthorised shipment of hazardous waste (OECD, 1994).

Polish environmental movements

In spite of many drawbacks a number of ecological movements and even political parties emerged, or intensified their activities, after 1989. In the 1991 parliamentary election, three separate ecological party coalitions took part: the Polish Ecological Party and the Polish Greens Party; the Polish Ecological Union and the Independent Ecological Federation; and the Polish Ecological Party-Greens. On paper, it seemed a great chance for the Green parties to influence governmental policies through parliamentary representation, and to be able to reflect the growing ecological fears of Polish society. Unfortunately, poor electoral campaigns and infighting amongst the Green parties meant they failed to gain any seats, but did obtain 251,000 votes, namely 2.26 per cent of the total cast.

Besides ecological parties, some 120 national and regional ecological associations/groups came into existence. These organisations covered the broad spectrum of environmental protection, mainly with the aim of reducing the ecological effects feared by the impact of a new market economy. Unfortunately, they have so far not used this opportunity to great effect and, like the ecological parties, have proved to be only of minor significance in Polish political life. Meanwhile, the idea behind official ecological policy was based on the concept of gradual implementation, each phase having clear objectives. The early short-term phase (three to four years) aimed at eliminating those changes to the natural environment which threatened the health and life of the population. The mid-term phase (up to 2000) stressed the need to reverse any adverse tendencies responsible for pollution, enabling Poland to achieve those environmental standards set by the EU for prospective membership. Long-term priorities up to 2030 assume that the safety of society will be achieved through the complete fulfilment of eco-development principles and environmental protection.

These goals have to be achieved through modern pro-ecological production methods, repair of any previous environmental damage, pro-ecological economic restructuring, and incorporation of cultural and ethical values towards the environment (Anon., 1990).

While the contribution of environmental groups towards solving ecological problems has remained limited in Poland, their importance is steadily growing. Arguments behind the environmental benefits of environmental groups in Poland can be highlighted by looking at the achievements of one of its largest environmental organisations. For example, the Upper Silesian region, with its dour environmental predicament, has attracted much NGO attention. Vigorous environmental concern is epitomised in Katowice, headquarters of the Polish Ecological Club (Upper Silesian Branch) since its foundation in 1980 in Kraków. As early as 1992 it was involved in promoting atmospheric protection from pollution (Poborski, 1993), and developed links with most environmental schemes in the region through its thirteen local branches. The club is now a national institution which has forged contacts with many NGO communities abroad. Nationally, it is involved in a number of projects; these either involve the club directly, as with the setting up of a distribution network for selling pollution-free food, or indirectly, through producing a database of public institutions and NGOs for people interested in improving the environmental situation (Slocock and Sowiński, 1996).

Another example is the Institute for Sustainable Development founded in the early 1990s by members of the Polish Ecological Club. It participated in a project completed in 1992 comprising a development strategy for northeast Poland, an area known as the 'Green Lungs of Poland' – the cleanest and least contaminated part of the country (Cygler, 1993). The study was well received abroad and the institute was asked by the Council of Europe to prepare a similar report on the 'Green Lungs of Europe' which included parts of the Baltic Republics, Belarus, Poland, Russia and Ukraine. The institute is an independent, non-governmental and non-profit-making organisation. It focuses on policy issues collaborating with the Polish government at central, regional and local levels, with ecological NGOs as well as the business and development sector. Its main activity is to prepare in-depth reports and reviews, and to organise workshops/seminars, conferences and training courses. It advocates environmental issues to parliamentary commissions and government agencies together with keeping in contact with other NGOs and representation on mass-media environmental discussions. Financial support comes mainly from the USA (e.g. Ford Foundation and Rockefeller Brothers Fund) as well as *ad hoc* grants for specific schemes from diverse Polish sources and from abroad.

By the late 1990s, Poland's various environmental movements remained broadly located yet segmented and split. Unfortunately, scant attention has been given to environmental issues by the major political parties, and there has also been a move away from ecological affairs by the general public, which is often too busy with the everyday economic problems of Polish life (Millard, 1998). Perhaps greatest efficiency and success amongst Poland's numerous environmental groups is to be found at the local level. Here, organised opposition against damage from some local pollution sources in the community has led to the closure or liquidation of the problem. Unfortunately, such successes have little influence on state environmental policy and the broader canvas of national ecological degradation. Even so, environmental movements still have an important role to play in Poland's future nationally, largely through questioning government ecological policy and possibly offering Polish society sensible alternative solutions.

Biodiversity and national parks

Biodiversity protection in Poland is guarded by the 1991 Law on Nature Conservation (Anon., 1994). This has legally demarcated many types of nature protection, particularly national parks, nature reserves, landscape parks, areas of landscape protection and biosphere reserves. Linked through ecological corridors, these sectors provide the basis for a protective system for ecological areas (Tederko, 1997). The aims of this system include, amongst others, conserving ecological processes and securing the ecosystem. It endeavours also to sustain biodiversity, support the survival of species and biocenoses and save Poland's inorganic natural heritage. In addition, besides these noble aspirations, the law established other nature conservation elements, including areas set aside for ecological use, natural-landscape groups and collection centres for documentary material (Denisiuk, 1994). Individual categories advocated in the system of protected areas included in the Nature Conservation Law of 1991 have experienced considerable growth (Table 9.4).

Since the 1991 Nature Conservation Law, the Boards of National Parks have been subordinate to the Ministry of Environment Protection, Natural Resources and Forestry. By the end of 1997, Poland had twenty-two national parks covering a total area of 3054sq.km, or 0.97 per cent of state territory. Three new national parks were added in 1995 including the largest, Biebrza Marshes

Table 9.4 Poland: protected areas, 1990–97 (area/ha; number)

Category	*1990*	*1995*	*1996*	*1997*
National parks	165,933 (17)	270,103 (20)	301,056 (22)	305,401 (22)
Nature reserves	116,952 (1,001)	121,303 (1,122)	128,001 (1,183)	130,379 (1,204)
Landscape parks	1,215,445 (68)	1,971,533 (102)	2,136,468 (106)	2,187,748 (109)
Monuments of nature	– (18,876)	– (26,423)	– (30,205)	– (30,811)
Protected forests	2,679 –	3,312 –	3,354 –	3,235 –

Source: As in Table 9.1, pp.51–53.

(592sq.km), together with Magurski (199) and the Stołowe Mountains (63); two more were established in 1996 – Narwiański (73sq.km) and Bory Tucholskie (47). The law also suggested the possibility of collecting entry fees to the parks, but this has not yet been universally adopted.

Like other sections of the country, Poland's national parks have also suffered from serious environmental degradation (Mazurski, 1997). Considerable deterioration of conditions in the parks can be attributed to local influences which can often be restored at reasonable cost, as in the Kampinoski (385sq.km) and Bieszczady (278) national parks. More serious however, is the practice of intensive agriculture around several villages in the Kampinoski National Park near Warsaw, which upsets nature's normal pattern; also increasing visitor numbers have resulted in extensive trampled areas, as well as the ubiquitous litter. In the Bieszczady National Park unlawful camping is commonplace and often accompanied by unchecked camp fires causing harm both to grass and trees. Most national parks are also subject to the illicit collection of wild fruit and flowers.

Poland's higher national parks suffer seriously from air pollution and modern development. Air pollution continues to be the scourge of national parks (e.g. the Świętokrzyski National Park (7sq.km) near Kielce in central-eastern Poland). Most pollution originates from local/national sources, imperilling the parks' ecology. The major offender is SO_2, but contributions are also made by CO_2, NO_x and such heavy metals as copper, lead and zinc. International pollution affects the Karkonoski and Babiogórski (33sq.km) national parks which suffer from transboundary pollution emitted by industrial enterprises and vehicles in the northern part of the Czech Republic and East Germany. Their pollution emissions have weakened the vegetation and obscured landscape panoramas in these parks.

The need to modernise has resulted from the growth of national and international tourism. For example, in the Karkonoski National Park (55sq.km) the growth of a small urban centre at the park's entrance (Karpacz, with a population of 6000) is now, ironically, endangering the park's ecology (Borys *et al.*, 1994). A similar problem has arisen in Kampinoski National Park where second homes, mainly owned by Warsaw's richer inhabitants, have been constructed. Zakopane (population 28,800 in 1997) in the Tatra Mountain foothills has experienced an urbanisation boom resulting from ski-run expansion in the Tatra National Park (211sq.km), which has placed a serious risk on the park's ecology.

A different scenario faces Poland's lowland parks. They are mainly affected by diffuse, non-point source pollution and depleted ground water-tables, the result of poor management at nearby farms (e.g. at the Poleski National Park (97sq.km) near Lublin in eastern Poland, with the conversion of wetland areas into fields for agricultural use). Other national parks are frequently plundered for their river stones, timber, and any other conceivable valuable resource. It is not unknown for some more unscrupulous visitors to catch or poach some of the national park's rarer species, including eagles and bears.

The 1991 law defined a nature reserve as an area which contained natural or little changed ecosystems, particular species of plants/animals, or immovable elements having significant value for scientific, natural, cultural or landscape purposes. In 1997, 1204 such reserves covered a total of 1303sq.km. Most have been established to prevent the extinction of certain species or change to the landscape pattern judged to be typical of a certain area or region. The most important are the 'strict' nature reserves which enjoy total state protection and are only used for scientific purposes.

In 1992, three new biosphere reserves (Eastern Carpathian – including Bieszczady National Park and the Ciśniansko-Wetliński and San Valley Landscape Parks – Tatra and Karkonosze) were established. These reserves were located in wider protected bilateral/trilateral areas stretching beyond Poland's national frontiers. This was part of a larger programme in which Poland co-operated in the creation/operation of twelve Euroregions. Euroregions, like national parks, suffer many common ecological problems, but on a large transborder scale (Mazurski, 1999). One area particularly affected is the Nysa ('Neisse') Euroregion. Situated within the 'Black Triangle', it has suffered greatly from industrial pollution. But this is not due to the local situation since industrial development is limited to Bogatynia and Zgorzelec apart from the large lignite mine and 'Turów' thermal power station at the former site. Rather the Euroregion has suffered mainly from transborder pollution; in the past this was largely from the German side, but now more seriously from the Czech Republic. This has affected the ecological balance of the Polish sector over a large part of its area (Borys and Mazurski, 1993; Mazurski, 1996a, 1996b; Kulczycka, 1997). Clearly, the major common task is for closer borderland co-operation in environmental research, hopefully leading to a better understanding of the future ecology for this Euroregion.

Natural monuments are registered under separate categories in Poland and their protection is the responsibility of individual voivodships. They largely comprise old, historic trees, stands of trees, erratic glacial blocks, etc. Between 1990 and 1997 their number increased by 11,935 (58 per cent) from nearly 19,000 to over 30,000. However, the percentage of 'individual trees' grew to over three-quarters of the total at the expense of all other categories, in spite of their greater numbers. This perhaps reflects the cheaper cost of protecting individual trees for voivodship budgets in comparison to the other, more complicated groups.

The hectarage of protected forest in Poland rose by one-fifth from 2679ha in 1990 to 3235 in 1997 (20.8 per cent). The need for forest protection, however, is growing. Poland's policy has been prompted by the country's population increasing faster than its timber resources, in spite of greater reafforestation; increasing social demand for recreation areas, putting more pressure on forest resources; the country's changing climate and greater soil contamination/acidification threatening greater eutrophication, leading to negative effects on biocenosis and forest ecosystems; the need for better adaptation of Polish forests to change if local species are to be preserved; and a rise in forest management costs, especially for resource protection, leading to self-financing problems (Szujecki, 1995).

Conclusion

Poland has some of the most ecologically valuable ecosystems in Europe, but they are complemented by some of the worst areas of pollution. Even so, certain regions still possess a very high degree of undamaged ecology and biodiversity. In fact, ecologists have estimated that 8.5 per cent of state territory exhibits nearly pure, uncontaminated environmental conditions. Another 19 per cent contains biological complexes that are run on solid, sustainable and ecological principles. Thus, over a quarter of Poland's area embraces a resource that would be the envy of many industrial nations. Nevertheless, there is still much to be done, but it is gratifying to learn that recently the EU gave a positive response to Polish efforts in the environmental field (Anon., 1997c; Jesień, 1997). Certainly external pressure, particularly from the EU, will encourage the country to keep on the right environmental track, but if Poland wants to be successful rather than just survive it needs firm government – none more so than in the environmental field.

Since 1989, Poland has made progress in cleaning up industry and improving water and air quality while the restructuring of the coal, steel and energy industries will boost the anti-pollution drive (Anon., 1999a). Use of new technology in recent years has meant a decline in SO_2 by about 42 per cent and nitric oxide (NO) by 25 per cent, as well as a considerable drop in industrial dust emissions and other harmful substances. However, increasing transport use has meant that both vehicular gases and dust continue to be above the EU norm. Half a million metres of untreated sewage are recorded in the River Vistula daily, in spite of the construction of around 300 sewage purification plants annually and a further 1300 currently under construction (Anon., 1999b).

Poland must increase investment to improve its environment. In order to meet EU environmental standards imposed by 2010, Poland must raise its annual spending by US$550mln. Unfortunately, environmental problems have declined as a topic of interest and debate in the Polish Parliament (Sejm). There is also a dichotomy between wishing, on the one hand, to increase its national income and on the other to produce a cleaner environment. If these two aims are to be realised, then there must be greater integration between national plans for the political economy and regional plans for more localised development with regard to environmental protection. The Polish Ecofund Foundation, based on exchanging debt for environmentally friendly investments, could be critical in this respect (Pudlis, 1999). It is becoming ever more apparent that the EU will have to grant Poland more time than the 2003 membership entry deadline to meet environmental legislation. About 4 per cent of Polish investment in environmental protection during the 1990s was provided from EU funds, rising to a higher percentage after 1999. Hopefully, Poland's well-known image for its neglect of the environment under communism will finally be dispelled during the first decade of the twenty-first century.

References

Agenda 21 (1998) *Sprawozdanie z realizacji w Polsce w latach 1992–1998* (Narodowa Fundacja Ochrony Środowiska), NFOS, Warsaw.

Anderberg, S. (1999) 'Economic and Environmental Development in the Black Triangle Region', in J. Linnerooth-Bayer, R. E. Löfstedt and G. Sjöstedt (eds) *Transboundary Risk Management* (Risk, Society and Policy Series), Earthscan, London, pp.10–21.

Anon. (1990) *Polityka Ekologiczna Państwa* (Ministry of Environmental Protection), Warsaw, pp.9–11.

Anon. (1994) *Protected Areas in Poland*, Institute of Environmental Protection, Warsaw, 35pp.

Anon. (1995a) 'P-10ZA Automatic Gravimetric Dust Meter', *Poznań Fair Magazine* 15(150), p.36.

Anon. (1995b) 'An Ecological Challenge in the Construction Industry', *The Coastal Times* 10(28), Sopot, p.13.

Anon. (1995c) 'Podhale Geothermal and Environment Project (Poland)', *PPC Project Fact Sheet*, Project Preparation Committee Secretariat, London, 3pp.

Anon. (1995d) 'Low NO_x Emission Burners to Protect Environment', *Poznań Fair Magazine* 15(150), pp.32–3.

Anon. (1997a) 'A Bright Light in Poland', *Network 21* 8, p.12.

Anon. (1997b) *Alternative Transport Policy in Poland* (Institute for Sustainable Development), Warsaw, 36pp.

Anon. (1997c) 'Ability to Assume Obligations of Membership', Ch.3 in *Agenda 2000 – Commission Opinion on Poland's Application for Membership of the European Union* (DOC/97/16 pt.B) (15 July 1997), Brussels/Strasbourg, p.36.

Anon. (1999a) 'Poland Must Speed up Environmental Spending', *New Europe* (1–7 March), p.8.

Anon. (1999b) 'Ekologia-kula u nogi', *Dziennik Polski* (26 February), p.3.

Bialic, A. (1995) 'Energy Efficiency Project: Measuring Snugness by Computer', *The Green Voice. Polish and Regional Ecological Review*, Warsaw (6 August), p.3.

Bilkowski, E. (1995) 'Pro-Ecological Activities in Bełchatów Power Station', *Poznań Fair Magazine* 15(150), p.31.

Borys, T. and Mazurski, K. R. (1993) 'Technical and Economic Possibilities of Limiting Air Pollution in the Border Areas of the "Neisse" Euroregion: Initial Situation Study', *Air Pollution in the Border Areas of the Euroregion 'Neisse'* (Sudety), Wrocław, 23pp.

Borys, T., Mazurski, K. R. and Kachniarz, K. (1994) 'Uwarunkowania środowiskowe rozwoju turystyki w Karkonoszach', in *Turystyka jako element rozwoju gospodarczego regionów przygranicznych*, (Fundacja Rozwoju Demokracji Lokalnej; Centrum Samorządu Lokalnego), Jelenia Góra, pp.8–14.

Chełstowski, D., (1997) 'Na spalonym: Czy na warszawskim Targówku powstanie pierwsza w Polsce spalarnia śmieci?', *Wprost* (16 February), p.50.

Clarke, J. and Cole, D. H. (1998) *Environmental Protection in Transition: Economic, Legal and Socio-political Perspectives on Poland*, Ashgate, Aldershot.

Cygler, M. (1993) 'The Project "Green Lungs of Poland" – An Implementation of the Sustainable Development', *Umwelt in Osteuropa* (Beiträge der Akademie für Natur- und Umweltschutz Baden-Württemberg), Stutgart, pp.66–85.

Czaja, S., Fiedor, B., Graczyk, A. and Jakubczyk, Z. (1996) *Opłaty Produktowe i Depozyty Ekologiczne* (Państwowy Instytut Naukowy – Instytut Śląski w Opolu), Opole, 96pp.

Denisiuk, Z. (1994) 'Poland: Country Profile', *Newsletter: IUCN Central and East Europe* (IUCN East European Programme), 9, 7, pp.3–5.

Elvingson, P. (1997a) 'Polish Emissions: Explaining a Decline', *Acid News* 2, pp.1–4.

Elvingson, P. (1997b) 'European Forests: Still Many Trees Damaged', *Acid News* 4–5, pp.14–15.

Elvingson, P. (1998) 'Damage Has Steadily Increased', *Acid News* 4, p.14.

EUC (1993) *Finance Memorandum P9312*, European Union Commission, Brussels.

Forowicz, K. (1997) 'Nadzwyczajne skażenia przemysłowe: Strefa zagrożenia obejmuje 2/3 obszaru Polski', *Rzeczpospolita* 104(4661) (6 May 1997), p.23.

GUS (1997) *Environmental Protection, 1997*, Główny Urząd Statystyczny, Warsaw.

GUS (1998a) *Statistical Yearbook: The Republic of Poland*, Warsaw.

GUS (1998b) *Mały rocznik statystyczny, 1998*, Główny Urząd Statystyczny, Warsaw, pp.56, 47, 485, 477, 44, 40, 32.

Hanneberg, P. (1996) 'Poland Heading in the Right Direction', *Enviro* 20, pp.10–12.

Jankowski, W. (1997) 'The Reasons for, and Ecological Conclusions from, the 1997 Flood in Poland', *Newsletter: IUCN Central and Eastern Europe* (IUCN, East European Programme), 12(25), p.7.

Jesień, L. (1997) 'Is Poland Ready for EU Membership?', *Studien des Institutes für den Donauraum und Mitteleuropa* (November), Vienna, 29pp.

Johannesson, M. (1998) 'Environmental Space: Making Pollution Personal', *Acid News* 2, pp.1, 3, 20–1.

Karaczun, Z. M. (1994) 'Air Pollution in Poland', Ch. 2 in *Report No. 6: Policy of Air Protection in Poland* (Part III: 1 January to 31 December 1993), Institute for Sustainable Development, Warsaw, pp.10–14.

Karaczun, Z. M. (1997) 'GEF Project on Conversion of Coal-boilers into Gas', Annexe 4 in *Report No. 4: Policy of Air Protection in Poland* (Part IV: 1 January 1994 to 31 December 1996), Institute for Sustainable Development, Warsaw, pp.119–22.

Kulczycka, A. (1997) 'Diagnoza stanu środowiska w Euroregionie Nysa oraz działania na rzecz jego poprawy', *Vědecká Pojednání* III(1), pp.140–8.

Maciążek, A. (1995) 'Science for Environment: Institute of Meteorology and Water Economy', *Poznań Fair Magazine* 15(150), p.22.

Marszał, T. and Nalewajko, J., (1995) 'Environmental Protection in Poland and its International Context', in T. Marszał (ed.) *Planning and Environmental Policy*, Łodż University Press, Łódż, pp.25–37.

Mazurski, K. R. (1994) *Zagrożenia środowiska Dolnego Śląska*, Sudety, Wrocław.

Mazurski, K. R. (1996a) 'Główne problemy ekologiczne Euroregionu Nysa', *Védecká Pojednání* II(1), pp.46–53.

Mazurski, K. R. (1996b) 'Ecological Problems of Border Areas in the Neisse Euroregion and a Trial of its Improvement', in J. Wotte, W. A. Halang and B. J. Kräkier (eds) *Environmental Engineering and Pollution Prevention*, Kluwer Academic, The Hague, pp.131–41.

Mazurski, K. R. (1997) 'Poland's National Parks: Problems and Prospects', *Forum for Applied Research and Public Policy*, Summer, pp.112–14.

Mazurski, K. R. (1999) 'Environmental Problems in the Sudetes, Poland', *GeoJournal* 46(3), pp.271–7.

Millard, F. (1998) 'Environmental Policy in Poland', in S. Baker and P. Jehlička (eds) *Dilemmas of Transition: The Environment, Democracy and Economic Reform in East Central Europe*, Frank Cass, London/Portland, Oregon, pp.145–61.

MOSZNIL (1999) *Ministerstwo Ochrony Srodowiska*, Zasobow Naturalnych i Lesnictwa, Warsaw.

National Environmental Policy of Poland (1991) Ministry of Environmental Protection, Natural Resources and Forestry, Warsaw.

OECD (1994) *Environmental Performance Review of Poland*, Central Statistical Office, Paris, pp.71, 122 (Figure 8.1), 121, 43, 37 (Table 2.2).

OECD (1997) *Environmental Data, 1997*, Central Statistical Office, Paris, p.5.

PHARE, (1997) 'Infrastruktura terenów wiejskich dla gospodarki wodnej: P9312/06', *Opis Programu*, Instytut Melioracji i Użytków Zielonych, Falenty, Warsaw.

PIOŚ (1998) *Stan srodowiska w Polsce*, Raport Panstwowej Inspekcji Ochrony Środowiska w Polsce, Warsaw.

Poborski, P. S. (1993) *Air Pollution in Upper Silesia: Report*, Information Centre for Air Pollution, Katowice.

Pudlis, E. (1995a) 'Small Really is Beautiful – and Effective', *The Green Voice: Polish and Regional Ecological Review* (2 July), p.G1.

Pudlis, E. (1995b) 'Giving Financial Incentives to Energy Conservation', *The Green Voice: Polish and Regional Ecological Review* (31 December), p.G8.

Pudlis, E. (1995c) 'Capital's Plight', *The Green Voice: Polish and Regional Ecological Review* (12 November), p.G4.

Pudlis, E. (1999) 'Environment Investment: Waste Not, Want Not', *Warsaw Voice* (11 April), p.7.

Pyka, J. L. (1996) *Klimat Dolnego Śląska* (Prace Instytutu Geograficznego: Seria C, Meteorologia i Klimatologia, Tom III), Wydawnictwo Uniwersytetu Wrocławskiego, Wrocław.

Rytkowski, K. (1993) 'Lasy', in R. Andrzejewski and M. Baranowski (eds) *Stan środowiska w Polsce*, Państwowa Inspeckya Ochrony Środowisku Centrum Informacji o Środowisku GRID-Warszawa, Warsaw, pp.43–52.

Salay, J. (1996a) *Electricity Production and SO_2 Emissions in Poland's Power Industry* (IMES/EESS Report No.23), Department of Environmental and Energy Systems Studies, Lund University/Lund Institute of Technology, Lund.

Salay, J. (1996b) *Energy Use and Air Pollution in a Transitional Industrialized Economy: The Case of Poland*, Department of Environmental and Energy Systems Studies, Lund University/Lund Institute of Technology, Lund.

Slocock, B. and Sowiński, M. (1996) 'Regional Management of Industrial Pollution in Eastern Europe: Air Quality in Katowice Province, Poland', *European Environment* 6, pp.21–7.

Szczygielski, L. (1997) 'Z czym do Unii Europejskiej. Rozwoj infrastruktury technicznej wsi w perspektywie roku 2010', *Stowarzyszenie Inzynierow i technikow wodnych i melioracyjnych*, Sekcja glowna Inzynierii Sanitarnej Wsi, Falenty.

Szujecki, A., (1995) 'Forests and Forestry in Poland at Threshold of 21st Century', *Poznań Fair Magazine* 15(150), pp.27–9.

Tederko, Z. (1997) 'River Corridors in Poland; An Assessment of the Natural Values of the Odra and Vistula Rivers', *Newsletter: IUCN Central and East Europe* (IUCN, East European Programme) 12(10), pp.11–12.

Warner, J. (1999) 'Poland: The Environment in Transition', *Geographical Journal* 165(2), pp.209–21.

Węcławowicz, G. (1996) *Contemporary Poland: Space and Society*, UCL Press, London, p.86.

Wiśniewski, R. J. (1993) 'Wody śródlądowe w Morzu Bałtyckim', in R. Andrzejewski and M. Baranowski (eds) *Stan Środowiska w Polsce*, Centrum Informacji o Środowisku GRID-Warszawa, Państwowa Inspekcja Ochrony Środowiska, Warsaw, pp.72–9.

Wiszka, A. and Hindson, J. (1991) 'Protecting a Polish Paradise', *Geographical Magazine (Analysis)* LXII(6), pp.1–3.

Witek, T. (1993) 'Gleby', in *Stan środowiska w Polsce*, Centrum Informacji o Środowisku GRID – Warszawa, Państwowa Inspekcja Ochrony Środowiska, Warsaw, pp.58–65.

Wojciechowski, J. (1995) 'Laboratory of Applied Catalysis SWINGTHERM', *Poznań Fair Magazine* 15(150), p.37.

Zawitaj, J. (1995) 'Poland in European Ecological System', *Poznań Fair Magazine* 15(150), p.30.

10 Slovakia

Vladimir Drgona and David Turnock

With an area of just over 49,000sq.km Slovakia is a diverse Central European country, with the high ground of the Carpathians in the northern and central areas grading through hill country to the Danube lowlands in the south (Figure 10.1). The country is well settled and quite highly urbanised, the population of the capital (Bratislava) exceeding 450,000 while large towns of over 80,000 (Banská Bystrica, Košice, Nitra, Prešov and Žilina) account for most of the other seven county centres. Small towns are numerous, especially in former mining areas which are of considerable historical interest. But they are also of importance in connection with mineral and thermal springs which have great value for tourism. Meanwhile Slovakia has almost 2900 villages, though in some mountain areas the rural settlement is highly fragmented. Intensive human activity in the primary sector as well as energy, manufacturing and tourism has generated many conflicts in the way resources are used. Environmental problems have often emerged, especially in the more devastated areas, and these are now being more widely

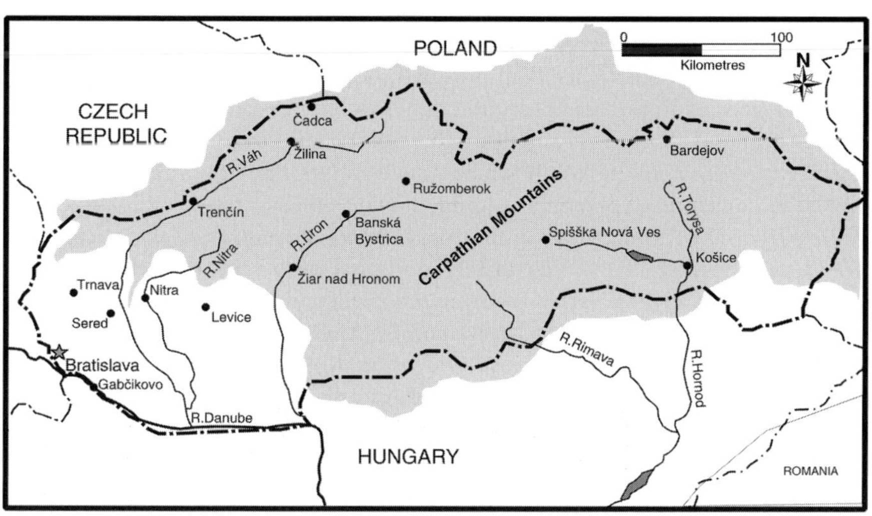

Figure 10.1 Salient features of Slovakia

recognised and addressed in the context of sustainability and with reference to all relevant interests (Huba and Ira 1996): thus reservoirs provide water supplies but also recreation facilities and a habitat for migratory birds. This chapter will consider the damaging effects of industry and circulation both on the physical environment (woodlands and wild life) and human health; while agriculture is in an ambivalent position, being both the victim of industrial pollution and also the cause of erosion and soil degradation in some areas of intensive cropping. As is the case elsewhere in the region there is a need for speedy action to cope with the legacy of socialism but also to regulate today's private enterprise and facilitate EU accession, which requires compliance with the environmental *acquis*.

Air quality

The air is polluted by sulphur dioxide, nitrogen oxides, carbon monoxide and particulates associated with industrial plants, power-stations and domestic heating systems while traffic is responsible for much of the heavy metals pollution. All these pollutants impact negatively on soil, vegetation and human health. Since 1985 the situation has been monitored by the Slovak Hydro-Meteorological Institute (SHMU) with the information published in a register of emissions and air pollution sources ('Register emisii a zdrojov ovzdusia' (REZZO)) with reference to stationary units of fuel combustion (large pollution sources exceeding 5MW capacity REZZO 1; medium sources of 0.2–5MW REZZO 2; and small sources REZZO 3) and all mobile sources (REZZO 4). While the figures for 1989 to 1997 are still high in comparison with Western Europe, there has been an improvement in acidification with reductions of 81.2 per cent for particulates, 65 per cent for sulphur dioxide, 46 per cent for nitrogen oxide and 29.5 per cent for carbon monoxide arising from desulphurisation of fuels and various technical measures, including the installation of filters. The relatively low reductions in nitrogen oxide and carbon monoxide arise due to the greater importance of traffic as the source rather than power-stations and factories.

The hazards of air pollution for human health have been moderated as has the damage to farmland and forests, though acidification levels are still high in the main Carpathian zone especially where gneiss and granite are present (Lehotský 1990). Relatively high levels of sulphur dioxide emissions were still evident in 1997: Košice (74.7t p.a./sq.km), Bratislava (64.9), Prievidza (47.9), Michalovce (31.1) and Spišská Nova Ves (13.2) due to power-stations, refineries and metallurgy. The highest levels for nitrogen oxides, linked with manufacturing and circulation, are recorded in Košice (66.3), Michalovce (20.0) and Bratislava (16.4); while particulates are most evident in Košice (45.1) and Michalovce (11.2) – linked with power-stations and metallurgical works – and carbon monoxide pollution – caused mainly by the chemical and metallurgical industries – occurs especially in Košice (346.5), Žiar nad Hronom (21.3) and Žilina (11.7), due to chemical and metallurgical industries. Transboundary pollution is a significant phenomenon with Slovakia exporting 193,000t of sulphur dioxide and 131,400t of nitrogen oxide while importing 156,000t and 88,400t respectively in 1995, according to estimates by the Meteorological Synthesising Centre in Oslo reported by the Slovak Environmental Ministry in 1998. There is a possible significant impact from urban pollution in Bratislava as well as Banska Bystrica, Kosice and Trnava where local meteorological stations report pollution levels relating to industry and traffic on an hourly basis.

Water pollution

Being part of the Danube basin, the water resources of the Carpathians are important for Slovakia and for some neighbouring countries. Pollution hazards arise through the discharge of organic matter, phosphorous amd nitrogen by the population at large (through untreated sewage) and also by agriculture (especially nitrogen) and such industries as cellulose, chemicals, paper and textiles. Water quality is monitored at 256 sites with reference to indicators concerned with oxygen (Group A); chemicals: both basic (Group B) and additional (Group C), biology and microbiology (Group D); while six other sites monitor heavy metals (Group E); and radioactivity (Group F). Five water quality classes are recognised from Category 1 (very clean) to Category 5 (heavily polluted). The most polluted rivers are the Dudvah, Hornád, Nitra and Váh. Important actions are linked with the Danube Pollution Reduction Programme of 1991: international cooperation was agreed in 1994 in order to improve water quality and reduce accidental discharges of pollutants having in mind not only the Danube itself but the need to protect Black Sea aquatic geosystems (Vaisher 1998). Meanwhile there has been monitoring of ground water resources since 1982 involving 291 monitoring stations distributed across twenty-seven areas. Particular importance is attached to the Danubian region of Žitny ostrov (the largest ground water reservoir in Central Europe): serious contamination has been revealed around Bratislava through chlorinated hydrocarbons, phenols, sulphates and specific organic substances.

Sustainable management of water resources is essential to protect both the environment and human health, especially in view of increasing public demand for clean water. Piped water supplies are becoming more widely available (80.9 per cent of the population in 1997 compared with 77.8 in 1993 as the pipe system grew from 20.4 to 21.7th.km) although the east is less well served than the central and western areas (Table 10.1). The number of sewage disposal plants increased by nearly 50 per cent during 1993 to 1997 from 208 to 299 (daily capacity from 1.63 to 1.97mln.cu.m), but only 53.7 per cent of the population is connected, with the western part of the country in greatest need.

Soils and vegetation

Soil is a basic but fragile natural resource. Much has been lost by the development of settlements, industries and transport systems in the valleys and basins (accounting for 218.6th.ha in 1997 or 4.4 per cent), while agricultural land (2444.6th.ha – 49.9 per cent) and woodland (1996.4th.ha – 40.7 per cent)

Table 10.1 Public water supply and sewerage systems by regions 1994–97

Region	*Share of population supplied from water pipe systems (%)*				*Share of population connected up to public sewerage systems (%)*			
	1994	*1995*	*1996*	*1997*	*1994*	*1995*	*1996*	*1997*
Bratislavský kraj	94,3	94,7	95,2	93,3	81,6	81,8	82,3	80,7
Trnavský kraj	76,2	77,6	77,0	80,4	40,9	41,1	41,3	42,5
Trenčiansky kraj	83	83,7	84,0	83,8	54,3	53,7	54,1	52,5
Nitriansky kraj	72,4	73,9	74,5	81,1	42,1	42,2	42,3	44,5
Žilinský kraj	82,3	82,4	82,7	83,2	46,9	47,3	47,5	48,0
Banskobystrický kraj	79,9	80,9	80,2	82,2	52,6	53,2	53,5	56,6
Prešovský kraj	69,3	70,6	71,8	73,6	49,1	49,2	50,8	50,8
Košický kraj	75,2	76,2	76,6	76,8	54,9	56,0	56,0	56,7

Source: MESR 1998.

have been damaged in a number of ways: the remaining surfaces comprise water bodies (93.3th.ha – 1.9 per cent) and other categories (150.5th.ha – 3.1 per cent). Damage is monitored by twenty-one key monitoring sites where readings are taken each year and a basic network of 650 rural sites (338 in forests and 312 on agricultural land) where investigations are made every five years in connection with international programmes such as the International Cooperative Programme on Forest Health Evaluation. Altogether some 19,250 plots are involved. Fortunately 69.5 per cent of soils are not contaminated, while 28.7 per cent are 'exposed' (with contamination roughly equivalent to reference values) and 1.8 per cent are contaminated (0.4 per cent heavily so). There is contamination with heavy metals (including arsenic, cadmium, chrome, copper, lead, mercury and zinc) linked with industries (chemicals and metallurgy: for example, the former nickel smelter at Sered), power-stations, sewage sludge, waste incineration and traffic, as well as contaminated irrigation water and manure, mineral fertilisers and pesticides (MESR 1998). There is also the irreversible phenomenom of acidification (which reduces soil fertility) linked with nitrogen and sulphur emissions and fertiliser use.

One solution is to reduce the application of fertilisers and pesticides which were first used on a large scale in the 1960s and increasingly so in the 1970s. Then a low point was reached in 1992 to 1993 when some 95th.t of industrial fertilisers (64.9th.t of nitrogen, 16.5 phosphate and 13.7 of potassium) were used (41.6kg/ha) followed by a rise to almost 133 (57kg/ha) in 1996 to 1997 (88th.t for nitrogen, 24.5 for phosphate and 20.5 for potassium) along with 854.2th.t of lime and 7961th.t of organic manure: increasing again due to the decline of industrialised animal husbandry. Meanwhile the use of pesticides fell from 19,017t in 1980 to 3513 in 1997.

Reference should also be made to soil erosion by water and wind which affects 18.3 per cent of agricultural land in an extreme form (Drgona 1996; Izakovičova 1992) especially in the Bardejov, Čadca, Považska Bystrica and Trenčin areas

Table 10.2 Soil erosion 1997

Erosion	*t/ha p.a.*	*Water erosion in '000 ha*	*% from agricultral land*	*Wind erosion in '000 ha*	*% from agricultural land*
Low	0–4	1065.4	45	2213.7	93.5
Middle	4–10	474.5	20	113.6	4.8
High	10–30	426.2	18	9.5	0.4
Extreme	over 30	402.5	17	30.8	1.3

Source: MESR 1998.

(Juráň 1990) (Table 10.2). Wind erosion is a hazard on large fields of 100–200ha which were created by land consolidation in the socialist period, especially in the Podunajskă nižina, Východoslovenská nižina and Záhorska lowlands and hilly areas of Nitrianska, Pohronska, Trnavska and Žitavska. These problems can be controlled through reducing exposure by planting underseed and by extensification, including the conversion of cropland to pasture on steep slopes through 'greening' programmes. Non-agricultural land or pastures within intensive farming zones are now being protected in the lowlands mentioned above.

Some 40.6 per cent of Slovakia is forested (1.9mln.ha), a high proportion, exceeded only by Austria (46 per cent), Sweden (69 per cent) and Finland (77 per cent); 1.30mln.ha (67.5 per cent) are used commercially, while 0.33mln.ha (17.4 per cent) serve special purposes (aesthetic and recreational) and 0.29mln.ha (15.1 per cent) are for the protection of soil and water. However, the condition of the woodlands has deteriorated in the industrial areas through pollution which has affected the trees both directly and indirectly (through damage to the soil), but northwest winds also bring pollution from Upper Silesia which has damaged some national parks and protected landscape areas (Drgona *et al.* 1993; Huba 1997). However, there has been a slight improvement in forest health: in 1988, 7 per cent of woodlands were damaged to a medium or heavy extent and 32 per cent slightly, whereas the comparable figures for 1997 were 3 per cent and 28 per cent respectively. The lower level of pollution damage evident in the 1990s may be attributed largely to deindustrialisation (Oszlanyi 1997). However, there is a need to restore the original forest structures – following the tendency in the nineteenth century to use foreign seed (mainly Austrian) to reafforest bare lands after wind and insect casualties (Voloscuk 1998).

Other aspects of pollution: noise and waste

Noise pollution is a problem in cities and industrial areas. Traffic is an important component, especially road traffic, although railway traffic has been monitored in Trnava. Noise pollution is considered to be especially serious in forty-six towns and villages, affecting an estimated 1.36mln people. All these places were monitored in 1997, and 19.7 per cent of the population suffered road traffic noise of more than 55dBA compared with 1.3 per cent for a similar level with regard to rail traffic. At over 65dBA the proportions are only 6.5 and 0.5 per cent and over 75dBA 0.1 and 0 per cent, according to the MESR (1998). People may well be affected in other places but the problems are not considered to be severe enough to justify routine monitoring.

Waste management

Some 19.8mln.t of waste was collected in 1997, combining agricultural, industrial and municipal sources along with sewage sludge and dredged materials (MESR 1998): 9.7mln.t is special waste (including 1.8mln.t of hazardous waste in the form of acids, heavy metals and oil) and 10.1mln.t ordinary waste. The trend has moved sharply downwards from 34.0mln.t in 1992 (9mln.t of special waste and 25mln.t ordinary waste), although the improvement is overstated because there has been a harmonisation of statistics with EU methodology as well as a new system of waste collection from enterprises. Most waste (including a quarter of the special waste) is disposed of by landfill at 540 different sites. However, only 124 sites fully comply with the technical standards laid down: some old hazardous waste sites pose a threat through contamination of soil and ground water, including several places formerly associated with the Soviet military (Levice, Nové Zámky, Ružomberok and Zvolen-Sliač). However, the situation is now more tightly regulated under the waste management programme launched in 1997. New landfills have opened at Brezová pod Bradlom, Cerova, Dolný Bar, Pezinok and Zlkovce, while others are being prepared (Michalovce, Slavosovce, Spišská Bela, Žilina and Zlaté Moravce). According to data supplied by the Regional Information System on Waste (Regionalny informacny system o odpadoch), recycling accounts for only 5.1mln.t of waste, though it covers 54.6 per cent of the total volume of special and hazardous waste: scrap iron is being used by VSZ in Košice and at the Podbrezova Ironworks, while 115.6th.t of waste-paper is processed in the mills of Harmanec, Ružomberok, Štúrovo and Žilina. Waste separation accounted for 42.3th.t of recycled material (Table 10.3). Through this process material is collected for the Skloobal glassworks at Nemšova while lead is recovered from old batteries at Šala.

The Slovak Environment Agency has developed a Waste Information System over recent years, supported by PHARE. The system collects data on the generation and movement of waste, especially hazardous waste, while formulating and supervising waste management programmes for local administrations. In 1997 a waste management centre based in Bratislava opened the first Basel Convention regional training centre for waste managers, linked with technology transfer. The Convention, which was ratified in 1991, aims to reduce the movement of hazardous waste across state borders and to provide for disposal in the country of origin. Nineteen Central and East European countries are involved in this UN Environment Programme (Program OSN pre životne prostredie) which is especially concerned with environmentally friendly methods of managing hazardous waste. During 1998 to 2000, Slovakia has been adopting new framework laws on waste, as well as air and water.

Table 10.3 Waste separation in 1997 (tonnes)

Type of waste	*Amount*	*Being used as*		
		Secondary raw material	*Energy generation material*	*Other methods of recycling*
SR total	42,379.6	39,162.7	774.2	2,442.7
Paper	8,140.9	8,021.9	52.1	66.9
Glass	12,127.0	12,090.5	0	36.5
Textiles	238.8	229.1	6.0	3.7
Plastics	752.7	670.2	0	82.5
Metals	12,575.3	12,423.8	0	151.5
Bio-waste	7,963.6	5,579.6	610.7	1,773.3
Hazardous comp.	103.9	80.6	23.3	23.3
Others	477.4	67.0	105.4	305.0

Source: MESR 1998.

A special concern is the nuclear programme at the power-stations of Jaslovské Bohunice (opened in the communist period) and Mochovce (1998–99). Both stations comprise two 440MW generators and there are particular concerns over safety at Jaslovske Bohunice (supplying 29 per cent of the country's power in 1997). The first reactor (Al) was inaugurated in 1973 but closed after accidents in 1976 and 1977. Now the plant is in a decommissioning phase with regard to the two remaining 440MW VVER reactors which are to operate until 2006/08 (Szollos 1998). Slovakia would like to retain the capacity for longer (until 2010/12) because after first (1994) contemplating closure in 2000 on safety grounds, Euro188mln has been spent on safety upgrades to the point where Slovakia considers that the plant fulfils international requirements. But Germany has made the issue crucial for the attempts by the new government to accelerate membership of the EU. Austria also intends to block Slovakia's progress towards accession until an acceptable closure plan is announced. Meanwhile, radioactive waste generated by this plant will be stored at Mochovce which has a capacity for 7200 containers (22,320cu.m). Slovakia has planned the shipment of spent nuclear fuel from Bohunice to Mayak in Russia, but the deal has been suspended indefinitely on the grounds of cost (US$6mln), even though the alternative of dry storage in Slovakia will be even more expensive at US$1.7bln.

Environment policy

Stricter policies have become necessary on account of the deterioration of the environment and the costs in terms of quality of life (Pavlinek and Pickles 1999). In particular, health researchers are becoming increasingly concerned about the long-term effects of environmental degradation on human health through inducing life-threatening conditions (Tolba 1992). While the leading cause of death relates to the circulatory system (54.7 per cent) linked with hypertension, a high chloresterol level and smoking, cancers and tumours account for 21.7 per cent, and 85 per cent of these are caused by environmental factors such as ionising radiation and carcinogenic chemicals in air, food and water as well as alcohol, drugs (including chemotherapeutic agents) and smoking. An increasing number of people are affected, including many under 65 years of age. In 1992 life expectancy on average was 68.9 years for men and 76.7 for women (below the European level by some 6.5 and 4.5 years respectively). However, there seems to be no clear correlation between life expectancy and pollution at the local level. The highest life expectancy is in the polluted regions of Bratislava, Kosice and Prievidza where there are large male populations with high school and university education, while the lowest levels largely apply to non-polluted areas with a male majority and mainly basic education qualifications.

A more sustainable approach to environmental management has been evident since the Rio Conference and its Agenda 21 which brought an almost immediate response in Slovakia through the National Environment Policy of 1993, recognising the need for environmental policy to prevent further damage, enhance biodiversity, improve public health and safeguard resources for succeeding generations. The 'polluter pays' principle was accepted but it was also recognised that progress towards sustainability under the new democratic situation would depend on synergy between government and the organisations of civil society, working through NGOs and the education system. Change has been helped by economic restructuring with greater emphasis on light industry and services, along with energy efficiency.

The Environment Ministry has prepared a national environmental action programme (1996) and strategy for the application of Agenda 21 at regional and district level. Policy has been assisted by outside finance from EU PHARE (totalling ECU4.43mln during the four years 1994 to 1997) with scope for air and water protection, waste management, nuclear safety, energy conservation and health protection. There is also support for environmental education and public awareness, including management courses and media campaigns linked with the government's environmental education conception of 1997 (covering schools, public information and training for administrative staff). The World Bank's Global Environmental Facility (GEF) has provided US$1.96mln over the same period for such projects as biodiversity protection in the Poloniny and Tatra national parks and the Záhorie protected landscape area. Help has also been forthcoming from individual European countries: mainly from Denmark (ECU7.19mln) and Switzerland (1.93), but also from Austria (0.80), The Netherlands (0.53), Norway (0.40), Luxemburg (0.20) and from Belgium, France and the UK (0.20).

Approximating the EU environmental acquis

Slovakia has only a limited programme in the massive task of approximation, although an updated National Environmental Action Plan has been adopted by Parliament to ensure that environmental issues are integrated with other policies in future, and public input has been facilitated by a free phone line to the authorities. But acceleration of the alignment process is needed. Air quality is being safeguarded by the adoption of some emission quotas covered by EU Directives and requirements for the gathering of evidence on pollution sources. Standards for noise measurement have been adopted, especially in the case of vehicles and machinery, and the Slovak Nuclear Regulatory Authority has planned for breakdowns and accidents and has regulations covering radioactive waste shipment. Preparatory steps have been taken in the field of nature protection and the transposition of the Habitats and Birds Directive. However, acts dealing with water and waste are not yet finalised and progress with industrial pollution and risk management is awaited; likewise chemicals and genetically modified organisms. Shortcomings apply not only to legislation but also to the provision of finance and administrative capacity: improvements in the availability of trained personnel are needed everywhere but especially for state bodies concerned with industrial pollution, nature, waste and water and the environmental inspectorates.

Regional classification: the Upper Nitra Valley

The Environment Agency has produced a regional classification of environmental quality (Table 10.4). While 75.1 per cent of the area and 44.5 per cent of the population enjoys satisfactory or high-quality conditions, 10 per cent of the area and 37.5 per cent of the population has a deteriorating or heavily deteriorating environment. There are nine 'endangered areas' with a total area of 4671sq.km and a population of 1.78mln where the latter (Group IV–V) conditions predominate: Middle Zemplin (1081sq.km; 207th population); Bratislava(710sq.km; 491th); Košice (658; 267); Upper Nitra (519; 179); Middle Gemer (401; 56); Trnava-Galanta (384; 162); Middle Pohronie (371; 170); Middle Spiš (364; 87); and Upper Považie (183; 162). The Upper Nitra Valley (between Topolčany and Prievidza) was designated in 1993, especially on account of the conditions around Nováky and Zemianske Kostolany where a power station was built in 1950. It is a good example of an area with severe environmental problems in view of the long-standing importance of coal-mining, electricity generation and

Table 10.4 Regional environmental classification

Quality level	*Specification of quality*	*Population of SR*		*Area of SR*	
		number	*%*	*km²*	*%*
I. and II.	High-quality environment, satisfactory environment	2,379,000	44.5	36,824	75.1
III.	Moderately deteriorated environment	964,000	18.5	7,306	14.9
IV.	Deteriorated environment	752,000	14.0	2,795	5.7
V.	Heavily deteriorated environment	1,262,000	23.5	2,109	4.3
Total		5,357,000	100.0	49,034	100.0

Source: MESR 1998.

the chemical industry. Emissions are still heavy although there were significant reductions between 1993 and 1997 thanks to the Environment Agency's Regional Action Programme which has led to the desulphurisation of power station emissions. However, there is still a problem at the chemical plant through the emission of carbide and calcareous dust, chlorine, ethylene/propylene oxides and vinyl chloride (Holikova 1998). Meanwhile the Nitra River is classified as Category V – through contamination with As, Cu, Hg and Pb – at the Chalmová station (immediately downstream of the Nováky power station and chemical plant). Further pollution in this area arises from lignite-mining (e.g. Baňa Nováky) and from sewage generated in both Nováky and Prievidza. Arsenic from the Nováky power plant pollutes the soil with concentrations reaching 61mg/kg-1, and the ash reservoir is particularly dangerous.

The relief of the Nováky area in the Nitra Valley has been affected by lignite-mining through the appearance of waste tips and the effects of subsidence which disturbs the natural drainage and ground water system and creates swamps and lakes at the bottom of the depressions. Meanwhile the power station generates large amounts of ash which is transported by water and dumped in sludge beds, situated first beside the power station at Zemianske Kostolany (where the dike broke in 1965, necessitating a temporary alternative site on river alluvium) and now amidst a karstic landscape at Chalmová with continuing harmful effects on the surrounding landscape including a rise in the temperature of the local mineral waters and danger to swimming (Turnock 1996). Reference should also be made to sludge emanating from the chemical works and quarries developed to produce construction materials. Some devastated land has been restored to agriculture and this must continue to be a priority in the future.

The first environmental initiatives began in 1992 when Prievidza District applied for funding under the OUVERTURE Programme and a Green Action project was implemented in the Handlova coalfield. The old miners' colony was rebuilt in the 1970s and 1980s but much wasteland remained untreated and was used for illegal dumping, although conversion to residential use was intended. The project was implemented by the Prievidza Business and Innovation Centre (PBIC), working with local government, the district labour office, companies and financial institutions along with partner administrations in several other countries: Greece (Stavroupoli, Thessaloniki), Slovenia (Ljubljana), Spain (Murcia) and the UK (Edinburgh). An environmental action plan has been produced as well as an urban architecture study with a view to converting the remaining wasteland into an industrial estate.

Slovakia is also placing its own expertise at the disposal of Third World countries through Project NEWS (North East West South) which was set up in Prague in 1992, with finance from The Netherlands, to disseminate know-how in local government for the particular benefit of Nicaragua. Since 1993 the project has continued both in the Czech Republic and Slovakia; in the latter case through the organisation Cooperation NEWS in Slovakia based at Gelnica. Gelnica itself has links with The Netherlands (Gennep) and Nicaragua (San Pedro de Lovago); likewise Nitra (with Zoetemeer and Jinotega) and Trnava (with Dordrecht and Granada) with the primary aim of providing aid for Nicaragua. The result of this cooperation is the implementation of specific projects in Nicaragua, such as the reconstruction of the Casa de Los Tres Mundos cultural centre in Granada, help in building and refurbishing a children's centre, pensioners' club and library in Masaya, equipping a laboratory in San Pedro de Lovago, regulating the Ducuali basin in Jinotega and building access roads to farms and primary schools in remote areas. The project also facilities contacts between NGOs and cultural exchanges. Meanwhile, a regional development agency has emerged as a public–private partnership to make the best use of natural and human resources. Complementing the activities of the PBIC, it is particularly interested in agriculture, food processing and tourism.

Other regional programmes

The Environment Agency has a cooperation agreement with Japan's International Cooperation Agency (JICA) in respect of a Regional Environmental Management Plan for the revitalisation and optimum resource use in the Hron basin, including the installation of monitoring equipment linked with GIS technology. Under this agreement other Far Eastern organisations are involved (Pacific Consultants International and Oyo Corporation) and Slovakia's Forest Research Institute, Geological Survey, Hydrometeorological Institute and Soil Fertility Research Institute as well as regional environmental offices in Banska Bystrica and Nitra. All municipalities in the area have responded to questionnaires which will also contribute to the final evaluation (JICA 1999).

The work of environmental NGOs

There is also the record of success by NGOs beginning with the Slovak Union of Nature and Landscape Protection (Slovensky zvaz ochrany prirody a krajiny (SZOPK)) formed in 1969 to oppose environmentally harmful projects – even during the communist years of the 1970s and 1980s – and work for alternative energy, rural land use, water management and tourism strategies (Huba 1998). An important initiative taken in the 1970s was the proposal for a Podunajsko national park on a trilateral basis (including Austrian and Hungarian territory) in an area of natural and cultural value in the Danube floodplain threatened by the Gabčikovo dam and navigation complex (discussed below). However, there was renewed activism in 1987 by SZOPK over conditions in Bratislava. This organisation also became involved over Gabčikovo-Nagymaros and the leading Slovak political party at the time of the Velvet Revolution (Public Against Violence (VPN)) included former SZOPK members. But ecology took a back seat and the Slovak Green Party (Strana Zelenych na Slovensko) split on the Gabčikovo issue (Šnajdr 1998). Huba (1997, p.258) refers to a certain amount of amateurism in the environment movement at the time but emphasises the fear of mass unemployment as the overriding concern.

However, in 1994 Greenpeace Slovakia staged a demonstration at Mochovce with the aim of getting the EBRD to cancel a loan to finance completion of the station; this attracted a lot of media attention and the first ecology camp (ekotabor) was held to train new members for direct action. SZOPK has also been perpetuated as the Society for Sustainable Living (Spoločnost pre trvalo-udržatelny život) (Šnajdr 1998). This organisation together with several other NGOs has campaigned against the idea of holding the Winter Olympics in the High Tatra. It was felt that there would be too much pressure on such a small area and that sharing the meeting with Zakopane would be a better idea. However, the 2006 games have gone to Torino and there is general scepticism over the feasibility of a further bid for 2010 given the scale of funding required. However, after the Velvet Revolution a number of local groups split away from the main body of the Slovak Union: organisations such as the Friends of the Biele Karpaty Mountains (Karpatske ochranarske zdruzenie altruistov), concerned with various projects in this area, The River Vah Union (Važska únia), protecting the river ecosystem against pollution, and the WOLF Forest Protection Movement (Lesoochranarske zoskupenie VLK) based at Tulčik, which seeks better woodland management through avoidance of large clear-fells in the Polana Biosphere Reserve (1998). Reference should also be made to the Centre for Sustainable Alternatives (Sosna-Centrum pre trvaloudržatelné alternatívy) which seeks alternative agriculture by enhancing farmers' awareness of the importance of grassland biodiversity in the Carpathians, with special reference to a 20,000ha area which is both a protected landscape area and a biosphere reserve.

The Daphne Centre for Applied Ecology was established in 1993 to restore and conserve natural areas. It builds on monitoring carried out in the Tatra since the 1960s: the 500 species currently present are only a quarter of the stock of a century ago. Animals are being killed by poaching while breeding is disrupted by tourism (climbing and skiing). Again, the Stražovske Vrchy needs restorative work after damage from tourism developments since the 1970s (also the limestone quarries of Jabloňove since 1949). Daphne has compaigned to stop the dumping of highly toxic waste from the Slovnaft oil refinery in Bratislava since 1963 which has accumulated at Devinská Nová Ves within the karstic and tectonically complex Male Karpaty Protected Landscape Area and close to an urban settlement. It is also heavily involved in the the Morava floodplain in collaboration with other organisations such as the Worldwide Fund for Nature (WWF). It has taken action to halt the degradation of forest in Záhorska nižina (lowland) where 522sq.km of lowland forest to protect soil on sand-dunes have been poorly managed.

A range of other organisations play a role in conservation work, in training and in providing information for the public. The Society for Sustainable Living (Spoločnost pre trvaloudržatelny život) was established in Bratislava in 1993 by professional people (ecologists, geographers, lawyers and philosophers) concerned about human survival. The organisation supports sustainable technologies through local and regional centres, and seeks to enhance public awareness and participation. Water and Life (Ludia a voda) is helping the inhabitants of the Torysa Valley in the east of the country to find a more sustainable future through the project for a Village of the Third Millennium (Kravčik *et al.* 1998). Previously this area was associated with the Javorina military training ground in the Levočské vrchy mountains and a controversial dam project for the Tichý Potok River. The Association of Environmentalists in North-Eastern Slovakia (Združenie ochranárov severovychodneho Slovenska) is a major regional group and mention should also be made of the Amaryllis Environmental Protection Movement (Amaryllis: ochrana životneho prostredia), the Association for Harmonised Living (Pospolitost harmonického života), Flora and Fauna (Flora a fauna) and the Gentian Club (Klub gentiana). There is also a conservation element in the work of such organisations as the animal rights organisation Freedom for Animals (Sloboda zvierat) which campaigns against fishing, hunting and the exploitation of animals in business (including laboratory use); and the Healthy Schools' Civic Association (Zdravé školy) seeking better physical and educational development of the students through a supply of information on such matters as ethnicity, unemployment and other social problems. The Association for the Healthy Cities of Slovakia (Asociácia zdravých miest Slovenska) has twelve full members committed to improving health through better housing, working and leisure conditions. There are also contributions from the Scout Movement (Slovenský skauting), the Slovak Ecological Society (Slovenská ekologická spoločnost), the Slovak Geographical Society (Slovenská geografická spoločnost) and the Slovak Ornithological Society (Slovenská ornitologická spoločnost).

The major international organisations are also represented. Friends of the Earth Slovakia (Spoločnost priatelŏv Zeme) was established in 1997 to seek economic and environmental justice and enhancement of living standards including environmental quality. Priorities are the protection of indigenous woodland, help for endangered communities, retention of public transport and reduced waste. The World Conservation Union in Slovakia (IUCN Slovakia) prepares programmes for environmental protection (especially projects in involving former Soviet military bases), monitors nature conservation work (especially the ecological and economical evaluation of fishponds, the sustainable development of the Kysuce region and biodiversity conservation in Polana PLA). It has helped to implement a national ecological network, while offering training and administering funds to provide small grants for nature conservation projects. Finally, the organisation Ecological Bricks for our Common House in Europe is active over transboundary aspects of nature conservation including the parks of Pieniny, Tatra, Slovakian Karst and the Danube–Morava floodplain.

NGOs are aided by the government through small grants to help their conservation work projects and they benefit from an environmental training programme established in 1995 to help implement Agenda 21 in Slovak cities, especially the project for sustainable cities (Trvaloudržatelné mestá). Nevertheless, the NGOs have retained their independence and have frequently criticised the government. For example, Mikuláš Huba (Society for Sustainable Living in the Slovak Republic), Lubica Trubiniova (Greenpeace Slovakia) and Katarina Simončičova (Slovak Union of Nature and Landscape Protectors, Bratislava) feel that the government is not doing enough for the environment while taking the credit for voluntary work. Further help for NGOs has come for external sources. An Environmental Training Project (administrated by a consortium led by the Hubert H. Humphrey Institute for Public Affairs at the University of Minnesota) facilitates environmentally sound economic restructuring through grants to provide training for Slovak NGOs over negotiation and conflict resolution, finance, policy analysis and the conduct of public awareness campaigns. ECOPOLIS (formerly the Environmental Partnership for Central Europe) has also supported environmental NGOs over technical and organisational problems.

Nature and landscape: the Lower Morava

The legal framework for nature and landscape protection was revised in 1994 (Table 10.5). There are seven national parks (Tatra with 74.1th.ha; Nízke Tatry 72.8; Poloniny 29.8; Malá Fatra 22.6; Slovenský raj (19.8) and Pieninský (3.7), plus Poloniny (29.8) and Muránská planina (20.3) designated in 1997 (Figure 10.2). There are also fifteen protected landscape areas (PLAs) and in descending order of size these are: Štiavnicke Mountains 77.6th.ha, Horna Orava 70.3, Malé Karpaty 65.5, Kysuce 65.5, Velká Fatra 60.6, Biele Karpaty 43,5, Ponitrie 37.7, Slovenský kras 36.2, Strážovske vrchovina 31.0, Záhorie 27.5, Východné Karpaty 26.0, Polăna 20.1, Cerova vrchovina 16.3, Latorica 15.6 and Vihorlat 4.4. There are also protected sites and a range of nature reserves and natural monuments: 881 in 1993 and 1008 in 1997, involving 101.5th.ha and 101.8 respectively (Figure 10.2). The total number of conservation sites has increased (from 902 in 1993 to 1030 in 1997), although the total protected area has fallen over the same period from 961.7 to 943.6th.ha (Table 10.5). This situation has arisen due to inadequate funding and is quite unusual because

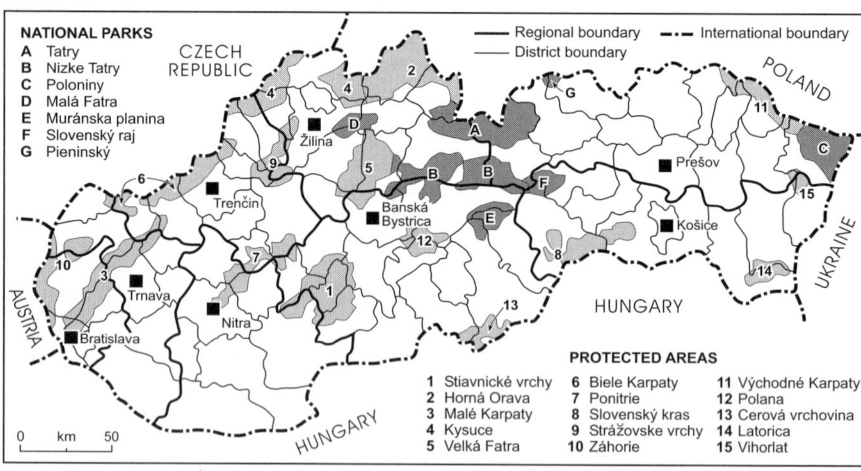

Figure 10.2 Slovakia: administrative divisions, major towns and protected areas

Source: Administrative maps.

Table 10.5 Preserved areas 1993–97

	Number of sites					*Total area (th.ha)*				
	1993	*1994*	*1995*	*1996*	*1997*	*1993*	*1994*	*1995*	*1996*	*1997*
National parks	5	5	5	5	7	199.7	199.7	199.7	199.7	243.2
Protected landscapes	16	16	16	16	15	660.5	660.5	660.5	660.5	598.6
State nature reserves	448	448	.	.	.	90.6	90.6	.	.	.
Nature reserves	.	.	559	559	575	.	.	92.7	92.9	96.0
Natural monuments	.	.	248	254	259	.	.	1.4	1.3	1.4
Other preserved areas	433	453	172	174	174	10.9	11.0	8.8	4.3	4.4
Total	902	922	1000	1010	1030	961.7	962.2	963.0	958.7	943.6

Source: MESR 1998.

Note
In 1994 there was a reorganisation of nature reserves and a new category of natural monument was introduced. However, reorganisation continued over subsequent years to deal with anomalies over the boundaries of protected areas for waters, springs and cultural monuments. As a result the total protected area has continued to decline.

the normal trend is for protected areas to be increased during transition. However, since the number of areas has increased it must be assumed that some marginal areas have been given up. Legislation on landscape protection was overhauled in 1994 when a new classification system was introduced: many other types of preserved area were categorised as natural monuments.

The World Heritage Committee lists some 550 sites including four in Slovakia: Spišský Castle and related monuments (Spišská Kapitula Historic Town Reserve, Spišské Podhradie Historic Zone and Žehra's Church of the Holy Ghost), Banská Štiavnica Historic Town Reserve and associated technical monuments, the Vlkolinec Folk Architecture Reserve, and the caves of Slovensky kras karst and Aggtelek kras karst. There is also GEF for Tatra and Dolné Pomoravie (biodiversity of forest ecosystems) and Východne Karpaty biosphere reservation. Meanwhile, Spišiak and Sčasná (1998) see the Lower Morava as a model area for cross-border cooperation between Slovakia, Austria and the Czech Republic, with potential for tourism (especially agrotourism), agriculture, small industries and agrotourism (cycling, fishing, swimming and walking, with additional tourist facilities including artificial ski slopes). The result of this cooperation is that agriculture should concentrate on vegetable growing for the Bratislava region, while industries include food (Gajary), construction (Velké Levare) and engineering (Kúty, Plavecký Stvrtok, Záhorska Ves and Zohor). Sewage plants are envisaged for Gajary (1.08mln.ECU) and Zohor (0.64mln.ECU), while geothermal energy will be used for heating in the villages of Šastín-Stráže, Lakšárska Nová Ves and Borsky Mikulas (costing ECU4mln). However, while the international traffic corridors provide opportunities for local development, there are environmental concerns, and special attention is being given to the Dolne Pomoravie area comprising seventeen villages in a protected landscape area on the Austrian–Slovak border in the Záhorie district (Huba and Ira 1999).

However, the meadow environment is of great importance and the protection previously afforded by the Iron Curtain has now been removed, although the Marchegg Nature Reserve on the Austrian side, supporting an important white stork colony, is safeguarded as part of Natura 2000. NGOs from the three countries, including Distelverein–Union for the Morava River (Austria), Veronica (Czech Republic) and Daphne (Slovakia), along with the government of Lower Austria and the environment ministries of the three countries, have been encouraged by WWF to conserve some 44sq.km of Morava floodplain meadow (the largest such complex in the whole of Central Europe) which is threatened by new projects for the Bata Channel and the Danube–Morava–Oder–Labe Canal (Slavik and Halás 1998). Various bridge projects (e.g. to connect Hohenau with Moravský Ján and Záhorska Ves), and increasing traffic generally, are also likely to have a significantly negative impact upon the area. So conservation is becoming a stronger concern and GEF money is now assisting this project. Historically the meadows covered 161sq.km but the landscape has been transformed by river regulation and conversion of intensive arable farming, along with gravel quarrying, inappropriate forest management, overfishing and water pollution.

A general lack of scientific or economic appreciation of biodiversity in the area stems in part from its marginalisation on the frontier of the communist world (though since 1989 the meadows have become a popular recreation area and failure to set aside sufficient areas for recreational use is now one of the root causes of biodiversity loss). An integrated approach to river basin management should include designation of a trilateral international Ramsar site (which now seems unlikely) and a series of sustainable resource use scenarios to secure a safer future for existing biodiversity. Detailed economic assessment of ecosystem functions and values is required in order to properly understand and value the true costs and benefits of different possible development and resource use scenarios, with local participation through 'strategic rural partnerships'. Progress should be facilitated by the emerging environmental policy climate within the EU and the candidate countries (which include the Czech Republic and Slovakia) and appears to offer significant opportunities for more sustainable management of natural resources, while the Danube River Protection Convention (DRPC), which came into force in 1998, offers a framework for international cooperation and more integrated river basin management which was previously lacking.

More than 850 species of plant have been recorded in the Morava meadows (twelve species of meadow plant and four species of water macrophyte are included on the Slovak Red List of endangered plants) which also provide important sources of food and habitat for a wide range of birds, including many rare and endangered species. Floodplain forests are dominated by willow and poplar species, forming a wet softwood association which is among the most threatened ecosystems in Slovakia. The wetlands are also important breeding and migration sites for fish: forty-eight species have been recorded and forty were recently confirmed, of which thirty-six are endangered on a European scale. Much of the floodplain area is included in the Záhorie Protected Landscape Area (PLA), which was designated in 1989. There are also two national nature reserves (NNRs) which accord the highest level of protection. Horný les NNR

(543ha) was established in 1981 and includes floodplain forests and meadows. The same year Dolný les NNR (186ha) was also designated, including valuable alluvial softwood forest and oxbow lakes.

Since 1993, the study area has been a Ramsar Site. The area is also recognised as one of the 'Ecological bricks of our Common House of Europe' and there has been regular trilateral discussions since 1995, with the Morava-Dyje accepted as a common natural and cultural heritage. However, protection and sustainable management of the floodplain meadows and forests remains problematic. Although the PLA administration aims to promote environmentally sensitive management of the area, problems of land ownership (land restitution is far from complete), conflicting legislative frameworks and illegal activities effectively nullify the conservation measures. Mowing and traditional management of meadows is extremely valuable in terms of biodiversity but is increasingly difficult to maintain. The floodplain forests have suffered in the past from the invasion of alien species and outbreaks of disease among elm and oak. In the Czech Republic dams and reservoirs have been built on the Dyje – but at Nove Mlyny the lake has been 'ecologised' by lowering the level 85cm to reveal a chain of small islands to form a biological corridor (though water managers want to raise the level again); such corridors (territorial systems of ecological stability) are now central to conservation in the Czech Republic and Slovakia. There are also trilateral concerns over water quality, revitalisation of watercourses disturbed by canalisation – Levarsky Channel meanders are being opened up (Valachovič 1998) – restoration of natural floodplain forests and grasslands, and the development of fish populations (including spawning sites and migration routes) (Vlašín 1999).

International relations: the Gabčikovo–Nagymaros controversy

While consensus has been achieved on the Morava, the project to build two dams on the Danube in the section where the river marks the frontier between Hungary and Slovakia has probably been the most controversial environmental issue of the 1990s as regards relations between two former communist states. The plan was given serious thought in the 1950s under Comecon auspices, but was shelved in favour of an integrated electricity grid. Interest was rekindled on a bilateral basis in the 1970s with particular enthusiasm in the then-Czechoslovakia, firmly recentralised after the Prague Spring of 1968 yet still dedicated to regional development projects for the benefit of Slovakia (Ostry 1988). The new plan was organised on a joint basis with Hungary in 1977 and work had started in both countries by 1980. The USSR was broadly supportive, through the prospect of an upgraded continental waterway (with the section from Bratislava to the Black Sea effectively freed from the historic problems of low water) and a perceived political advantage in linking Hungary more closely with her more reliable northern neighbour. However, opposition on environmental grounds surfaced in Hungary in the early 1980s and, despite renewed commitment to the project in 1985 (when financial problems were overcome),

Hungary later suspended its contribution and subsequently the Antall government decided to withdraw from the project. At a time of political change, public opinion considered that farmland and wetland would be lost, while earthquake damage might precipitate a major flood disaster.

Meanwhile, Czechoslovakia (and after 1993 Slovakia) claimed that the Gabcikovo dam, along with an artificial 25km canal at a level 11–18m above the natural river, would protect the Žitny Ostrov (Rye Island) zone from flooding and would allow water to be released into the natural drainage system to protect the wetlands. This would also safeguard Hungary's drinking-water which is obtained by seepage from the Danube by means of a complex process of filtration into the water-table. Navigation would benefit, and the electricity generated (eight 90MW turbines) would mean less dependence on coal-burning power-stations and so reduce gaseous and solid emissions. There was also a wider issue because of the deepening incision of the river on its natural course, attributed to dam construction in Austria in the 1950s, holding back gravel that was formerly carried downstream to the delta. As it carried less material, the river eroded more effectively. The impact was exacerbated by dredging carried out by the Slovaks at Bratislava (to avoid flooding and improve the port) in the 1970s and 1980s; also through work to straighten the channel and increase flood rates. Both projects were seen as unavoidable as erosion made navigation increasingly difficult: indeed some engineers argued that a further dam would eventually be needed to the south of Budapest.

However, while the abrogation of the treaty in 1989 resulted in a stoppage at Nagymaros near Visegrad (where a 160MW power station was planned) where the Danube lay wholly within Hungarian territory, it was possible for Czechoslovakia to proceed unilaterally in 1991 with the variant of the Gabčikovo project which diverted water into an artificial channel at Cunovo – instead of Dunakiliti further downstream – where that state controlled both banks of the river (Fisher 1993). The scheme was completed, much to the dismay of Hungary on account of the ecological risks and also the blow to national pride arising from the diversion of the river into Slovak territory which was a perceived 'attack on the country's borders' (Fisher 1993; Okolicsanyi 1992). Hungary was also sensitive to the fact that members of Slovakia's minority Hungarian population bore the brunt of the inconvenience. However, the Gabčikovo scheme is now complete and the environmental situation appears satisfactory in view of the arrangements to release appropriate amounts of water to sustain the floodplain woodlands, part of which have now been safeguarded through the Protected Landscape Area of Dunajske luhy (12,284ha) (Kovačovský and Rychlik 1994). Meanwhile, accessibility has improved for the 'island' between the two channels (Paričková and Brožiková 1994; Peter and Mirko 1994). Some people who feared a 'disaster' but chose not to move can now appreciate some potential for rural tourism (Schraggeova and Roškova 1994).

Hungary challenged Slovakia through the International Court of Justice between 1994 and 1997 (during which time the work at Gabčikovo was completed) and, although the court ruled that Slovakia should not have proceeded unilaterally, it also upheld the legality of the original agreement and Hungary's obligation to proceed at Nagymaros (Williams 1995). The former Horn government was prepared to negotiate on this basis, but his main coalition partner successfully pressed for more impact studies, and a new Hungarian government has now rejected any notion of obligation and the matter remains in dispute. It is certainly true that the project is a survival from the communist past and the Hungarians are captive to a hasty commitment made in 1977 'based on the environmental paradigm of the 1950s and energy supply projections of the mid-1970s' (Ostry 1988, p.22). On the other hand there is a school of thought which argues that since the regulation of the Danube has started in German and Austrian territory, the adjustments to the bed make it essential for similar works to continue downstream. However, from the environmental angle the interest lies in the contrast between the pressure groups in the two countries. Perhaps because there are differences in national interest, the Hungarian greens have been much more effective than their Czechoslovak and Slovak counterparts. The Slovak ecologists were ignored and the project went ahead, destroying 30sq.km of floodplain woodland in the process, while creating an artificial river system which now needs a permanent energy input despite the scope for floodplain management.

Conclusion

Slovakia's record on environmental protection is one of substantial progress over the past decade. Large-scale industrial development ceased in 1989 and management is now being placed more strongly on sustainability principles. There have been improvements in air quality and the quantity of waste is now under control but much more recycling is still needed, as well as better environmental information, since there is no tradition of free access to such information. But the shift is towards EU norms and important in this respect is the 1998 Access to Environmental Information Act which should improve the flow of information for the public sector, educational institutions (including universities and research institutions), NGOs, policy-makers and journalists. It is also relevant to point out that until the end of the 1990s Slovak politics was dominated by a nationalist rather than a European agenda. But the present government wishes to accelerate the process of integration which will require greater priority for the environmental elements of the *acquis*, despite what has been achieved by PHARE so far. In this task Slovakia has the resources of a substantial voluntary sector which has been robust in its championing of environmental issues in the past. The country is also well integrated in terms of cross-border arrangements which will channel in valuable expertise in the years ahead.

References

Drgona, V. 1996, 'Ecological problems arising from intensive agriculture in western Slovakia'. *GeoJournal* 38: 213–18.

Drgona, V., Baráth, J. and Kramárekova, H. 1993, 'Regional analysis of air contamination in the Slovak Republic', in V. Drgona (ed.) *Geographical Studies* (Nitra: University of Education), pp.97–114.

Fisher, S. 1993, 'The Gabčikovo–Nagymaros dam construction continues'. *Radio Free Europe/Radio Liberty Research* 2(37): 7–12.

Holiková, V. 1998, 'Environmenalny akčny program pre Hornú Nitru'. *Enviromagazin*, mimoriadne cislo 18–19.

Huba, M. 1997, 'Slovak Republic', in J. Klarer and B. Moldan (eds) *The Environmental Challenge for Central European Economies in Transition* (Chichester: Wiley), pp.227–70.

Huba, M. 1998, 'Mimovladne environmentalne organizacie na Slovensku ako iniciatori a realizatori rozvojovych alternativ životneho prostredia'. *Životné prostredie* 32(4): 206–10.

Huba, M. and Ira, V. 1996, O koncepcii trvalej udržatelnosti vo vztahu k niektorým geografickým aspektom vyvoja Slovenska, Geografický časopis 48(3–4): 285–97.

Huba, M. and Ira, V. 1999, 'Border region and its development from the sustainability perspective: the case of the Lower Morava river region, Slovakia'. *Moravian Geographical Reports* 7(2): 32–9.

Izakovičová, Z. 1992, 'Ekologicke rizsiká v rozvoji polnohospodarstva'. *Zivotne prostredie* 27(6): 311–14.

Jakál, J. 1998, 'Antropická transformácia relíefu a jej odraz v krajine Hornej Nitry'. *Geograficky casopis* 50(1): 3–22.

Japan International Cooperation Agency (JICA) 1999, *Suhrn priebežnej spravy-regionalny plan environmentalneho manazementu pre pvodie Hrona* (Banská Bystrica: JICA).

Juráň, C. 1990, 'Erózne procesy na územi Slovenska a perspektiva protieroznej ochrany polňohospodarskej pódy', in Anon. (ed.) *Póda-najcennejsi zdroj: výskumny uśtav pódnej urodnosti* (Bratislava: Výskumny uśtav pódnej úrodnosti).

Kovačovský, P. and Rychlik, I. 1994, 'Selected avifauna components as indicators of environmental changes on the area of water dam construction on the Danube'. *Acta Environmentalica Universitatis Comenianae* 3: 108–19 (in Slovak with an English summary).

Kravčik, M., Těsliar, J. and Lisoňová, M. 1998, 'Predstavujeme región hornej Torysy'. *Životné prostredie* 32(4): 211–14.

Lehotsky, M. 1990, 'Hodnotenie tlmivej schopnosti pód Slovenska voči antropogenemu zakyslovaniu'. *Geogralický Časopis* 42(4): 357–74.

Ministry of the Environment of the Slovak Republic (MESR) 1998, *State of the Environment Report* (Banska Bystrica: Environment Agency).

Nadacia IUCN 1996, *Aspects of Implementation of the National Network* (Bratislava: Nadacia IUCN).

Okolicsanyi, K. 1992, 'Slovak–Hungarian tension: Bratislava diverts the Danube'. *Radio Free Europe/Radio Liberty Research* 1(49): 49–54.

Ostry, D. 1988, 'The Gabčikovo–Nagymaros dam system as a case study in conflict of interest in Czechoslovakia and Hungary'. *Slovo: Journal of Contemporary Soviet–East European Affairs* 1(1): 11–24.

Oszlanyi, J. 1997, 'Forest health and environmental pollution in Slovakia'. *Environmental Pollution* 98: 389–92.

Páričková, I. and Brožiková, M. 1994, 'Gabčikovo water dam in thoughts and experiences of inhabitants of its hinterland'. *Acta Environmentalica Universitatis Comenianae* 3: 61–6 (in Slovak with an English summary).

Pavlinek, P. and Pickles, J. 1999, 'Environmental change and post-communist transformation in the Czech Republic and Slovakia'. *Post-Soviet Geography and Economics* 40: 354–82.

Peter, A. and Mirko, B. 1994, 'The impact of Gabčikovo water dam on changes of spatial relations in the derivation channel and inundation of the Danube'. *Acta Environmentalica Universitatis Comenianae* 3: 120–30 (in Slovak with an English summary).

Schraggeova, M. and Roskova, E. 1994, 'Perception and evaluation of a local environmental change (Gabčikovo water dam)'. *Acta Environmentalica Universitatis Comenianae* 3: 73–81 (in Slovak with an English summary).

Slavik, V. and Halas, M. 1998, 'Cezhraničné vázby a cezhraničná spolupraca SR a CR na priklade okresu Skalica', in: A. Dubcova (ed.) *Úlohy regionalnej geografie Slovenskej a Českej republiky v podmienkach transformujucich sa ekonomik, Geografické Informácie 5* (Brno: Katedra Geografie PF MU/Nitra: Katedra FPV UKF), pp.65–75.

Snajdr, E. K. 1998, 'The children of the greens: new ecological activism in post-socialist Slovakia'. *Problems of Post-Communism* 45(1): 54–62.

Spišiak, P. and Sčasná, S. 1998, 'Obnova ruralneho priestoru Delneho Pomoravia', in A. Dubcova (ed.) *Úlohy regionálnej geografie Slovenskej a Českej republiky v podmienkach transformujúcich sa ekonomik, Geografické Informácie 5* (Brno: Katedra Geografie PF MU/Nitra: Katedra FPV UKF), pp.85–91.

Stanners, D. and Bourdea, P. 1996, *Europe's Environment 1995* (Copenhagen: European Environment Agency).

Szöllos, J. 1998, 'Alternativne rozvojove perspektivy energetickeho systému Slovenska'. *Životné prostredie* 32(4): 191–6.

Tolba, M. K. 1992, *Saving our Planet: Challenges and Hopes* (London: Chapman and Hall).

Turnock, D. 1996, 'Environmental issues in the Nitra valley of Slovakia', in F. W. Carter and D. Turnock (eds) *Environmental Problems in Eastern Europe* (London: Routledge), pp.188–205.

Vaishar, A. 1998, 'Danube pollution reduction programme'. *Moravian Geographical Reports* 6(1): 61–2.

Valachovic, D. 1998, 'Obnova riecňeho systému v podmienkach povodia rieky Moravy'. *Daphne* 5(1): 8–13.

Vlašin, M. 1999, 'Trilateral floodplains of the March and Thaya rivers'. *IUCN Newsletter Central and Eastern Europe* 32: 6–7.

Volosčuk, I. 1998, 'Ecological stability in the Tatra Mountain forests'. *Ecology* 17: 39–48.

Williams, K. 1995, 'The foreign relations of an independent Slovakia'. *Slovo: Journal of Contemporary Soviet–East European Affairs* 8(2): 87–108.

11 Slovenia

Dušan Plut, Metka Špes and Valentina Brečko

Introduction

Seen from a geographical viewpoint, Slovenia – with 20,256sq.km of territory, 1.98mln population and GDP in 1998 equalling US$9700pc – is a distinct Central European transit country, located at the contact of four macro-regional units: the Alps, the Mediterranean, the Pannonian lowland and the Dinaric Mountains (Gams, 1991) (Figure 11.1). Due to this location, its physico-geographical features are intensely intertwined, resulting in an exceptionally diverse landscape and intensely patterned geographical mosaic of small landscape units. Its landscape and biotic diversities, together with abundant forests, essentially contribute to Slovenia's highly regenerative capacity. Yet for some important and widespread

Figure 11.1 Salient features of Slovenia

landscape degradation processes (air and water pollution) the regenerative capacity of individual landscape ecosystems and landscape-forming elements is low and varies intensely throughout the year. In Slovenia, geographical constants and variables together with the impact of environmental and other processes in its vicinity are reflected in the intensity of landscape degradational processes and degree of environmental degradation (Gams, 1991).

With the demise of intensive environmentally unfriendly socialist industrialization in the late 1980s, the consequences of industrial and energy pollution on the landscape's ecology have been long-lasting, very specific and partly irreversible in certain individual regions (Plut, 1987). In the mid-1990s, Slovenia was ranked a moderately polluted European country by empirical environmental indicators, although environmental degradation was higher than that attested by the achieved economic development level and the income per inhabitant (Figure 11.2). In spite of some individual cases of focused pollution reversal, Slovenia's environmental condition has not changed over the past decade; namely, the additional environmental pollution has been faster than the successful environmental revitalization. It is estimated that by the mid-1990s annual losses from environmental pollution totalled about 4–6 per cent of GDP, while less than 1 per cent of GDP (state budget, private funds) was allocated to environmental protection and nature (Plut, 1997). Empirical investigations in Slovenian towns show that environmental quality has already become one of the social differentiation factors. In addition, the environmental awareness of people living in the most polluted Slovenian towns is rather low, which reflects a social crisis (Špes, 1998).

Pollution of water resources

The extent and differential degrees of water resource pollution (sources, streams, lakes, sea, ground water) in Slovenia result from diverse factors. These include water-ecological sensitivity of landscape-ecological units, the drainage areas of individual water sources, and the quantity and composition of waste water discharged into water sources. As a result of the large karstic surface area of head water zones, specific seasonal oscillations in water discharge are significant. These, together with the prevailing low river gradients, thin layer of weathered material above ground water areas, greater water-ecological sensitivity of shallow Alpine lakes and the enclosed character of the North Adriatic, all contribute to the seasonal regenerative capacity. This means that in many cases water resources are very moderate, especially in summer.

The manufacturing industry, agriculture and households are the main producers of waste water pollution. Slovenian industry still discharges nearly half (48 per cent) of its waste water into the environment unfiltered, while the amount of unprocessed waste water from urban and rural areas is even larger. By the mid-1990s, the register of water polluters and diverse filtering-process efficiencies was not yet complete. Between 1987 and 1993, the quantity of organic pollution (measured in population equivalents) declined, partly due to recession/diminished production and reduced quantities of waste water. At the beginning of the 1990s most filters gave only modest results; a mere one-tenth of waste water

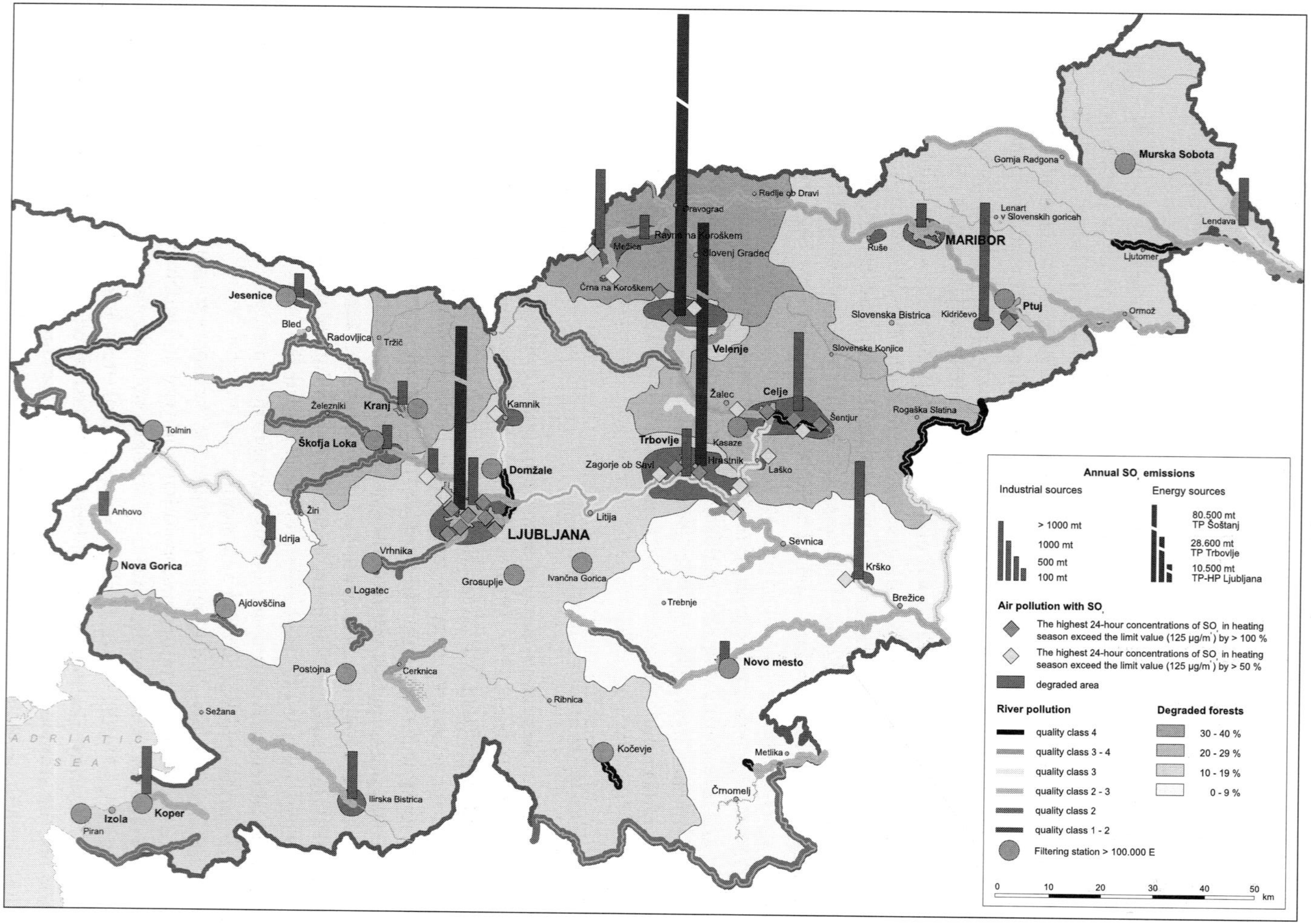

Figure 11.2 Environmental pollution in Slovenia in 1995

Source: National Atlas of Slovenia. Reproduced with permission from the Director, Institute of Geography, Ljubljana, Slovenia.

was processed efficiently through physical-chemical-biological treatment. By the mid-1990s, only half (48 per cent) of Slovenia's population was connected to a sewerage system, and less than one-fifth (18 per cent) to filters. Thus from the late 1980s to the mid-1990s, although the quantity of waste water diminished due largely to industrial decline, the percentage of untreated waste water ranked significantly above the average of EU countries. In the mid-1990s, manufacturing accounted for three-fifths of the pollution from toxic substances (metals, pesticides, organic compounds), whilst agricultural sources and the sewerage system each totalled one-fifth (Environmental Performance, 1997).

Slovenian rivers are abundant in water throughout almost the entire year. This results from fairly high average annual precipitation (about 1600mm), considerable relief energy, and the Drava and Mura rivers flowing in from Austria. Slovenia's rivers have a high self-purifying capacity. Also typical for Slovenia is the dense network of rather short and less abundant streams with minimum summer discharge. During the second half of the 1970s and the first half of the 1980s, heavy pollution of Slovenian rivers began to abate (Radinja, 1979; Plut, 1997). It has not increased since 1990, largely due to a rather sharp decline in industrial production and the resultant decrease in industrial waste water. Nevertheless, by the mid-1990s, pollution in Slovenian rivers was still above average, especially during periods of low rainfall in summer. Water pollution was highest in smaller rivers during the mid-1980s and early 1990s. This coincided with a period when the disparity between low self-purifying capacity and the shortness/weakness of tributary streams led to a decline in efficient natural purification coupled with additional over-polluted waste water. The only unbroken area of relatively pure rivers (i.e. Class 1–2 quality but not Class 1) occurs in Alpine areas where water quality is also influenced by higher rainfall and the increasing role of tourism/recreation facilities.

Between 1970 and 1995, data on water quality in Slovenian rivers revealed that individual sections had long been organically and inorganically critically polluted; river sediments were found to have been polluted by toxic elements and compounds. River pollution by lead, cadmium, zinc, chromium, nickel and other metals is only gradually diminishing. For example, in spite of closure of the Idrija mercury mine in the Idrijca sub-Alpine region, mercury still continues to be a pollution source. However, thanks to a systematic reduction of pollution in Austria between 1987 and 1994, water quality in the Mura and to a lesser extent the Drava rivers improved. Slovenia's largest river, the Sava, suffered highest pollution in the 1980s, followed by a slight reduction, but remained high in the mid-1990s (i.e. Class 2–3 quality on average). Perhaps significant here is the fact that most large urban centres, including the capital, Ljubljana, and Celje in the Sava drainage basin, lacked any waste water treatment facilities.

By the mid-1990s, water quality varied in Slovenia's natural and artificial lakes. Particularly worrying was the decline in water quality in high mountain lakes. At the same time Bled's Alpine lake, which is an important tourist attraction, was ranked with other moderately polluted (mesotrophic) lakes. The self-purifying capacity of Lake Bled, with its artificial inflow of fresh water and outflow of polluted lake water from the bottom layers through a pipe, has also increased, as has the inflow of polluted lake water into the Sava Bohinjka river downstream. In contrast to Lake Bled, over the past few years water quality in Lake Bohinj has gradually declined. During the mid-1990s Lake Bohinj was ranked among mesotrophic lakes largely due to its average concentration of nitrogene compounds, and among oligotrophic unpolluted lakes according to the annual concentration of phosphorus and chlorophyll. Excessive input of nutritive and organic matter is apparent during summer stratification due to the lowered oxygen concentration at the bottom of the lake.

Environmental pollution in Slovenia first influenced the quality of surface waters, and later ground waters, i.e. those in non-consolidated sediments and karstic areas. Since 1990, ground water pollution has not declined. This represents an acute landscape-degradational process reflecting the diverse self-purifying capacities of Slovenian regions and water resources, and excessive environmental pollution. Otherwise, the distribution of rich drinking-water resources, of which only one-third is exploited, is uneven. Their pollution already endangers drinking-water supplies in certain areas such as the karstic, sub-Mediterranean and sub-Pannonian regions. Rather small water-supply systems prevail in Slovenia, pumping water from numerous, less abundant sources. Such 'fragmentation' makes control over potable water quality more difficult, as well as their protection against pollution. During the mid-1990s, about 90 per cent of Slovenia's population was supplied with water from public systems. Ground water provided over half (52 per cent), Karst water sources 45 per cent, and surface waters only 3 per cent. In 1994, households consumed up to 86.5 mln.cu.m (i.e. about 100 l/pc daily) of drinkable water, followed by various other economic activities, 56mln.cu.m, and manufacturing/mining, 33mln.cu.m (Urad Republic, 1998).

In the fluvio-glacial deposits of the Mura, Drava, Savinja, Sava, Soča rivers and some of their tributaries, there are large quantities of ground water which total over one-third of Slovenia's reserves. Ground water areas cover nearly one-fifth (18.4 per cent) of Slovenia and differ considerably in their landscape sensitivity. Water quality is regularly monitored at eighty-four sample points in eighteen areas. Ground water provides drinking-water sources in state-monitored areas, and is especially important in the sub-Pannonian region. Slovenia lacks more abundant ground water resources in areas of solid rock, and in the Ljubljana basin with its heavy consumption of drinking-water (Brečko, 1998a). Ground water areas at the bottom of basins and valleys comprise one-sixth of Slovenian territory, and coincide with areas of intensive agriculture and urbanisation. As well as waste water from settlements and production activity, ground water is also polluted by rubbish dumps, abandoned gravel pits and the numerous livestock farms in northeast Slovenia. Intensive agriculture is also a source of pollution: hop production in the Lower Savinja valley, sugar beet, maize and other fodder plants in eastern Slovenia, and vegetables in all ground water areas. These have greatest impact on quality in shallow ground water areas where the water-table is less than five metres deep, as in Apaško, Mursko, Prekmursko and Ptujsko polje, together with the Lower Savinja valley. Here during the first half of the 1990s, ground water was heavily polluted with nitrates and pesticide residues. Ground water in Dravsko Polje, the most abundant water source in northeast Slovenia, was also polluted with pesticide residues; yet this not only resulted from agricultural pesticides but, above all, from dumping solid and liquid waste from a pesticide manufacturing plant into a gravel pit.

The pollution of ground water from agriculture is also present to a lesser degree in some other areas such as Krško-Brežiško and Soško polje and the Kamniška Bistrica valley. Ground water areas in central and western Slovenia have greater self-purifying capacities, with the water-table mainly over 15m deep, having a faster resupply of water and greater reserves. Thus they provide better quality water in the Ljubljansko, Sorško and Kranjsko polje together with the Ljubljansko Barje, in spite of numerous environmental pollution sources. In these areas, damage to water quality is caused mainly by industrial pollutants and waste water from settlements (Brečko, 1998b). In 1996, the permitted nitrate concentration level in drinking-water (50mg/l) was at least exceeded occasionally in half of Slovenian ground water areas; the stricter standard imposed on atrazin concentration (0.1 microgram/l) was exceeded everywhere, except in the rim areas of the Ljubljansko Barje, the Čateško and Soško polje, and the Vipava and Hudinja valleys. In 1995 chlorinated and halogenated organic solvents were found to be present in the ground water of the Kamniška Bistrica and Lower Savinja valleys, as well as the Soško, Sorško, Krško-Brežiško and Čateško polje. Of the trace metals, zinc was most frequently found (MOP, 1997a).

Over two-fifths (44 per cent) or about 9000sq.km of Slovenia's territory is karst; its water resources have a moderate self-purifying capacity. Half of Slovenia's drinking-water comes from karstic sources, and their extensive and cavernous hydrographic catchment areas are very vulnerable ecologically. In western and southern Slovenia (i.e. the Alpine and Dinaric areas), karstic water sources are often polluted, but because alternative sources are lacking they are very important for supplying drinking-water. State monitoring of water quality is provided by about seventy source points, where pollution of water and sediments is periodically controlled. However, regular monitoring of water quality occurs at larger endangered karstic sources used for supplying drinking-water. The quality of karstic sources is most heavily affected by unprocessed waste water from settlements, while numerous illegal dumps in their catchment areas pose a delayed threat. By the mid-1990s, water pollution also suffered from increased concentrations of lead, cadmium, zinc, chromium and other metals found in sediments. In the mid-1980s, a specific form of chemical pollution occurred in a karstic source at Krupa in southern Slovenia near the Croatian border. This resulted from a long period of dumping used capacitors from a nearby plant into karstic sink holes in the Krupa catchment area. The extremely high concentration was reflected in the water source, comprising polychlorinated biphenyls (PCB), which was over a hundred times higher than the permitted concentration level (Plut, 1987). Industrial PCB pollution of the Krupa River has gradually declined, but in 1996 it was still ranked as having only Class 4 quality.

Part of the North Adriatic coast also belongs to Slovenia. It is 46.6km long and the sea here is shallow, making its marine ecosystem ecologically vulnerable from the heavily polluted solid/liquid waste of about 15mln inhabitants. Its geographical location along the rim of the calm and shallow North Adriatic, where depths rarely exceed 50m, together with the geological/hydrological conditions of the hinterland, curtail its self-purifying capacity. The inflow of hinterland water, as well as solution from interior rocks supply rather large amounts of nutritive soils, which means the sea here is inclined towards eutrophication (Malej, 1994). Further danger arises from dense coastal settlement, intensive agriculture, heavily industrialised hinterlands, and a long legacy of *laissez-faire* attitudes towards environmental protection. The pollution of Trieste Bay is mainly caused by large quantities of waste water supplied by the Po River. Its 75,000sq.km drainage area contains polluted material from the industrialised, intensively farmed and densely inhabited region of North Italy. The Po River contributes a major part of pollution into the northwest Adriatic Sea. This polluting process is on a large scale; its macro-regional dimension leads to the meso-regional pollution of Trieste Bay, where local pollution from Slovenia also adds its contribution to this part of the sea (Radinja, 1990). About three-quarters of all nutritive soils containing nitrogen (73 per cent) and phosphorus (74 per cent) come from the Po River (Malej, 1994). The major autochthonous pollutants of the coastal area are industrial and communal waste water sources from littoral settlements as well as heavy maritime transport (Koper's port), representing a considerable threat particularly from the transport of crude oil. In summer months tourist facilities significantly increase the amount of waste water. Over 400,000 people live along Trieste Bay where daily water consumption reaches 300 l per inhabitant, a proportional amount of which is sewage. The so-called 'flourishing' of Trieste Bay has resulted from the excessive growth of algae. This coalesces into dense, slimy masses which are pushed by winds and tide towards the coastal shallows where a layer of algae, seaweed and waste half a metre thick can develop. In the long and extremely hot summer of 1994, the localised death of bethic fauna was noticed in the bottom layers of Trieste Bay. This was mainly due to lack of oxygen and the poor hygienic quality of coastal waters, in certain areas heavily impaired by proximity to the inflow of waste water from the littoral towns.

Air pollution

The geographical diversity and mosaic-like form of Slovenian regions strongly influence the extent and degree of environmental degradation, above all air pollution. Areas of heavy and weak environmental degradation are intertwined resulting in regional and/or valley and basin air pollution. The most polluted Slovenian regions lie in sub-Alpine and Alpine Slovenia, and are limited to basins and narrow and deep mountain valleys. The enclosed landform characteristics and related meteorological and other features further intensify the negative effects caused by environmental pollution and contribute to increased degradation. In turn, this also causes significant disparities between emissions and concentration of air pollutants (CoAP) (Špes, 1998).

The tranquillity of Slovenia is particularly unfavourable from an air pollution viewpoint. Temperature inversions are often formed, causing a concentration of polluted air over a rather limited space covered with an inversion cap, a frequent phenomenon during the colder half of the year. This period is also very detrimental because of communal winter emissions from residental heating sources which directly influence annual degradation patterns through specific maximums during the colder months. Annual air pollution fluctuations are intensified, especially in densely inhabited areas with their predominantly communal housing emissions; industrial emissions are more evenly spread over the year.

Slovenia's most polluted CoAP areas are not evenly or concentrically distributed around the main emission sources; typical are the horizontal and vertical asymmetries, most evident on damaged vegetation. Intensified negative landscape effects caused by environmental pollution are also found in areas where silicate rocks occur beside carbonate ones because the former poorly neutralise acid precipitation. The distinctiveness of a degraded environment in the most polluted CoAP areas (Celje basin, Zasavje region, Mežiška dolina valley) is surprising, because they contain only a few small towns. Their smaller emission amounts and emission synergy with industry do not occur, so that the negative effects of single-source emissions prevail in these areas. Moreover, none of these sources is very large, neither from industrial nor energy-producing plants. In the past the old, technologically outdated and ecologically unfriendly industrial plants, with their origins dating back to the mid-nineteenth century, contributed most of these emissions. Most of them were not based on domestic raw materials; more importantly their origins were based on craft traditions, favourable transport locations and proximity to energy sources, all found in valley bottoms and basins. From the ecological viewpoint, this was an unfavourable situation, made more acute by planned postwar industrialisation. This meant that Slovenian industrial towns suffered heaviest environmental pollution from the end of the 1960s to the mid-1980s, when it was important to overcome economic backwardness rapidly with planned industrial expansion. Greatest sufferers in this process were towns located in narrow, deep valleys and basins, where possibilities for neutralising harmful emissions were limited (Špes, 1998). At the beginning of the 1970s, Trbovlje, Celje, Žerjav, Mežica, Črna and Hrastnik had the highest air pollution concentrations of all major Slovenian towns, especially for SO_2 and smoke. High CoAP values were found in the annual, monthly and daily averages, while critically high values mainly occurred as half-hourly values, and certain places even exceeded 5mg/cu.m (Paradiž, 1994).

In contrast to other countries in the region, a survey of Slovenia's heaviest polluted areas revealed that it was not the largest urban centres which were affected most, but hinterland towns near to thermal plants and major industrial units, usually situated in areas of very unfavourable meteorological conditions. The most polluted areas of Slovenia are still the Zasavje region (Trbovlje, Hrastnik, Zagorje) where urban, industrial and only indirectly thermal plant emissions have shared the consequences; the rim around the Šaleška dolina valley due to thermal plant emissions; towns in the Mežiška dolina valley and Celje with their industrial and communal housing emissions. However, experts have also noted an increase in summer ozone concentrations, which often exceed recommended maximum levels for urban centres. In larger Slovenian towns, individual streets with intensive transport activity are further subjected to excessive pollution of NO_x at street level; all other transport pollutants occur only in narrow tiers along these transport routes (MOP – Environmental Report, 1996).

A recent survey of CoAP values, especially SO_2 as the main air pollutant in Slovenia, has revealed a gradual pollution decrease over the first measurements made in the mid-1980s. Maximum and average winter concentrations have especially declined; thus the frequency of the measurements performed at the time when the maximum or limit/critical values were well above agreed norms has been reduced. However, old industrial towns which were among the most polluted over the twenty years of measurement still remain the most degraded, even though their industrial plants have considerably reduced the amount of harmful emissions. This confirms the belief that it is the share of communal housing emissions which has actually increased, although the industrial sources are still considered the main cause. The annual amount of SO_2 emissions in Slovenia increased up to 1986, but since then they have gradually declined, thanks to technological innovation in industry and attempts at ecological improvement. Production in outdated, non-competitive industrial plants has ceased due to the economic crisis and social changes, which often led to a reduction in environmental pollution. In addition, numerous industrial branches have either substituted poor-quality domestic coal with better imported varieties (e.g. coal from Indonesia, Bohemia), or replaced it with other energy sources like gas/fuel oil. More than one-third of Slovenia's population lived in urban settlements containing a large share of residences heated by natural gas in 1996. Even so, the amount of natural gas utilised for general consumption is still relatively low. Reduced concentration of pollution in recent years has resulted from milder winters, more favourable winds and therefore fewer temperature inversions. The implementation of occasional revitalisation programmes and/or banning the burning of high-sulphur coal in towns such as Maribor has also helped.

Slovenia still produced the following emission amounts in 1996: 109,700t of SO_2, 70,144t of NO_x, 95,371t of CO, and 15.1mln.t of CO_2. Over half of these emissions remain in the air over Slovenia, while the other half drift away along air streams to other European countries. Between 1990 and 1996, SO_2 emissions in Slovenia dropped by 44 per cent, or from 97kg/pc to 55. According to 1996 data, all but 5 per cent of SO_2 emissions came from industrial plants; the leading offenders were the pulp-and-paper mill at Krško, the zinc factory at Celje and the oil refinery at Lendava. Over 85 per cent of SO_2 came from three thermal plants (TP); their contribution oscillated slightly from year by year, according to the temperature and length of winter, together with the amount of nuclear energy produced, its price, related economic saving, etc. Of the total TP emissions, the Šoštanj plant contributed 67 per cent, Trbovlje 24 per cent, and the heating plant at Ljubljana 9 per cent. Emission data revealed that all three

TP plants exceeded the permitted limit for emission concentrations. About 6 per cent of SO_2 was generated through residential heating, in which the quality of market-available fuels was quite significant (Environmental Performance Reviews, 1997). By 2010, Slovenia will hopefully have reduced SO_2 emissions by 70 per cent over their total in 1980, when they amounted to 235,000t. In contrast to SO_2, annual emissions of NO_x have increased during the past decade; between 1990 and 1996 they grew from 28.3kg/pc to 35.2, largely as a result of the increased density in road transport. In 1995, this rise was curtailed due to the larger number of cars with catalytic converters. Road transport contributed almost two-thirds of all NO_x emissions. Thermal plants contributed about one-fifth in the mid-1990s, compared with 31 per cent in 1980. Even so, they were the main intermittent source of emissions in the mid-1990s (Environmental Performance Reviews, 1997). According to EU demands, NO_x emissions in Slovenia had to be reduced by 35 per cent in the year 2000 compared with 1985; this will certainly not be achieved easily, considering actual conditions and rapid growth of motorised transport.

Transport also contributed over 94 per cent of all CO emissions, which between 1990 and 1996 increased by almost one-fifth. The most rapid increase in emissions came from CO_2 (principal gas in the greenhouse effect); over three-quarters (76 per cent), was registered in Slovenia between 1971 and 1989, and directly attributable to the growth of road transport. In 1996, CO_2 emissions reached 15.1mln.t, mainly coming from thermal power plants and road transport. Any increase in annual CO_2 emissions in Slovenia is contrary to plans for EU membership. According to the 1997 agreement, annual emissions should be reduced by 15 per cent by 2010 compared with 1990. As regards CO_2 emissions from fossil fuels per inhabitant, Slovenia had 7900kg/pc (Plut, 1998).

During the summer months of vegetation growth the limits of CoAP ozone values were also exceeded at most measuring sites, even in mountainous areas. In the 1970s and 1980s, the average concentration of SO_2 in the most polluted Slovenian towns (Trbovlje, Hrastnik, Zagorje, Črna, Celje, Ljubljana, Krško, Maribor) amounted to 0.20–0.45mg/cu.m of heat during the colder seasons. At the beginning of the 1990s, the concentrations of SO_2 in the heating season in the above urban centres decreased to 0.08–0.2mg/cu.m; during non-heating seasons, they amounted to 0.01–0.03mg/cu.m, and the annual average was 0.6–0.11mg/cu.m. During the mid-1990s, the average annual concentration in the most polluted towns totalled 0.04–0.06mg/cu.m (Ciglar, 1997).

In the first half of the 1990s, air pollution (especially with SO_2 and smoke) decreased, especially in urban areas. Nevertheless, the maximum limits and critical values were still exceeded in certain places such as the Zasavje, Šaleška dolina valley and its rimland, the Upper Mežiška dolina valley, Ljubljana, Škofja Loka, Celje, Maribor, etc.; most of these places transcended the maximum SO_2 limit, as well as those for suspended particles and ozone. In non-urban areas, the main pollutants occur as high concentrations of ozone and photo-oxidants, and polluted precipitation, especially in areas around thermal power plants with their high SO_2 emissions.

Forest degradation

Over half (54 per cent) of Slovenia is covered by forests which represent, along with water, the country's most important natural resources. They are most extensive in the Alpine and Dinaric karstic regions, where their proportion is even higher due to the overgrowth of farming areas, especially pastures. According to data from the Forestry Institute for 1996, their large area and abundant wood reserves totalled 210cu.m/ha, with a current annual growth of 5.5cu.m/ha. Nevertheless, their degradation is proving a worrying factor as a result of both man-made and natural influences (Ferlin, 1998). Foresters estimated that around 17 per cent of trees suffered from over 25 per cent defoliation in 1995. This resulted both from the impact of air pollution and natural causes, such as drought, pests and arboreal diseases. Furthermore, more than a quarter of trees (coniferous 37.6 per cent; deciduous 17.6 per cent) were above the degradation threshold of less than 10 per cent defoliation (Bogataj, 1997). Most heavily affected were firs, chestnuts and oaks. The least defoliated trees grew in the Alpine region, while higher defoliation occurred in the sub-Mediterranean (especially beech) and Dinaric karstic regions (particularly fir). Defoliation has slightly decreased over the past few years, largely due to the marked improvement in the condition of coniferous trees, but defoliation among deciduous trees has increased slightly (Table 11.1).

As to the degree of degradation, Slovenian forests were similar to the Central Europe average; even so, tree degradation is still high in spite of certain improvements. From a more general viewpoint, the most endangered trees in Slovenia are those damaged genetically because their capacity for natural reproduction has been reduced. These are primarily located in areas of high emission sources generated from industrial and energy-producing plants in such areas as the Upper Mežiška, Šaleška, Dravska and the Upper Sava valleys, the Zasavje, the surroundings of urban centres like Celje, Ljubljana, Maribor, Kranj, Škofja Loka and Krško, and also Žirovski Vrh due to its distinctive character. From an ecological viewpoint, forest stability has diminished in recent years, largely due to increased natural disasters after 1993, and continued air pollution which still has a strong impact on their survival. Currently, acid rain does not influence the degradation of Slovenian forests due to the neutralising capacity of carbonate surfaces.

Table 11.1 Defoliation index (>25%) 1987–96 (%)

Tree species	*1987*	*1989*	*1991*	*1993*	*1996*
Coniferous	43	39	31	31	27
Deciduous	9	8	6	8	13
Total	24	23	16	18	19

Source: Bogataj, 1997.

Soil pollution

Soil/vegetation pollution in Slovenia has been systematically studied since the early 1990s. The most extensive research has been done in industrially polluted areas near to larger urban centres, along heavily used transport routes and areas of intensive agriculture. Monitoring of soil pollution between 1989 and 1996 was carried out by the Centre of Pedology and Environmental Protection. Soil analysis from over 800 samples at 267 chosen locations throughout Slovenia, together with the analysis of 200 samples of plant pollution, was undertaken (Poročevalec, 1996). Studies revealed that soil/vegetation pollution was most affected by transport emissions, while agriculture had only a minor impact. For example, to the east of Celje, soil is heavily polluted by zinc from industrial emissions; in Ljubljana, by lead from car exhausts along transport routes; in the Jesenice area from heavy metals and emissions from its ironworks; in the Mežiška dolina valley, by lead from industrial production, etc. In areas of intensive agricultural production (Dravsko, Ptujsko, and Krško polje) research did not reveal any significant pollution, except for pesticide residues which are omnipresent. Investigations also confirmed how soil pollution affected plants, especially vegetables. Among the most polluted plants were salad and root crops. Experts have also anticipated the harmful effects of pollution on other agrarian cultures and forest vegetation.

Waste management

Almost half the total waste consisted of waste from agriculture, forestry and the food-processing industry (40 per cent); this was followed by the building industry (26.3 per cent), energy-production (13.7 per cent), and human settlements (10 per cent). The lowest amount (9.7 per cent) came from manufacturing industries. There are very few natural possibilities for utilisation as waste dumps in Slovenia. This is due to the extensive karstic topography and ecologically sensitive ground water areas. Organisation of a waste collection system is further hindered by the variety of landforms and prevailing dispersed pattern of settlement. The data on the amount of generated municipal waste are only approximately known. In 1995, municipal waste amounted to some 850,000 to 1,023,000mln.t and accounted for about 10 per cent of total waste generation (450–512kg/pc p.a.) (Radej *et al.*, 1999). Over time, the proportion of people whose waste is collected regularly has increased, reaching three-quarters of the population by 1995. Naturally, the amount of waste has rapidly increased (7 per cent annually), leading to a growing problem of waste treatment and management. Nine-tenths of waste is dumped at prescribed locations in Slovenia, with only one-tenth usefully recycled (Plut, 1997). Of the country's fifty-three municipal communal waste dumps, twenty-nine (i.e. 55 per cent) are illegal; there are even more dumps that lack proper supervision, in particular heavily polluting water and soil resources. Most of these dumps are expected to be full within the next decade. Selective waste collection is organised in Slovenia, but mainly in a restricted number of settlements, or at certain specific dumps which restrict acceptance of waste quantities in order to defer complete infill for several years. One can divide Slovenia into two parts regarding regular waste collection. Most municipalities in the western half of the country provide households with a regular collection, but in the eastern part only half of the residents enjoy such facilities. Between 10 and 20 per cent of communal waste is currently taken to 10–15,000 illegal dumps. If one takes into consideration the smallest dumps, the number of illegal dumps would exceed 50,000. Such smaller illegal dumps are very numerous and active in the vicinity of settlements lacking organised waste collection, while abandoned and unrestored dumps are present throughout the country.

Besides communal waste, Slovenia produced 19,000t of hazardous waste in 1996. From then onwards, an annual register was kept on hazardous and special waste. Between 1991 and 1995, thirty-three enterprises were established to perform waste treatment services. A total of thirteen sites exist for collecting special waste along with two waste incineration units. Pharmaceutical waste is burned at the Lek plant in Lendava, and phytopharmaceutical rubbish at the Pinus plant at Raće; both plants are located in northeast Slovenia. Radioactive waste generated at the Krško nuclear plant is disposed of on site. By 1997, 2599cu.m of low and medium radioactive waste had been eliminated, while 470 burnt-out rod elements were disposed of in special water-filled pools containing boric acid. Another temporary dumping point for low and medium radioactive waste is the research institute at the Nuclear Reactor Centre in Podgorica, near Ljubljana. It disposes of waste from smaller sources such as hospitals, institutes/research institutions and manufacturing industries. There is also an abandoned uranium mine located at Žirovski Vrh and covered by overburden.

Degradation of the urban environment

More than half (50.5 per cent) of Slovenia's inhabitants now live in urban areas; if one includes the outskirts, this totals three-quarters of the population. Two decades ago, urban dwellers were lower than two-fifths. Immediately after 1945 intensive demographic growth began due to rapid urbanisation and a marked desertion of agrarian activities. Fastest growth took place during the 1960s, largely due to some urbanised areas increasing their inhabitants by one-fifth or more from 1961 to 1971. More generally, demographic conditions stabilised during the 1970s and 1980s, but the number of inhabitants in the vicinity of urban areas began to grow rapidly, so that suburbanisation spread to cover almost one-third of the country. After 1981, house-building on the urbanised outskirts became higher than in towns.

Urban areas in Slovenia suffer from the most polluted landscape ecosystems. Air and water pollution are the main offenders, followed by soil pollution (especially heavy metals), landform degradation due to mining, disorganised communal and other waste dumps, and the effects of increased corrosion of materials due to acid rain, louder noises, etc. As a general rule, the more heavily degraded urban areas lie in the Alpine and sub-Alpine regions. Their locations at the bottom of valleys and basins intensify the lack of space, leading to the physiognomic heterogeneity of individual urban quarters. Thus industrial, public utility and other urban space functions are heavily entwined. The regime of annual degradation with its winter maxima is clearly evident. At first glance, the distinctiveness of the degraded environment is surprising because it is usually found in smaller towns with lower quantities of communal housing emissions. However, they are still too high for the natural capacity of their ecologically sensitive basins and valleys where such towns are located.

It is no coincidence that towns with century-old industrial centres are the most polluted. From the 1970s to the mid-1980s air pollution was the dominant factor in urban areas such as Ljubljana, Celje, Štore, Hrastnik, Trbovlje, Zagorje, Jesenice, Medvode, Ravne and Črna with their profuse industrial and communal housing emissions, although less from transport. In the mid-1970s, about 15 per cent of Slovenia's population (i.e. 36 per cent of urban inhabitants) resided in an urban environment containing critical air pollution levels, while a further 14 per cent lived in an excessively polluted environment. After 1980, the share of industrial emissions began gradually to decrease largely thanks to increased use of quality fuels; there was also greater utilisation of filters and, especially over the past decade, closure of certain industrial plants and reduced production. Nevertheless in 1991, about a quarter of the country's inhabitants, i.e. 44 per cent of the urban population, still resided in areas having excessively polluted air. Air pollution continues to be an acute problem in Slovenian towns; parallel with the reduction of SO_2 and smoke emissions has come increased concentrations of nitrogenous oxides, ozone, etc. as a result of increased urban road transport. A typical feature of these emissions is the lack of specific annual oscillations, because of constant transport flows and the prevailing pollution of the bottom air layers, which occur fairly evenly throughout the year.

Pollution of urban water sources depends on the self-purifying capacity of individual streams and quantity/composition of waste water from urban areas. The regenerative capacity of Slovenian rivers where most urban settlements are sited is 8 12km for Class 1 purification. Yet major polluters such as urban centres and industrial settlements are more than twice as densely located along rivers on average as elsewhere, thus exceeding a stream's regenerative capacity (Radinja, 1979). With regard to the amount of pollution in urban water sources, at least three-quarters of Slovenia's urban population reside near to moderately or heavily polluted streams (i.e. Class 2–3 and Class 3 quality), whil over one-seventh of urban dwellers live in settlements with critically polluted water sources.

Environmental pollution was taken into indirect consideration only during the postwar boom of Slovenian towns. First, the built-up areas became denser within the urban district framework, often located in the worst polluted environments. This expansion further contributed to the annual amount of communal housing emissions, especially in winter. Some new districts were even enlarged in a polluted environment. When urban space became scarce, towns were largely expanded outwards to areas which were previously regarded as less attractive for habitation. This was due to their peripheral location and remoteness from transport facilities, although they were sited in less polluted environments, a decisive factor when originally built. Generally speaking, environmental degradation is not considered much of a limiting location factor when residential districts are expanded. More recently, increased construction of the latter has meant that they now reach into the ecologically more favourable outskirts. In fact, non-ecological reasons were decisive in this expansion process, although on occasion they do coincide with ecological justifications as well (Špes, 1994).

Social ecology

Research has taken place into people's attitudes towards the environment in Slovenia's most polluted regions and their reactions to it. While economic, social and also partly political factors were noted, the greatest need was to solve basic existing problems (Špes, 1994). It was also proved that a relatively low ecological awareness by inhabitants existed. From these Slovenian investigations, it also appears that ecological awareness increases when analogous to general education. It is therefore necessary to expand not only ecological education but also the overall general educational level, to enable a broader understanding of causes/consequences of environmental pollution. Therefore, ecological awareness should also play a part in forming ideas about the role of the living environment as a broader, general concept.

According to certain socio-economic indicators, individual population groups already understand their residential and wider environmental situation, along with ecological problems which have elicited similar responses. Their spatial distribution also effectively reflects that of the degradational structure. Education and economic powers of the population are among the most meaningful indicators of differentiation in sections of the diversely polluted landscape. The difference is that the educational structure emphasises indirect influence such as planned or proposed changes, while economic power has a direct impact. The higher the education level, the stronger the ecological awareness and ability to distinguish the real ecological problems; this applies both at residential level and in a wider global context, and these people make the most suggestions for improving the environment. In this group, people's potential migrations are most likely to be ecologically conditioned. Analysis of income data has also confirmed that a larger number of economically weak groups live in the heaviest polluted environments; in contrast, less polluted areas enjoy a gradual increase in average buying power. Typical in the Slovenian context, especially regarding ethnic structure, are the differences that exist between statistical and survey data. Survey data reveal that members of other ethnic groups react more or less equally to environmental pollution, or have attitudes similar to the locals. This suggests that they have attitudes to the environment similar to those of their Slovenian neighbours. Similar findings related to the large number of inhabitants of the most polluted urban regions of Slovenia who came from other former Yugoslav republics. Data on the population's ethnic structure appear to be quite a reliable indicator of urban differentiation in areas with the worst residential environments. Result have shown that a degraded environment is most passively accepted by the lowest socio-economic groups, which ensures its preservation or even degeneration.

It has been generally proved that due to flawed economic conditions, people are more tolerant in evaluating degraded environments than they were a decade ago. However, they are more realistic when it comes to emissions; they not only blame manufacturing industries, but also recognise the negative influence of residential heating emissions generated mainly during winter, especially from low-quality fuels. It is unfortunate that the most polluted urban areas have the lowest average socio-econmic structure, and the inhabitants are least likely to identify the main environmental pollution source. Instead they prefer to blame manufacturing or transport as the main culprits, rather than the heating of their homes. This deters the introduction of environmentally friendly installations and improved heating methods. In general therefore, ecological awareness is lowest in areas where people have been directly dependent on employment in the manufacturing/mining sector over several decades. Ecological awareness is higher in rural environments not only because of closer links and dependence on nature, but also because there is less dependence on manufacturing for employment (Špes, 1998).

Regional dimensions of Slovenia's pollution

In Slovenia, the following landscape–ecological types can be established from an environmental protection/vulnerability viewpoint: Alpine (mountainous), Dinaric (karstic), sub-Pannonian, sub-Mediterranean (littoral), and basin/valleys. A detailed landscape–ecological division of urban settlements showing the highest landscape vulnerability confirms that urbanised areas in parts of narrow valleys are the most sensitive and polluted by emissions. These are followed by pollution from emissions in basin areas in the sub-Alpine and sub-Pannonian regions. Problems are further accentuated by long-term negative human activity on the landscape, especially in basin/valley ecosystems, together with – and more explicitly – several recent decades of emission pollution from the manufacturing and energy sectors. High landscape sensitivity is typically found in narrow sub-Alpine valleys and basins, especially head-water areas, enclosed landforms, areas of poor ventilation and frequent temperature inversions, where excessive emissions affect only the modest self-purifying capacity of the region (Špes, 1994).

The basic geographical/regional elements for environmental protection planning and decision-making are clearly the predominantly mountainous and forest-covered territory, and to a lesser extent the bottoms of the larger valleys/basins and karstic fields where a high percentage of land has only shallow soils. Other areas needing attention are those which possess a predominantly dispersed settlement pattern. The influence of factors such as over-intensive use of the cultural landscape, emigration from mountainous areas, intense postwar industrialisation and the uncontrolled degradation of air/water sources, forest areas and valley/basin landscapes all demand greater planning application by future generations.

Environmental movements

A new, post-independence Slovene legal system has not yet been achieved. During the transitional phase all previous laws remained in force, providing that they did not conflict with any part of the newly emerging legal system. It is therefore expected that several environmental laws will continue to be applicable for some time to come. Slovenia is paying special attention to harmonising its new environmental legislation with that of the EU. The Environmental Protection Act (EPA) of 1993 remains the basic law for environmental policy and management. Since independence it has contained the first comprehensive collection of aims, principles and rules for environmental management. The EPA is based on Agenda 21, environment-related EU directives, and Slovene experience of environmental management. It provides the basis for preparing other laws. Regulations, directives and standards are being prepared at different government levels, depending on the law being formulated (Economic Commission for Europe, 1997).

Since 1990, the number of environmental non-governmental organisations (NGOs) in Slovenia has increased rapidly. More than eighty such organisations now exist. In terms of their membership, Slovene NGOs are relatively small (up to fifty members). Larger membership numbers are to be found in the twenty-two Slovene Federation of Environmental Protection Associations, the Slovene Natural Sciences Association (800 members), the Slovene Environmental Movement (310 members) and the Association for the Observation and Examination of Birds (300 members). Slovene NGOs function at the local, regional, national and international levels. The range of NGO activity is contextually very extensive. They cover all important environmental activities, especially through education and promoting public awareness. They also draw attention to government offices for environmental problems. This involves proposing potential environmental solutions and legislative initiatives, as well as promoting permanent developments and helping to monitor the state of the environment, etc. NGO activity can broadly be characterised by the following: three organise educational and training programmes, nature protection, ecological summer camps and campaigns. For example, in some of the larger cities there are crusades to collect bulky waste free of charge. Three are involved in environmental policy-making. Three have participated in all major international meetings since the United Nations Conference on Environment and Development in Rio de Janeiro, and on occasions NGOs have held alternative meetings simultaneously with governmental conferences.

In 1993, on the basis of EPA provisions, the Slovene Parliament established the Council of Environmental Protection, a body of researchers and scientists.

The Council provides descriptions of environmental conditions and analyses special topics, such as the National Highway Network Construction Programme, waste disposal and treatment problems, and a strategy for the effective use and supply of energy. In 1996, the first volume of New Circumstances and Horizons was issued, entitled *Views on Nature and Environmental Protection in Slovenia*. It has been widely distributed, and a second volume on chemicals and the environment was published in 1997 (Economic Commission for Europe, 1997).

Conclusion

In the 1990s, Slovenia had to face the challenges of sustainable development against a backcloth of exhausted natural resources, degraded valley/basin ecosystems, detrimental air pollution and problems associated with forested areas, water resources and unorganised dumps. All these difficulties came at a demanding time in the transition to an independent market economy after a decade of declining levels in the standard of living (Plut, 1997). As a medium-ranked polluter amongst European countries, Slovenia had built up its economic development during a period of socialism based on exploiting natural resources and polluting the landscape, especially the air, water and forests. Even after independence in 1990, the environmental basis of economic development continued to be neglected, if only in a rather more moderate form. In Slovenia, as in other post-socialist transition countries, the danger exists of too strict an adherence to market laws at the expense of the environment. It is therefore very important to enforce principles of sustainable development. However, the search for accord between development and environmental protection is not easy. Slovenian regions have different landscape vulnerability and pollution intensity. Account must be taken of their original landscape features when subjected to environmental and developmental restoration. Unfortunately, environmental awareness in the heaviest polluted settlements is rather low, as many local inhabitants are employed in environmentally polluting plants (Špes, 1994). Moreover, the social crisis which followed changes in the political system meant that environmental protection was rather neglected. Unfortunately, there is a belief amongst politicians and the public that environmental protection hinders economic growth and secure employment.

The introduction of sustainable development principles from a spatial, environmental and regional aspect conflicts with economic, regional, agricultural and transport policies. Yet the latter have top priority in Slovenia's attempts to adapt to the European concept of development. Irrespective of the EU decision on Slovenia's application for membership, the country has already integrated itself into the European economy. It now has a two-thirds share in goods exchange and enjoys a certain degree of cooperation with EU members. These factors actually forced Slovenia to adapt its economic, developmental, infrastructural, social, environmental and partly demographic policies to basic strategic and developmental EU directives. Adaptation of Slovenia to EU norms in the environmental field will only be possible in the medium term. This is dependent on promulgating plans and enforcing environmental legislation, provision of a rapidly executed national environmental programme and creation of a sound development strategy. Unfortunately, creation of an efficient legislature capable of embracing EU directives in all individual environmental fields, such as management of urban waste water, drinking-water, waste disposal, will only be possible in the long term. It will also demand an essential increase in environmental investment and administrative capacity (Commission of European Union, 1997). Failure in the state budget to provide increased finance from 0.5 per cent to about 1.5 per cent GDP by 2003 for protection of the environment and nature will mean that Slovenian enterprises will further lag behind ecologically friendly industrial processes adopted in other developed countries. This, in turn, will diminish their own medium/long-term competitive capacity. After Slovenia's independence there were attempts to intensify and integrate the economic, political, safety and cultural aspects of the country into the newly formed European space. Its rapid adaptation to contemporary processes of sustainable development makes such moves even more important and, in the end, inevitable.

References

Bogataj, N., 1997, *Degradacija gozda s posebnim ozirom na metodološko problematiko popisa* (magistrsko delo), Oddelek za geografijo, Filozorska fakulteta, Ljubljana.

Brečko, V., 1998a, *Vpliv pokrajinskoekoloških dejavnikov na vodno oskrbo Ljubljane* (magistrsko delo), Oddelek za geografijo, Filozorska fakulteta, Ljubljana.

Brečko, V., 1998b, 'Pokrajinska občutljivost območij podtalnic z vidika kmetijstva', *Kmetijstvo in okolje*, Kmetijski inštitut, Ljubljana, pp.49–56.

Ciglar, R., 1997, 'Kemikalije v zraku. Kemizacija okolja in življenja – do katere meje?', *Slovensko ekološko gibanje*, Ljubljana, pp.295–316.

Commission of European Union, 1997, *Agenda 2000 – Commission Opinion on Slovenia's Application for Membership of the European Union*, DOC 97/19, Brussels.

Economic Commission for Europe, 1997, *Slovenia*. Economic Commission for Europe, Committe on Environmental Policy. United Nations, New York/Geneva.

Environmental Report, 1995, Republic of Slovenia – Ministry of the Environment and Physical Planning, 1996, Ljubljana.

Ferlin, F., 1998, 'Uspešnost novega sistema sonaravnega gospodarjenja z gozdovi v Sloveniji', *Gozdarski vestnik* 2, Ljubljana, pp.81–96.

Gams, I., 1991, 'Nekatere geografske stalnice Slovenije', *Geografski vestnik* 63 (Zveza geografskih društev Slovenije), Ljubljana, pp.7–24.

Malej, A., 1994, 'Vidiki varstva morja', *Okolje v Sloveniji*, Tehniška založba, Ljubljana, pp.289–294.

Ministry of the Environment and Physical Planning, *Environmental Report 1995, 1996*, Ljubljana.

MOP, 1997a, *Kakovost voda v Sloveniji v letu 1995*, 1997, MOP–Hidrometeorološki zavod Rep. Slovenije, Ljublana.

MOP, 1997b, *Onesnaženost zraka v Sloveniji v letu 1996*, 1997, MOP–Hidrometeorološki zavod Rep. Slovenije, Ljubljana.

Paradiž, B., 1994, 'Posledice onesnaževanja zraka v lokalnih in planetarnih razsežnostih', *Okolje v Sloveniji*, Tehniška založba Slovenije, Ljubljana, pp.330–344.

Plut, D., 1987, 'Slovenija – zelena dežela ali pustinja?', *Krt* 19, Ljubljana.
Plut, D., 1988, *Varstvo geografskega okolja*, Filozofska fakulteta, Ljubljana.
Plut, D., 1997, *Slovenija na križpotju*, Mihelač, Ljubljana, p.386.
Poročevalec, 1996, *Predlog poročila o stanju okolja 1995*, Državnega zbora, 29 February, Ljubljana.
Radej, B., Pirc Velkavrh, A., and Globevnik, L. 1999, 'Indicators on environment and development', *Urad R Slovenije za makroekonomske analize in razvoj*, Ljubljana.
Radinja, D., 1979, 'Onesnaženost slovenskih rek in njene pokrajinske značilnosti', *Geografski vestnik* 51 (Zveza geografskih društev Slovenije), Ljubljana, pp.3–17.
Radinja, D., 1990, 'Dimenzije Tržaškega zaliva in slovenskega morja ter njihov regionalni pomen', *Primorje*, Zbornik (Zveza geografskih društev Slovenije), Portorož, pp.13–20.
Špes, M., 1994, *Degradacija okolja kot dejavnik diferenciacije urbane pokrajine (na izbranih slovenskih primerih)*, (doktorska disertacija), Oddelek za geografijo, Filozofska fakulteta, Ljubljana.
Špes, M., 1998, 'Degradacija okolja kot dejavnik diferenciacije urbane pokrajine', *Geographica Slovenica* 30, Inštitut za geografijo, Ljubljana.
Statistične Letopis R. Slovenije, 1998, *Urad Republike Slovenije za Statistico*, SLRS, Ljubljana (and earlier editions).

Part III

Country studies

Southern group

12 Albania

Derek R. Hall

Introduction: history of a low level of development

Although only the size of Wales or the state of Maryland, Albania is notable for the beauty and variety of its natural landscapes (Figure 12.1). The country's mean altitude of 708m is twice the European average. Varied relief and climatic conditions, significant water and mineral reserves, a diversity of fauna and flora, and often stunning Ionian and Adriatic coasts characterize one of the hitherto lesser-known corners of Europe. Environmental pressures from economic exploitation and development were relatively low prior to the Second World War, although reclamation of some of the mosquito-infested lowland wetlands had begun.

Pre-war industry was restricted in scale, scope and location. With no large industrial establishments, processing and manufacturing was little more than a cottage industry. Eighty-eight per cent of all industrial enterprises employed fewer than fifteen workers, and by 1938 only 4.5 per cent of national income was derived from industry (Prifti 1978, p.52). It was mostly limited to food processing, with nine diesel-powered modern flour mills, two olive oil refineries (Elbasan, Vlora), a fish-preserving plant at Shkodra, a brewery (at Korça, built by the Italians) and four raki distilleries. Tobacco processing was also undertaken and cigarette manufacture took place at Durrës, Shkodra, Elbasan, Berat and Vlora, each of which also contained a soap factory. Building materials were represented by brickworks at Elbasan, Korça and Tirana, and cement plants at Shkodra and Elbasan, the latter begun in 1940. In all cases, manufacturing was limited to domestic requirements and was concentrated in a handful of towns. Power was usually provided by internal-combustion diesel engines, except for a small hydroelectric plant serving Korça. Two Italian companies provided electric lighting for Tirana and Durrës.

Oil was first raised in 1918, and by the end of the 1930s exploration and development of some 400 wells was largely under Italian auspices. The crude was piped to the coast near Vlora for direct export to Italy. The capacity of the pipeline was said to be 5000 barrels a day in 1937, and special wharves were constructed to handle two million barrels p.a. The first shipment took place in 1935. Just as there was no oil refining capacity, so too Albania possessed no

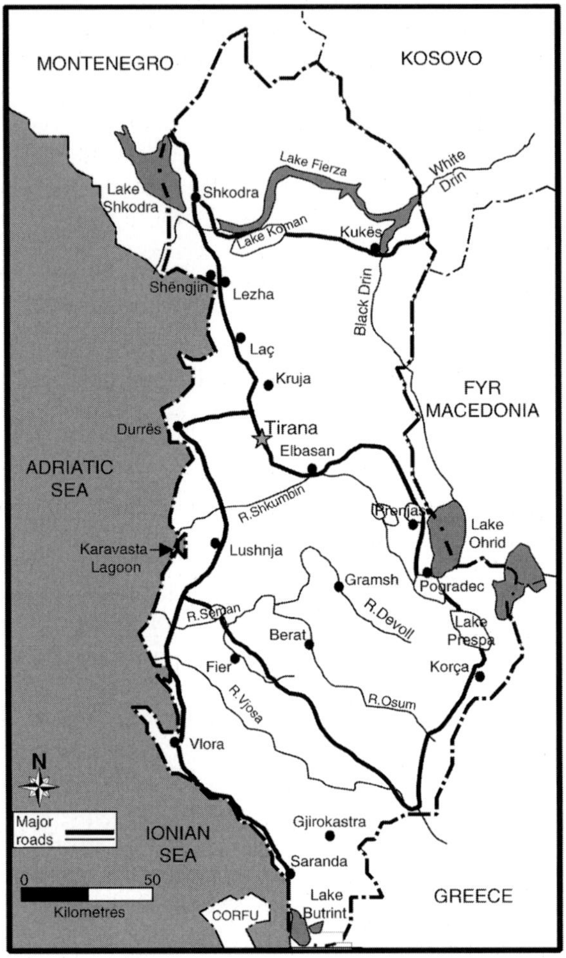

Figure 12.1 Salient features of Albania

facilities for the processing of the country's copper and chromium resources. Lignite was worked to a limited extent for domestic fuel and for brick making and lime burning. Tourism was limited (Hall 1991, 1994): at Durrës, latterly the focus of both domestic and international tourism, there were just fifteen cottages with seventy-three rooms and one hotel with twenty-eight rooms in 1945 (Bërxholi 1986, p.27).

A public health service was first created in 1922, although it was hampered by lack of funds and personnel. At that time tuberculosis was very common, with a particularly high incidence of adolescent pulmonary tuberculosis among the Islamic female population. High mortality levels in the Gjirokastra, Shkodra and Kruja districts were notable. Syphilis was widespread in rural areas. Intestinal parasites were common, and undernourishment was apparent everywhere. Cancer and cardiovascular diseases were reported to be rare. Anthrax was the major epizootic affecting humans, being widespread in the coastal plain. Most notably, malaria was endemic in all areas below 1200 metres, and hyperendemic in parts of the coastal plain and in the river valleys. The country's census of May 1930 enumerated a total population of just over one million (1,003,000). In December 1933 the Director-General of Health reported 250,000 cases of malaria (McKinley 1935) – one-quarter of the total population. Under the auspices of the Health Committee of the League of Nations, surveys were undertaken in 1923 and 1924 to prepare plans for the eradication or control of malaria (Haigh 1925). While the inaccessibility of many of the country's villages hampered the spread of infectious diseases, this situation was gradually being broken down during the inter-war period, with increasing seasonal and more regular migrations and movement between upland and lowland regions. Methods of recruitment for the army also contributed to this process (Mason *et al.* 1945, pp.119–22). Average expectation of life at birth in 1938 was just 38.3 years (KPS 1989, p.37).

Some draining of marshlands was actually undertaken by the Rockefeller Institute, and this was said to have reduced the incidence of malaria. After their invasion in 1939, the Italians had grandiose plans for reclamation schemes, involving much of the coastal plain, Lake Maliq and the Viosa valley. Of the country's total land area in 1938 only 11 per cent was given over to arable farming or fruit growing: largely confined to upper courses of rivers, upland plateaux and basins. Thirty-one per cent was pasture, 36 per cent woodland, 11 per cent was potentially productive but uncultivated, and the remaining 11 per cent barren and unproductive. It was recognized, however, that the malarial lowlands were potentially very fertile. Indeed, an Italian company obtained a concession of 3000ha at Sukth, near Durrës. Italian settlers drained and irrigated the land for arable cultivation to produce a model farm (which was to become a model state farm for visitors after the war). This enclave flourished virtually outside of the Albanian domestic economy. Albanian sources claim that after the Italian invasion of 1939 plans were laid for 'tens of thousands' of settlers to establish themselves on the coastal plains. Overall, Albania was to have absorbed at least a million settlers from southern Italy. Between 1933 and 1934, when the Albanian government had entered into secret negotiations with France in an effort to circumvent economic subservience to Italy, the French government had demanded exclusive rights to exploit the Albanian subsoil and sought control of much of the coastal plain for French settlers (Pollo and Puto 1981, pp.207–8, 225).

Upland erosional problems were exacerbated by the considerable herding of sheep and goats. In 1938 the value of wool exports was the highest for any single item of trade; cheese was also significant. Goats were bred for mohair and milk. By the late 1930s it was recognized that many forest areas, containing all of the best European hardwoods as well as conifers, had been devastated by reckless cutting for fuel or charcoal burning. Walnut had almost disappeared. In the late 1930s about 390,000cu.ft were exported annually, mostly to Italy. Preservation and planting had not yet been considered (Mason *et al.* 1945, pp.246–7).

The country is rich in energy sources and minerals. While oil and natural gas, bitumen, some brown coal and hydroelectricity resources are found in the south, most minerals and water-power potential are located in the northern half of the country. Extracted metals include manganese, chrome, bauxite, copper and nickel, and non-metallic minerals are phosphorus, calcium, clay, gypsum and quartz. In the late 1980s annual output amounted to just over 2mln.t of coal and just over 1mln.t each of chromite, ferrous-nickel ore and copper ore (Bërxholi 1990). The northern rivers have been harnessed such that hydropower provides some 80 per cent of the country's electricity needs; the 285km-long River Drin provides half the total. Brown coal is burned in a number of power-stations, with deleterious atmospheric and health consequences, as, for example, at Korça. Consideration of long-term energy provision stimulated some studies into the potential for low-cost solar, wave, wind, geothermal and photosynthetic means of energy generation (Karadimov 1989), but these have not been followed up.

Albania's low level of economic development had been exacerbated by the dogmatic manner in which autarky was pursued by the country's Stalinist leadership (1944–91). Article 28 of the 1976 Constitution, drawn up as the rift with the country's then patron, China, was widening, and not abolished until 1990, stipulated that no credits or loans could be taken from foreign governments or organizations (PSR of Albania 1977, p.17). Foreign aid, joint ventures and other foreign interests in the Albanian economy were summarily precluded, with the result that much Albanian trade was based either on barter or through the maintenance of an import–export equilibrium.

The Stalinist emphasis upon heavy industrialization, with belching smokestacks viewed as symbols of socialist modernization to be extolled and emulated, was combined by Albania's post-war leadership with dogmatic notions of 'self-reliance' (Gurashi and Ziri 1982). This developmental distortion reached its zenith in the late 1970s with the construction of the Elbasan metallurgical complex, the country's largest industrial plant. Begun by the Chinese, this project was abandoned by them when Sino-Albanian political relations were broken off in 1978. Apparently denied access to the original plans and to pollution control technology, the Albanians subsequently laboured over a number of years to complete the project, with some assistance from India during the 1980s. The plant directly employed a 10,000-strong workforce, and had some 300 other enterprises dependent upon it. Including the substantial atmospheric pollution damage extending across the valley in which Elbasan lies, it was estimated that the plant cost the Albanian exchequer the equivalent of US$40mln p.a. (Atkinson *et al.* 1991). At the end of the communist period, renovation to bring it up to an acceptable standard was estimated at US$100mln (BBC 1991). As a consequence, the Elbasan plant was closed, starkly symbolizing Albania's changing global circumstances. Such legacies of its Stalinist past and the environmental consequences for the country – both physical and human – have, nonetheless, tended to cast a long shadow over the country's experience of the 1990s. Indeed, the Elbasan plant was subsequently restarted.

Although the country's post-war birth rate declined from Europe's highest at over thirty-five per thousand in the mid-1960s to a 1988 figure of 25.3, a very explicit pro-natalist policy, with highly constrained access to contraceptive methods, remained until 1991. As a result of a virtual halving of the death rate since 1960, the country's annual natural increase of 2 per cent (2.01 per cent average 1979–89) – about five times the European average – placed severe pressure on the country's ability to feed itself. At 111 inhabitants/sq.km in 1989 (compared to just twenty-eight in 1923), Albania had the highest population density in the Balkans (Vejsiu 1990). The low average age of population (27 years), with 43 per cent of the total under 20 years old, gradually built up pressure which by 1990 saw a large body of young people coming to the forefront of demands for change within the country.

The economy had stagnated in the 1980s (Sandström and Sjöberg 1991), and in the early 1990s, the Albanian population of 3.3 million represented a very weak consumer society. A low consumption level coupled with an emphasis upon import substitution and recycling resulted in relatively little material waste. The individual ownership of motor vehicles, for example, remained forbidden from the 1950s until 1991, by which time they still numbered fewer than 10,000 (Atkinson *et al.* 1991). Typical examples of pragmatic substitution within an economy of shortage included the use of recycled strips of pneumatic tyre as bindings in hothouses, potentially lethal multiple sections of glass used to replace broken windows in buses, and the use of rubber tyres for fuel in small bitumen plants.

Policies for protection and conservation

Early examples of Albanian environmental legislation include those concerning control of hunting (November 1951) and forestry protection (October 1963) (IUCN 1967; IUCN-EEP 1990). National parks were created under the auspices of these laws. Until the early 1970s, however, most ecological effort was concentrated on the protection of rare animals and plants. Protected areas have been classified into three categories (Atkinson, *et al.* 1991; IUCN-EEP 1990):

- *National parks*: dual role of providing areas for public access, recreation and education, and of protecting the landscape. Within them, a total protection regime is imposed, with no permanent human occupation or exploitation. Hunting and ancient grazing rights are prohibited, although villagers from neighbouring settlements are permitted to gather dead wood. Tourism is encouraged, with limited motor access, and each park has a forest lodge for accommodation. There are currently six internationally recognized national parks (Dajti, Divjaka, Llogora, Lura, Thethi and Tomori, extending in size up to 4500ha. In addition to those parks listed in Table 12.1 Albanian sources also refer to the 1350ha Bosdovec National Park at Drenova (Bërxholi and Qiriazi 1986, p.38; Hoda 1991), noted for its black pine and beech forests.
- *'Integral' reserves*: areas strictly protected for nature conservation; public entry is prohibited.

Table 12.1 Protected areas of Albania

Name	*Conservation management category*[a] *(ha)*[b]	*Year designated*	*Characteristics*
1 National parks			
Dajti	II – /4,000	1966	Beech and Bosnian pine, mixed woodland flora with Mediterranean maquis
Divjaka	II 1,194/4,000	1966	Aleppo and Stone pine forest, salt marsh and dune flora on the Adriatic coast
Llogora	II 1,040/3,500	1966	Black pine forests
Lura	II 1,300/4,000	1966	Bosnian, Macedonian, Scots and black pine, beech and spruce; Alpine meadows
Thethi	II 2,700/4,500	1966	Pine forest
Tomori	II – /3,000	1956	Beech forests and Bosnian pine; alpine meadows
2 Major nature reserves (1000ha or larger)			
Berzan	IV 1,000		
Cangonj	IV 3,000		
Fushë-Senje	IV 4,200		
Karaburun	IV 12,000		
Kuturman	IV 4,000		
Pishë-Poro	IV 5,500		
Rrushkull	IV 1,800		
3 Special protected areas			
Divjaka[c]	SPA 1,000	1965	Pelican reserve within the coastal national park

There are a further eleven Category IV protected areas of less than 1000ha and six Category VIII areas.

Sources: Atkinson *et al.* 1991, pp.11–14; Basha 1986, p.22; Bërxholi and Qiriazi 1986, pp.37–8, 1990; Çobani 1989, p.49; IUCN 1990, p.23; IUCN-EEP 1990; Hoda 1991; Polunin 1980, pp.108–11; author's field observations.

Notes

a For definitions of the conservation management categories see IUCN 1990, pp.9–14.

b Sizes are taken from Bërxholi and Qiriazi ('national parks') and IUCN/IUCN-EEP respectively. Çobani refers to 'about 8400ha of national parks'; Bërxholi and Qiriazi cite 18,000ha.

c 1993+ EU Phare programme funding for protection of the Karavasta Lagoon.

- *'Orientated' reserves*: traditional human activities such as fishing are permitted. Dual role of providing areas for education and of protection of the landscape and wildlife.

Protected animals include the brown bear, lynx and chamois, while bird colonies on the islands of Kuna (in Lezha district) and Ada (Shkodra district), as well as the pelican colony in Divjaka, are also protected. Overall, some 350 types of bird are found within the country. Artificial winter feeding has been organized for roe-deer, chamois and wild boar (Bërxholi and Qiriazi 1986, pp.39–41). By contrast, frogs have been exported in large numbers to France for culinary purposes.

During the 1990s, with successive phases of political and social upheaval, the administrative framework for enforcing and overseeing environmental protection has been weakened. A Central Environmental Protection Commission, established in 1979, operated from within the Council of Ministers, and included representatives of ministries and research institutes. Its major role was to monitor the work undertaken by ministries, local authorities, institutions and farms. Later reorganization saw the establishment of a Ministry of Public Economy and Protection of the Environment, which was demoted in 1991, for reasons of streamlining governmental structure, to committee status. A watchdog role was supposedly undertaken by the Ministry of Health's Department of Hygiene and Environmental Protection. A number of ministries have had their own sections for environmental protection, although a major responsibility for nature conservation fell within the purview of the Ministry of Forest and Water Economy. Under the communists, each administrative district's executive committee had one person notionally responsible for environmental matters, although rarely were they trained for the purpose.

Several scientific institutions have directly and indirectly contributed to environmental protection work:

- The Academy of Sciences, established in 1972, derived from the Committee for Science and other university institutes located in Tirana, comprising three sections: social sciences (humanities), natural sciences, and technical sciences. The Institute of Hydrometeorology, Seismological Centre, Centre for Geographical Study and Laboratory for Hydrological Research comprise the latter. A national system of 220 meteorological and 210 hydrological stations was established in 1949, and membership of the World Meteorological Organization has been held since 1975. A system of seismological stations, now numbering thirteen, was inaugurated in 1968, and contributes to UNESCO Balkan projects. Zoological sciences have been weakly represented, as the structure of the Academy has tended to favour the humanities, although the Centre of Biological Research within the natural sciences section has played an important role (Lloshi 1990).
- The Natural Sciences Faculty of the State University, established in 1957 (Anon. 1985b).
- The Natural History Museum, which belongs to the University.
- The 14ha Botanical Gardens, which are jointly supported by the University and Tirana city authorities.
- The Institute of Scientific Research in Hygiene and Epidemiology in Tirana, whose work includes analysis of the physical and chemical properties of drinking-water, protection from environmental pollution, the study of noise and vibration, dietary nutrition and occupational diseases (Cikuli 1984, p.77).

During 1991 higher education was reorganized: the Luigj Gurakuqi University in Shkodra (a centre of pre-war intellectual life) was reopened, as was the Tirana Agricultural University. The former 'Enver Hoxha' State University in Tirana was divided into two institutions: Tirana University and the Polytechnic University. Nine higher education institutions emerged from this reorganization.

The communists' penchant for scarring the landscape with hastily constructed polluting factories and poorly finished apartment blocks was only marginally offset by positive attitudes towards urban conservation and protection. Many religious buildings were destroyed or rebuilt for non-religious functions during and after the cultural revolution of 1967, and many old town centres were destroyed through comprehensive redevelopment based on structure plans of the 1970s (Hall 1987, p.51). Urban green and open spaces were also emphasized in such redevelopment programmes, however. The southern city of Vlora, for example, had over 50ha of such space by the mid-1980s, with an aim to provide 10sq.m of greenery for each of the city's 65,000 inhabitants by 1990 (Bërxholi 1985, p.41). Such green spaces provided important open air sites for humans and wildlife alike. In the open spaces of central Tirana, for example, nightingales and scops owls could be regularly heard up to the early 1990s. Subsequent growth in motor cars (to about 68,000 by 1994) and firearms use, and the spontaneous development of bars and cafés, as well as migrants' temporary homes on urban open space, has rendered most of such environments far less inviting to wildlife.

Selective urban conservation, or at least protection, was embodied in the designation of protective zones in much of the old residential districts of Berat and Gjirokastra ('museum cities') (Riza 1978; Strazimiri 1987), and in the historic parts of Durrës and Kruja (Bërxholi 1986; Karaiskaj and Baçe 1975), while restoration was carried out on an individual site or street basis in a number of other urban centres such as Korça, Shkodra, Tirana, Vlora and Elbasan (Bërxholi 1985, p.27; Miho 1987; Thomo 1988). Major archaeological sites, notably at Apollonia and Butrint, were also under the protection of the state (Adhami 1988; Baçe and Çondi 1987; Zarshati 1982), although, as communism collapsed, this did not prevent the theft of fifteen third-century BC sculptures from the former site in November 1991 (McDowall 1991). In 1992 UNESCO declared the Butrint archaeological site (Hodges *et al.* 1998) and later Gjirokastra as World Heritage Sites (WHSs). At the time of writing, the Butrint Foundation, a UK charitable organisation working in tandem with the Albanian government to help restore the site as an archaeological and visitor attraction, was seeking to enlarge the boundaries of the Butrint WHS following damage and vandalism during 1997 (Martin 1999).

International cooperation and agreements

Until the death of the country's post-war Stalinist leader Enver Hoxha in April 1985, the country tended not to participate in any international gatherings or treaties, arguing that each was dominated by at least one of the superpowers.

However, participation by viral disease experts at an international congress in Cairo that very month indicated a swift change of direction and a desire to participate in scientific gatherings (Anon. 1985a). Subsequently Tirana acted as host to a number of international conventions, such as the 1987 inaugural meeting of the Balkan seismology working group, and in May 1990 the first gathering of Balkan energy ministers, which agreed to establish working groups for environmental protection.

Most notably for environmental matters Albania was a party to the February 1988 Belgrade foreign ministers' communiqué on tourism, transport and protection of the environment, and there was significant Albanian participation at the Balkan Scientific Conference on environmental protection in the area, held in Varna, Bulgaria in September of that year (e.g. Gjiknuri 1988; Puka 1988). In 1989 the country signed the World Heritage Convention (IUCN 1990, p.227). This was followed in 1990 by Albania becoming the last Mediterranean country to be a party to the UNEP Regional Seas Programme (Barcelona Convention), proposing the inclusion of six coastal sites under the convention. In October 1991 Albania gained access to the World Bank and the IMF, and to the European Bank for Reconstruction and Development (EBRD). The country also became a signatory to the 1971 Convention on wetlands of international importance (Ramsar Convention), which has been important in supporting conservation measures in Karavasta Lagoon, and has helped to secure PHARE programme funding. A coastal area management programme was also developed in conjunction with UNEP and the World Bank (CEP 1998d), and Albania became a member of the Black Sea Economic Cooperation Zone group in 1994. Yet despite establishing the framework of international multilateral cooperation agreements, only a handful of intergovernmental and interministerial bilateral agreements have been secured. However, cooperation has been undertaken with neighbours at a regional level: Pogradec with Struga and Ohrid in Macedonia for monitoring pollution in Lake Ohrid and undertaking joint studies for its protection. Local cooperation has also taken place with Montenegro over protection of Lake Shkodra and the Buna River (CEP 1998d).

Changing attitudes towards the environment

Article 20 of the 1976(–1990) Constitution pointed to environmental protection as being a duty of the state, all institutions and citizens. Yet in practice, these statements carried little force given the state's 'Stalinist' approach to heavy industry. Within an economy of shortage, however, programmes against waste, continuing low levels of consumerism and a policy of recycling materials had also long characterized the Albanian approach, if not primarily for altruistic environmental reasons.

A broader dimension to Albanian environmental concern nevertheless emerged after the 1989 upheavals in Eastern Europe (Hall 1990b, 1990c), marking the outward sign of a change of approach. Some pronouncements, reflecting divisions within the ruling elite and unsupported by legislative process, began to express criticism of previous practice. Early in 1990 the Minister of Health pointed to 'imperious' development processes as the cause of substantial deforestation and atmospheric pollution. Later that year, in one of the then most explicit official comments on the country's environmental situation, the party newspaper *Zëri i Popullit* carried an article by the chairman of the Foreign Economic Co-operation Committee, arguing that the coal industry, mining 2.4mln.t annually, was both uneconomic and environmentally harmful, and that it was unacceptable and irresponsible for the country's leadership to permit the industry to continue in its present manner. This was just the tip of a large iceberg. Ironically, however, just as in agriculture, the poor production level of the mining industry had recently stimulated increased state subsidies whereby prices paid would be a function of production costs – which, while attempting to boost production, could only be a recipe for further state-sanctioned environmental degradation, as more difficult and sometimes fragile environments were more intensively worked.

June 1990 saw a rather hollow mass environmental protection campaign instigated under the title of 'Week of the Mediterranean Environment', with lectures held in the capital on anti-pollution measures. People in the rest of the country were mobilized for 'cleaning and restoration' activities (BBC 1990). However, the hastening of political and economic change from the end of 1990, and the drafting of a new constitution, provided the impetus for a much wider public debate of environmental concerns than was possible hitherto, although this was overtaken and negated by the near-anarchic condition of the country during the second half of 1991.

The legalization of opposition political parties in December 1990 saw the establishment of an Albanian Ecology Party. In the country's first post-war multi-party elections, held in March 1991, the party's three manifesto pledges were environmental improvement, national unity and peace, but it failed to win any seats in the new Parliament. Unlike the major opposition Democratic Party, however, which appeared to alienate the collectivized rural peasantry by declaring a wish to privatize the land, the Ecology Party opposed rural land sales. Later that year, the Ecology Party was joined by the Green Party, while a Society for the Preservation and Protection of the Environment was established largely by members of the scientific establishment.

Set up in 1991, the Committee of Environmental Protection (CEP) was restructured and strengthened in 1993, with twenty-five staff, and coming under the auspices of the Ministry of Health and Environmental Protection. Other ministries also had responsibilities for aspects of environmental protection. A basic law on environmental protection was passed in 1993 which acted as the umbrella for further legislation, the first of which embraced hazardous wastes and residues, forestry, the protection of flora and fauna, and hunting. It soon became clear, however, that non-enforcement of environmental legislation was a major problem. A World Bank-funded national environmental strategy plan was drafted in 1993, and in the following year the National Environmental Action Plan (NEAP) was approved by the government. Amongst its provisions, the plan lays down that every ministry and municipality should have its own structure and policies to address environmental problems. As a consequence, regional environmental agencies (REAs) were established at the prefectural level, although a number of local authorities opposed this development, and Tirana City, Fier, Pogradec and Lushnja districts, contrary to the law, refused to accept REA inspectors' decisions (CEP 1998c).

Demographic dimensions

As noted above, until the late 1980s, acknowledgement of environmental problems was subordinated to economic growth and increased food production. This, together with Albania's notorious reticence in publishing accessible statistical data (see e.g. Sjöberg and Sandström 1989), saw very little published material, and even less public debate, on environmental 'problems' and their human consequences. The last three full years of communism did, however, see the availability of published annual statistics (KPS 1988, 1989, 1990), and this continued into 1991 (Ministry of Economy 1991). Up to this point no data on any form of pollution had been published, and no released demographic data, for example, had ever alluded to birth defects, occurrence of premature births or other possibly environmentally induced problems. Similarly, observations on public health only referred optimistically to living conditions, recreational opportunities and mass immununization programmes, citing the eradication of malaria and syphilis and the 'appreciable reduction' of tuberculosis (from 296 new cases per 100,000 population in 1951 to thirty-five in 1981). By the early 1980s, the most commonly encountered diseases, in descending order, were said to be tracheo-bronchitis, angina, influenza, localized pneumonia, and diseases of the heart and blood vessels (Cikuli 1984, pp.33, 73).

Although health care under communism was unimpressive (Hall 1994), it was regarded as a state obligation. In the 1990s, the health care system deteriorated, particularly in rural areas, as it has become starved of funds. Destruction and looting of facilities between 1991 and 1992 exacerbated problems. Most notably, there was an increase in both infant mortality and mothers dying in childbirth. The latter was estimated at thirty-five per 100,000 live births for 1996, and infant mortality had increased to forty-nine per 1000 live births compared to a 1990 figure of twenty-eight (EIU 1999b; Smeets *et al.* 1997).

This is partly due to the growth of spontaneous settlements following regional migration away from northern upland villages (Hall 1996). Additional unemployment pressures and social tensions have heightened the country's rural and urban social problems (de Waal 1997). Some estimates put Tirana's population as high as 450,000 by 1996 (Standish 1996), with the more alarmist predictions suggesting some 1.2mln people might be living in the capital by the turn of the millennium – a figure approaching one-third of the country's likely total population. The spontaneous settlement of Bathore, near Kamëz, to the north of Tirana, for example, has attracted over 30,000 settlers. The 'town' has little in the way of infrastructure – no drinkable water or sewage system, electricity or roads. Many children do not attend school, and up to 90 per cent of young children and many adults suffer from disease, including rickets and skin infections. In August 1995 an attempt to evict Bathore's residents failed, and the authorities subsequently legalized the settlement, promising a 2km asphalted link to the main road and several drinking-water fountains. The large numbers of people involved and the buildings being erected pose health problems within the sites themselves and present potential threats to neighbouring settlements, as witnessed by a cholera outbreak in 1994 and a resurgence of polio in 1996 (Anon. 1997).

Data sources and problems

A wide range of research-based environmentally related literature was produced under the constrained conditions of communism, although often with a very limited circulation, such as that of the country's Academy of Sciences. The Institute of Hydrometeorology, for example, has produced works on Albania's climate and hydrology, including a climatic atlas, and has attempted to publish monthly and annual bulletins since 1956, with a 'Studies' series since 1960 *(Studime meteorologjike dhe hidrologjike)*. The Centre for Biological Research has produced an annual journal *(Vjetari punimeve shkencore te QKB)*. The Seismological Centre has published works on seismic zone definition, an appraisal of the 1979 earthquake, translated into English (Dede 1983), and a catalogue of the country's earthquakes, as well as its own monthly and annual bulletins (in Albanian and English). Since 1986 the Laboratory of Hydraulic Research has produced the occasional *Kerkime dhe studime hidraulike* (Hydraulic research and studies). Soil studies and the production of pedological maps of the country have been undertaken since 1954 by the Institute of Pedological Studies and its predecessors (Skarço 1984, pp.66–7). Other environmentally related research publications include the Geography Institute's series on geographical studies (Kabo *et al.* 1985, 1987, 1988), the Institute of Folk Culture's French-language journals *Culture Populaire Albanaise* (published since 1982) *and Ethnographie Albanaise* (since 1979), and the Archaeology Institute's periodical *Illiria* (from 1971). The State University's important Natural Sciences faculty has published its *Bulletin i Shkencore te Natyrei* (Journal of Natural Science) since the 1950s. The Institute for Monuments of Culture has produced the twice-yearly *Monumentet*, containing archaeological and architectural studies of the country's cultural monuments.

More pedagogical studies include a range of texts on the geography of Albania (e.g. Bërxholi *et al.* 1988; Ostreni 1972; Xherahu and Baruti 1975), together with a popularized foreign-language version (Bërxholi and Qiriazi 1986), a monumental work on Albanian flora encompassing more than 2500 line drawings (Demiri 1983), studies of plants with medical and other applications (e.g. Ahmataj and Ceku 1984; Nano 1987; Piperi and Kajno 1990), and atlases (e.g. Anon. 1971; Bërxholi 1990a), as well as a 1:200,000 physical map of the country (Bërxholi and Qiriazi 1990). The Academy of Science's Centre for Biological Research, the University's Natural Sciences Faculty and the Higher Agricultural Institution collaborated on a five-volume work on the country's flora (Lloshi 1990, pp.41–2).

Table 12.2 International publications

Institutions(s)	*Topic*	*Reference*
Ministry of Food (Division of Fisheries) Tirana	Conservation of freshwater fish	Rakaj and Flloko 1995
Seismological Institute, Tirana	Earthquake incidence and factors	Muco and Puka 1993; Muco 1995, 1998a, 1998b
Geology and Mining Faculty, Polytechnic University, Tirana	Geophysical mineral exploration	Frashëri *et al.* 1995
Chemistry Department, Natural Science Faculty, University of Tirana	Lead exposure; soil pollution; heavy metals marine pollution	Baraj *et al.* 1995; Saraci *et al.* 1995; Allabashi *et al.* 1996; Babi *et al.* 1998
Environment and Health Department, Institute of Public Health, Tirana	Lead exposure	Tabaku and Panariti 1996
Institute of Public Health and Veterinary Research Institute, Tirana	Lead exposure	Panariti and Bërxholi 1998

Some regional monographs have been produced in popular form in foreign languages (Basha 1986; Bërxholi 1985, 1986), while a number of districts have had their own *Almanak*, a wide-ranging occasional compilation of papers on local, including environmental, matters (e.g. Kraja 1976; Toçka 1985), albeit centrally edited and published. Generally, however, the Albanian literature on environmental concerns has been fragmented and uncoordinated, ranging over specialist periodicals such as *Probleme Ekonomike* and *Bujqesia Shqiperisë* (Albanian Agriculture) and the now defunct glossy and hitherto bland periodicals produced for foreign consumption, such as *New Albania* (e.g. Pano 1990) and *Albania Today* (e.g. Çobani 1989). Optimistic works were published under communism on the material culture and man-made environment, especially in the areas of archaeology and architecture (e.g. Adhami 1983; Meksi 1983), or on human transformation of the environment, such as through agricultural development (e.g. Anagnosti 1985; Skarço 1984).

Post-communist work has seen both (1) collaboration with foreign academics and environmentalists, and (2) publication in foreign, often English-language journals (Table 12.2), rather than continuing with Albanian-language material and publications. This partly reflects the emigration from Albania of a number of leading academics, the internationalization of Albanian work, both in terms of its quality and recognition by those working in other countries, and the significance of international academic cooperation programmes, such as those developed within the EU's TEMPUS framework. There is also a developing critical literature on the role of aid and 'assistance' (e.g. Mema 1998).

Atmospheric pollution

Urban and industrial sources

Under the communists, official pronouncements argued that constant attention coupled with the clear separation between industrial and residential zones guaranteed a minimum of atmospheric problems in Albania's major settlements. Reality was self-evidently otherwise in a number of industrial centres, and by the late 1980s it was officially admitted that pollution from various sources was increasing in all urban centres. For example, work on the climatic factors influencing the content and transport of atmospheric pollutants urged that levels of soot and sulphur dioxide concentrations receive particular attention (Çobani 1988). Carbon dioxide and nitrogen oxide were also problematic.

Since 1988 there has been a requirement for new industrial developments to be constructed at least 4–5km from residential areas. But pre-existing industrial concentrations within the built-up areas of such major settlements as Tirana, Shkodra and Korça posed major atmospheric, visual, noise and olfactory pollution problems. A handful of polluting industries were removed from locations close to residential areas (Karadimov 1989). But atmospheric pollution technology was virtually unknown (Ellenburg and Damm 1989), and early Western assistance was to address this deficiency with, for example, the installation of flue gas desulphurization plant.

Major atmospheric polluters (see Figure 12.1) include: (1) the Elbasan metallurgical complex, which, although closed for a time in the early 1990s is now leased by a Turkish company to raise steel production from 20,000 to 80,000t per annum. Since the late 1970s this plant has been a major source of untreated pollutants (SO_2, Hc, CO, $NO_{x)}$), directly impacting on the town and surrounding forest land; (2) cement works at Elbasan, Kruja, Vlora, Tirana and elsewhere: 746,000t of cement were produced in 1988 compared to 16,000t in 1950 (KPS 1989, p.75); (3) oil- and lignite-burning power-stations such as those in Tirana and Korça employing locally mined coal: that mined near Korça having a notably high ash and sulphur content. Post-war lignite production grew dramatically from 41,000t in 1950 to 2.18mln.t in 1988 (KPS 1989, p.75); (4) petrochemical complexes at Ballsh and Fier emitting harmful levels of H_2S and SO_2 (the Cërrik oil refinery is, however, idle); (5) the nitrate fertilizer plant at Fier, emitting NO_x and NH_3 gases; (6) copper-smelting plants at Kukës, Rubik and Laç, producing considerable SO_2 emissions (CEP 1998b); (7) battery factories at Berat and Gramsh; (8) the copper wire factory at Shkodra, with 11,600t production in 1988 compared to 3100t in 1970 (KPS 1989, p.75); (9) PVC and tannin plants at Vlora; (10) small yet locally heavily polluting bitumen plants, such as those in Rrogozhina, near Saranda, and in Tepelena.

A number of major settlements are situated in basins and valleys, or are located adjacent to high ground, such that pollutants tend to become trapped. This feature is often viciously exacerbated by temperature inversions, as experienced in the country's highest city Korça (at 800m) on calm evenings, when the centre and major residential areas of the town may become literally covered with the sulphurous outpourings of the lignite-burning power station.

Although the level of motorized vehicles remained low until private ownership was permittd in 1991, the use of low-grade, sulphurous fuel, coupled with poor maintenance of often long time-expired Chinese and Soviet bloc machines, did produce relatively high concentrations of vehicle emissions within a number of towns, apparently unmeasured. In rural areas and some urban back streets, poor road maintenance or lack of hard core continues to produce high levels of atmospheric dust, especially through the long dry summers. Motor car numbers peaked in 1994 at 68,000, although the 34,000 buses and lorries operating in the country in 1995 had increased in number by only a thousand since 1991 (EIU 1999b).

The domestic burning of wood and low-grade coal which continues to be mined adds to urban pollution problems, particularly in winter and under conditions of temperature inversions. The natural gas is piped to a number of urban homes (Hall 1990a); wood-burning ovens have been produced in quantity within the country and are domestically ubiquitous.

Solid and liquid waste

Considerably increased post-communist consumption of imported items meant that from a low base, solid waste generated by discarded packaging – glass, plastic, cardboard and metal – rose dramatically without an adequate system in place for its collection, transport and safe disposal. In 1995 a PHARE programme was begun to examine urban and industrial waste management with a view to identifying appropriate disposal and treatment sites for eight major centres (CEP 1998e). Albania entered its post-communist phase with a complete absence of sewage treatment plants, with obvious deleterious effects on water, land and human health. Subsequent PHARE and World Bank programmes have set about addressing this problem (CEP 1998e).

The post-communist globalization of Central and Eastern Europe has seen cross-border dumping of toxic waste, notably from the former East German territories. Albania was drawn into this process when a German playing-card manufacturer, Schmidt-Cretan, shipped 480t of hazardous chemicals to Albania in five consignments between 1991 and 1992. The toxic contents of these deliveries included toxaphene and phenyl mercury acetate, both of which had been banned in the EU since 1983. These shipments were made under the guise of humanitarian aid for Albania's agricultural sector. With East Germany's entry to the EU by virtue of German unification, the chemicals' toxic status now meant that disposal within Germany would cost between $4800 and $6600 a tonne (*Industrial Waste in Albania* 1996). Unable to return the materials to Germany, the Albanian authorities had stored them in various locations. In Bajze, close to the Montenegrin border, the toxic liquids had seeped into the soil in the surrounding area, and gases were being emitted into the air. Drinking-water wells nearby were endangered, as was Lake Shkodra. In Milot, leaking containers had been placed next to a school. These and other similar cases underscored the urgency of returning the pesticides to Germany, since Albania appeared not to be suitably equipped to handle the disposal of such materials.

This case was used as a means to bring the problem of waste transport from Western Europe to poorer East Central Europe on to the global stage. In March 1994, the Basel Convention in Geneva focused on the dumping of waste from OECD countries in non-OECD countries. The convention meeting resulted in agreement on stricter measures, including a provision to prohibit the movement of hazardous waste for final disposal from OECD to non-OECD countries with immediate effect. The loophole which allowed disposal under the designation of 'recycling' would be closed by 1997 (*Industrial waste in Albania* 1996).

However, as part of the EU's PHARE programme a further 214t of pesticides were transported to Albania as 'essential supplies', and the World Bank funded a project named the 'Critical Imports Project', which led to an additional 160t of pesticide imports. The pesticides shipped in both cases were highly toxic, possible carcinogens, which posed a potential threat to the water supply. In addition, there was no provision of training for the Albanian farmers who had never been exposed to such materials before. Indeed, the average farmer proceeded to handle these dangerous chemicals without the proper protective clothing or training necessary to prevent accidents. By 1993 Albania already possessed 3000t of pesticides in store rendering additional imports unnecessary, and the poor state of the economy meant that few farmers could afford to buy imported pesticides anyway.

Water pollution and management problems

Albania is particularly rich in water resources and has extremely active hydrological systems (Ciavola *et al.* 1999). Average annual precipitation in the upland areas is the highest in Europe at around 1500mm, and the average annual water flow is 42bn.cu.m (Frashëri 1988). Drainage water management is only a recently discovered tool (Grazhdani *et al.* 1996), and studies have been pursued on the effectiveness of subsurface drainage to reduce soil and nutrient erosion and surface run-off. The sources of Albanian rivers are invariably above 1000m. Rivers flow fast over relatively short courses to the sea and provide substantial hydroelectric potential. Until large storage lakes were developed in conjunction with the HEP schemes (Koman, Fierza), navigation of most Albanian waterways was virtually impossible.

By the early 1980s only 56 per cent of the country's villages were supplied with drinking-water (Shkodra and Ganiu 1984, p.36). In 1986, the Party resolved to ensure that by 1990 sufficient quantities of pure water would be available for all the country's settlements. Subsequent attention was centred on the construction of aqueducts to supply towns and villages (Frashëri 1988). But poor domestic and industrial plumbing and maintenance of pipes, aqueduct irrigation canals, village taps and standpipes results in interruptions to supply and an enormous albeit inadvertent waste of water. This also permits bacteria and other harmful substances to enter domestic water supplies, posing a continuous threat of infection.

Marine pollution

The impact of untreated urban domestic and industrial waste, offshore discharges, and pollution originating in Italy, Yugoslavia and Greece have begun to take their toll on Albania's marine environment. The polluted nature of water adjacent to the ports of Shëngjin, Durrës, Vlora and Saranda, where waste and spillages from vessels are locally problematic, has increasingly impacted upon nearby resorts. This is exacerbated by the discharge of polluted rivers reaching the Adriatic coast, notably the Buna on the Yugoslav border, the Mat south of Lezha, and the Shkumbin, Seman and Viosa between Durrës and Vlora. At the latter, for example, high concentrations of heavy metals have been found (Babi *et al.* 1998). Pollution has adversely affected fisheries and doubtless other fauna and flora. Some 260 kinds of fish may be found in Albanian waters. Sardines are caught in the Bay of Vlora, and at Durrës and Shëngjin; grey mullet is caught in coastal lagoons, and other varieties fished include white perch, barbel, cod and tuna (Bërxholi and Qiriazi 1986, p.43).

In 1991, the country's first 'National Symposium on the Coastal Space' was held in Tirana, with Italian attendance, and soon after, alongside Italian, Greek, Yugoslav and EC representatives, Albania became a signatory to the Declaration on the Adriatic Sea in Ancona. An initiative was launched to clean up the northern stretch of the Adriatic and to keep the Albanian coast pollution free. Yet in the extreme south of the country, careless local recreational activity coupled with flotsam from the many Greek and Italian vessels passing through the Corfu Straits has already contributed to a substantial littering of the coast with waste material at such locations as Ksamil. Offshore oil and gas exploration have been taking place, and although no commercially exploitable deposits had been found by 1999, impacts from these activities will need to be carefully monitored.

Tourism along Albania's varied Ionian and Adriatic coasts has been concentrated in and around Durrës Beach, which can accommodate up to 60,000 largely domestic holiday-makers per month in summer (Bërxholi 1986, p.30), and a handful of other centres such as Vlora and Saranda (Albturist n.d., 1989; Hall 1984a, 1984b, 1990e, 1991). With a number of Albanians arguing that tourism was the country's only short-term salvation, plans to render tourism a major element of the Albanian economy have seen a number of joint-venture contracts being pursued for substantial developments along the coast and inland, following a joint plan drawn up with the EBRD to recognize tourism investment priority zones (Hall 1994, pp.260–4; Ministry of Tourism and EBRD 1993). Subsequent bouts of civil unrest within the country have tended to inhibit both inward investment in this area and international tourism development, such that most tourism development has been domestically driven, through the erratic development of a domestic market and privatization process which has seen the rise of a limited number of entrepreneurs.

As stability returns to the country, increasing pressures, particularly along the relatively unspoilt Ionian coast, cannot but bring environmental deterioration in their wake. Sewage, solid waste and other refuse disposal, degradation of habitats through recreational pressures, substantially increased traffic and the need for improved transport infrastructures and other utilities will all take their toll, imposing adverse impacts upon the very marine and landward environments which can provide the basis for relatively high-volume tourism expansion.

River pollution

In the coastal plains, the more important rivers are visibly polluted by industrial waste. In 1993, oil extraction activity in the south of the country was responsible for the discharge of over 800,000cu.m of waste liquids, while the chemical complex at Laç in the north was responsible for just under 800,000cu.m (CEP 1998b). In the following year, oil extraction waste was down to 700,000cu.m and Laç had ceased operation. The oil-processing complex at Ballsh is recognized as one of the worst industrial polluters in the country, discharging large quantities of waste water into the Gjanica river, while the Fier nitrate fertilizer plant discharges significant amounts of ammonia waste waters (CEP 1998b). In the latter case limited industrial activity has significantly reduced pollution levels. In the lighter industrial sectors, leather and textile factories – growth areas for Italian and Greek funded outward processing exploiting low labour costs – discharge untreated waste waters which include sulphur, chromium and organic components (see e.g. Piperidou *et al.* 1996). With little waste treatment, untreated discharges continue to pose a threat both to aquatic life and human health. Run-off from increasing use of chemical fertilizers is also a contributory factor. The lower reaches of many streams and rivers in the central part of the country may already be organically dead. Even in upland areas careless mining activity has damaged water bodies, such as the Shkumbin running red downstream from the ferro-nickel workings of the Librazhd and Prenjas areas. It has been acknowledged for some time that waste from paper mills (such as near Lezha) and plants associated with non-ferrous metals has damaged aquatic flora and fauna along a number of rivers and estuarine areas, although the former are now recycling paper. Some of the consequences of industrial closure and production reduction are reflected in improving river water quality (Table 12.3).

Lake pollution

Albania's lakes are recognized as wetlands of international importance. They have acted as breeding and feeding areas for white and Dalmatian pelicans with eleven pairs of the latter being located in 1985 (and five birds at Divjaka in 1989), the pygmy cormorant, a large colony of which is found on Lake Shkodra, various herons (Karadimov 1989), and several other waterfowl species. The precarious position of some of these is a matter of concern: wetland modifications are thought to have caused recent decline. Lake Ohrid has particular interest

Table 12.3 River water quality (in mg/litre)

River	*Year*	*Dissolved* O_2	NBO_5	NH_4	NO_2
Kir	1989	0.05	25.50	0.20	0.15
	1994	8.51	6.71	0.06	0.15
Shkumbin	1989	5.85	14.85	1.30	0.30
	1994	11.84	2.96	0.64	0.20
Tirana	1989	5.40	37.50	3.50	0.35
	1994	6.28	7.00	0.20	15.09

Source: Institute of Hydrometeorology, Tirana, cited in CEP 1998b, pp.4–5.

due to a rich endemic and relict fish and aquatic fauna (Carp 1980): for example, twenty-four varieties of carp are unique to the lake (Bërxholi 1986, p.30). In 1999, German funding was secured for the development of a sewage disposal scheme at Pogradec in order to help prevent deterioration of the water quality of Lake Ohrid. The water of the Prespa lakes has seen little significant change in nutrient status and an absence of eutrophication, although tourism pressures are threatening terrestrial fauna (Catsadorakis and Malakou 1997; Hollis and Stevenson 1997).

The country's four major natural lakes are shared with Montenegro (Shkodra, 368sq.km), Macedonia (Ohrid, 362.6sq.km), Greece (Little Prespa, 23.5sq.km) and both Macedonia and Greece (Prespa, 285sq.km). While this raises questions of transboundary pollution, particularly with increasing urban and industrial development in the vicinity of Lake Ohrid, it also provides opportunities for regional cooperation. The smaller lakes of the northern mountains, such as Lura, appear largely to have retained their character, although domestic tourism pressures are increasing. In the extreme southwest of the country, the larger Lake Butrint (16.3sq.km) has been developed for mussel cultivation. However, diversion of some of the lake's freshwater sources for irrigation purposes has increased salinity to the detriment of lake fauna.

Soils, flora and fauna

Albania can be divided topographically into a sub-Alpine and Alpine zone, situated north of the River Drin, with peaks attaining heights of over 2000m; a mountainous zone of alternating deciduous montane forests of beech and cultivated basins, between the Drin and Osum valleys; a broad Mediterranean and transitional deciduous forest zone with scattered patches of Mediterranean pine south of the Osum valley featuring limestone mountains and valleys; and a coastal plain of phrygana, maquis and cultivation, dissected by numerous rivers (Grimmett and Jones 1989). Two-thirds of the country contains sedimentary deposits and one-third igneous rock. Soils developed on serpentine rocks cover a large area containing a range of industrial minerals. Research undertaken at eight sites across the country suggests that high concentrations of heavy metals are present in soil and vegetation (Shallari *et al.* 1998). The coastal zone consists mainly of alluvial material: of 250,000ha, 220,000ha have been reclaimed, mostly from mosquito-infested marshland (Kusse and Winkels 1990).

Of exploitable land, almost half is covered by forest and bush, some 20 per cent by natural pasture and 30 per cent by cultivation. Such an ecosystem relationship favours floral and faunal diversity: the country contains some 3500 species of flora, with several hundred endemic and sub-endemic species. These are well documented within the country (see e.g. Demiri 1983), and have attracted attention in the West (Almond 1988a, 1988b; North 1990; Polunin 1980; see also Turrill 1929). Some 865 specimens are claimed to have curative properties, and Albania has been Europe's leading exporter of medicinal herbs, supplying around 20,000t p.a. worth US$20mln.

The Albanian fauna is rich and varied, with Mediterranean and European characteristics. Officially, hunting has been severely restricted but since 1991 has gained greater significance as a means of accruing hard currency. In practice, it appears to have been widespread and locally safeguarded. By the end of 1993 some 300 foreign hunters had entered the country, often facilitated by Italian travel companies. Marine fauna and ecosystems have been adversely affected by the use of dynamite to catch fish. A National Council of Fauna and Hunting was established in 1994 with the objective of regulating hunting activities (CEP 1998e). With increasing human and economic pressures, deforestation may threaten the habitats of such rare animals as the brown bear, of which there were an estimated 500 to 800 in the northern mountains in 1987 (Atkinson *et al.* 1991).

Farming and food practices

Agricultural production generates 60 per cent of GDP and continues to employ about half of the workforce. Albania had maintained the lowest level of urbanization in Europe – just 35.5 per cent of the population were recorded as living in towns in the 1989 census: an increase of 2 per cent over 1979. This reflected an explicit rural-led development programme and the maintenance of a higher natural increase amongst the rural population (Hall 1987; Sjöberg 1989, 1991). In the forty years after 1946 cultivatable land increased by 240 per cent, but between 1950 and 1986 the population increased by 250 per cent, with the amount of arable land per person decreasing by 10 per cent. This process accelerated during the 1980s, and exerted pressure on the communist authorities to rethink their pro-natalist approach, although this was eventually overtaken by events. Arable land comprises 700,000ha, 450,000ha (64 per cent) of which is on sloping ground. Expansion of arable acreage has hitherto entailed an improvement in desalination methods (although salination remains a problem), reclamation from coastal and lakeside lagoons, and an extension of cultivation in upland areas, often through extensive terracing (Kusse and Winkels 1990). Collectivized agriculture absorbed a rapidly growing workforce during the 1970s and 1980s, but with privatization and uncertainty, the 70,000 new recruits who entered the national job market every year under communism have been deflected through rural to urban migration, new forms of petty trading, grey economic activities and emigration.

In the waning months of communism in 1990 a programme of encouraging the development of private plots was begun, particularly aimed at stimulating vegetable production. For the first time peasants were allowed to rear cattle on their own plots, and cooperatives were asked to transfer some of their stock to their members for this purpose: in hilly and mountainous areas 2000sq.m to each member in addition to their existing private plots, and in lowland areas peasants were to be given up to 2000sq.m including their plots (EIU 1990a, p.36). While clearly intended to raise food productivity and incomes, prolonged drought tended to undermine the programme. Nonetheless, the domesticated animal population increased dramatically although a significant growth in the number of sheep and particularly goats was also evident, which, through indiscriminate grazing, exacerbated conditions for renewed erosion in upland areas.

New land and private property laws in 1991 dissolved all cooperatives but forbade the purchase or sale of land. In the consequent rush to regain land held before the war, former rural land-owning families were unilaterally claiming and fencing off territory, even destroying public buildings such as schools and installations such as oil wells in the process. By 1994, 94 per cent of farmland had been privatized (Giorgio 1994). This structural change, coupled with international aid programmes and substantial remittances from Albanian émigrés, saw agriculture become a key motor for economic recovery. In the process of attempting to bring some equity to the redistribution of cooperative land, rural ownership became extremely fragmented. Some three-quarters of the country's arable acreage was given back to peasant families, producing over 379,000 holdings of between 0.8 and 2ha (Deslondes and Sivignon 1995), but with significant local and regional variations.

Prospects for conflict or subversion of distribution tended to increase where several problematic factors coexisted, such as the combination of long-time resident pre-communist landowners and large numbers of newcomer families in a village. The likelihood of conflict was increased further in villages where hilly or mountainous terrain exacerbated demographic pressures on the land. Throughout the country's rural land privatization programme, failure of government authorities to enforce the law was a recurrent theme. Local communities were thereby given, often by default, a degree of flexibility to shape events in ways that suited their needs and minimized contact with corrupt officials (Lemel 1998). This related both to the land distribution process and to events such as the proliferation of spontaneous settlements.

Only in August 1995 was legislation passed enabling the buying and selling of land as one means of encouraging consolidation. Credit was made available to farmers through an agricultural development fund, set up with support from the World Bank and other international donor agencies. The EU PHARE programme accorded Euro6mln to support the creation of a functioning market in agricultural land and other agricultural reforms. Ironically, consolidation was assisted by some farmers selling their land in order to invest in the country's high-yielding pyramid 'investment' schemes which subsequently collapsed during 1997, once again bringing chaotic conditions to the country.

Many tasks have reverted to being labour intensive, undertaken by human and animal power, often using improvised implements and vehicles. Reversion to smaller private holdings has exacerbated this situation. On the other hand, whole areas of upland agriculture, particularly that which enjoyed substantial investment of financial and labour resources in the 1960s and 1970s, have been abandoned as a result of rural to urban and upland to lowland migration. While a shedding of surplus rural labour might have appeared to present an ideal opportunity for agricultural mechanization, difficult topography coupled with capital shortages has further constrained such an application.

With funding from the Soros Foundation and the UK Government Know-How Fund, the Plunkett Foundation has been helping to develop self-supporting rural associations, including seed, sapling and flower growers, bee keepers and herb cultivators, with the aim of improving the quality and presentation of local upland products for niche marketing in Albania and abroad (Lucey 1999). A major potential source of rural investment capital, émigré workers' remittances, was largely diverted to pyramid schemes in 1996 and lost in the subsequent schemes' collapse. Germany has made efforts to encourage emigrants to return to Albania by offering start-up credit for those wanting to open small businesses. However, the flood of Kosovar refugees to Kukës and a number of cities during 1999 further complicated ethnic Albanian mobility patterns as well as exacerbating local socio-economic problems. A further constraint on agricultural amelioration, particularly in upland northern regions, has been the resurgence of blood feuds. Government statistics suggest that 1200 people were killed in such feuds between the end of 1991 and June 1998, while it is claimed that members of some 760 families cannot leave their homes and till the land through fear of reprisal (Cockerell and Peacock 1999).

Problems resulting from the increased use of chemical pesticides and fertilizers in agriculture brought a recognition of the need for control (Hoxha 1988). Insecticides have been used on a large scale: about 800t in 1960, 5900 in 1970, and 10,200 in 1980. Application of chemical fertilizers grew considerably in the post-war period, rising overall from just 1000t in 1950 to 96,500t in 1988, representing an increase from 2.6 to 132.2kg per hectare of tilled land (KPS 1989, p.89; Papajorgji 1989, p.43). Soil studies undertaken in 1993 in two important agricultural regions found significant heavy metal contamination and high concentrations of persistent organochlorine pesticide residues – notably DDT and DDE (Pilidis *et al.* 1996; Saraci *et al.* 1995). In February 1996 and April 1997 a total of four people died and twenty-six exhibited signs of lead intoxication in rural areas after eating flour contaminated from the repair of a mill's cracked grindstone with metallic lead (Panariti and Bërxholi 1998; Tabaku and Panariti 1996). Studies of children living in Berat, the location of the country's battery plant, have also found high blood lead concentrations (Allabashi *et al.* 1996; Baraj *et al.* 1995; Tabaku *et al.* 1998).

Forestry and erosion problems

Forests in Albania cover just over 1mln.ha or 36 per cent of land territory. Some evidence of the effects of acid rain exists, particularly in the country's northern forests (Atkinson *et al.* 1991). International collaboration would appear to be indispensable in this as in other aspects of environmental protection. USAID and the US Peace Corps helped to develop strategies for the transfer of management rights of state-owned forests to private and commune ownership, and in the development of a forest extension service to provide agro-forestry advice to the new managers. Funding has been provided by the World Bank (USAID 1996).

Deforestation is of major significance, and, while historically rooted, continues on an official and unofficial basis, not least to meet the large demands of domestic wood-burning ovens. Several valuable tree species were extracted by the Italians in the 1930s, although lowland forests have been declining for centuries because of fuel wood extraction and clearance for agriculture. While some of the country's best timber remains relatively inaccessible in mountainous areas, much lowland destruction was wrought in the early 1990s both as part of the outburst of anti-state violence within the country and a desperate search for fuel. Trees in orchards or for roadside or cattle shade were taken for firewood or rough construction. In 1990 there were 8.3 million fruit trees, but by 1993 this figure had been reduced to 3.2 million (EIU 1999b, p.25). Other woodland damage has resulted from fires and the overgrazing of goats (Green and King 1996) and sheep. In 1994, for example, 509 forest fires were recorded, affecting some 400ha (CEP 1998e).

Partly to meet increasing food demands for a rapidly growing population, vast areas of hillside were dogmatically terraced in the 1960s and 1970s (Anagnosti 1985). A slogan of the late 1960s was 'Let us take to the hills and mountains and make them as beautiful and fertile as the plains' (Skarço 1984, pp.63–4). Over 100,000ha of virgin land were broken in for cultivation and irrigation during that plan period and a further 92,000ha between 1971 and 1975. However, many of the hillsides should never have been terraced, the topsoil being too shallow and the scrub too fragile for working. In many cases, bedrock soon appeared. Terracing might have been more effective had the terraces been properly built up with stones and then planted immediately. Fruit trees were often planted in areas that were too high and dry. Olive planting had better results, but ground vegetation was too sparse to prevent soil erosion, rapidly leading to gullies between the trees (Atkinson *et al.* 1991). A number of problems resulted from the deployment of mass mobilization, particularly the involvement of untrained 'volunteer' urban youth in the terracing work. Much of this work was abandoned in the early 1990s and, with substantial upland to lowland migration, remains one of the sadder disdained and discredited symbols of the Hoxha era.

Over a wide area, disuse and misuse has been the fate of the results of communist rural mass action. Irrigation systems in hilly regions have deteriorated and been damaged; channels have been clogged up, erosion on slopes has considerably increased and small stone dams have been destroyed as a result of the removal of materials for residential construction purposes. A number of new, often unauthorized quarries have been opened up, adding further to erosion and visual amenity problems (CEP 1998a).

Afforestation affected almost 200,000ha between 1951 and 1988, with a further 467,000ha of forestry improvement during that period (KPS 1989, p.90). This programme helped to stabilize some areas of hitherto fragile conditions. The planting up of susceptible areas with stabilizing trees has usually involved Pinus *pinea* and *P. radiata*. The districts of Puka, Shkodra, Librazhd, Kukës and Fier were the main focus of this activity. In the Vlora district, for example, by the mid-1980s 4500ha of forest had been improved and 4000ha of denuded land reforested (Bërxholi 1985, pp.31–2). Fruit and olive tree planting has latterly involved individual rather than contour terracing, and this policy could be seen being implemented in such districts as Saranda and Pogradec. All afforestation had stopped by the early 1990s. The country now faces a major need for reforestation and soil conservation as gully and rill erosion has accelerated on many sloping lands (Zdruli *et al.* 1997).

The impacts of irrigation and reclamation

Since 1940 major programmes of wetland drainage have been undertaken as part of the drive against malaria, with most marshes being converted to agricultural lands irrigated through a series of reservoirs. Rivers, especially in the coastal belt, were regulated and canalized. Between 1946 and 1983 some 9 per cent of the country was drained, and, between 1966 and 1970 at the height of the programme, 20 per cent of all capital investment was devoted to major hydrological projects. Inland, drainage of the Maliq Lake and marshes started in 1946 resulted in a loss of some 60,000ha of wetlands by 1974 and the improvement of a further 170,000ha for agricultural production. The area became Albania's main sugar beet source.

With rainfall unevenly distributed throughout the year – Albanian agriculture often suffers from a shortage of water between April and September – water improvement projects have been closely associated with irrigation, which by 1988 had been extended to some 58.2 per cent of all cultivatable land (compared to 47.4 per cent in 1970 and 10.5 per cent in 1950), representing an irrigation capacity of 416,800ha, the figure for 1970 being 283,800ha and for 1950 39,300ha (KPS 1989, p.89; Papajorgji 1989, p.43).

Other environmental problems

Visual pollution

During and following the period of worsening relations with China in the mid-1970s, national defensive measures entailed two forms of physical installation.

Air raid shelters were built into urban residential and industrial areas. These are often distinguished by heavy metal doors above small elevations at ground level, although they tend to be unobtrusive. Defensive concrete pillboxes, however, numbering several hundred thousand, were scattered throughout the country, visually scarring the landscape and encroaching on valuable agricultural land. They were particularly numerous in 'strategic' areas such as near borders, ports and along the key Durrës–Tirana trunk-road. The Democratic Party pledged to dismantle them at the earliest opportunity (*Newsweek*, 8 April 1991, p.52), although many still remain. Indeed, in the northeast of the country their original purpose was realized in repulsing border incursions and responding to sniper fire from Serb forces during the conflict over Kosova in 1999. On a lighter note, miniature bunkers carved from soapstone have been available in Tirana as tourist souvenirs.

Insensitively located rural apartment blocks, usually prominent symbols of the communist era, associated with the administrative centres of state farms, can often be found in otherwise harmonious landscapes, as in the coastal village of Lukova, where three four-storey units dominate a ridge some 200 metres above the Ionian Sea. Slightly less prominent are the eleven four-storey blocks of the nearby coastal village of Piqeras. A greater visual blight under the communists was imposed by officially sponsored hillside graffiti, often marked out in bare limestone, extolling the virtues of the party and 'Comrade Enver'. These confections were usually put together by members of the armed forces (see Gardiner 1976, pp.61–2), although this practice was by no means unique to Albania. One of the most grotesque and persisting examples is an enormous 'ENVER' prominently splayed across five ridges of Mt Shpirag facing the town of Berat and extending to a height of some 60 metres. Twice during the 1990s the army attempted to destroy this 'monument' with dynamite and napalm, with little success.

High levels of rural–urban and interregional migration have resulted in large numbers of spontaneous settlements in urban open spaces – notably in Tirana and Durrës, and along the beaches of coastal resorts such as Golem near Durrës, Ksamil near Saranda, and the fragile Karavasta lagoon area. Illegal construction of bars and kiosks was one of the first visual symbols of post-communism, and represented the innocent belief that market freedom could be interpreted literally. The Socialist Party government elected to power in 1997 began a concerted effort to clear the country's major urban green spaces of these constructions towards the end of 1998. In January 1999 the Prime Minister took a personal role in a drive to clean up Tirana: the owners of illegal 'permanent' buildings were given six months to dismantle them, and construction companies were told to remove waste and to reinstate pavements.

Conclusions

An economy of shortage has continued to prevail in Albania. Innovation, improvisation, import substitution, recyling and low levels of consumerism and material waste have, from necessity, persisted into the unstable post-communist period. Entry into the world political and economic arena, transformation to elements of a market economy and programmes for privatization are familiar in much of East Central Europe. In Albania, however, these processes follow some distance behind those of the rest of the region, reflecting the deeper ideological nature and inertia of the old Albanian regime, the country's hitherto relative isolation, and its low level of economic development and technological sophistication. Periods of domestic disruption between 1991 and 1992 and since 1997, together with the impact of events in Kosova, have further constrained the efficacy of post-communist restructuring processes. Apart from war-ravaged Bosnia and Hercegovina, Albania endured the greatest contraction in industrial production in ECE. In 1998 industrial output was less than 30 per cent of 1989 (EIU 1999c). Positive environmental benefits have included the full or partial closure of heavily polluting and uneconomic plants which were developed only as the result of a combination of a dogma of self-reliance and poor transport and distribution systems. However, an infusion of foreign capital and aid has been only slowly assisting the implementation of pollution and waste control technologies and strategies.

The pursuit of sustainable development projects is important in raising environmental awareness, but policies may be contradictory. The promoters of 'sustainable' tourism schemes may emphasize that the lack of local infrastructure offers a pristine, uncommercialized environment which is attractive to niche rural tourists. Yet residents participating in such projects may reluctantly offer their support on the assumption that tourism will help improve their local infrastructures and services. This has been emphasized, for example, in social surveys undertaken in the Karavasta lagoon area south of Durrës, a Ramsar site since 1994 and a potential important focus for nature tourism development. A PRA (participatory rural appraisal) approach, involving trained teams from local villages in this, one of the most valuable wetland environments on the eastern Adriatic, indicated that a major priority for locals was an improvement in their basic infrastructural services for access, water supply, electricity and telecommunications (Runca 1996).

Spatial restrictions on personal freedom, such as the imposition of exclusion zones for environmentally sensitive areas or the prohibition of fishing and other pursuits in particular areas, is politically sensitive in view of the half century of communist imposition. While Albania's 1993 Law on Environmental Protection provides the legislative framework for imposing exclusion and proscription, a major problem at Karavasta is that the lagoon supports up to 500 fishermen, most of whom are locals who pursue their activities illegally. Any sustainable form of lagoon fishery would be unable to support more than the eighty members of the current exclusive fishing cooperative. As fishing contributes substantially to the local community, yet conflicts with the conservation and tourism aspirations of the area, difficult questions of inclusion and exclusion, protection, diversification and participatory decision making have been posed (Runca 1996).

Increased consumerism coupled with a development strategy emphasising tourism growth requires substantial infrastructural investment. Inevitably a number of lowland and coastal environments will be further modified, while improved access to upland areas will accelerate processes of change there. Increased waste, a reduction in recycling, improvisation and import substitution, greater use of motor vehicles with attendant atmospheric pollution and congestion problems all represent negative environmental consequences of change. Albania could, and should, learn from the mistakes of both East and West: its environment is fragile, but its resources, both physical and human, are potentially great (Hall 1999).

References

Adhami, J. (1983) *Rrugë dhe Objekte Arkeologjike në Shqipëri*, Tirana, 8 Nëntori.

Adhami, S. (1988) 'Historical monuments, their restoration and use', in Anon. *Legacy of Centuries*, Tirana, 8 Nëntori.

Ahmataj, H. and Ceku, K. (1984) *Bimët Aromatike*, Tirana, Botimi Shtëpisë së Propagandës Bujqësore.

Albturist (n.d.) *The Tourist Map of Albania*, Tirana, 8 Nëntori.

—— (1989) *Hotels and Tourist Centres in Albania*, Tirana, 8 Nëntori.

Allabashi, R., Cullaj, A., Tabaku, A., Babi, D. and Celo, V. (1996) 'Atom absorption spectrotomic determination of lead in whole blood of differently exposed collectives in Albania', *Fresenius Environmental Bulletin* 5(3–4): 126–32.

Almond, M. (1988a) 'A Balkan journey: Part 1', *The Rock Garden* 20(4): 459–66.

—— (1988b) 'A Balkan journey: Part 2', *The Rock Garden* 21(1): 76–81.

Anagnosti, V. (1985) *The Terraces of Lukova*, Tirana, 8 Nëntori.

Anon. (1971) *Atlas Gjeografik i Shqipërisë*, Tirana, Hamid Shijaku.

—— (1985a) 'Albania comes in out of scientific cold', *New Scientist*, 30 May, p.8.

—— (1985b) *The Enver Hoxha University of Tirana*, Tirana, 8 Nëntori.

—— (1997) 'Polio outbreak – Albania 1996', *European Journal of Pediatrics* 156(1): 77.

Atkinson, R. T., Bouvier, M., Hall, D. and Prigioni, C. (1991) 'Albania', in IUCN-EEP *Environmental Status Reports 1990, Volume 2, Albania, Bulgaria, Romania, Yugoslavia*, Cambridge, IUCN.

Babi, D., Celo, V., Cullaj, A. and Pano, N. (1998) 'Evaluation of the heavy metals pollution of the sediments of Vlora Bay Adriatic Sea, Albania', *Fresenius Environmental Bulletin* 7(9–10): 577–84.

Baçe, A. and Çondi, D. (1987) *Buthrot*, Tirana, 8 Nëntori.

Baraj, B., Nati, R., Celo, V., Babi, D., Tabaku, A. and Robinson, D. (1995) 'Determination of lead in the blood of some children in the town of Berat, Albania', *Fresenius Environmental Bulletin* 4(9): 527–32.

Barnes, G., Moyer, D. D. and Gjata, G. (1994) 'Evaluating the effectiveness of alternative approaches to the surveying and mapping of cadastral parcels in Albania', *Computers, Environment and Urban Systems*, 18(2): 123–31.

Basha, H. (1986) *Tirana et ses Environs*, Tirana, 8 Nëntori.

BBC (1990) 'Albania: environmental protection campaign', *Summaries of World Broadcasts*, 21 June, EE/WO133 A/6.

—— (1991) 'Albania: need for caution in closing unprofitable enterprises', *Summaries of World Broadcasts, 18 April*, EE/WO175 A/4.

Bërxholi, A. (1985) *Vlora and its Environs*, Tirana, 8 Nëntori.

—— (1986) *Durrës and its Environs*, Tirana, 8 Nëntori.

—— (ed.) (1990a) *Atlas Gjeografik i RPS të Shqipërisë*, Tirana, Shtëpia Botuese e Librit Shkollor.

—— (1990b) 'On the surface – flowers and leaves, underground – silver and gold', *New Albania* 5: 6–7.

Bërxholi, A. and Qiriazi, P. (1986) *Albania: A Geographic Outline*, Tirana, 8 Nëntori.

—— (1990) *Republika Popullore Socialiste e Shqipërisë: Harte Fizike*, Tirana, Shtëpia Botuese e Librit Shkollor.

Bërxholi, A., Memisha, O. and Konomi, F. (1988) *Gjeografia e Shqipërisë*, Tirana, Shtëpia Botuese e Librit Shkollor.

Carp, E. (1980) *A Directory of Western Paleartic Wetlands*, Cambridge, IUCN-UNEP.

Catsadorakis, G. and Malakou, M. (1997) 'Conservation and management issues of Prespa National Park', *Hydrobiologia* 351: 175–96.

Committee for Environmental Protection (CEP) (1998a) *Damages to the Geographical Terrain*, Tirana, CEP.

—— (1998b) *Impacts of the Industrial Productive Sector*, Tirana, CEP.

—— (1998c) *Institutional Strengthening*, Tirana, CEP.

—— (1998d) *International Cooperation*, Tirana, CEP.

—— (1998e) *The Situation of Forests and Protected Areas*, Tirana, CEP.

—— (1998f) *Urban Environment Pollution and Damage*, Tirana, CEP.

Ciavola, P., Mantovani, F., Simeoni, U. and Tessari, U. (1999) 'Relation between river dynamics and coastal changes in Albania: an assessment integrating satellite imagery with historical data', *International Journal of Remote Sensing* 20(3): 561–84.

Cikuli, Z. (1984) *Health Service in the PSR of Albania*, Tirana, 8 Nëntori.

Çobani, J. (1988) 'Problems of state in protection of the atmospheric environment', in Balkan Scientific Conference on Environmental Protection in the Balkans, *Abstracts*, Varna.

—— (1989) 'Conservation of the environment in Albania', *Albania Today* 109: 47–9.

Cockerell, W. and Peacock, P. (1999) 'Political asylum or economic migration?', *Besa* 3(1): 26–7.

de Waal, C. (1997) 'Socio-economic trends in post-communist Albania', in S. Bianchi and M. Uvalic (eds) *The Balkans and the Challenges of Economic Integration – Regional and European Perspectives*, Ravenna, Longo Editore.

Dede, S. (1983) *The Earthquake of 15 April 1979 and the Elimination of its Consequences*, Tirana, 8 Nëntori.

Demiri, M. (1983) *Flora Ekskursioniste e Shqipërisë*, Tirana, Shtëpia Botuese e Librit Shkollor.

Deslondes, O. and Sivignon, M. (1995) 'Albanian agriculture – from the cooperative to subsistence farming', *Revue d'Études Comparatives Est–Ouest* 26(3): 143–57.

Diamanti, E., Ibrahimi, B., Tafaj, F., Mezini, E., Dodbiba, A., Dobi, V., Catone, S., Genovese, D., Simeoni, P. and Fiore, L. (1998) 'Surveillance of suspected poliomyelitis in Albania, 1980–1995: suggestion of increased risk of vaccine associated poliomyelitis', *Vaccine* 16(9–10): 940–8.

Economist Intelligence Unit (EIU) (1990a) *Romania, Bulgaria, Albania: Country Report No. 3, 1990*, London, EIU.

—— (1999b) *Albania: Country Profile 1999–2000*, London, EIU.

—— (1999c) *Albania: Country Report Second Quarter 1999*, London, EIU.

Ellenburg, H. and Damm, K. (1989) *Albanien 1989*, Berlin, Institut für Geographie, Technische Universität.

Fiore, L., Genovese, D., Diamante, E., Catone, S., Ridolfi, B., Ibrahimi, B., Konomi, R., van der Avoort, H. G. A. M., Hovi, T., Crainic, R., Simeoni, P. and Amato, C. (1998) 'Antigenic and molecular characterization of wild type 1 poliovirus causing outbreaks of poliomyelitis in Albania and neighbouring countries in 1996', *Journal of Clinical Microbiology* 36(7): 1912–18.

Frashëri, A., Lubonja, L. and Alikaj, P. (1995) 'On the application of geophysics in the exploration for copper and chrome ores in Albania', *Geophysical Prospecting* 43(6): 743–57.

Frashëri, M. (1988) 'Hydric resources in the context of Albania's environmental protection problems', in Balkan Scientific Conference on Environmental Protection in the Balkans, *Abstracts*, Varna.

Gardiner, L. (1976) *Curtain Calls: Travels in Albania, Romania and Bulgaria*, London, Duckworth.

Giorgio, S. (1994) *Farm Survey*, Tirana, Republic of Albania Ministry of Agriculture and Food and United Nations Food and Agriculture Organisation.

Gjiknuri, L. (1988) 'Fauna and some protection problems', in Balkan Scientific Conference on Environmental Protection in the Balkans, *Abstracts*, Varna.

Grazhdani, S., Jacquin, F. and Sulce, S. (1996) 'Effect of subsurface drainage on nutrient pollution of surface waters in south-eastern Albania', *Science of the Total Environment* 191(1–2): 15–21.

Green, S. and King, G. (1996) 'The importance of goats to a natural environment: a case study from Epirus (Greece) and southern Albania', *Terra Nova* 8(6): 655–8.

Grimmett, R. F. A. and Jones, T. A. (1989) *Important Bird Areas in Europe*, Cambridge, International Council for Bird Preservation.

Gurashi, A. and Ziri, F. (1982) *Albania Constructs Socialism Relying on its Own Forces*, Tirana, 8 Nëntori.

Haigh, W. E. (1925) *Malaria in Albania*, Geneva, League of Nations Health Organisation.

Hall, D. R. (1984a) 'Foreign tourism under socialism: the Albanian "Stalinist" model', *Annals of Tourism Research* 11(4): 539–5.

—— (1984b) 'Tourism and social change: reply to Romsa', *Annals of Tourism Research* 11(4): 608–10.

—— (1987) 'Albania', in A. H. Dawson (ed.) *Planning in Eastern Europe*, London, Croom Helm.

—— (1990a) 'Albania', in J. A. A. Sillince (ed.) *Housing Policies in Eastern Europe and the Soviet Union*, London, Routledge.

—— (1990b) 'Albania: the last bastion?', *Geography* 75(3): 268–71.

—— (1990c) 'Change closes in on Europe's last communist bastion', *Town and Country Planning* 59(9): 251–4.

—— (1990d) 'Introduction: geographic dimensions of change', *Geography* 75(3): 239–44.

—— (1990e) 'Stalinism and tourism: a comparative study of Albania and North Korea', *Annals of Tourism Research* 17(1): 36–54.

—— (1991) 'Albania', in D. R. Hall (ed.) *Tourism and Economic Development in Eastern Europe and the Soviet Union*, London, Belhaven Press.

—— (1994) *Albania and the Albanians*, London, Frances Pinter.

—— (1996) 'Albania: rural development, migration and uncertainty', *GeoJournal* 38(2): 185–9.

—— (1999) 'Representations of place: Albania', *Geographical Journal* 165(3): 161–72.

Hoda, P. (1991) Lecturer in Botany, Faculty of Natural Science, University of Tirana, Personal interviews, April/August.

Hodges, R., Martin, S. and Saraci, G. (1998) *The Butrint Foundation: The Rediscovery of Butrint*, London, The Butrint Foundation, http://www.albania.co.uk/main/page17.html

Hollis, G. E. and Stevenson, A. C. (1997) 'The physical basis of the Lake Mikri Prespa systems: geology, climate, hydrology and water quality', *Hydrobiologia* 351: 1–19.

Hoxha, D. (1988) 'Assessment and protection of land', in Balkan Scientific Conference on Environmental Protection in the Balkans, *Abstracts*, Varna.

Industrial Waste in Albania (1996) http://gurukul.ucc.american.edu/ted/ALBANIA.HTM

IUCN (1967) *Liste des Nations Unies des Parcs Nationaux et Reserves Analogues*, Brussels, IUCN.

—— (1990) *United Nations List of National Parks and Protected Areas*, Cambridge, IUCN.

IUCN-EEP (1990) *Protected Areas in Eastern and Central Europe and the USSR*, Cambridge, IUCN.

Kabo, M., Kristo, V., Qiriazi, P., Krutaj, F., Gonda, G., Meçaj, N. and Bërxholi, A. (eds) (1985) *Studime Gjeografike*, Tirana, Akademia e Shkencave e RPSSH.

Kabo, M., Bërxholi, A., Krutaj, F., Gonda, G., Meçaj, N., Qiriazi, P. and Kristo, V. (eds) (1987) *Studime Gjeografike 2*, Tirana, Akademia e Shkencave e RPS të Shqipërisë.

—— (eds) (1988) *Studime Gjeografike 3*, Tirana, Akademia e Shkencave e RPS të Shqipërisë.

Karadimov, I. (1989) *Balkan Mayors on the Environment*, Sofia, Sofia Press.

Karaiskaj, G. and Baçe, A. (1975) Kalaja e Durrësit dhe sistemi i fortifikimit përreth në Kohën e vonë antike, *Monumentet* 9: 5–33.

Komisioni i Planit të Shtetit, Drejtoria e Statistikës (KPS) (1988) *Vjetari statistikor i R. P. S. të Shqipërisë*, Tirana, Komisioni i Planit të Shtetit.

—— (1989) *Vjetari statistikor i R. P. S. të Shqipërisë*, Tirana, Komisioni i Planit të Shtetit.

—— (1990) *Vjetari statistikor i R. P. S. të Shqipërisë*, Tirana, Komisioni i Planit të Shtetit.

Kraja, E. (ed.) (1976) *Shkodra Almanak*, Tirana, 8 Nëntori.

Kusse, P. J. and Winkels, H. J. (1990) *Remarks on Desalination and Land Reclamation in the Coastal Area of the People's Socialist Republic of Albania*, The Hague, Dutch Ministry of Agriculture, Nature Management and Fisheries.

Lemel, H. (1998) 'Rural land privatisation and distribution in Albania: evidence from the field', *Europe–Asia Studies* 50(1): 121–40.

Lloshi, X. (1990) *Academy of Sciences of the PSR of Albania*, Tirana, Publishing Council of the Academy of Sciences of the PSR of Albania.

Lucey, T. (1999) 'The Plunkett Foundation', *Besa* 3(2): 14.

McDowall, L. (1991) 'New prime minister installed as 32 die in Albanian food riot', *Guardian*, 11 December.

McKinley, E. B. (1935) *A Geography of Disease*, Washington, DC, The George Washington University Press.

Martin, S. (1999) 'The World Heritage Site of Butrint and the work of the Butrint Foundation', *Besa* 3(1): 21–3.

Mason, K., Myers, J., Winterbotham, H. S. L., Longland, F., Davidson, C. F., Turrill, W. B., White, N. and Mann, S. E. (1945) *Albania*, London, Naval Intelligence Division.

Meksi, A. (1983) *Arkitektura Mesjerate në Shqipëri*, Tirana, 8 Nëntori.

Mema, F. (1998) 'Did Albania really need operation "Alba"?', *Security Dialogue* 29(1): 59–62.

Miho, K. (1987) *Trajta të Profilit Urbanistik të Qytetit të Tiranës*, Tirana, 8 Nëntori.

Ministry of Economy (1991) *Statistical Year Book of Albania*, Tirana, Ministry of Economy.

Ministry of Tourism and EBRD (European Bank for Reconstruction and Development) (1993) *Albania: Investing in Tourism*, Tirana and London, Albanian Ministry of Tourism and EBRD.

Muco, B. (1995) 'The seismicity of Albanian earthquakes and cross-correlation with rainfall', *Physics of the Earth and Planetary Interiors* 88(3–4): 285–91.

—— (1998a) 'Catalogue of M-L = >3.0 earthquakes in Albania from 1976 to 1995 and distribution of seismic energy released', *Tectonophysics* 292(3–4): 311–19.

—— (1998b) 'Twenty years seismic monitoring of induced seismicity in northern Albania', *Pure and Applied Geophysics* 153(1): 151–62.

Muco, B. and Puka, L. (1993) 'Periodic features of seismic activity in Albania', *Pure and Applied Geophysics* 141(1): 59–69.

Nano, T. (1987) *Bimet Mjaltese*, Tirana, 8 Nëntori.

North, C. (1990) 'Into Albania', *The Rock Garden* 22(1): 63–71.

Ostreni, A. (1972) *Gjeografia e Shqipërisë*, Tirana, Shtëpia Botuese e Librit Shkollor.

Panariti, E. and Bërxholi, K. (1998) 'Lead toxicity in humans form contaminated flour in Albania', *Veterinary and Human Toxicology* 40(2): 91–2.

Pano, N. (1990) 'Ecological balance in the Mediterranean Sea', *New Albania* 5: 28.

Papajorgji, H. (1989) 'Triumph of the correctness of the Party on the countryside', *Albania Today* 109: 42–6.

Peja, N., Sarigul, G., Siki, M. and Crivelli, A. J. (1996a) 'The Dalmation pelican, Pelicanus crispus, nesting in Meditarranean lagoons in Albania and Turkey', *Colonial Waterbirds* 19(1): 184–9.

Peja, N., Vaso, A., Miho, A., Rakaj, N. and Crivelli, A. J. (1996b) 'Characteristics of Albanian lagoons and their fisheries', *Fisheries Research* 27(4): 215–25.

Pilidis, G., Ioannidou, A. G., Saraci, M. and Stalikas, C. (1996) 'Determination of organochlorine pesticides and selected heavy metals in Albanian soils', *Fresenius Environmental Bulletin* 5(9–10): 551–6.

Piperi, R. and Kajno, K. (1990) *Flora Mjekësore e Korçës*, Tirana, Drejtoria e Arsimit Shëndetësor.

Piperidou, C. I., Chaidou, C. I., Saraci, M., Stalikis, C. and Pilidis, G. (1996) 'Determination of phenolic compounds, PAHs and heavy metals in surface waters in Albania', *Fresenius Environmental Bulletin* 5(9–10): 557–62.

Pollo, S. and Puto, A. (1981) *The History of Albania*, London, Routledge and Kegan Paul.

Polunin, O. (1980) *Flowers of Greece and the Balkans*, Oxford, Oxford University Press.

Prevots, D. R., Atti, M. L. C. D., Sallabanda, A., Diamante, E., Aylward, R. B., Kakariqi, E., Fiore, L., Ylli, A., van der Avoort, H., Sutter, R. W., Tozzi, A. E., Panei, P., Schinaia, N., Genovese, D., Oblapenko, G., Greco, D. and Wassilak, S. G. F. (1998) 'Outbreak of paralytic poliomyelitis in Albania 1996', *Clinical Infectious Diseases* 26(2): 419–25.

Prifti, P. R. (1978) *Socialist Albania since 1944*, Cambridge, MA, MIT Press.

Protected Areas Database Unit (PADU) (1990) *Information from Protected Areas Database*, Cambridge, World Conservation Monitoring Centre.

PSR of Albania (1977) *Constitution of the People's Socialist Republic*, Tirana, 8 Nëntori.

Puka, V. (1988) 'Physical-chemical qualities of river water', in Balkan Scientific Conference on Environmental Protection in the Balkans, *Abstracts*, Varna.

Rakaj, N. and Flloko, A. (1995) 'Conservation status of fresh-water fish of Albania', *Biological Conservation* 72(2): 195–9.

Riza, E. (1978) *Gjirokastra: Museum city*, Tirana, 8 Nëntori.

Runca, E. (ed.) (1996) *Karavasta Lagoon – Wetland Management Project*, Brussels, European Commission Phare Programme Project Summary.

Sandström, P. and Sjöberg, Ö. (1991) 'Albanian economic performance: stagnation in the 1980s', *Soviet Studies* 43(5): 931–47.

Saraci, M., Cullaj, A., Robinson, D. and Damo, B. (1995) 'Analytical investigation of soil from a tree plantation in south-east Albania', *Fresenius Environmental Bulletin* 4(10): 624–9.

Shallari, S., Schwartz, C., Hasko, A. and Morel, J. L. (1998) 'Heavy metals in soils and plants of serpentine and industrial sites of Albania', *Science of the Total Environment* 209(2–3): 133–42.

Shkodra, G. and Ganiu, S. (1984) *The Well-being of the Albania People and Some Factors and Ways for its Continuous Improvement*, Tirana, 8 Nëntori.

Sjöberg, Ö. (1989) 'A note on the regional dimension of post-war demographic development in Albania', *Nordic Journal of Soviet and East European Studies* 6(1): 91–121.

—— (1991) *Rural Change and Development in Albania*, Oxford, Westview Press.

Sjöberg, Ö. and Sandström, P. (1989) *The Albanian Statistical Abstract of 1988: Heralding a New Era?* Working Paper 2, Department of Soviet and Eastern Studies, Uppsala University.

Skarço, K. (1984) *Agriculture in the PSR of Albania*, Tirana, 8 Nëntori.

Smeets, H., Caushi, N. and Alikaj, E. (1997) 'Infant and maternal mortality in Albania', *European Journal of Public Health* 7(3): 279–83.

Spawforth, A. (1994) 'British tycoons help to unearth ancient city', *Observer*, 20 November.

Standish, A. (1996) 'Albania stumbles on road to democracy', *The European*, 4 April.

Stanfield, D. (1993) 'A proposal for creating a land market in Albania', *Computers, Environment and Urban Systems* 17(3): 263–8.

Stanfield, D. and Raco, M. (1994) 'Land markets, information and a property registration system in Albania', *Computers, Environment and Urban Systems* 18(2): 133–41.

Strazimiri, G. (1987) *Berati: Qytet Muze*, Tirana, 8 Nëntori.

Tabaku, A., Bizgha, V. and Rahlenbeck, S. I. (1998) 'Biological monitoring of lead exposure in high risk groups in Berat, Albania', *Journal of Epidemiology and Community Health* 52(4): 234–6.

Tabaku, A. and Panariti, E. (1996) 'Lead intoxication in rural Albania', *Veterinary and Human Toxicology* 38(6): 434–5.

Toçka, J. (ed.) (1985) *Librazhdi Almanak 1*, Tirana, 8 Nëntori.

Turrill, W. B. (1929) *The Plant Life of the Balkan Peninsula*, Oxford, Clarendon Press.

USAID (1996) 'The Albanian Private Forestry Development Program', *ENR Update*, August/September, 1–2: 4 (http://eni-environment.net/newsltr/AUGSEP96.htm)

Vangeluwe, D., Beudels, M. O. and Lamani, F. (1996) 'Conservation status of Albanian coastal wetlands and their colonial waterbird population (Pelecarniformes and Ciconiiformes)', *Colonial Waterbirds* 19(1): 81–90.

Vejsiu, Y. (1990) 'The vitality of a population', *New Albania* 6: 14–15.

World Monuments Watch (1997) *Butrint Archaeological Site*, http://www.worldmonuments.org/butrintprogf2.html

Xherahu, Q. and Baruti, V. (1975) *Gjeografia e Shqipërise*, Tirana, Shtëpia Botuse e Librit Shkollor.

Zarshati, F. (1982) *Monuments of Culture in Albania*, Tirana, 8 Nëntori.

Zdruli, P., Eswaran, H., Almaraz, R. and Reich, P. (1997) 'Developing the prerequisites for sustainable land use in Albania', *Soil Use and Management* 13(1): 48–55.

13 Bosnia and Hercegovina

Richard Clarke

Introduction

Bosnia and Hercegovina declared sovereignty and seceded from the residue of the Socialist Federal Republic of Yugoslavia (SFRJ) in October 1991, following similar action, first by Slovenia, then by Croatia and after a plebiscite, boycotted by many ethnic Serbs, in which a majority of those voting backed independence (Figure 13.1). The following April, Bosnia and Hercegovina (BiH)[1] was recognised as a legal entity by the EU and USA. A month later it was admitted to the UN. With secession came internal conflict and external aggression, fomented by nationalists in the Croat and Muslim as well as Serb communities. The war left a quarter of a million people dead, maimed or traumatised, and the economy, infrastructure and physical and social fabric of the country in ruins. It seems fitting to dedicate this chapter to the large numbers of Bosnians (of all nationalities as well as none) who tried their utmost to prevent the war and who continue today to work for a multi-ethnic, democratic and environmentally healthy Bosnia.

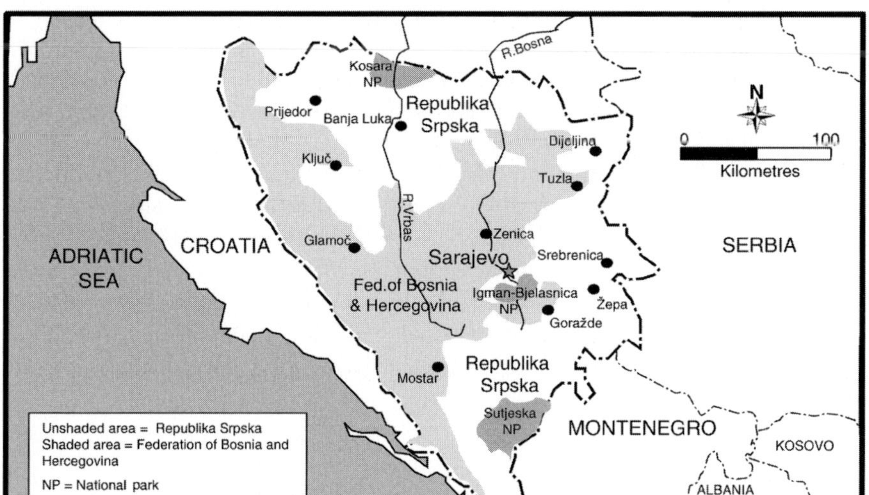

Figure 13.1 Salient features of Bosnia and Hercegovina

In particular, it is dedicated to those who remained in its capital Sarajevo throughout its siege by those who hoped to destroy both the city and the ideals that it represented.

Any discussion of current environmental issues in Bosnia and Hercegovina must be framed by three considerations: (1) the biophysical characteristics of the region and its socio-cultural history and heritage, including the economic and political legacy of BiH's status as a former republic of SFRJ and its inheritance of environmental problems (and achievements) of the socialist era (Jančar 1987; Fisher 1990; IUCN 1991a, b; Terselič and Juras 1991); (2) the immense human, material and environmental cost of the 1992 to 1995 war which followed Yugoslavia's collapse and which all but destroyed the economy and the physical and social infrastructure of BiH; (3) developments subsequent to the war (including patterns of economic and socio-political transition), some environmentally negative, others positive, which provide pointers for the future.

Physical and ecological features

Bosnia and Hercegovina is a small country of 51,200km^2. A small 'stem' of territory joins it to the Adriatic at Neum (interrupting the southernmost coast of Croatia just south of the Neretva delta, some 50km north of Dubrovnik) and provides a coastal strip of a mere 23km. Of Bosnia's 1500km of terrestrial borders (some half of which is demarcated by rivers), roughly two-thirds adjoins Croatia to the northern and southwestern sides of the triangle, while the remainder (on the eastern side) abuts Serbia and Montenegro (Figure 13.1).

Over 55 per cent of the land surface of BiH is defined as mountain (700m and above). A similar proportion consists of dinaric limestone and dolomites, much of it characterised by karstic features. The whole territory is subject to occasional and sometimes destructive earthquakes. The parallel orientation of coastal mountains leads to great climatic variation within a small area. For example, average annual rainfall varies from 730mm (Bijeljina) to 1975mm (Ljubinje); winter temperatures of 10 or 12°C are common on the coast at the same time as heavy snow blankets (Sarajevo), only 120km inland.

Vegetation follows climate and is Mediterranean towards the coast, continental inland. Forest covers almost 50 per cent of the land surface: included in this total are some significant areas often cited as 'virgin' climatic-climax communities. The 1400ha of Peručica between the mountains of Maglič and Volujak represent the largest area of near-natural forest complex in southeast Europe. Anthropogenic vegetation types in Bosnia range from small fields and alpine meadow inland, to maquis, Mediterranean scrub and wetlands towards the coast, to Alpine pasture in the highest regions.

Economic and social background

A characteristic of Bosnia (which it shares with other southern republics and in which it is a mirror of the Balkans as a whole) is ethnic and cultural diversity.

Of its population of just over 3.2mln (the population at the 1991 census was 4.36mln), no one group forms an absolute majority. Since the end of Turkish rule the largest group has been the Muslims (with between 40 and 44 per cent of the population), followed by Serbs (31–40 per cent) and then Croats (17–22 per cent). Estimates of numbers of different ethnic-national groupings vary according to source as well as date. Only 5 per cent of the population at the 1991 census described themselves as 'Yugoslav' (i.e. repudiated any ethnic affiliation). Patterns of settlement were very mixed so that pre-conflict ethnic maps resembled a patchwork quilt or (like the countryside itself) a landscape of small fields.

Despite eruptions of populist nationalism, declared ethnicity was for many people an arbitrary matter, depending on which parent's origin was followed. For most it was a nominal category, significant in the years prior to Yugoslavia's collapse only because the alternative – to declare oneself a Yugoslav – had become increasingly unfashionable by the mid-1980s. Moreover, such affiliations were cultural rather than religious. Many individuals (at least in towns, as well as those from partisan families or villages) were secular; many had little time for the established religious institutions. It is only since the conflict that religion has become, for many, a way of declaring their national affiliation. As virtually all knowledgeable commentators (including those hostile to Yugoslav socialism) have observed (e.g. Donia and Fine 1994; Malcolm 1994; East and Pontin 1997), contrary to popular perceptions of 'ancient ethnic hatreds' the causes of Bosnia's tragedy came not from the prejudices of its people but from the scheming of opportunist nationalists (Serb, Croat and Muslim), and from the fatal miscomprehension and interference of outsiders. Newly created political divisions do not in general conform to any natural geophysical areas (a particular problem today in catchment areas where source and consumption of water may be in different entities); in the context of this chapter their prime significance is their debilitating effect on united action for protection and remediation.

A second characteristic of BiH which it shares with other southern Balkan states is its historically low level of socio-economic development; it was one of the less developed republics of pre-1991 Yugoslavia. Nevertheless, from 1945, as in the rest of the SFRJ, urbanisation and industrialisation were the major determinants of environmental change. BiH is rich in mineral resources, including coal, iron ore, bauxite, manganese, copper, chromium, lead, zinc and silver. Much post-war industrial development was promoted by cross-federal transfers of investment from the developed 'north' to the poorer south of the former Yugoslavia; this was motivated as much by political considerations as by logistic factors to do with the location of raw materials, energy supplies or transport routes. A particular feature of post-1945 development in BiH was the establishment of military bases and munitions production facilities. The Republic had some 40 per cent of former Yugoslavia's armaments production capacity including tank and aircraft manufacture.

In the first three decades of the SFRJ, the population of BiH grew by 40 per cent and the economy by an average of 6.2 per cent p.a., although the rate of both population and economic growth slowed in the 1980s. The single most apparent consequence for the landscape was the establishment of several large industrial complexes and a rapid (and in some towns, chaotic) growth of suburbs at the expense of the countryside. Much of the suburbanisation has been 'planned' (usually high-rise) development by city authorities or individual enterprises; much of it however has simply grown (without planning permission or proper sewerage or transport systems) as privately built incremental sprawl. In addition to large (often internationally financed) industrial plant, the unique Yugoslav system of self-managing enterprises coupled with lax planning controls encouraged numerous small-scale enterprises. One of the foci of post-1945 investment, Zenica is commonly held to have been one of the worst polluted cities in the former Yugoslavia. In some other cities, including Sarajevo, the geographical spread and location of industry has been constrained by mountainous terrain, though this has also exacerbated the consequences of pollution as exemplified by Sarajevo's frequent temperature inversions. Prior to independence BiH as a whole had 20 per cent of the surface area of former Yugoslavia, 18 per cent of its population and 12 per cent of its GDP, but contributed 32 per cent of its atmospheric pollution (Independent Bureau for Humanitarian Issues 1988).

The obverse of post-1945 urban development has been the 'undevelopment' of the countryside. The influx of young families to the towns left innumerable farmsteads to be cared for by ageing parents or (in marginal areas) abandoned altogether. At the same time as many of the older houses in the countryside became derelict, new ones have been built, often close by, sometimes by the families of the original owners, sometimes by newly affluent town dwellers as weekend cottages ('vikendica') and holiday homes. The result in some areas has been a significant conversion of much farmland and woodland to concrete and tarmac, while in other areas fields have reverted by natural succession to scrub and secondary woodland.

The war and its aftermath

In addition to the large numbers of dead (5–7 per cent of the population; estimates from different sources vary widely) the human costs of the war include over 200,000 wounded, of whom 13,000 are permanently disabled. Very few people continued to live in their own homes throughout the war, and some 2.6mln people (55–62 per cent of the population) were displaced. Since the Dayton Accords many of these have returned home (though under 10 per cent of returnees have gone back to places where they were part of an ethnic minority), but there remain over one million refugees and displaced persons about equally distributed between the two entities. Estimates of material destruction and damage are unreliable since returns by property owners are certainly inflated by hopes of reparation; however, the level is certainly very high. It is claimed that some 63 per cent of all dwellings in BiH sustained at least some damage and that as much as 18 per cent were totally destroyed. Whole residential districts of Sarajevo have been reduced to rubble. Throughout the whole of BiH productive installations, communications and energy utilities were often as significant as military targets as the focus of destruction. As much as 78 per cent of the industrial capacity of BiH may have been destroyed, and production is only now beginning to recover from its 1994 low of 4 per cent of the 1990 output.

Most parts of the communications infrastructure (including fifty-nine bridges) have been damaged by military action or lack of maintenance. Over three-quarters of Bosnia's electrical generating capacity was damaged; some is still out of operation. Water and sewerage systems in some localities remain non-functional. Educational facilities have been severely damaged and enrolments at all school levels in some areas are estimated at around 50 per cent below eligibility. Hospital facilities (as measured by the number of available beds) are less than two-thirds of their pre-war levels (World Bank 1998). War damage in the Republic was estimated in 1996 by Hans Van den Broek, EU Commissioner for Central and Southeast Europe, to amount to between $30 billion and $50 billion (Simpson 1996). Physical destruction has been accompanied by economic collapse. 'Real' unemployment is currently some 80 per cent; annual per capita income has fallen to around $500 (compared to $1900 in 1990). Such statistics provide a backcloth against which a concern for the environment can appear almost an indulgence. It is certainly the case that most foreign aid has been directed, initially, towards humanitarian assistance, then subsequently (and inadequately) towards economic rehabilitation. However, the environment is important not only in itself but because it has the capacity to provide a focus for action which can transcend ethnic differences and the continuing division of the country into two separate 'entities'.

That division is (next to the human suffering and the physical destruction which is immediately apparent to any visitor) the single most significant legacy of the war. The wartime territories which emerged as a consequence of the shifting conflicts and alliances between Bosniak, Croat and Serb forces – namely the Muslim/Croat 'Federation of Bosnia and Hercegovina' (FBiH or 'Federation', covering 51 per cent of the territory) and Republika Srpska (RS, covering some 49 per cent) – were recognised in the November 1995 Dayton Accords (and confirmed in the subsequent December 1995 Paris Peace Treaty) as separate, semi-autonomous entities within one legal state to be governed by a bicameral National Assembly. The reality is that whilst the Federal government in Sarajevo often aspires (in the absence of any functioning governmental structures at state (i.e. BiH) level) to speak for the whole of BiH, Republika Srpska understands itself to be effectively independent. Significant decisions affecting FBiH, RS, or BiH as a whole are taken in the Office of the UN High Representative (OHR) and by other members of the OHR Advisory Council (including the OSCE, UNHCR, SFOR, IBRD and IMF). The spring 1999 NATO bombardment of Serbia has, if anything, hardened the resolve of Serb nationalists to have nothing to do with an integrated BiH. In addition, much of the Croat part of the Federation is, or was (until the February 2000 elections in Croatia replaced the ruling nationalist party with a broader left coalition which renounced claims on the 'Croat' part of BiH), effectively ruled from Zagreb (under the Tuđman administration, many Hercegovinian Croats occupied positions of power and influence in Croatia). Inevitably, official environmental pronouncements often have a political subtext. In the context of this chapter, one consequence is that good information for the whole of BiH is difficult to come by. Much is unpublished and/or consists of individual estimates; some is merely anecdotal. All statistics relating to the area should be treated with caution and those in this chapter should be regarded as 'best estimates' only.

A major environmental problem (unique in Europe to BiH until recent conflict in Kosovo) is the continued hazard of land-mines. Estimates of the numbers of mines laid (by all 'sides') during the war range from 350,000 to 35 million. A realistic estimate is perhaps 2–3mln mines, in some 16,000 minefields. Despite efforts at clearance, aided by foreign governments (Austria has given DM3.5mln for satellite surveillance of mined lands) and companies (such as Caterpillar International) which have exploited opportunities for the sales of mechanised de-mining equipment, only a fraction of minefields have to date been cleared. Those that have are mainly in and around inhabited areas, and on some of the settled front lines where the densest minefields were laid and their locations are known. In remoter areas and in regions where front lines continually changed (for example, in the Muslim enclaves around Srebrenica, Žepa and Goražde) many minefields still exist, their location is unclearly defined and injury and death still regularly occur, particularly to children (United Nations High Commission for Refugees 1996). Some rural areas are regarded as 'no-go' and it is estimated that some 20 per cent of the wooded area of BiH may still be mined (Hadziabdić 1997). Even in Sarajevo the Jewish cemetery remains heavily mined and there has yet to be any significant attempt to clear it. What impact the continuing – or past – presence of mines may have or may have had on wildlife is unknown; there has been and is no monitoring. What is clear is that for many years to come, mines will remain a major environmental hazard, and their removal will be a precondition for effective environmental management.

Legislation policy and planning

The formal principles of environmental protection in BiH derive from the former Yugoslav Federal constitution whereby all natural resources with the exception of private holdings of up to 10ha were defined as social property. In principle this might have permitted a higher degree of social awareness and care for the environment than in (for example) the UK, where land is privately owned and environmental protection requires limitations on property rights, often involving compensation. In practice, social ownership often meant that 'what is everyone's is no one's' and financial mechanisms for securing environmental objectives were lacking. Particularly in the case of industrial undertakings, environmental costs (for example, of air and water pollution) were, even when recognised, perceived as 'externalities' and were often disregarded in policy decisions. In this respect the situation in SFRJ differed little from that in Western European countries, in that only where there was specific legislation backed by mechanisms for monitoring and enforcement was protection effective. However, the high levels of devolution and decentralisation in the former Yugoslavia combined with a generally lower level of socio-economic development. The consequence was that although environmental pressures were often of lower intensity than in more 'developed' areas in the north, lower levels of environmental protection meant that their effects were often much greater.

One of the features of socialist Yugoslavia was its unique system of workers' self-management, with enterprise autonomy (and a degree of market discipline) superimposed on an essentially socialist economic base. This was coupled with a high degree of decentralisation, a high level of political awareness and debate, and openness to 'Western' ideas, influence (and investment). In all but the most essentially federal matters (such as defence), legislation and policy-making was devolved to the constituent republics. Within each republic, implementation and decision-making was often devolved to a local level. Day-to-day environmental regulation and management was generally the responsibility of municipalities or individual enterprises, subject to general policy guidance (and funding) from the responsible ministries, together with advice and/or regulation via a parallel system of semi-autonomous institutes (for example, water protection, natural and cultural heritage protection, urban planning). This broad situation remains unchanged by independence but the war has inevitably disrupted communications and policy implementation, which in general remains distinct within the two entities. There were no provisions in the Dayton Accords for any all-BiH structures of environmental management, with the exception of a Commission to preserve national monuments, which has so far not been instituted. In the Federation, the principal ministries (of Physical Planning and Environment (*Ministarstvo za prostorno planiranje i okoliš*) and of Agriculture, Water Management and Forestry (*Ministarstvo za Poljoprivredu, Vodaplaniranje i Šumarstvo*)) try to retain at least a 'watching brief' across the whole of the former Republic. In practice however they have very little contact with their counterparts in Republika Srpska.

Environmental law both in the Federation and RS is therefore still based largely on former Yugoslav environmental regulations but policy and implementation within the two entities has diverged. By January 1999 Federal BiH had drafted new framework laws on Environmental Protection (*Zakon o Zaštiti Okoliša*) and on Nature Conservation (*Zakon o Zaštiti Prirode*), both closely based on those recently adopted in Slovenia and Croatia. RS, in turn, had drafted its own environmental law, which is a copy of the law currently in force in Serbia. This led to concern on the part of both OHR and international bodies, and of environmentalists within the respective entities, that, once adopted, the very different approaches of this draft legislation would be yet a further hinderance to future collaboration. In July 1998 a Memorandum of Understanding was signed at the Regional Environmental Centre (REC) in Szentendre, Hungary. The Szentendre Agreement is one of the most far-reaching agreements concluded to date between the two entities. It accepts the fact that environmental issues extend across the border between the two states and provides for the institution of an Environmental Steering Committee (ESC) to generate common approaches to environmental issues across the two entities. The Committee comprises eight members (four from each of the two entities), plus representatives of the Office of the High Representative, USAID, the European Commission, the World Bank and other bodies as non-voting members. Its remit includes harmonisation of environmental legislation and standards, environmental monitoring, information systems and plans for emergency situations, physical planning, and relations with the international donor community (Csagoly 1999). Subsequent work, financed by the Japan Special Fund, resulted in the employment of local coordinators to work in the REC offices in Sarajevo and in Banja Luka and has focused on the development of pilot national, regional and local environmental action plans (EAPs, with the possible objective of a NEAPP for the whole of BiH), and on legislative coordination between the two entities. Early progress included the initiation of a process of drafting a new environmental law applicable across the whole of BiH. This involved a commitment by the responsible ministries in both entities to withdraw the existing draft laws from further parliamentary process. The intention (accepted by both Ministries) was to institute a new and transparent process with opportunities for public and NGO involvement, resulting in new environmental legislation across the whole of BiH. This could also have led to BiH participation in international conventions including the Basel, Danube and Aarhus conventions (REC and Japan Special Fund 1999). At the same time, international agencies (including the OHR) expressed significant doubts as to whether such an 'internal' process could yield concrete results within a reasonable timespan. An EU-funded project was therefore established in early 2000 to draft an all-BiH framework law using independent contractors. Any harmonisation of environmental legislation across the two entities will be rendered more difficult, however, by their very different governmental structures. In particular, FBiH is characterised by a high degree of devolution of decision-making to its ten constituent cantons (80 per cent of Federation income is spent at canton or municipal level) which contrasts sharply with the high degree of centralisation in RS.

The Szentendre Agreement also accepts that many environmental issues involve cooperation with neighbouring states and should therefore, as with foreign policy in general, fall within the competence of the common institutions of BiH. Currently, the governments of both entities in BiH formally regard themselves as bound by a number of international conventions, by succession from the former Yugoslavia. These include the 1973 (Washington) Convention on International Trade in Endangered Species (CITES), the 1979 Ramsar Convention and the 1987 (Montreal) Convention on Substances that Deplete the Ozone Layer. However, under the Dayton Accords the entities are not themselves able individually to enter into any international agreements, which can only be done at state (BiH) level; and in the absence of effective state structures BiH is unable to accede to new agreements and is therefore not party to major conventions such as those arising from the 1992 Earth Summit, including the UN Framework

Conventions on Climate Change and the UN Biodiversity Convention. Nor is it effectively party to other international agreements or programmes (such as the Pan European Biological and Landscape Diversity Strategy, or the Aarhus Convention) drafted subsequent to 1990; however, there is considerable interest in both academic and government circles, as well as NGOs in doing so, and representatives from both entities attend international meetings.

In addition to obvious problems of enforcement and monitoring of legislation, there is an additional issue of transparency. Partly as an inheritance from the previous culture of official institutions, and partly as a reaction against constant external examination, there is a reluctance to open the working of organisations to external scrutiny. This applies even to environmental data which have no sensitivity in a political or commercial context. The limited data that exist are not regarded as appropriate for the public domain, and this creates a disincentive to the collection of additional information which might be useful to make effective assessments of environmental quality. For example, the Institute of Water Resources in BiH is unable to provide information on river quality which might help to assess the consequences of war, or the effectiveness of remediation policies. Accession to the June 1998 Aarhus Convention (formally, the Convention on Access to Information, Public Participation in Decision-making and Access to Justice in Environmental Matters signed by the EU and thirty-five governments) would be a significant step because the Convention links humanitarian concerns and civic rights (including freedom of information) to environmental protection.

Urban development

One of the features of the siege of Sarajevo is the way that despite the war, 'ordinary life' continued for many people under the most difficult circumstances. This was reflected even in policies for environmental improvement. A major scheme for the post-war revitalisation of Sarajevo was produced by the Institute for City Planning 'Zavod za Planiranje Razvoja Grada Sarajeva' at the height of the war (February 1993 to June 1994). Proposals included a visionary programme of tree planting to create a system of 'green corridors' throughout the city, each with its own distinct species assemblage related to microclimatic features of particular areas and each linked with forest parks (Trebevič, Bulozi) and nature parks (Bjelašnica, Ozren, Jahorina) on the city periphery (Zavod za Planiranje Razvoja Grada Sarajeva 1994).

Subsequent to the conclusion of the Dayton Accords, priority has been given to repairs to (or clearance of) damaged property and the recommissioning of industrial enterprises. Numerous reconstruction and development projects have begun, many of them financed with external assistance. Significant donor bodies include the World Bank, USAID and several national governments. Industrial production has begun to recover and is currently some 15–20 per cent of pre-war levels. Prospects for environmental improvements (including implementation of wartime plans for a 'green Sarajevo') have taken second place to these priorities and have also been much frustrated by new political conflicts.

One difficulty is that the situation regarding property restitution and privatisation for public properties remains confused. This has so far obstructed the implementation of management proposals for Sarajevo's cemeteries, which form such a significant feature of the city's landscape and which were established to a large degree on land nationalised in the process of agrarian reform in pre-socialist Yugoslavia between 1919 and 1939. It seems likely that property restitution will be limited to land appropriated subsequent to 1945. However, this process will be much more complex than even in other former Yugoslav republics because of the major shifts of population that have occurred since 1991.

The elaboration and implementation of environmental policies is constrained not only by the division of Bosnia into its two constituent 'entities' but also by conflicts of authority within these entities, both at government and municipal level. For example, agreed plans to plant trees in Sarajevo's largest central park have been halted by parallel proposals for building developments. Planning issues outside the central area are dogged by friction between the city and the canton – the latter swelled by an influx of refugees from rural areas. One example of such conflicts in an environmental context is a proposal for new urban development over an important neolithic archaeological site in Ilidža, where the conservation objectives of the city authorities are directly opposed by recently arrived 'New Sarajevans' whose physical circumstances mean that jobs and accommodation are priority issues.

The ambient environment

One of the consequences of the war was the collapse of monitoring programmes, the severing of links between individuals in university, government and NGOs, and the physical destruction of records, including scientific data that had been painstakingly collected over a long period. There is currently little ongoing environmental monitoring. It is clear however that the consequences of the war for the ambient environment have been both mixed and geographically uneven. In some places damage to processing and storage facilities has produced pockets of intense local pollution; elsewhere, however, economic collapse has meant that industrial pollution has been (temporarily, at least) much reduced.

Generation of industrial waste plummeted during the war from an estimated 250mln.t in 1989 to virtually zero in 1992 (Federal Ministry of Physical Planning and Environment 1998). However, new dumps and stores of wastes of all categories accumulated during and since the war and these remain largely unrecorded. This wartime waste includes an estimated 800t of unused medicines (from some 35,000t sent as humanitarian aid). Some 90 per cent of these medicines have expired and are unusable, and are stockpiled, particularly in Mostar (FBiH) and Prijedor (RS). Effective means of disposal do not exist within BiH and the price of incineration abroad is estimated at between DM2000 and 50,000 per tonne (Independent Bureau for Humanitarian Issues 1988). Scattered but largely unrecorded military wastes together with the legacy of unexploded mines and munitions present a major continuing hazard.

As industrial production develops, the provision of appropriate technical facilities with administrative coordination, backed by legislation, will be critical. Responsibility for waste management in the Federation is split between the Ministry of Physical Planning and Environment, and the Ministry of Agriculture, Water and Forestry. However, waste disposal is in general undefined by legislation at a Federal level, except in the most general terms by the Law on Physical Planning (for internal transports of waste) and the Law on Customs (for transborder transports). Even this is largely ineffectual because of poor or uncoordinated border controls, with the result that illegal movement and disposal of environmentally hazardous wastes is largely uncontrolled.

Effective structures exist only at municipal level. Most domestic solid wastes are disposed of by landfill. There is no independent monitoring of procedures or quality. Hazardous wastes present a particular problem and are largely dealt with on an *ad-hoc* basis. There are no operating separation plants or high-temperature disposal facilities for toxic or unstable wastes. Disposal in power plants and cement factories is minimal and, where it exists, is performed by poorly trained personnel and is unmonitored. The Federal Government has begun the process of ratification of the Basel Convention but little is happening on the ground with respect to infrastructural and organisational implementation.

The most critical environmental issue in BiH is water cycle management. In terms of pollution of surface and subsurface waters, BiH is generally accepted to have been the most polluted Republic of the former Yugoslavia. Pollution was particularly bad in several areas along the valley of the River Bosna (as well as the lower parts of the Neretva valley) where most industrial production was located. Zenica and Tuzla were particular blackspots for water quality with chemical processors and other facilities discharging waste directly into watercourses. Throughout the country, many rivers were of poor quality due to lack of waste water and sewage treatment plants. In urban areas only around 70 per cent of households were linked to a sewerage system and these were often inefficient. No tertiary treatment plants existed anywhere. The general recognition of this problem led to the production, just before the war, of a major twenty-year plan for improvement of the water supply (only about 45 per cent of the country by area was connected to a piped water supply) and of waste water disposal facilities of the whole country.

The 1992 to 1995 war seems to have led to some local improvements in surface water quality following temporary cessation of industrial operations. For example, water in much of the Neretva is reported to be of near drinkable quality. At the same time, however, the war has caused a considerable local worsening of the situation particularly with regard to domestic wastes following destruction or disrepair of many municipal facilities. Since the end of the war, all 2200km of major rivers in Federation areas have been surveyed by the Hydro-Engineering Institute. Although only a relatively limited number of sampling stations were used, these were selected to represent significant catchments. Recorded parameters include suspended and total solids, dissolved oxygen and BOD, pH and bacterial contamination (biological indicators appear not to have been used). The results show only about 3 per cent of all rivers to be totally free of pollution: almost 30 per cent of rivers fall into the category of eutrophic or severely eutrophic (Federal Ministry of Physical Planning and Environment 1998). The worst pollution seems to be of the River Bosna downstream of Zenica which is claimed by environmental activists to be little more than an industrial channel (Omanović 1999). Much international assistance from bodies such as the World Bank, USAID and individual member governments is directed to the re-establishment of drinking-water and sewage systems in RS (Government of the Republic of Srpska 1999) as well as in Federation areas. One of the current features of post-war reconstruction is the entrepreneurial activity of foreign firms (for example, that of Atkins in promoting small-scale combined heat and power CHP stations). Aid-related infrastructural development may not always be necessarily environmentally benign, however. A proposal to dam the Neretva at Konjic to create a hydro-electric power station has been vigorously contested by environmental campaigners. More generally, the lack of requirements for environmental assessment means that the environmental effects of many proposed developments are unknown.

The situation with respect to air pollution shows some similarities to that of water quality. It is claimed that with 20 per cent of former Yugoslavia's surface area, 18 per cent of its population and 12 per cent of its GDP,, BiH contributed 32 per cent of its air pollution (Independent Bureau for Humanitarian Issues 1988). Pre-war 'blackspots' for atmospheric pollutants included Zenica and Tuzla. During the war SO_2 and NO_x emissions fell considerably though peaks of other pollutants recorded during the war included particulates from domestic wood-burning in Sarajevo (Table 13.1). These have now in turn been replaced by other pollutants including hydrocarbons and lead from motor vehicles. Such pollutant levels have been exacerbated by other forms of 'aid' including the importation from Germany since the war of some 80,000 secondhand cars which do not conform to EU emission standards. There are no regulations on automobile exhausts in either entity of BiH and the use of unleaded petrol is still uncommon. In other areas the re-establishment of industrial production has restored pollution to its pre-war levels or even above. Annual production of pollutants from Zenica steelworks has been estimated at 35,000t of SO_2,

Table 13.1 Air quality in Sarajevo

	SO_2 concentrations (mg/m^3)		*Particulate matter (mg/m^3)*	
	1990	*1995*	*1990*	*1995*
Annual average	84	12	38	38
Winter average	143	18	62	54
Maximum (24h)	486	66	235	628
98th percentile (24h)	382	51	175	304

Source: Meterological Institute Bjelave 01 station: city background.

11,000t of dust, and 85 × 10^9cu.m of smoke (Omanovič 1999). The location of urban and industrial development in valleys where air circulation is often poor means that atmospheric pollutants are often localised, but concentrated. This is particularly apparent in hot weather in Sarajevo and Zenica where temperature inversions are common, resulting in occasional but high pollutant levels.

The rural landscape

Over 40 per cent of the land surface of BiH is classified as agricultural land (20 per cent arable, 25 per cent meadow and pasture: 2 per cent permanent crops). However, agriculture accounted for only 9 per cent of GDP in 1989 and, even before the conflict, the country produced less than 50 per cent of its food requirements. The lower lands of northern Bosnia support some wheat and corn as well as vineyards, orchards and livestock. However, leaching and soil erosion in mountain regions and on slopes, caused by high rainfall and the short inland growing season, often leads to reduced agricultural productivity. Moreover, farm holdings are mostly small and were in private ownership, often managed by older people whose income was supplemented by that of their children who had moved to the town.

During the war there was mass movement off the land. In rural areas, where they have been permitted to do so, some families have now returned to their farms and are re-tilling their fields; however, ethnic cleansing and the continued division of the country, coupled with low per capita incomes and the chronic shortage of domestic capital, mean that many farms and houses still lie abandoned and derelict. Others have been occupied by refugees, but the uncertain (and possibly temporary) nature of their tenure means that even where there is the money to do so (which is rare) there is little permanent investment and many areas have an unkempt appearance. Larger areas of once-farmed countryside remain uninhabited because of continuing threats to returnees, the absence of habitable buildings, or because of the threat of landmines.

Approximately 48 per cent of BiH is covered by forest. The long history of forest management embodies Austro-Hungarian forestry traditions. These include an emphasis on continuous cover techniques aimed at maintaining sustainable yields. The forest area of BiH is distributed fairly equally between the Federation and RS. It consists of approximately 58 per cent hardwoods (mainly beech, with some oak). The remaining softwoods are principally spruce and fir, with some pine.

Responsibility for woodlands of BiH remains that of regional Forestry Economic Area Administrations (*Šumsko Privredno Područje*) (SPP), each responsible for between 20,000 and 85,000ha of forest including state-owned and private areas. Prior to the war, each of the eighty-five SPPs combined regulatory and production functions in accordance with a comprehensive ten-year forest plan approved by the Ministry. The plan specifies forest maintenance, reforestation and silvicultural development and an annual allowable cut within which logging quotas and cut schedules for each woodland area are determined. In reality, inadequate regulation meant that forest plans were often patchily or poorly applied. As in all other areas of environmental management and policy, the war has caused much greater disruption. The SPPs remain in existence: forty-three in Republika Srpska and forty-five in the Federation, of which five, in Hercegovina, operate outside Ministry control (Citizens Democracy Corporation 1998). Each SPP is also a commercial organisation, cutting timber and marketing logs, and relatively free of state control. The SPP can sell timber (cut according to licences it issues to itself) to any internal purchaser. Exports (currently to Germany, Italy, Croatia and Slovenia) are approved by the Ministry which sets minimum domestic and export prices. It is likely that, as in other areas of the former Yugoslavia, the future will see the regulatory activities of the SPPs separated from their executive and commercial functions and that these latter will be privatised.

It is claimed that the war has led to significant deterioration in forest quality. However, the nature and distribution of damage seems to vary considerably in different areas, and proper inventories have yet to be compiled. In general, the forests of Republika Srpska seem to be in better shape than those of the Federation. The major causes of damage include uncontrolled cutting (there were significant exports of timber from both sides throughout the war); lack of management (including a neglect of reforestation and of selective felling for improved growth); and direct damage from military action. Large areas of woodland have been affected by uncontrolled fires. It is estimated that between 15 and 20 per cent of the forest area remains mined and inaccessible; shell and bullet fragments embedded in trees in some areas have lowered the grade of timber and have increased the cost of harvesting it.

The mere 23km of coastline might at first suggest that coastal issues are not significant in BiH. The point is often made, however, that the inland catchments draining into the Adriatic (some 25 per cent of the country's land area), especially the Neretva delta, are part of the coastal system. BiH is signatory (by succession) to the Mediterranean Action Plan (MAP) and a number of Federation environmental projects are related to a Coastal Area Integrated Management Plan produced with EC life funding. Several of these projects address land-based pollution, particularly in the Neretva delta region.

Natural and cultural heritage protection

As with the ambient environment, legislation and administration for nature conservation varied in detail between each of the Republics of the former Yugoslavia (Jančar 1987; Fisher 1990; European Parliament Directorate-General for Research 1991; IUCN 1991b; Terselič and Juras 1991). In outline, however, the situation in Bosnia corresponded to that elsewhere, in that day-to-day management of protected sites and areas was delegated at a local level (to municipalities and often to individual enterprises) and overall control was exercised through a tripartite relationship between local authorities, the responsible ministries and the quasi-autonomous Institute for the Protection of Cultural, Natural and Historic Heritage (*Zavod za Zaštitu Kulturnog, Historijskog i Prirodnog Naslijeđa BiH*).

Inevitably, there is an absence of empirical data on the effects of the war on wildlife. Effective monitoring has yet to begin. One significant achievement has been the publication of a national botanical Red List covering the whole of BiH. This was commissioned in 1990 by the Ecological Society of BiH, and compiled under siege conditions during the war using literature and herbarium sources; it was finally published in 1996 (Šilić 1997). The list covers all vascular plant groups (pteridophyta and spermaphyta) and comprises 678 species registered according to the old IUCN threat categories, including eight Extinct or probably so; forty-three Vulnerable, 289 Rare and fifty-two Insufficiently known. Preparatory work has also begun on vertebrate groups.

Only 281km^2 of the country receives any legal protection as conservation sites and protected landscape. This is equivalent to 0.55 per cent of the land surface. This is the lowest proportion of all the former republics of SFRJ. The greater part of this total – just over 250km^2 – comprises just five areas (the national parks of Sutjeska and Kozara, the Jahorina and Trebević nature parks, and the Peručica Primeval Reserve). The remaining 250-odd sites are individually very small.

As in other former Yugoslav republics, the two national parks were declared by the Republic Assembly under their own acts, *Zakon o Nacionalnom parku Sutjeska* and *Zakon o Nacionalnom parku Kozara*. However, of the two national parks (both now in RS), only Sutjeska had a formally constituted managing authority (and staff). Lower-tier areas such as regional parks and nature reserves were established by lower bodies, latterly under the the 1985 Framework Act for the protection and use of cultural historic and natural heritage (*Zakon o zaštiti i korišćenju kulturnoistorijskog i prirodnog nasleđa*), by regional and municipal authorities in collaboration with the Institute for Protection of Cultural and Natural Monuments and Natural Rarities (683/54).

Regional and town authorities were also able to declare lower levels of protected landscape. For all these categories, levels of protection procedures were variable. Management and administration, although nominally the responsibility of municipal authorities, was in practice often delegated to scientific or commercial enterprises. The natural resource of some areas was also commercially exploited and protected status used as a vehicle for this process. An example is Hutovo Blato Nature Reserve in the Neretva delta which was formally administered by an experimental farm belonging to the Ministry of Agricultural Economy (Singleton 1985). Hutovo Blato was also let annually by the municipality of Čapljina via a tourist organisation to Italian bird hunters (Terselić and Juras 1991). In addition to these locally run areas, a small number of strict (non-intervention) reserves, mainly of *Picea omorica Panc*, were designated by the Forest Authority.

All these areas were in existence prior to 1991, since when legal protection has been largely in abeyance, administrative structures have collapsed and little effective management or monitoring has taken place. Major problems include the lack of funding, of qualified individuals capable of undertaking the work, of effective communication between entities and of access to critical sites. For example, the most recent survey data for Peručiča forest – arguably the most significant natural area in BiH – are from 1984.

However, there are signs of at least some activity in both entities. Both national parks (Kozara and Sutjeska) are in RS and have been included in a DM5mln appeal for funding for reconstruction and repair of places of culture, with sums identified as necessary for rehabilitation, of DM774,500 for Sutjeska and DM532,000 for Kozara (Government of the Republic of Srpska 1999).

Within the Federation there has been a revival of proposals for the extension of the protected area system already under discussion before the war: these include two riverine areas (the source and upper reaches of the rivers Una and Sana) and several mountain areas (Treskavica, Prenj, Čvrsnica and Vranica) (Terselič and Juras 1991). If all these pre-war proposals were to be realised, the total, although double the present area, would still be little over 1 per cent of the whole country. Under a new Law on Physical Planning *Zakon o prostornom uredjenju*, a spatial plan for the Federation to the year 2000 was produced under which it was proposed to designate up to 16 per cent of the land surface. One proposal is for a new national park of almost 1000km^2 or just under 2 per cent of the land surface of BiH, focused on Prenj, Čvrsnica and Čabulja. However, there is some scepticism as to whether any of these proposals will be implemented, at least in the short term.

In addition to national parks and nature reserves, several other areas have (or had) the status of recreational and conservation areas under city structural plans. For the most part these are merely formal designations, inherited from the former, pre-1991 system. In Sarajevo however they formed part of the plans for post-war regeneration developed during the siege which have recently been given new impetus by the institution of an Olympic Committee to prepare a bid for Sarajevo to host the 2010 Winter Olympics. The proposal includes the Igman and Bjelašnica mountains, the site of the 1974 Winter Olympics where skiing resumed again in 1998. Here, the woodlands remain largely unfelled, despite the chronic shortages of fuel during the war. Jučan, Trebevič and Jahorina were by contrast heavily felled by soldiers during the war and the wood sold to city dwellers at high prices. The consequence has been major erosion and landslips. Jahorina and Trebevič remain heavily mined. Since the proposed area for the new park straddles the two entities (and on the Republika Srpska side is the area from which the bombardment of Sarajevo was controlled) the significance of these proposals is as much political as environmental (and, if achieved, would be as great as the obstacles to doing so).

A significant advancement is that the governments of both entities are already beginning to consider the development of tourist strategies. In RS, attempts are being made to raise funding for the rehabilitation of skiing centres on Mount Jahorina, and of spa facilities in Višegrad, Teslič, Bijeljina, Dubica, Laktaši and Srebrenica, to attract game hunters in state-designated hunting areas and to use other natural features (especially fishing and rafting in the upper Drina) as tourist attractions. Logistic difficulties, as well as the all too obvious destroyed houses and other reminders of ethnic cleansing around the Muslim 'safe haven' of Srebrenica, make it unlikely that these attempts will result in significant revenue – either from aid agencies or tourists – in the near future. However, in Sarajevo and (to a lesser extent) in Mostar, recreational commercialism has already begun. In addition to services provided for foreign aid personnel and their families there has begun a significant but steady trickle of more 'conventional' tourists (a significant proportion of them, it seems, attracted as spectators of the destruction of the heritage of these cities).

One of the features of the conflict was the destruction – often deliberate – of cultural and historic buildings. The destruction of property was not just collateral damage; architecture, as one of the more enduring manifestations of cultural heritage, was often deliberately targeted by both sides (Herscher 1998). The wanton and unnecessary destruction at the height of the fighting by Croat irregulars of the ancient bridge in Mostar (which had no military significance but did serve as a footbridge for civilians between the 'Croat' west and the 'Muslim' east of the city) has become an enduring image – captured in film footage – of the war. In Serb (as in many Croat) nationalist areas there has been a systematic destruction of mosques; in many Muslim areas, Orthodox churches have been a target of fire and stand empty and derelict. An unpublished 1995 report issued by the Institute for the Protection of Cultural, Historical and Natural Heritage in Sarajevo documents damage to 1115 mosques, 309 Catholic churches, thirty-six Serbian Orthodox churches (this last figure is almost certainly an underestimate) and 1079 public buildings. In a few cases this has 'merely' undone restoration to buildings which were already severely damaged during the Second World War. Many cultural monuments however have been irreparably destroyed. Such is the case with the Jewish Library, one of the most significant cultural assets of pre-war Sarajevo. In the case of some of the other most notable public buildings and monuments (for example, the National Library in Sarajevo) repairs have already begun, usually funded by outside organisations.

In Sarajevo too, whilst most of the suburbs (particularly around the airport) remain empty, the old central bazaar area of the town shows almost no signs of three years of shelling. Almost all the mosques have been repaired (and several new ones built), clearly with no shortage of external funding. Within a limited area of perhaps a square kilometre the market area of Sarajevo has become a tourist centre again, replete with trinket shops, though the tourists are almost all foreign military, humanitarian and advisory personnel. New 'heritage' interpretation (Carter *et al.* 1995) is a feature of reconstruction. Pavement grenade blast scours have been levelled with red plastic filler to mark the spots where people died from Serb nationalist shelling from the surrounding hills, which overlook the city. Meanwhile, Gavrilo Princip's footprints on the Latin Bridge (where the spark which ignited the First World War was struck) have been removed and the bridge renamed, as have many other streets. At the time of writing a major outcry has prevented the proposed removal of a monument (untouched by shelling during the siege) to the liberation of Sarajevo by partisan troops in 1945.

Ironically it is damage to the religious artefacts, much more than to the living places and economy of Bosnia, that has caught the attention of (and stimulated action among) international funding bodies. Some of the destruction suffered has been compounded by 'wholly inappropriate new construction by NGOs using concrete blockwork and red tile roof finish' (Barakat 1997). At a deeper level some of the (doubtless equally well-intentioned) activity of outside bodies reveals the historic complexity beneath the symbolic simplicity of cultural heritage. An example is Počitelj, a small historic town on the banks of the Neretva. Although there was little collateral damage from actual fighting, after the exclusion of the Muslim population (and their replacement by Croat refugees) the mosque was deliberately blown up. The mosque and town have become the focus of an action-research project based at the Post-War Reconstruction and Development Unit (PRDU) at the University of York (UK) Institute of Advanced Architectural Studies, funded by the World Monuments Fund, and the (US) Samuel H. Kress Foundation (Revitalisation of Počitelj 1997). A vacation study of the town by students on the University's MA course in Post-war Recovery Studies focused on the need to safeguard the remains of the mosque (and of other traditional Islamic buildings) pending possible reconstruction (Barakat and Wilson 1997). The probable existence of much older (Orthodox Christian) remains below the mosque was seen as a 'complicating' factor, and one of less importance than the continuing occupation of the town by Croats, who would be as unlikely to regard with favour the rebuilding of the mosque as they would any repair and display of the earlier Orthodox remains. The destruction to the cultural heritage has raised more acutely in Bosnia than elsewhere the perennial problem of all 'heritage' reconstruction, that it is always in part 'an arbitrary exercise in cultural memory, is always a question of what to save, what to put back, what to take apart' (Wilson 1992, p.235). Although the institution of a Commission on the Protection of National Monuments was one of the requirements of the Dayton Accords, there has to date been no action.

Civil society and non-governmental organisations

A feature of former Yugoslavia including BiH was the strong tradition of non-governmental organisations (NGOs). These included scientific societies and amenity organisations as well as sporting societies and (most importantly in an environmental context) the Alpine clubs, which were active at a variety of different levels from local and municipal groups to regional and national organisations. The work of many of these NGOs was supported, both directly and through support of particular projects and initiatives, by the BiH government and/or municipal administrations. The more 'scientific' NGOs, in particular, often had close links with universities as well as sometimes with ministries.

Important groups that were active at a local level included the Country Museum *Zemaljski muzej* and other organisations particularly involved in cultural heritage and species protection. More explicitly political bodies included, notably, the River Una Protection Society, *Društvo za stvaranje kulture čuvanja i zaštite rijeke Une*, based in Bihać. The latter, known popularly as the 'Emeralds of Una' (*Unski Smaragdi*) claimed a membership of some 112,000 immediately prior to the war. A feature of the 'Emeralds' was its work with young people, for which it was awarded the UN Global 500 Charter medal. This it returned during the war in protest against the UN's ineptness and inertia, particularly with regard to the suffering of children who bore the brunt of the conflict. The Greens of Zenica (*Zeleni Zenice*), another local body, also continued to work during the war, but their activities were restricted to the local level and had little effect (Gjigas 1997). More recent environmental campaigning has been initiated by loosely formed local groups on specific issues. One example is local opposition to proposals to dam the Neretva River to create a hydro-electric power installation near Konjic. Another is action by residents of Doboj who organised a petition (1999) on pollution of the River Spreča (from which the Banja Luka water supply also comes) arising from factories in the Tuzla region (REC and Japan Special Fund 1999).

Subsequent to the end of the war, other bodies such as the Environmental Protection Union, *Savez za zaštitu i unapredjenje čovjekove okoline BiH*, and the Green Alliance, *Savez Zelenih BiH*, have been formed, but activities are largely restricted to large towns in Federation areas. The most influential of currently active organisations, the Foundation for Sustainable Development and Quality of Life (*Fondacija za podsticanje uravnote zenog razvoja i kvaliteta zivota*) (FONDEKO), was founded in 1994 to initiate Agenda 21 initiatives. Its management board includes prominent academics as well as businesses and utilities; it has already produced several educational texts for high school pupils and has had a regular thirty-minute radio slot. In 1996 FONDEKO founded a quarterly colour magazine, *Fondeko svijet* (Fondeko World) as a 'scientific-popular revue on nature, people and ecology', excellently produced with all articles including a short English summary (see Abadžić 1997); in early 1988 it organised a conference on the future of protected areas in the Republic. The organisation (like several other NGOs) has its own website. Other significant organisations include the Sarajevo-based ECO-BiH, an umbrella body coordinating a number of local organisations, and BETA, the Bosnian Environmental Technology Association, whose activities have included the encouragement of organic farming (with a certification scheme and demonstration fields in FBiH and RS), and a summer camp for students on Mount Vranica, where participants worked on plans for sustainable use of local natural and human resources. Usually these organisations depend on the initiative and energy of a handful of individuals who may have academic affiliations; thus ECO-BiH is associated with the Centre for Ecology and Natural Resources at the Science Faculty, and BETA with the Faculty of Agriculture at Sarajevo University. Other organisations may be more political in nature; for example, in mid-2000, the Banja Luka-based Eco-Movement RS provided the basis for the establishment of the RS Green Party.

Overall, the number and level of activity of NGOs in BiH is, understandably, low but is nevertheless growing. A survey carried out in late 1996 by the Sarajevo office of the Regional Environmental Centre (Gjigas 1997) secured only ten completed questionnaires distributed to thirty-eight NGOs in the Republic. The number of active environmental NGOs was estimated at fifteen, around half of them based in Sarajevo and all but three formed after 1990. Most respondents defined themselves either as grass-roots organisations or as associations of environmental professionals. None had any full-time paid staff. About half of the groups surveyed were either dormant, defunct, or had moved with no forwarding address. By mid 2000, REC estimated that some hundred-plus NGOs were active across BiH and RS, with more than 150 funded projects. Such growth is only partly to be attributed to the funding activity of external agencies (including REC); it is also a measure of a growing level of environmental awareness and of the potential of environmental activism as a non-partisan focus for civic action.

Conclusions: the future

Continuing political and ethnic conflicts and divisions in Bosnia make it difficult to avoid partisan analyses even of environmental issues. It has not been the purpose of this chapter to analyse the causes of the recent conflict in BiH, much less to allocate blame between the protagonists. Two things are clear, however: first, that the war itself was a tragedy not only for the country's people, but also – directly and indirectly – for its environment, natural and cultural; second, that its consequences will dominate the future for many years to come. Nationalism, the lack of a developed civil society and effective oppositional politics in BiH comprise the flip side to poverty, dependency and isolation; both are obstacles to environmental progress. At the same time, the environment has provided a focus for some of the most positive post-conflict developments; within stability pact arrangements for Southeast Europe, the Regional Environmental Reconstruction Programme (RERP) now plays a prominent role, linking environmental remediation to the (re)construction of civil society. Recent growth in NGO activity, the likelihood in the near future of all-BiH framework environmental legislation and a widespread awareness within and without BiH of the quality of its natural resources provide hopeful signs for the future.

Acknowledgements

Acknowledgement and thanks are due for help and information provided by colleagues at the universities of Sarajevo (FBiH) and of Banja Luka (RS), and within the ministries of the two entities. I am particularly grateful to Nesad Seremet and Djordje Stefanović of the Regional Environmental Centre, to Ing Djuro Fukarek and to other friends in Sarajevo, particularly Zdravka Grebo of Job 22 and Rade Jevtić of Radio Zid for their hospitality during visits to BiH. Much of the research for this chapter was carried out during study visits in 1997 and 2000 with the permission and assistance of Birkbeck College, London University.

Note

1 The territory of Bosnia and Hercegovina is commonly referred to as 'Bosnia' by outsiders as well as by many former Yugoslavs, but in fact consists of two entities, one of which provided the basis for the short-lived attempt to set up a Croatian entity of 'Herceg-bosna' during the war.

References

Abadžić, N. 1997: 'Šta Hoćemo?', *FONDEKO Svijet* 1 (1), spring, p.3.

Barakat, S. 1997: 'Analysis of Cultural Heritage Damaged by War in Bosnia-Herzegovina', *Revival* 8, August, p.4, Supplement.

Barakat, S. and Wilson, C. 1997: *The Revitalisation of the Historic Settlement Počitelj.* York: PRDU, Institute of Advanced Architectural Studies.

Carter, F. W., Hall, D. R., Turnock, D. and Williams, A. M. 1995: *Interpreting the Balkans.* Geographical Intelligence Paper 2. London: Royal Geographical Society with the Institute of British Geographers.

Chandler, D. 2000: *Bosnia. Faking Democracy after Dayton.* London: Pluto Press.

Citizens Democracy Corporation. 1998: 'Forestry in Bosnia and Hercegovina', http://www.cdc.org/Bosnia Forestry.

Csagoly, P. 1999: 'Federation, Republic in B&H Sign Historical Pact', *REC Bulletin* 8 (1), pp.18–19.

Donia, R. J. and Fine, J. V. A. J. 1994: *Bosnia and Hercegovina; A Tradition Betrayed.* London: Hurst and Company.

East, R. and Pontin, J. (eds) 1997: *Revolution and Change in East and Central Europe.* London: Pinter.

European Parliament Directorate-General for Research. 1991: *Nature Conservation in Austria, Finland, Norway, Sweden, Switzerland, Bulgaria, Czechoslovakia, Hungary, Poland, Romania, Yugoslavia and the Soviet Union.* Environment, Public Health, and Consumer Protection Series 17. Luxemburg: European Parliament.

Federal Ministry of Physical Planning and Environment. 1998: Untitled.

Fisher, D. 1990: 'Environmental Politics in Yugoslavia', *Environmental Policy Review* 4 (2), July. The Soviet Union and Yugoslavia.

Gjigas, O. (ed.) 1997: *NGO Directory: A Directory of Environmental Nongovernmental Organisations in Central and Eastern Europe* (3rd edn). Szentendre, Hungary: Regional Environmental Centre. See also http://www.rec.org/REC/Publications.

Government of the Republic of Srpska. 1999: 'Reconstruction and Development 1998', http:\\www.srpska.org 11 November 1999.

Hadžiabdić, S. 1997: 'Rat i Šumarstvo Bosne i Hercegovine', *FONDEKO Svijet* 1 (1), spring, pp.42–4.

Herscher, A. 1998: 'Remembering and Rebuilding in Bosnia', *Transitions*, pp.76–81. March.

Independent Bureau for Humanitarian Issues. 1988: *Human Development Report; Bosnia and Herzegovina.* Sarajevo: IBHI and United Nations Development Project.

IUCN. 1991a: *Basis of the Long-term Prospects of the Conservation and Management of the Natural Heritage of Yugoslavia.* Gland, Switzerland: International Union for the Conservation of Nature. Unpublished Report.

—— 1991b: *Environmental Status Reports No. 2: Albania, Bulgaria, Romania, Yugoslavia.* Cambridge: International Union for the Conservation of Nature.

Jančar, B. 1987: *Environmental Management in the Soviet Union and Yugoslavia: Structure and Regulation in Federal Communist States.* Durham, NC: Duke University Press.

Malcolm, N. 1994: *Bosnia. A Short History.* London: Macmillan.

Omanovič, M. 1999: 'Environmental Problems of the Town of Zenica', http://www.zenica.net/ekologija/mustafa.omanovic 11 November 1999.

REC and Japan Special Fund. 1999: 'Support for the Development of Environmental Law and Policy in Bosnia and Herzegovina', Status report. July 1998 to March 1999.

The Revitalisation of Počitelj; a war-damaged historic settlement in Bosnia-Herzegovina. 1997: *Revival* 8, August Supplement.

Šilić, C. 1997: 'Spisak biljnih vrsta – Pteridophyta i Spermatophyta – sa crvenu knjigu Bosne i Hercegovine', *Glasnik Zemaljskog Muzeja Bosne i Hercegovine Prirode nauke, 31/1992–1995/1996*, pp.323–367 (in Bosnian).

Simpson, J. 1996: 'Business Opportunities in Bosnia . . . Including Environment', *OMRI Daily Digest* 142 (II), 24 July, pp.11–12.

Singleton, F. 1985: 'National Parks and the Conservation of Nature in Yugoslavia', Paper presented at the Third World Congress for Soviet and East European Studies, Washington, DC, 30 October to 4 November.

Terselić, V. and Juras. A. 1991: 'Nature Conservation in Yugoslavia', in Directorate-General for Research (ed.) *Nature Conservation in Austria, Finland, Norway, Sweden, Switzerland, Bulgaria, Czechoslovakia, Hungary, Poland, Romania, Yugoslavia and the Soviet Union*, Environment, Public Health, and Consumer Protection Series, 17. Luxemburg: European Parliament, Directorate-General for Research, pp.117–37.

United Nations High Commission for Refugees. 1996: 'Minefields in Bosnia-Hercegovina: Overall Risk Assessment', Office of the Special Envoy, External Relations Unit.

Wilson, A. 1992. *The Culture of Nature.* Oxford: Blackwell.

World Bank. 1998: *Reconstruction of Bosnia and Herzegovina: Priorities for Recovery and Growth.* Washington, DC: World Bank. http://www.org.html 11 November 1999.

Zavod za Planiranje Razvoja Grada Sarajeva. 1994: *Akcioni Program Revitalizacije Zelenila Grada Sarajeva 1993–2000.* Sarajevo: Zavod za Planiranje Razvoja Grada.

14 Bulgaria

John Pickles, Mariana Nikolova, Caedmon Staddon, Stefan Velev, Zoya Mateeva and Anton Popov

Introduction

By the end of the Second World War, 80 per cent of Bulgaria's workforce was still employed in subsistence agriculture. By the 1960s, the economy had been transformed into an agro-industrial export economy dependent on inputs, credits, and orders from the Soviet Union and other Comecon countries. Like other economies throughout the region, national economic development in mining, industry and agriculture was achieved by the rapid adoption of large-scale technologies and organizational practices in which environmental and health concerns were minimized; what Green has called 'the hubris of giganticism' (Green 1989, p.1). In many ways it was this very hubris that brought the environmental and health consequences of rapid industrialization, urbanization and environmental degradation to the forefront of public concerns in the 1980s, eventually erupting as a mass mobilization of political forces for 'ecological defence' (EcoGlasnost n.d.; Pickles 1992; Pavlinek and Pickles 2000). The commitment to national development through forced industrialization prompted new environmental legislation in the 1980s (Table 14.1), and resulted in little political room for manoeuvre for the Zhivkov regime when opposition forces organized around environmental and health rights, particularly in the context of parallel opposition to state violence against the country's Muslim minorities (see Pickles and Begg 2000).

In this chapter, we focus on efforts to reform environmental legislation and administration since 1989, with particular attention to the role of geographical scale and popular action in this process. Initial commitments to environmental reforms were, in large part, responses to the power of popular action and of ENGOs. But this power quickly diminished as political and economic conditions changed after 1989. The normalizing of environmental politics also weakened efforts to implement stronger forms of environmental regulation, a withdrawal from commitment that was signalled in the spring of 1995 by the scandalous post facto rewriting of the Environmental Protection Act to retroactively legalize the construction of the Djerman–Skakavitsa dam and by the subsequent suspension of talks with the government by the International Monetary Fund (IMF) because of its lack of commitment to environmental reforms. But the chapter

Table 14.1 Selected environmental legislation in Bulgaria

Enacted before 1989	*New or revised 1990–99*
Law on Mines and Quarries (1957, amended in 1986)	Law on Protection of Air, Water and Soil from Pollution (amended in 1991 and 1992)
Law on Forests (1958, amended in 1979)	Law on Protection of Nature (amended in 1991)
Law on Protection of Plants from Diseases and Insects (1960, amended in 1968)	Law on Regional and Urban Planning (amended in 1990 and 1991)
Law on Protection of Air, Water and Soil from Pollution (1963, amended in 1968, 1975, 1977, 1978, 1988)	Law on Public Health (amended in 1991 and 1995)
Law on Protection of Nature (1967, amended in 1977, 1978, 1988)	Law on Ownership and Use of Agricultural Land (1991, amended in 1995)
Law on Regional and Urban Planning (1973, amended in 1974, 1977, 1979, 1980, 1984, 1985, 1986, 1987, 1988)	Environmental Protection Law (1991, amended in 1992 and 1995)
Law on Protection of Arable Land and Pastures (1973, amended in 1977, 1981, 1989)	Law on Liquid Fossil Fuels Tax (1996)
Law on Public Health (1973, amended in 1976, 1983, 1985, 1986, 1988, 1989)	Clean Air Act (1996)
Law on Hunting (1982)	Law on Restricting the Harmful Impact of Wastes (1997)
Fisheries Act (1982)	Law on Forests (1997)
Law on the Use of Atomic Energy for Peaceful Aims (1985)	Law on Protected Areas (1998)
Law on Marine Environment (1987)	Law on Power Industry and Energy Efficiency (1999)
	Waters Act (1999)
	Hunting and Protection of Wildlife Act (Pending)
	Access to Public Information Act (pending)

also focuses on the ways in which, in responding to the demands of reform and transformation, everyday activities are also changing environmental conditions and practices. The chapter focuses explicitly on the impacts on environmental conditions of changing regulatory and social practices, with specific attention to regulatory reform, environmental hazards, agriculture, forestry, water resources, protected landscapes and industry.

Reforming the regulators

Efforts to rewrite state socialist environmental legislation had already begun before the street demonstrations for 'ecological defence' that occurred in Sofia and other cities in November 1989 (Angelov 1989; Pavlinek and Pickles 2000). But the mass demonstrations mobilized political opposition to the Zhivkov regime and placed environmental health at the head of reform priorities (see Pickles (1992) and Koulov (1998) for extended discussion of these events). Environmental awareness was high and the role of environmental social movements in bringing about political reforms was strong (Popov and Baikov 1992; see also Genov 1994).[1] In 1990, the new Ministry of Environment (MoE) was established and the reform of environmental legislation began in 1991 (Table 14.1).[2] Its aim was to create market-oriented legislation that would be enforceable (Mindjov 1995, p.26). The 1991 Constitution gave citizens the right to a 'healthy and friendly environment' and obliged them to protect the environment (Article 15). Citizens also had the right to be informed about environmental conditions affecting them (Article 41) (Tzvetkova 1996, p.25). The new Environmental Protection Law was enacted in 1991 and has been amended nearly every year since. Among other measures, the law introduced the polluter pays principle, the prevention and precautionary principle, and the public right of access to environmental information. The law also introduced charges for the use of natural resources, environmental impact assessment procedures, and established the National Environmental Protection Fund and Municipal Environmental Protection Fund. Detailed regulations for the law on Environmental Impact Assessments were passed in 1992 (Georgieva 1993, pp.74–7; Mindjov 1995, p.26; OECD 1996, p.28), followed by laws dealing with protected areas, waste, protection of medicinal plants, land protection, marine environment, water, air and mineral resources (OECD 1996, p.29).[3]

Despite these reforms, the development of new environmental laws has been slower in Bulgaria than in other ECECs such as the Czech Republic and Slovakia (see Pavlinek and Pickles 2000). Indeed, by the mid-1990s some observers even argued that there had been little change in the approach towards the environment since 1989. In the words of Georgieva (1993, p.73), 'the first and the second communist (socialist) dominated post-totalitarian governments followed the past tradition of wishful thinking by developing an ambitious, completely unrealistic, and not related to legal and institutional changes strategy for environmental improvements.'

Since 1996 existing environmental monitoring units have been upgraded and new units established to gather information on environmental conditions, pollution and health, partly with financial and technical support from EU-PHARE (see MoE 1996a, 1996b; Ministry of Environment and Waters 1998). Institutional coordination problems remain, the laws regarding property rights as they apply to environmental and health data are unclear, and established user practices still tend to foster data-hoarding and data-hiding instead of data-sharing (see Mikhova and Pickles 1994a, 1994b; Pickles and Mikhova 1998). However, gradually environmental information is being seen as public information, particularly under the influence of European organizations.

In cases where national and regional jurisdictions are weak, local 'obstini' authorities have stepped in to exercise their constitutional right to regulate more rigorously than does national legislation. Obstini with environmentally active mayors or administrators – especially those who found some political advantage from working with environmental forces after 1989 – have, in some cases, been able to develop their own policies and regulations for local food supplies and agricultural practices (Paskaleva *et al.* 1998). As a result, at national, regional and

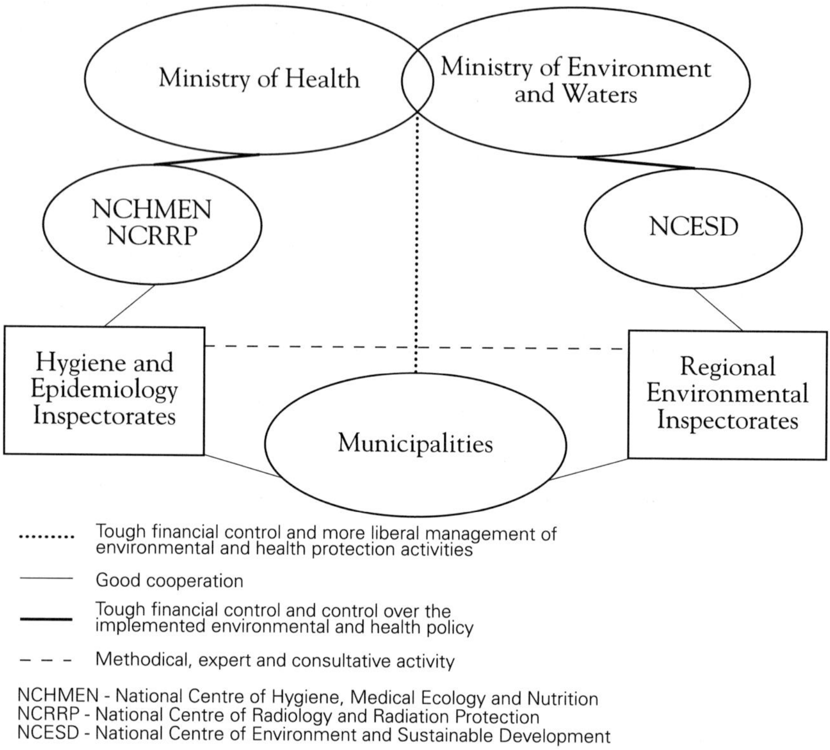

Figure 14.1 Institutions of environmental protection and regulation in Bulgaria

Source: Ministry of Environment and Waters, Sofia.

local levels new systems of administration, licensing, coordinating and registering of activities with potential environmental impacts have been introduced, along with economic measures (sanctions, taxes, customs duties and tariffs) that regulate the environmental behaviour of citizens and enterprises.

Overall, institutions dealing with environmental issues have been substantially restructured since 1989 and the practices of these institutions seem to have changed markedly, especially in regard to the provision of information since 1997. But coordination problems remain, staff members are often under-qualified and poorly paid, and economic difficulties facing those they must regulate leave little room for real enforcement. The result is an uneven map of institutional alignment and effectiveness (Figure 14.1).[4] Under these circumstances any assessment of the state of the environment might appear premature, but in fact changes on the ground have run far ahead of formal practices of policy and regulation. In what follows we focus on the role of climate change in affecting environmental hazards, on the degree to which the environmental legacies of the centrally planned economy have been ameliorated, and on the environmental consequences of transition.

Figure 14.2 Map of Bulgaria

Source: See acknowledgement.

Environment and environmental hazards

It goes without saying that fundamental to any understanding of environmental quality is the actual condition of the environment. However, in one regard at least, environmental conditions have been overlooked in assessments of ecological transitions. Recent collaborative work by Knight *et al.* (1995) and by Velev (1996, 1997, 1998) has shed light on the importance of climate change in understanding ecological transformations in Bulgaria. Specifically, this work has focused on the prolonged drought conditions between 1982 and 1994.[5] The drought was most marked in southern Bulgaria and in the high mountain zone of Vitosha, Rila, Pirin and Osogovo, but affected all but the northeast of the country to some degree (Figure 14.2). Between 1982 and 1994 precipitation decreased by 10 per cent in northern Bulgaria, by 20 per cent in southern Bulgaria, and by 30–40 per cent in the high mountains. Since 1994, no droughts have been observed and annual precipitation totals have been around or above normal.

Researchers have been trying to explain why the drought has been so pronounced in the high mountains and why it has not been accompanied by the kind of warming trend expected by global models of greenhouse warming. The explanation seems to have to do with changes in atmospheric circulation over the past fifteen to twenty years. The number of Mediterranean cyclones passing over the Balkan Peninsula decreased from an average of eighteen per year during the 1960s (Martinov 1970) to seven per year in the 1980s. The number of days with anticyclones has increased, and in most cases these are blocking anticyclones which remain over eastern and southeastern Europe for ten or fifteen days, or even longer.[6] The broader consequences of the 1982 to 1994 drought are not clear. Precipitation declines in 1992 to 1994 influenced river run-off in southwestern Bulgaria. During these years the stations in Kiustendil, Sandanski, Pazardzhik, and on the Cherni Vrah and Mousala mountain peaks recorded fifteen to twenty successive months with precipitation totals about 50 per cent below normal. Water flow from the heavily affected high mountain areas within the Iskar River basin declined and (exacerbated by ageing infrastructure) led to significant water supply problems in 1993, approaching crisis proportions in Sofia by the end of 1994 (see Vasilev *et al.* 1994; Knight *et al.* 1995; Staddon 1996, 1998; Velev 1996). River flow and reservoir levels (including that of the Iskar reservoir supplying drinking-water to the Sofia metropolitan area) dropped to dangerously low levels. The water crisis caused several problems for institutions responsible for providing drinking-water, and social unrest emerged in areas where people felt their water was being mined to compensate for the insatiable demands of Sofia and to cover for the poor planning of government ministries (Staddon 1996, 1998). The drought altered Alpine grass complexes in the high mountains and, if it had persisted, the bio-productivity of coniferous forests and other highland forests might have been affected. In non-mountain areas in southern Bulgaria the drought caused diminished river run-off, a drop in ground water levels and reduced moisture content in soils (Popov and Sarafov 1996). Assessing the drought impact on crop yields is particularly difficult. During the period of drought, socio-economic restructuring and major changes in the patterns of cropping were widespread. As a result, it is not possible to make distinctions between yield decline resulting from reduced fertilizer and irrigation use and drought-related declines.

Between 1995 and 1998, precipitation patterns began to normalize, even rising above normal in some regions. In turn, these increases resulted in flooding and spills from often poorly maintained reservoirs, culminating in serious flooding in northeastern Bulgaria on 3 September 1999.[7] Already, the wetter years have brought higher frequency and intensity of thunderstorms, heavy rains and snowfalls, and these have appreciably increased damage from hail, floods, landslides and avalanches.

One result of the Bulgarian transition is increased risk from natural events. Hurricanes, earthquakes, floods or particularly harsh winter storms can be particularly dangerous and damaging to health and property if people do not have the resources (or incentives) to maintain the safety of basic industrial, service or household infrastructure. For example, as a result of the winter storms in December 1996 and subsequent torrential rains, the water level of the Varbitsa River rose by ten metres, causing two deaths and material losses amounting to three billion leva. In Zlatograd alone 150 families lost their homes. Damages were higher than might otherwise have been the case because many people had been unable to maintain their property in weatherproof condition after five years of decline in real income. For the same reason, the National Fund for the Civil Defence Against Disasters and Accidents is currently able to compensate individuals for only one-fifth of their losses. The ability of state and local governments to finance hazard mitigation efforts has also diminished. The number of hydro-meteorological stations in the country has been reduced. The existing ten Hail Risk Reduction Sites, which protected about one-quarter of all arable land, were closed in 1996 because the Ministry of Agriculture could not afford to provide the US$183,000 needed for their operation in the 1997 agricultural season. As a result, 1997 hail damage in agriculture in the 'protected areas' amounted to an estimated US$2.89mln (Directorate of Hail Risk Reduction, Ministry of Agriculture). Moreover, with land restitution now effectively completed, the limited profitability of many smallholdings has meant that the proportion of land insured against crop damage has declined from 34 per cent in 1993 to 18.5 per cent in 1996 (State Insurance Institute 1997; Nikolova 1998a).

In addition, vulnerability to environmental hazards has increased because information about risk prevention is not available and policies for disseminating public service information are underdeveloped. Surveys carried out in 1997 in the Varna region of eastern Bulgaria after a series of major landslides indicate that only 5 per cent of those surveyed knew how to protect themselves from natural hazards, only 1 per cent actually relied on the safe warning system when exposed to danger, and only 3 per cent expected to receive compensation from the government in the event of a disaster (Nikolova 1998a, p.191). While these findings might suggest that citizens fail to respond rationally to risk, such responses in fact may be accurate assessments of current circumstances; in the 3 September 1999 flood in the same region in which six people drowned, no flood warnings were issued to residents, even though the severity of likely flooding had been known beforehand.

Ecological change in rural transitions

Agricultural transformation

Agriculture is another sector where transition policies have had ambiguous effects. Agricultural ecology suffered from the Communist Party's prioritization of rapid industrialization over environmental or other goals, creating incentives for farm managers and households to prioritise growth.[8] Substantial degradation of agricultural and other resources resulted. However, as Meurs *et al.* (1998) have shown, post-1989 agricultural policies have not been without impact on environmental quality. Privatization policies, for example, were expected to reduce erosion, excessive chemical applications and livestock pollution as new landowners (both individual farmers and members of new production cooperatives) sought to protect the value of their property. As prices adjusted to world market levels they were expected to encourage reduced applications of fertilizers and other chemicals, as well as the replacement of the giant Soviet machinery with smaller, more appropriate technology. Environmental benefits were also expected to accrue from decentralized decision-making, as centrally defined cropping patterns and fertilizer and pesticide applications were replaced with agricultural practices more attuned to local conditions and livestock management was decentralized. Some even argued that Bulgarian agriculture had the potential to develop a system of low-input agriculture, especially because locally developed alternative technologies such as Integrated Pest Management were available and West European markets for low-input agricultural produce were growing (Begg 1994). Others emphasized the possibility that pre-war Bulgarian traditions of sustainable, community-controlled agriculture might serve as a model for post-socialist agriculture (Bogdanova 1995).

Indeed, several of the benefits expected from decentralization of control and the emergence of markets have been realized. Fertilizer prices have increased to more than ten times their 1989 levels and fertilizer use has dropped drastically in response, from an average of 170kg/ha between 1985 and 1989, to an average of 108kg/ha between 1990 and 1992. Pesticide and herbicide use also dropped after prices jumped by a factor of ten in 1991, from 20,485kg/th.ha during the peak period between 1980 and 1985 to 10,807kg in 1992 (Begg 1994, p.15; CSO various years; World Bank 1991, p.13). Decentralization also resulted in some environmental improvements. Between 1989 and 1993, cooperative and individual farms slaughtered livestock: the cattle population fell by 39 per cent, pigs fell by 35 per cent, sheep by 44 per cent and poultry by 52 per cent. The majority of remaining livestock is now held by small, decentralized households (ranging from 46 per cent of pigs to 84 per cent of sheep) (Begg and Meurs 1998, pp.22, 24). As a result, there has been an overall reduction in waste and an increase in the use of natural fertilizers on small farms (Naidenova 1993). But the decentralization of livestock has also created some problems. Livestock were privatized to villagers living in closely clustered houses with attached barns designed to hold a few head of livestock for household use. EC-PHARE reports that the concentration of privatized livestock in these dwellings has created locally severe waste disposal problems, and such 'overstocking' is evident in many rural villages throughout the country (Douglass 1994, pp.5, 9).

On the other hand, marketization has also produced its own environmental problems, exacerbated by extreme uncertainty (Meurs *et al.* 1998). Legal title to newly restored land was slow in coming and many individual farmers and new cooperatives had to farm land under one-year leases or temporary use permits. Even where ownership has been established with relative certainty, conditions do not necessarily encourage owners to protect the long-run value of their 'investment'. Most of those initially receiving land did not plan to farm it themselves, and many did not plan to own it for more than a few years. At the same time, price instability, declining average incomes and the absence of land markets meant that owners had little basis for estimating land prices. Combined with the restitution of land in its fragmented pre-war boundaries, these conditions re-created pre-war pressures for households to deplete land in the interests of short-term survival and, partly as a result, 'soil-mining' has increased. Such farmers have been reluctant to invest in anti-erosion tree barriers or lined irrigation canals, or to experiment with soil-improving crop rotation. In 1993 and 1994, many new private farmers were not using any fertilizers at all, because they did not expect to be leasing the same land the following year. Reduced nitrate applications were certainly beneficial. Phosphorus applications could also be temporarily discontinued in certain areas with little reduction in yields because of the already high levels in the soil (Deets 1993, pp.4–5). But potassium was already seriously depleted in over 16 per cent of arable land and phosphorous was lacking in nearly 40 per cent (Nikolova 1992). As a result, by the mid-1990s yields began to decline and soils were degrading.

In the next few years these land use and natural resource management practices are likely to be challenged even further as the restructuring of forestry, surface water management, mining and fisheries transforms the structure of common property resources in rural areas. Current government policy is orientated towards establishing regulatory and management regimes for common property resources that are in line as far as possible with European Union standards. This will have to establish more 'eco-centric' and 'democratic' environmental practices while addressing widespread pressures for further property restitution (particularly of forest lands). The following sub-sections explore common property resources, with particular attention to forests, water and protected landscapes.

Forestry

Bulgaria's forests cover approximately 30 per cent of total surface area, 29.2 per cent of which is covered by coniferous species, which are largely concentrated in the mountainous areas of the Balkan, Rhodope, Rila and Pirin mountains. The remaining deciduous forests are much more species diverse. Oak-beech woodlands are well represented within lowland habitats, and have become commercially more attractive in recent years. The overall condition of both coniferous and deciduous resources is good, though many regions have significant problems with soil erosion caused by indiscriminate logging practices, tree pests and diseases and with foliage damage through environmental pollution (Staddon 1999b).

The present government recognizes the importance of forest privatization in its Program 2001, where the implications of forestry reform receives special attention – the only sector to receive specific discussion in the national environmental plan.[9] In fact, among the first legislation initiated by the Kostov government after April 1997 was the new 'Law on Forests' (Durzhaven Vestnik, 29 December 1997) and the 'Law for the Restructuring of Ownership of Forests and Lands in the Forest Fund' (Durzhaven Vestnik, 25 November 1997).[10] When the process is completed it is anticipated that approximately 15 per cent of total national forested land will have been restituted, though there will be considerable regional variations in levels of private ownership (Popov and Staddon 1999; Staddon 2000b).

Restitution of private ownership over forests poses several problems for continued access to forest-related resources. Many rural families have come to rely on informal resource extraction practices and are therefore highly dependent on the common property resources of public forest lands. These range from hunting to mushroom-collecting for sale to international markets to free-range grazing of sheep and goats to wood-gathering for heating, construction and small craft production (see Cellarius 1999; Staddon 1999a). A 1999 survey of a rural district of the mountainous southwest found that 100 per cent of respondent households depended upon locally collected firewood for heat, 50 per cent collected herbs, 42 per cent used 'common' forest lands for pasturing cattle and sheep, 27 per cent collected foodstuffs, and 21 per cent were involved in water harvesting (Staddon 1999c). The loss of such ancillary sources of household income could have serious consequences for the use of marginal lands and for social instability in regions where new private landowners attempt to restrict access and practices in areas long regarded as communal.

Freshwater resources

Though relatively poorly endowed with minerals, Bulgaria is blessed with an abundance of freshwater resources. Several major rivers rise within the country, including the Maritsa, Strouma, Mesta, Arda, Iskur and Yantra rivers, and Bulgaria shares more than 200km of Danube River frontage. Notwithstanding the ready availability of more than 2000cu.m/pc (by world standards an abundance) Bulgaria has struggled to provide water in adequate quantity and of satisfactory quality. Restructuring the communist era water management infrastructure has proven difficult, and has been made more difficult by restitution of agricultural and forest lands.

Ultimate responsibility for drinking-water provision is vested in the National Council on Waters, a cabinet-level body that reviews use and discharge permits of Bulgaria's freshwater resources (Knight *et al.* 1995; Staddon 1996). Decisions are passed on to various agencies, including the Regional Environmental Inspectorates (responsible for overall environmental quality monitoring), the Health and Epidemiological Inspectorates (responsible for monitoring safety of water supplies), the National Electric Company (responsible for utilizing water in hydro-power generation and for cooling nuclear and thermal power plants), and the Water Supply and Sewerage Companies (responsible for provision of drinking-water to point of use and management of domestic waste waters). Though some organizations such as the Water Supply and Sewerage Companies have been restructured into regionalized entities, there has been little commitment either to privatization or allowing new firm entry.

As early as the 1980s it had become evident that the water supply infrastructure was failing to provide sufficient quantities of potable water, a situation made more difficult by the drought. Before the rains returned in 1995 to 1998 many of Bulgaria's towns and cities faced periodic water rationing (Staddon 1996). With a return to normal or above-normal rainfalls the situation has eased, but relatively little has been done to improve the crumbling water supply and sewerage systems. Reservoirs formerly under the control of NEC, water supply and sewerage companies and other institutions have often been neglected and, where farming units have been disbanded, the management structures that were formerly responsible for maintenance and repair have often ceased to exist. As a result, in many cases it is not clear who is responsible for the safe use of these reservoirs, even where new organizations, such as the General Assembly of

Rightful Claimants (GARC), are formally charged with the legal responsibility for maintaining the formerly collectively owned reservoirs in safe condition. In practice, the assemblies have often rented out the reservoirs without also assigning responsibility for safe use and maintenance in the rental agreement. The national government still seems wedded to large-scale water projects, including the VK-Rila diversion scheme (resurrected in 1995) and the Gorna Arda hydropower plan, agreed jointly with Turkey in June 1999.

In the mid-1990s, a pilot project for Integrated River Basin Management in the Yantra River Basin, located in north-central Bulgaria, became the first real attempt to develop post-communist water resource management structures (Clark *et al.* 1993; Tsenova 1995). The project brought together water-users from all major settlements along the course of the Yantra River and from all user groups (domestic, industrial, recreational, etc.). These 'stakeholders' were charged with making collective decisions about uses, discharges and investment in water infrastructure throughout the basin, but thus far it has not been possible to assess its effectiveness.

Protected landscapes

Strong central control over natural resources after 1948 facilitated the relatively rapid expansion of Bulgaria's network of protected landscapes. The country now maintains five International Union for the Conservation of Nature Class 2 National Parks (Central Balkans, Vitosha, Strandzha-Sakar, Rila and Pirin), and a large network of smaller landscape parks, reserves and ecological zones. Even under central planning the area of protected forests was greatly expanded (from 8.4 per cent of the total forest area in 1955 to 39.8 per cent in 1995), and some of the largest forest reserves in Europe were created (e.g. Rila, 100,000ha and Central Balkan forests over 72,000ha). Bulgarian government authorities and ENGOs have worked with international institutions such as the US National Parks Service, the Swiss-based International Union for the Conservation of Nature, the UK-based Royal Society for the Protection of Birds and the Global Environmental Facility. As a result, the total area of protected land in Bulgaria doubled between 1992 and 1995, now totalling 4.5 per cent of land area (Cellarius 1998, p.62).

Bulgaria's protected areas face at least three challenges in the coming years. First, it is as yet unclear how dependency on foreign donor bodies for everything from expertise to legal reform to funding will be affected by the inevitable withdrawal of such support in the next decade. Second, there remain significant unresolved jurisdictional problems among different branches of government, especially between foresters and parks managers (Cellarius 1998). In some places, such as the Bansko Forest District in late 1999, this conflict has been resolved by assimilating forestry management into the park administration, but such a resolution seems unlikely everywhere. Third, a number of rural communities have begun to oppose the further expansion of park territories, arguing that a better balance must be struck between local economic development and landscape protection. Such community–park conflicts were exemplified by the public hearings held in July 1999 over the proposed new boundaries and regulations for Pirin National Park. In many cases, local residents continue their use of parklands for pasturing, gathering, fishing and hunting and in some cases have expanded clandestine logging activities inside park boundaries (Staddon 1999c). The public management of common property resources such as forests, freshwaters and protected landscapes will have to adjust to these practices and stakeholders in the years ahead. As we have seen, state regulatory systems have yet to complete their transformation from vehicles of central planning to arbiters of the tenuous and difficult balance between the public good and private property rights. Undoubtedly this transformation will require more time and resources, but this may be of little comfort to conservationists anxious about the short-term threats to particular natural resources or landscapes, or to rural communities still desperately seeking a new foundation for sustainable local economic development.

Transforming the industrial legacies of central planning

Despite uncertainties and reversals in environmental policy, environmental quality has improved since 1989. Pollution levels have rapidly declined and environmental indicators have improved. Citizens now have greater access to information about environmental pollution, and their political representatives have, to varying degrees, expressed support for and interest in environmental protections and rights (Pickles 1995; Pavlinek and Pickles 2000). Overall pollution levels have declined as industrial production has dropped and successive governments have taken steps to stop the worst cases of environmental devastation caused by state socialist industries (Figure 14.3). Some, however, fear that the real gains in environmental quality achieved in the early years of post-communist transformation may not be sustainable, but may be by-products of economic collapse, production

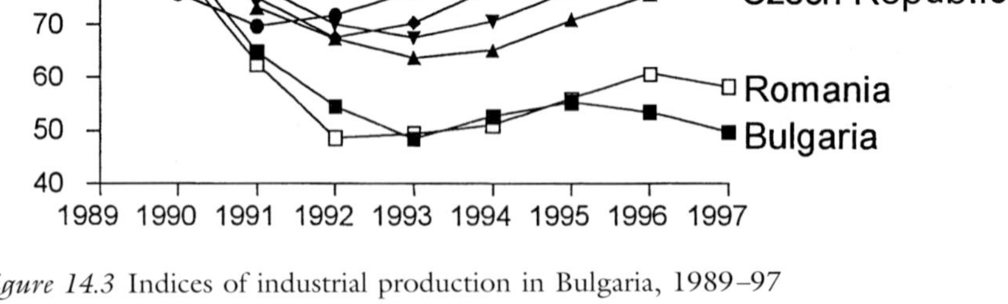

Figure 14.3 Indices of industrial production in Bulgaria, 1989–97

Source: Ministry of Environment and Waters, Sofia.

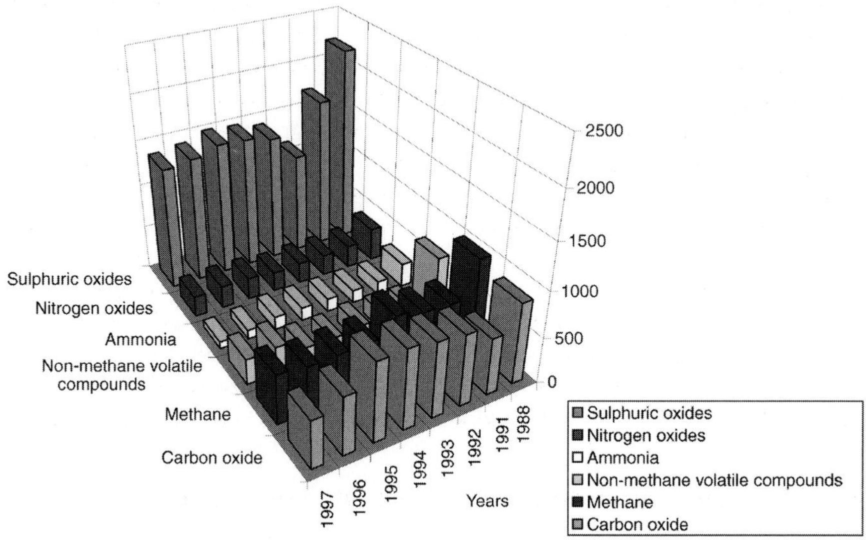

Figure 14.4 Emissions of atmospheric contaminants in Bulgaria ('000 tons)

Source: Ministry of Environment and Waters, Sofia.

declines, lowered inputs and 'end of pipe' upgrades (such as filters, scrubbers, long stacks, diffusers and de-sulphurization equipment). These may have resulted in reduced toxic and hazardous emissions, but have not restructured production along more efficient and less polluting lines (Roussev 1995a, 1995b). Air pollution still occurs around major industrial polluters, especially where topographic factors concentrate emissions. As economic conditions deteriorated after 1989 these 'hot spots' did decline in frequency and severity, but recently there is evidence to suggest that emissions of some pollutants have begun to increase again (Figure 14.4) (Mateeva 1995). Thus whether or not declines in pollution are sustainable will depend in large part on whether existing production facilities and technologies are put back into operation if old markets are re-established or new ones found. Whether that happens depends in large part on the availability of investment capital for new technology locally or through FDI.

Since 1989, however, there has been little investment in major technical restructuring in the worst polluting industrial sectors, and little direct mitigation of technological hazards or reduction of industrial accidents has occurred (*Standart*, 20 February 1997, p.14; see also Shapira *et al.* 1994; Begg and Pickles 1998; Nikolova 1998b; Paskaleva and Shapira 1998). Privatization and restructuring of the economy have been occurring very slowly, new investment in industry has been minimal, and there is little evidence that new, less hazardous processes of production are being adopted. The amount of public information about industrial disasters has increased and reports of them in the press have become commonplace, especially where they have had serious consequences for the health and lives of workers and people living in nearby communities. At the same time, new perspectives on the problems of technological risk have been introduced into Bulgaria and research on risk has been expanded (see Donchev 1994; Donchev *et al.* 1994; Ilieva 1994; Nikolova and Roukova 1994; Roukova and Nikolova 1994; Velev and Yordanova 1994; Ziapkov *et al.* 1994). We turn now to two examples of industrial transformation: Kozloduy nuclear power plant and manufacturing industries more generally.

Nuclear power, energy demand and technological hazards

The nuclear power plant (NPP) near Kozloduy is certainly the best example of an extreme technological hazard in Bulgaria and, because of the scope of the threat an accident would pose and the national and international attention and investment it has received, is unparalleled in the country. Other potential extreme technological hazards are posed by some of the 2200 facilities in Bulgaria that are ionizing radiation sources (including fault detection laboratories, scientific and educational institutions, facilities within the Ministry of Public Health, and industrial facilities), and some of these do pose serious potential risk to local populations. While the actual risk posed by Kozloduy is contested (Vasilev *et al.* 1997; see also Nikolova 1998b), 38.8 per cent of Bulgarians are concerned about potential radiation leakage (Marinov 1997).[11] In 1995, a series of popular protests throughout Bulgaria brought to a halt, for the second time, planning for the proposed nuclear power generating facility at Belene on the Danube River. Government plans to complete the construction of two 1000MW VVER power units with an estimated generating capacity of 1900MWe p.a.[12] on the Danube were immediately opposed by a coalition of domestic environmental groups, local protesters and European governments (Carter 1996; *Za Zemyata* 1999b). Similar coalitions have also pressed for the closure or scaling down of operations at Kozloduy.

Kozloduy provides approximately 35 per cent of total electricity demand and cannot therefore be easily or quickly phased out (IEA 1994). Moreover, 50 per cent of domestic demand is met by thermal power generation in seven major thermal power-stations (TPS) (IEA 1994; Committee on Energy 1998). Fuelled by highly polluting lignites, these plants also pose serious environmental, economic and health hazards and are being reviewed by Bulgarian and international energy experts for renovation, conversion or closure. Between 1991 and 1995, Bulgaria received Euro82mln for enhancing safety at the plant. Part of this programme included the preparation of thirty-eight regulatory and technical documents on nuclear safety and radiation protection, the retraining of inspectors and the licensing of operators in collaboration with the French nuclear power agency, and the upgrading of Bulgarian data to standards required by the Information and Public Relations Section of the International Nuclear Information System. As a result, safety has improved and risk of accidents has declined.

In 1995, a team of inspectors from the International Atomic Energy Agency (IAEA) and the Committee on the Use of Atomic Energy for Peaceful Purposes (CUAEPP) carried out three complete audits of operations at Kozloduy, finding no violations of the regulations and procedures (CUAEPP 1995). Also in 1995, following a great deal of national and international speculation about the safety of Unit 1 at the plant, Kozloduy submitted to a second inspection by the Inspectorate on the Safe Use of Atomic Energy (ISUAE). The statement by the investigating ISUAE Commission recommended that Unit 1 should be shut down for repair work, and this was carried out at the end of May 1996. Unit 1 was declared safe and capable of working for another eight years (four years longer than its projected operating time) after a follow-up investigation by experts from the European Union, Russia and Bulgaria in 1996. By the summer of 1999 all six units at the Kozloduy installation were once again in operation, generating approximately 3760MWe p.a.

The storage of hazardous wastes from Kozloduy (as well as from the other ionizing radiation sources) remains a problem. Kozloduy is not designed with facilities to treat 'radwaste' (any radioctive waste), and has facilities only for its interim storage. The quantity of low- and medium-grade radioactive waste has already exceeded 10–15,000t (Evstatiev and Kozhukharov 1995), and in recent years generated radioactive waste from used nuclear fuel (GRW) has been added to this waste because its processing in Russia has been discontinued. New elaboration technology has reduced by half the amount of solid radioactive waste produced (RAW), i.e. radioactive waste other than that generated by the nuclear power plant including those with low and medium levels of radioactivity (CUAEPP 1995, p.6). The remaining wastes remain untreated and are stored in the radwaste repository near v. Novi Khan, but this lacks appropriate long-term storage facilities. A national repository for final disposal of treated RAW will be needed if the RAW storage problem is to be solved, but plans for such a facility were put on hold in 1993 because of lack of funds.

Until 1990, spent fuel from Kozloduy was sent to the FSU for reprocessing under a bilateral agreement (Committee on Energy 1998). After the expiration of that agreement, spent fuel was stored on site in 'temporary' facilities. The on-site storage facility, which cannot even accommodate spent fuel from the plant's two 1000MW VVER units, was expected to exhaust its capacity in 1995. It has been reported that over 700t have piled up in this temporary storage facility, with between 10,000 and 15,000t of low and medium grade radioactive waste stored elsewhere in the country (Nikolova 1998b; *Za Zemyata* 1999b). In 1993, Russia offered to accept spent fuel for reprocessing again, but at a cost of US$1000/kg, paid in hard currency. In response, the Bulgarian Committee on Energy proposed construction of a spent fuel storage facility near the Kozloduy plant (Nikolova 1998b).

Bulgaria has now closed its domestic uranium mining and processing facilities and embarked on a long-term programme for site reclamation and environmental reconstruction. However, this work depends on financial assistance from EU and international aid programmes, including PHARE, THERMIE, JOULE II and USAID. From PHARE sources alone Bulgaria was expected to have received almost Euro100mln by the end of the 1990s specifically for nuclear power and radioactive waste management programmes.

Manufacturing industries, pollution and transformation

Other manufacturing industries, particularly heavy industries, have not fared as well in their efforts to attract investment capital or in completing fundamental technical upgrades. Between 1993 and 1995, the chemical and oil processing industry accounted for 34 per cent of the total number of disasters in all industrial enterprises in Bulgaria (Nikolova 1998b). Plants such as Neftochim-Burgas (chemicals, plastics and other oil derived products), Polichim in the industrial agglomeration of Varna and Devnya (primarily fertilizers), Neochim in Dimitrovgrad (chemicals, especially fertilizers and acids), and Agrobiochim in Stara Zagora (nitrogen fertilizers) continue to pose major pollution and environmental health hazards to local populations. Because of their size and importance to the national budget they are also difficult to regulate and costly to upgrade (Roussev 1998a).

Inward investment for industrial restructuring has remained low (see Begg and Pickles 1998). Especially in the large petro-chemical and chemical industries this greatly hinders the ability of enterprise managers to modernize equipment and reduce risks to workers and nearby residents. At least until the collapse of the lev in 1996, many of these plants generated substantial income for the central state, and as a result many have only recently been scheduled or listed for privatization. The slow speed of re-capitalization and technological restructuring can itself increase the degree of risk posed to the environment, workers and community, especially where enterprise managers and government officials have made decisions to run down production lines or even whole enterprises in preparation for cash privatization. Such a policy of running down to 'sell cheap' has had enormous short-term implications for environmental health and safety in and around such plants.

In some situations, transition itself has led to increased levels of pollution. In some cases, environmental degradation has been associated with new foreign investment and privatization, a risk long recognized by environmental activists in the region (French 1991), but one underestimated by governments eager to attract foreign investors. Poorly enforced environmental regulations make Bulgaria particularly vulnerable to such risks. For some Western investors, lower environmental, health and safety standards have been an added benefit of what is seen as a cheap but skilled labour market.

Environmental liability remains an important concern for many foreign investors. Rarely willing to pay for the environmental degradation caused by enterprises in the past and concerned about potential costs associated with changing environmental legislation (Klavens and Zamparutti 1995, p.6; Stanners and Bourdeau 1995, p.431), foreign investors have been proactive in shaping government policy. In some cases detailed environmental audits prepared by foreign companies prior to their investment have been used to negotiate lower prices for the purchase of enterprises. In other cases foreign investors have worked with government agencies to ensure that the costs of environmental liability will be met by the state. Some foreign investors prefer to avoid the issue entirely by adopting greenfield strategies for new investments, building their factories on previously unused sites (Reeves 1995, p.71). As a result, foreign capital is unlikely to lead to the clean-up efforts predicted in 1989, nor in the short term is it likely to be the vehicle it was expected to be to invest in new clean technologies.

At the same time, new investments are contributing to Western consumption patterns and these are bringing their own problems, particularly plastic packaging and increased domestic waste. Although ECECs are producing less than half as much domestic waste as Western Europe, the region is quickly catching up (Bisschop 1996, p.45). One of the most visible symptoms of transformation has been the rapid growth in car traffic resulting largely from imports of used cars from the West.[13] Increased car traffic has contributed to rapidly growing air pollution from carbon oxides, nitrogen oxides, lead and ozone as most cars are old and not equipped with catalytic converters. In Bulgaria, the share of car emissions in air pollution doubled between 1991 and 1994, and by 1994 cars accounted for 46 per cent of total NO_x emissions, 44 per cent of hydrocarbons and 47 per cent of carbon dioxide emissions (Georgieva and Moore 1997, pp.68–9).[14]

Prospects

Since 1989, environmental quality has been ameliorated in the worst hot spots, many of the most visible forms of pollution have been eradicated, and for many conditions have improved. However, it has been clear for some time that economic crises associated with transition have substantially weakened environmental movements throughout the region and that the 'normalization' of political life has effectively closed off spaces for a radical politics of civil or environmental action (although whether or not popular support for environmental programmes has also declined is unclear; see Pickles *et al.* 1998). As a result, with the exception of a few groups such as Borrowed Nature, the popular environmental politics of 1989 seem to have faded from political life (see Staddon 2000a). Management of the economy and cleaning up the environment now seem to be firmly in the hands of the central state, albeit at times influenced by proactive local governments. The deepening economic crisis has weakened the emergence of an independent policy arena and the roles played by environmental NGOs within Bulgaria, and has favoured a (re)assertion of a culture in which solutions are sought from the outside and from technical experts. In these circumstances, international agencies, corporations such as Monsanto, and foreign governmental bodies have been surprisingly influential in the formulation of public environmental policy.

In the long run it may well be the intense desire of politicians to enter into formal relations with the EU that brings about the biggest change in environmental policies and practices, as, for example, EU mandates on food quality and agricultural practices are met for export crops. At the opposite end of the spectrum, we simply do not yet know what the effects of severe economic crises will be on land management practices, particularly as common property resources are closed off to household economies. Will producers extend production on marginal lands leading to increased levels of degradation or will they switch crops or even move out of farming altogether? Will urban workers desperate to supplement the household table extend allotments into heavily polluted sidings, with consequent impacts on children's health? What effects will the rapid increase in resource extraction from public forest lands have on land quality as thousands of weekend visitors scour the forest floor for wood for burning, foodstuffs and medicines for the family, and mushrooms for export markets? What effect will forest privatization have on their livelihoods? And, as social and economic conditions for the mass of the population continue to decline or stagnate as a minority enjoy (and display) the fruits of rapid wealth accumulation, will environmental, health and social welfare concerns again begin to reconfigure the political landscape of local and national elections. Every day at every level of society issues such as these are being negotiated and resolved, with consequences for the shape of environmental practices and management that will result.

Acknowledgements

The authors would like to acknowledge assistance in translation from Rumiana Dobrina of the Institute of Geography, Bulgarian Academy of Sciences. Figure 14.2 was prepared by Richard Gilbraith of the Gyula Pauer Cartographic and Geographic Information Systems Laboratory of the Department of Geography, University of Kentucky. Figure 14.4 was prepared by Zoya Mateeva of the Institute of Geography, Bulgarian Academy of Sciences. Funding for research on which parts of this chapter are based was provided by the John D. and Catherine T. MacArthur Foundation and the National Science Foundation (grants SBR 9515244 and INT 9092190).

Notes

1 In the wake of strong criticism of the environmentally negative consequences of World Bank policies elsewhere in the world, and in light of the importance of environmental NGOs in the Bulgarian democracy movement, World Bank loans to Bulgaria were explicitly (and for the first time) tied from the beginning to the restructuring of environmental agencies and practices (World Bank 1991).

2 The Ministry was renamed the Ministry of Environment and Waters in 1997.

3 The Laws on Mines and Quarries, Protection of Air, Water and Soil from Pollution, Protection of Nature, Waters, Regional and Urban Planning, Protection of Arable Land and Pastures, Public Health and the Law on Marine Environment were all still in place in 1996, carry-overs from the old regime, but were in the process of being rewritten (OECD 1996, p.20).

4 Not all environmental problems can be addressed through institutional and regulatory reform. Some are importantly exogenous. The most important of these is cross-border transport of pollution, and the most extreme example of this was the Third Balkan War. In 1998, Serbian policies in Kosovo led to an extended bombing campaign by NATO across the territory of the former Yugoslavia, with targets often close to the

Bulgarian border and missiles occasionally landing in Bulgaria. By 19 May, five NATO missiles and one Serb missile had landed in Bulgaria, and these were causing alarm in the country, but also in international arenas. Greenpeace warned of the dangers of a nuclear accident, as missiles were landing 'dangerously close to Bulgaria's Kozloduy nuclear plant' (sixty miles east of the Serb border) (Kyriakidou 1999). Bombings destroyed petrochemical plants releasing thousands of tons of toxic chemicals, dioxins were produced from the burning of organic and chlorinate materials, and oil had been spilled into the Danube (Kyriakidou 1999). Bulgarian producer prices and industrial and retail sales plummeted as a result of deepening stagnation, and these were exacerbated by the loss of transport shipping, rail and road routes to major trading partners in Europe. As a result, many informal practices of natural resource extraction (for example, gathering, hunting and cutting in the forests, potato farming on marginal mountain lands, and extended cropping and grazing practices discussed earlier which had emerged with the economic stringencies of neo-liberal adjustment and had been enhanced by the 1996 economic crisis) were further encouraged.

Concern about the effects of bombing on environmental conditions prompted the government to respond strongly. The Environment Minister Evdokia Maneva insisted that 'monitoring of all elements of the environment continues along the entire border with Yugoslavia and everything necessary is being done to provide information to the population about the environmental situation' (BTA 1999). Data on conditions in the region for April and May from the National System for Environmental Monitoring were presented in Parliament, showing that permissible levels of controlled pollutants had not been exceeded, that there had been no change in radiation levels, and that measurements of organic and heavy metal pollution in soils showed increased heavy metal levels across a 142.9ha area near to the Timok River, but that these were associated with copper, arsenic, zinc, and lead pollution from the Bor mine.

5 Based on data from 1954 to 1996 from ten stations located at Sofia, Pleven, Kazanlak, Kiustendil, Vidin, Varna, Bourgas, Haskovo, Cherni Vrah and Moussala peak.

6 The issue of greenhouse gases and global warming scenarios is an important part of a broader debate about climate change and a corresponding geopolitics of blame. The scope and nature of these changes raises important questions about the current emphasis on greenhouse effects as an explanation of regional climate change and water regimes. According to global warming models, increased air temperatures should be most perceptible in polar regions due to pressure changes in the region. Pressure differences grow and zonal transfer accelerates. Under these conditions, the number of Mediterranean cyclones decreases – a fact observed between 1982 and 1994. But these changes in atmospheric circulation do not take into account the occurrence of blocking anticyclones, which are the second major cause for droughts in Bulgaria. Moreover, drought conditions are also related to the increased frequency of north-eastern humid air masses. For the past ten to fifteen years these invasions have not combined with Mediterranean cyclones from the south: precipitation in the high mountains decreases as a result of the poor vertical extension of cloud formation (particularly in southwest Bulgaria) which itself results from the lack of warmer air at altitude thrusting over the front.

The issue is important both scientifically and geopolitically. Scientifically, the debate is about whether and to what extent climate change in the region can be explained by greenhouse gas emissions (see Ministry of Environment and Waters 1998) or by other factors such as synoptic climate variation (Velev 1996, 1997, 1998). Geopolitically, the issue is important in terms of the ways in which Bulgarian institutions and policies are 'opened' to the discourse and policy recommendations and interventions of international agencies under the rubric of 'greenhouse gas emissions' and global climate change. That is, under the rubric of blaming past economic practices for current climate variation/change.

7 Seventy obstini in seven districts were affected, 1117 homes and public buildings were flooded, nineteen enterprises were completely destroyed, and the transport infrastructure was considerably damaged. At that point 70 per cent of the national disaster fund for 1999 had already been used to compensate losses from previous damages, precipitating urgent discussions in Parliament about the possibility of revising upwards the level of funding provision for infrastructural repairs (*Pari Vestknik*, 9 September 1999, pp.1–3).

8 Alongside collective and state farms, household production continued in all East and Central European countries. In some cases, as in Poland and Yugoslavia, many individual farms persisted. In other cases, including Bulgaria, individual farming was mainly limited to small household plots issued by collective farms to villagers.

9 In 1997, the Bulgarian government released its *Bulgaria 2001 Program* (Government of the Republic of Bulgaria 1997). The *Program* spelled out in strong, clear terms the state of environmental regulation and the difficulties that would face environmental policy-makers through to the end of the decade.

10 Enabling and implementation orders were passed in early 1998, including the 'Order for Implementing the Law on Forests' (*Durzhaven Vestnik*, 10 April 1998) and 'Order for Implementing the Law on Restitution of Ownership of Forests and Lands in the Forests Fund' (*Durzhaven Vestnik*, 13 March 1998).

11 There are no Chernobyl-style RBMK (graphite-moderated) reactors in Bulgaria (Nuclear Energy Institute 1996).

12 MWe denotes megawatts electrically generated, a figure that is usually between 5–10 per cent less than rated power plant capacity.

13 Today the size of the nation's motor fleet grows annually by 6.2 per cent, but only 20 per cent of imported cars are new.

14 A national Ozone Campaign was launched in summer 1998, coordinated by the environmental NGO Borrowed Nature.

References

Altvater, E. (1993) *The Future of the Market: An Essay on the Regulation of Money and Nature After the Collapse of Actually Existing Socialism.* London: Verso.

Angelov, T. (1989) *Vital Program for Ecological Education.* Expert Opinion, National Council for Education, Science and Culture of the People's Republic of Bulgaria. Sofia: Sofia Press.

Begg, R. (1994) 'Patterns and Paths: Agriculture and the Environment During the Bulgarian Transition', Working Paper, Indiana University of Pennsylvania, Department of Geography and Regional Planning.

Begg, R. and Meurs, M. (1998) 'Writing a New Song: Path Dependency and State Policy in Reforming Bulgarian Agriculture', in I. Szlenyi (ed.) *Privatising the Land.* London: Routledge, pp.245–270.

Begg, R. and Pickles, J. (1998) 'Institutions, Social Networks, and Ethnicity in the Cultures of Transition: Industrial Change, Mass Unemployment and Regional Transformation in Bulgaria', in J. Pickles and A. Smith (eds) *Theorizing Transition: The Political Economy of Post-communist Transformations.* London: Routledge, pp.115–146.

Bisschop, G. (1996) 'Optimism Wanes for a Prompt Clean-up', *Transition* 2(10), pp.42–45.

Bogdanova, S. (1995) 'Water Resources and Ecological Transitions', in *Ecology and Agriculture.* Sofia: Alia Publishing.

BTA (1999) Bulgaria-Environment, 2 June 1999 (as reported in http://www.db.online/ 2 June).

Carter, F. (1996) 'Bulgaria', in D. Turnock and F. Carter (eds) *Environmental Problems in Eastern Europe*. London: Routledge, pp.38–62.

Cellarius, B. A. (1998) 'Linking Global Priorities and Local Realities: Non-governmental Organization and the Conservation of Nature in Bulgaria', in K. Paskaleva, P. Shapira, J. Pickles and B. Koulov (eds) *Bulgaria in Transition: The Environmental Consequences of Economic and Political Transformation*. Aldershot: Avebury Press, pp.57–82.

Cellarius, B. A. (1999) 'Global Priority, Local Reality: Rural Communities and Biodiversity in Bulgaria'. Ph.D. Dissertation, Department of Anthropology, University of Kentucky.

Central Statistical Office (CSO) (various years) *Statisticheski Godishnik na Narodna Republika na Bulgaria*. Sofia.

Clark, M., Laredo, D. and Hogrewe, B. (1993) 'Water Quality Pre-Investment Studies in the Yantra Basin in Bulgaria'. WASH Field Report #408, USAID.

Committee on Energy (1998) *National Strategy for Development of Energy and Efficiency Until 2010*. Sofia.

Committee on Use of Atomic Energy for Peaceful Purposes (CUAEPP) (1995) *Annual Report*. Sofia: CUAEPP.

Deets, S. G. (ed.) (1993) *The Economics of Sustainable Agriculture: American and Bulgarian Perspectives*. Washington, DC: National Academy Press.

Donchev, D. (1994) 'Risk Situations in the Development and Location of Industry', *Proceedings of the Conference on Theoretical Problems of Geographical Knowledge*, 10–11 September 1993. Nesebar, pp.267–276.

Donchev, D., Yordanova, M. and Nikolova, M. (1994) 'Natural and Anthropogenic Factors and Prerequisites for Environmental Pollution in the Region of Varna-Devnya', *Proceedings of the Conference on Ecology, Economy, and the Living Environment in the Black Sea Region*, 3–4 June 1993. Varna, pp.194–203.

Douglass, P. (1994) 'Farm Buildings/Environmental Protection Expert Report: Report of Short Term Input', Technical Report No. 23, prepared for the Republic of Bulgaria Ministry of Agriculture and the Commission of the European Community PHARE Programme for Technical Assistance to Demonstration Private Farms Project, No. PMU 012/91.

EcoGlasnost (n.d.) *Policy Statement of the Social Movement EcoGlasnost*. Sofia, mimeo.

Evstatiev, D. and Kozhukharov, D. (1995) 'Concept of the Bulgarian Academy of Sciences for a National Radioactive Waste Repository (Geological Aspects)', *Journal of the Bulgarian Academy of Sciences* 3(4), pp.12–19.

French, H. (1991) 'Restoring the East European and Soviet Environments', in *State of the World 1991*. New York: Norton, pp.93–112.

Genov, N. (ed.) (1994) *Society and Environment in the Balkans*. Sofia: Bulgarian Academy of Sciences.

Georgieva, K. (1993) 'Environmental Policy in a Transition Economy: The Bulgarian Example', in A. Vari and P. Tamas (eds) *Environment and Democratic Transition: Policy and Politics in Central and Eastern Europe*. Boston, MA: Kluwer Academic, pp.67–87.

Georgieva, K. and Moore, J. (1997) 'Bulgaria', in J. Klarer and B. Moldan (eds) *The Environmental Challenge for Central European Economies in Transition*. Chichester: Wiley, pp.67–106.

Government of the Republic of Bulgaria (1997) *Bulgaria 2001 Program. 1997–2001*. http://www.govrn.bg/programme/index.html

Green, E. (1989) 'The Political Economy of Environmental Protection in the Soviet Union.' Paper prepared by the staff of the American Committee on US–Soviet Relations, No. 8, February. Washington, DC.

Ilieva, M. (1994) 'Risk and Risk Situations in Agriculture', *Proceedings of the Conference on Theoretical Problems of Geographical Knowledge*, 10–11 September 1993. Nesebar, pp.291–299.

International Energy Agency (IEA) (1994) *Electricity in European Economies in Transition.* Paris: OECD.

Klavens, J. and Zamparutti, A. (1995) *Foreign Direct Investment and Environment in Central and Eastern Europe: A Survey.* Washington, DC: The World Bank.

Knight, G., Velev, S. and Staneva, M. (1995) 'The Emerging Water Crisis in Bulgaria', *GeoJournal* 35(4), April, pp.415–423.

Koulov, B. (1998) 'Post-socialist Change in Eco-politics and Environmental Management: A Case Study of Bulgaria's Burgas Region', in K. Paskaleva, P. Shapira, J. Pickles and B. Koulov (eds) *Bulgaria in Transition: The Environmental Consequences of Economic and Political Transformation.* Aldershot: Avebury Press, pp.201–228.

Kyriakidou, D. (1999) 'NATO Bombing Hurts Balkan Environment – Greenpeace'. Reuters, 19 May. (Obtained from http://infoseek.go.com/)

Marinov, V. (1997) 'Radiatsionna zashtita v Republika Bulgaria: sastoianie, vazmojnosti, perspectivi', *Chetvarta nauchno-prakticheska konferentsia po zashtita na naselenieto pri bedstvia i avarii*, Book I, 5–6 November, Sofia, pp.55–71.

Martinov, M. (1970) 'Osobenosti na Premestvane na Tziklonite v Raiona na Balkanskia Poluostrov I Sredizemnomorieto', *Izvestia na IHM*, Vol. XVIII, pp.35–74.

Mateeva, Z. (1995) 'Climatic Conditions for Dispersion of Airborne Pollutants over Problem Areas in Bulgaria', *Atlas of the Environment of the Republic of Bulgaria*, Section 1, Theme 2. Sofia: ProGIS.

Meurs, M. and Begg, R. (1998) 'Path Dependence and the Evolution of Rural Institutions: The Bulgarian Case', in J. Pickles and A. Smith (eds) *Theorizing Transition: The Political Economy of Post-Communist Transformations.* London: Routledge, pp.115–146.

Meurs, M., Morrissey, M. and Begg, R. (1998) 'Village to State to Market: Agricultural Ecology and Transformation – the Bulgarian Experience', in K. Paskaleva, P. Shapira, J. Pickles and B. Koulov (eds) *Bulgaria in Transition: Environmental Consequences of Political and Economic Transformation.* Aldershot: Ashgate, pp.23–38.

Mikhova, D. and Pickles, J. (1994a) 'Environmental Data in Bulgaria: Problems and Prospects', *The Professional Geographer* 46(2), pp.228–235.

Mikhova, D. and Pickles, J. (1994b) 'GIS in Bulgaria: Development and Prospects', *International Journal of Geographical Information Science* 8(5), May, pp.471–477.

Mindjov, K. (1995) 'Bulgaria', in *Status of National Environmental Action Programs in Central and Eastern Europe: Case Studies of Albania, Bulgaria, the Czech Republic, Croatia, Hungary, Latvia, Lithuania, FYR Macedonia, Poland, Romania, the Slovak Republic and Slovenia.* Budapest: Regional Environmental Center for Central and Eastern Europe, pp.25–34.

Ministry of Environment (1996a) *National Action Plan on Environment and Health* (second version), September 1996. Sofia: Ministry of Health, Ministry of Environment.

Ministry of Environment (1996b) *Ecological Monitoring.* Sofia.

Ministry of Environment and Waters (1998) *The Second National Communication on Climate Change.* Sofia: Ministry of Environment and Waters, Energoproekt PLC.

Naidenova, E. (1993) 'Biological Control of Insect Pests in Agriculture', in Deets, S.G. (ed.) *The Economics of Sustainable Agriculture: American and Bulgarian Perspectives.* Washington, DC: National Academy Press.

National Statistical Institute (NSI) (various years) *Statistical Yearbook (1988–1998).* Sofia: National Statistical Institute.

Nikolova, M. (1992) 'Changes in Agrochemical Status of Bulgarian Soils', Pushkarov Institute of Soil Science and Programming, Sofia, mimeo.

Nikolova, M. (1998a) 'Hail Risk for Agriculture in Republic of Bulgaria (abstract of dissertation)'. Institute of Geography, Bulgarian Academy of Sciences, Sofia, p.11.

Nikolova, M. (1998b) 'Technological Hazards in Bulgaria', in K. Paskaleva, P. Shapira, J. Pickles and B. Koulov (eds) *Bulgaria in Transition: Environmental Consequences of Political and Economic Transformation*. Aldershot: Ashgate, pp.39–56.

Nikolova, M. and Roukova, P. (1994) 'Population and Health in the Region of Bourgas', *Proceedings of the Conference on Geoecology*, 9–10 April, 1994. Sofia, pp.148–153.

Nuclear Energy Institute (1996) *Soviet-Designed Nuclear Power Plants in Russia, Ukraine, Lithuania, Czech Republic, Slovakia, Hungary and Bulgaria*. Washington, DC: Nuclear Energy Institute.

OECD (1996) *Environmental Performance Reviews: Bulgaria*. Paris: Organization for Economic Co-operation and Development, Center for Co-operation with the Economies in Transition.

Paskaleva, P. and Shapira, P. (1998) 'State Industrial Enterprise and the Impacts of Restructuring in the Burgas Region, Bulgaria in Transition', in K. Paskaleva, P. Shapira, J. Pickles and B. Koulov (eds) *Bulgaria in Transition: Environmental Consequences of Political and Economic Transformation*. Aldershot: Ashgate, pp.163–200.

Paskaleva, K., Shapira, P., Pickles, J. and Koulov, B. (eds) (1998) *Bulgaria in Transition: Environmental Consequences of Political and Economic Transformation*. Aldershot: Ashgate.

Pavlinek, P. and Pickles, J. (2000) *Environmental Transitions: Transformation and Ecological Defence in Central and Eastern Europe*. New York and London: Routledge.

Pickles, J. (1992) 'Environmental Politics, Democracy and Economic Restructuring in Bulgaria', in J. O'Loughlin and H. van der Wustern (eds), *The New Political Geography of Eastern Europe*. London: Bellhaven Press, pp.167–185.

Pickles, J. (1995) 'Restructuring State Enterprises: Industrial Geography and Eastern European Transitions', *Geographische Zeitschrift* 2, pp.114–131. (Reproduced in J. Pickles and A. Smith (eds) *Theorising Transition: the Political Economy of Post-Communist Transformations*. London: Routledge, pp.172–196.)

Pickles, J. and Begg, R. (2000) 'Ethnicity, State Violence, and Neo-Liberal Transitions in Post-Communist Bulgaria', *Growth and Change* 31(2), spring, pp.179–210.

Pickles, J. and Mikhova, D. (1998) 'The Political Economy of Environmental Data in the Bulgarian Transformation', in K. Paskaleva, P. Shapira, J. Pickles and B. Koulov (eds) *Bulgaria in Transition: Environmental Consequences of Political and Economic Transformation*. Aldershot: Ashgate, pp.105–128.

Pickles, J., Pavlinek, P. and Staddon, C. (1998) 'Who Cares about the Environment? Popular Attitudes to Pollution and Environmental Reconstruction: Two Case Studies of Bulgaria and the Czech Republic', in K. Paskaleva, P. Shapira, J. Pickles and B. Koulov (eds) *Bulgaria in Transition: Environmental Consequences of Political and Economic Transformation*. Aldershot: Ashgate, pp.229–56.

Popov, A. and Baikov, B. (1992) 'Environmental Protection Legislation and Current Ecological Problems in Bulgaria, 1992', in *Annuaire de l'Université de Sofia 'Kliment Ohridski', Faculté de Geologie et Geographie*. Tome 85, Livre 2 – Geographie, pp.7–16.

Popov, A. and Sarafov, A. (1996) 'A Map of the Soil Moisture Deficit in South-eastern Bulgaria', presented at the National Conference on Cartography and Ecology. Sofia, pp.124–130.

Popov, A. and Staddon, C. (1999) 'Revision of Bulgarian Forestry Management Practices and Restructuring of the Wood-Working Sector: Case Study of the West Rhodope Mountain'. Unpublished mimeo.

Reeves, A. (1995) 'Towards Practical Environmental Provision in Hungary', *European Environment* 5, pp.70–75.

Roukova, P. and Nikolova, M. (1994) 'Geographical Aspects of Risk in the Living Environment', Proceedings of the Conference on Theoretical Problems of Geographical Knowledge, 10–11 September 1993. Nesebar, pp.431–434.

Roussev, M. (1995a) 'Problems during Transition to Sustainable Ecologo-economic Development in the East-European Countries', *Problems of Geography* 2, pp.17–28.

Roussev, M. (1995b) 'Ecological Aspects of Metallurgy in Eastern Europe', *Yearbook of Sofia University*. Sofia: St Kliment Ohridski University Press. Book 2 – Geography 87, pp.367–379.

Roussev, M. (1998a) 'Ecological Aspects of Chemical Industry in Eastern Europe', in *Yearbook of Sofia University*. Sofia: St Kliment Ohridski University Press. Book 2 – Geography 89, pp.427–437.

Roussev, M. (1998b) 'Challenges to Sustainable Development in Bulgaria', in *UNDP Capacity 21*. Sofia, pp.26–39, 49–56, 75–77.

Shapira, P., Paskaleva, K. and Roukova, P. (1994) 'Adjustment of State Industrial Enterprises in Burgas Region to the New Economic Conditions', in *Problems of Geography*. Book 1, Sofia, pp.45–54.

Sofia Press (1990) *Ecological Security for the Common European Home. Statements by Participants in the CSCE Environment Meeting and Representatives of International Non-Governmental Organizations and Movements*, 16 October to 3 November 1989. Sofia: Sofia Press.

Staddon, C. (1996) 'Democratization, Environmental Management, and the Production of new Political Geographies in Bulgaria: A Case Study of the 1994–95 Sofia Water Crisis'. Ph.D. Dissertation, Department of Geography, The University of Kentucky.

Staddon, C. (1998) 'Democratisation and the Politics of Water in Bulgaria: Local Protest and the 1994–5 Sofia Water Crisis', in John Pickles and Adrian Smith (eds) *Theorizing Transition: The Political Economy of Post-communist Transformations*. London: Routledge, pp.347–372.

Staddon, C. (1999a) 'Economic Marginalisation and Natural Resource Management in Eastern Europe', *The Geographical Journal* 165(2) (also forthcoming in Bulgarian, in *Problemite v Geografia* 1999, #1).

Staddon, C. (1999b) 'The Bulgarian Wood Products Sector in Transition'. Unpublished mimeo.

Staddon, C. (1999c) 'Restructuring Forestry in Post-communist Eastern Europe: Localities and Their Forests in Poland and Bulgaria', presented at the Arkleton Centre for Rural Development Research, University of Aberdeen, 30 November.

Staddon, C. (2000a) 'Environmental NGOs, Civil Society and Democratisation in Eastern Europe', in A. Warhurst (ed.) *Towards an Environmental Research Agenda*. London: Macmillan, in press.

Staddon, C. (2000b) 'Restitution of Forest Property in Post-Communist Bulgaria', *Natural Resources Forum* 24(3), pp.237–246.

Stanners, D. and Bourdeau, P. (eds) (1995) *Europe's Environment: The Dobris Assessment*. Copenhagen: European Environment Agency.

State Insurance Institute (1997) Personal Communication.

Tsenova, V. (1995) The Pilot Yantra River Basin Council, internal report for the Environmental Training Project (Sofia) and USAID.

Tzvetkova, K. (1996) 'Bulgaria', in *Approximation of European Union Environmental Legislation: Case Studies of Bulgaria, Czech Republic, Estonia, Hungary, Latvia, Lithuania, Poland, Romania, Slovak Republic and Slovenia*. Budapest: The Regional Environmental Center for Central and Eastern Europe, pp.21–39.

Vasilev, I., Georgev, B. and Marinkina, V. (1994) 'Aspekti na izmenenieto na klimata v niakoi raioni na Bulgaria i neobhodimost ot dopalnitelna nauchna informatsia za agronomicheskite nauki', *Bulgarian Geophysical Journal* 10(3), pp.38–48.

Vasilev, G., Bliznakov, V. and Hristov, J.(1997) 'Vlianie na NPP "Kozlodui" varhu okolnata sreda, raboteshtite i naselenieto', *Chetvarta nauchno-prakticheska konferentsia po zashtita na naselenieto pri bedstvia i avarii*. Tom I, 5–6 November, Sofia, pp.47–54.

Velev, S., (1996) 'Is Bulgaria Becoming Warmer and Drier?', *GeoJournal* 40(9), April, pp.363–370.

Velev, S. (1997) 'The Drought in Bulgaria During 1980–1994', *Problems of Geography Journal* 1–2, p.6.

Velev, S. (1998) *Tendentzii na Izmenenie na Temperaturata na Vazduha I Valejite v Bulgaria*. Sofia: St Kliment Ohridski University Press, pp.18–22.

Velev, S. and Yordanova, M. (1994) 'A Method of Anthropogenic Stress Assessment on a Certain Area and Results of its Application', *Proceedings of the Conference on Geoecology*. 9–10 April, Sofia: FIL-VEST, pp.30–35.

World Bank (1991) *Bulgaria: A Country Study*. Washington, DC: World Bank.

Za Zemyata (1999a) 'Radioactive Waste: Environmentalists Insist on Immediate Actions', Press Release, *Za Zemyata*, 27 January.

Za Zemyata (1999b) 'Fingerbooks: Kozludoy and Belene', *Za Zemyata*.

15 Croatia

Peter Jordan

Croatia's environment compared with other post-communist countries: similarities and differences

Croatia never complied with the general image of former communist countries, namely heavily polluted and suffering from environmental disaster. Air pollution was always moderate and confined to larger urban agglomerations and industrial centres like Zagreb, Osijek, Rijeka and Šibenik (Figure 15.1). Pollution of the principal rivers such as the Sava, Drava and Mura originated in other countries rather than in Croatia itself. The coastal waters of the long Croatian Adriatic coastline were always less polluted than those of the Italian counterpart, except for the west coast of Istria (Istra). It was only around larger cities and seaports like Rijeka, Šibenik, Split and Ploče that coastal waters displayed some pollution. Soil erosion by wind and water was advanced in larger areas, such as the karstic areas of the Dinaric zone with its thin soils (brown soils, lithosols, rendzinas, luvisols) on limestone exposed to heavy rainfall and winds like the Bora.

This rather specific environmental situation is partly based on Croatia's political and economic history, and partly on the country's physical background. First, Croatia's territory had been industrialised only to a minor extent before the First World War, in contrast to modern Slovenia or even Bosnia, where the availability of mineral deposits, iron and coal-mines encouraged early industrialisation. In 1910, Croatia's industry employed no more than 41,000 workers and was mainly composed of light industrial branches like timber processing, textile manufacture and the production of foodstuffs and beverages (Feletar 1996, p.75). This structure did not change significantly during the inter-war period, when Croatia had 128,000 industrial employees (1931) and industrial plant location became a little more dispersed (Feletar and Stiperski 1996, p.438). Thus Croatia did not have industrial agglomerations; when they arrived they were very much later than in other Central Europe countries but were an essential part of industrialisation during the communist period. Communist industrialisation did change Croatia's economic structure from being predominantly agricultural to industrial, but again in a rather dispersed way without creating larger industrial agglomerations. A more significant spatial effect was just 'litoralisation', i.e. the industrial opening up of the coast, formerly very much on the economic periphery. In 1993, industry

Figure 15.1 Salient features of Croatia

dominated as a contributor to gross national product in only five of the twenty-one Croatian counties (županje). These were the counties of Šibenik (Šibenska županja), Karlovac (Karlovačka županja), Sisak (Sisačko-Moslavačko županja), Krapina (Krapinsko-Zagorska županja) and Koprivnica (Koprivničko-Krizevažka županja) (ibid., p.442).

The second reason for Croatia's distinctive situation was the importance of tourism for the country especially along the coast during the post-war period. Former Yugoslavia was easily the first of the communist countries to open up its borders to Western commercial tourism. It had already done so in the 1950s and succeeded in developing it further between the 1960s and 1980s into a major European seaside attraction. It contributed crucially to Yugoslavia's foreign currency revenues. The main destination of Western tourists within former Yugoslavia was the Croatian coast. It was understood quite early on that success in tourism and further industrial development, especially of heavy industry, were competing goals there. However, within the communist political framework, industry had priority over tourism although tourism contributed to foreign currency revenues especially for the state budget. Unfortunately, decisions did not always fall in favour of tourism. In the early 1980s the location of a nuclear power station on the island of Vir, right in the centre of the Northern Dalmatian tourist region, was prevented thanks to local oppostion from the population and political authorities. However, at the same time a petrochemical complex was located on the tourist island of Krk which endangered the karst hydrological system and part of the island's water supply.

A third reason is linked to agricultural land use and tenure. These factors mainly explain the absence of larger-scale soil degradation as a result of agro-industrial use and the minor role of agriculture as a polluter of rivers and brooks. Along the coastal fringe as well as on the highlands and in the basins (polje) of the Dinaric Range (Dinarsko gorje) agricultural land use is very extensive. Besides unfavourable natural conditions this is mainly the result of depopulation due to a tradition of emigration since the late nineteenth century and of urban migration during the post-war period. In the more fertile and agriculturally intensively used interior of the country (Kordun, Banja, Hrvatsko zagorje, Slavonia, Syrmia (Srem)) environmental pressure on soil is relatively low due to small-scale private farming which persisted throughout the communist period (Taschler 1989). Contrary to other communist countries and also to other parts of former Yugoslavia (mainly Voivodina (Vojvodina)), agriculture in Croatia was collectivised only on a small scale. Just over one-fifth (22.2 per cent) of the agricultural land was transferred into the hands of self-administered agricultural-industrial enterprises (Poljoprivredno-industrijskikombinati(PIK)) or agriculturalproductioncooperatives (Poljoprivredne zadruge). The rest remained with private farmers (Malić 1996, p.89). Only the collectivised land was exposed to disproportional environmental pressure by mechanisation, intensive fertilisation and mono-cultivation, while private farming was carried out on small plots of which only 2.6 per cent were larger than 10ha (ibid., p.88). There was a very limited use of labour (about half worked part-time) and capital input (machines, fertilisers, etc.) while self-provision was the main target. Collectivised agriculture was more or less confined to the Baranja, East Slavonia and Syrmia, i.e. the eastern parts of Croatia's share in the Pannonian Basin with its fertile soils and favourable climatic conditions. Even there, soil degradation did not reach levels experienced in other communist countries.

The relatively low pollution of coastal waters southeast of Istria (Sekulić and Pečar-Ilić 1997, p.10) has, besides historical and economic factors, predominantly natural reasons. Historic and economic reasons are certainly important. For example, the peripheral coastal economic position and its hinterland up to the post-war period and its relatively low population and industrial density as well as agricultural intensity are significant even today. Even more important is the fact that a southeast–northwest direction prevails along the Croatian coast streams. They supply the coast with fresh water up to the Istrian peninsula from the open Mediterrenean Sea and prevent it from being too much affected by the heavily polluted River Po which is doing so much harm along the Italian coast. Adverse effects of the Po with its heavily industrialised and intensive agricultural catchment area in upper Italy are in Croatia confined to the western coast of Istria, situated opposite the Po delta. Further natural reasons for the low degree of pollution found in the coastal waters of the Croatian coast are that they are relatively deep, with the exception of the western coast of Istria, augmented by clean fresh water from the extended karstic regions of the hinterland by abundant submarine springs (vrulje). Here, apart from the moderately polluted Neretva, there are no major polluted rivers. This is an advantage compared to the western Italian coast, where large rivers carry quantities of polluted sediment and deposit it near to the coast, since submarine transport of sediments is weak in the Adriatic (Sekulić and Sondi 1997, p.101).

The state of the environment in Croatia, although it does not suffer from a single case of massive environmental devastation which would incite public interest at a European level, is nevertheless in a way characteristic of countries which have been under communist administration for decades. In Croatia, as in former Yugoslavia as a whole, environmental problems were neglected by the public authorities. Because of authoritarian political structures, there was also no civic movement in Croatian environmental degradation which encouraged sustainable development of natural resources. Violation of environmental standards was possible in Croatia almost without any legal consequences, whilst the country lacked enough capital to prevent the use of new technologies to reduce damage or repair it. This political and socio-economic situation caused and tolerated a variety of smaller environmental problems which were easly visible.

As in other post-communist countries, public discussion in present-day Croatia is certainly not dominated by environmental issues. This is not only true of the general public (television, daily newspapers), but also of professional circles like geographers. This may very well be explained by a generally favourable environmental situation which does not present any major problems. There is no developed sensitivity to minor problems due to the dominance over the decades of basic political and economic issues. Current economic difficulties overshadow many other concerns, especially the wars and violent conflicts in Croatia between 1991 and 1995 which resulted in the reconstruction tasks and specific development problems which have absorbed a lot of public attention right up to the present. When the state of Croatia's environment is on the agenda for public discussion, this is mostly connected with tourism. Tourism experts and managers like to hint at environmental problems; they are well aware of the fact that an excellent environmental situation along the Croatian coast could function as one of the strongest competitive advantages for the country's tourism.

It is characteristic that a Green party or movement was and still is playing no significant role at the national level. While in 1990 a Green Action (Zelena akcija) party had obtained one of the eighty seats with the first democratic elections to Parliament, there was no environmental movement represented after the 1995 elections. Croatia has no special Ministry for the Environment, only a State Directorate for Environmental Protection (Državna uprava za zaštitu okoliša). As a result, environmental investigations and publications are rare and data on the state of the environment are difficult to obtain.

The current environmental situation

Data from the mid-1990s show that a reduction in environmental damage was mainly due to a decline in industrial emissions. This resulted from a decrease of industrial production in Croatia itself as well as in the post-communist neighbouring countries Slovenia, Bosnia-Hercegovina and Hungary. This was a consequence of the economic transformation to market-oriented production, while in Bosnia it was mainly the result of war destruction and stagnation of the economy. Between 1986 and 1995, the total volume of industrial production was reduced by half (Statistički Zavod 1998). In some branches of heavy industry reduction had gone even further: for example, the manufacturing of iron and steel to 16 per cent, non-ferrous metals to 30 per cent, metal products to 30 per cent, machinery and equipment to 21 per cent and chemical products to 44 per cent of the production levels for 1986 (ibid.).

Pressure on environmental reduction between 1986 and 1995 was therefore the result mainly of decreased production in the most environmentally harmful branches of heavy industry and had much less to do with a successful restructuring of the economy, or with active measures against environmental pollution. The year 1995 experienced the only minimal increase of overall industrial production (0.3 per cent), which grew in 1996 to 3.1 per cent and in 1997 to 6.8 per cent. This positive trend also continued in 1998 (Feletar 1998, p.47). However, growth is not caused mainly by new, innovative, high-tech and non-polluting industries, but by traditional branches of light industry which were based on the use of cheap natural resources and cheap labour (Feletar and Stiperski 1996, p.440f.). Obviously, Croatia's industry was commerically successful only through using these competitive advantages. The assumption that pressure on the environment has re-increased at a similar rate is therefore justified.

Air pollution

The decline in industrial production is reflected by the drop in predominantly industrial emissions like sulphur dioxide. With regard to SO_2 from thermo-electric power-stations and oil refineries, in 1995 Croatia was almost a clean island between Slovenia, Bosnia and Hungary, with figures ranging from 1000–17,000t/150sq.km. Figures there ranged from 66,000 (parts of Slovenia) via 126,000 (NE Hungary) to 151,000 (NE Bosnia) (EMEP MSC-W Report 1/97). Nitrogen oxide emissions, typical not only for industry but also for road and air traffic, fluctuated from 18–25,000t/150sq.km. Croatia resembled Bosnia, but was ranked well below Slovenia, Hungary, Italy and Austria. The main industrial pollutant emitters of sulphur dioxide, and dust, are the metallurgical centres in Sisak, Šibenik (aluminium) and Dugi Rat near Omiš; chemical and petrochemical industries located in Zagreb, Osijek, Kutina, Sisak, Varaždin, Rijeka and Split; and the thermo-electric power-stations (each 280–950MW installed capacity) of Plomin (east Istrian coast), with a second (210MW) block nearing completion. In 1990 Plomin I emitted 43,991 tonnes of SO_2 (Blažević, 1996, p.136); Urinj, to the east of Rijeka (320MW) and Sisak (Figure 15.1).

The wider Zagreb region (including Sisak and Kutina) and the Gulf of Rijeka are most affected by industrial pollutants. Another case in point is the Croatian Baranja east of Osijek. However, industrial emissions there originate mainly from the Hungarian industrial and mining region Pécs-Komló. With reference to sulphur dioxide, about one-tenth of the pollution originates in Croatia. Most derives from Italy due to the prevailing westerly wind direction and, together with emissions from the Gulf of Rijeka, affects the higher parts of the northwestern

Dinaric Range, particularly Gorski kotar. Forest damage, especially to firs but also junipers, is most frequent there (Blažević 1996, p.137).

The burning of coal in households, and local traffic, create serious local meteorological situations, due mainly to local industrial emitters. Among Croatia's urban areas, air pollution is worst (i.e. excessively contaminated) in Zagreb, Rijeka, Split and Kutina, according to the State Directorate for Environmental Protection. Another classification based on sulphur dioxide and nitrogen oxide reveals the comparative concentration of air pollution in Croatian cities and towns having more than 50,000 inhabitants with the situation in other Central and Southeast European countries. In this classification, Zagreb, Rijeka and Pula are the most polluted places in Croatia and Category 2 (i.e. elevated pollution) within a scale of 1–6 (with 6 representing the worst) (Hartung *et al.* 2001).

River pollution

With regard to water quality of major rivers, improvements recorded by the mid-1990s compared to the late 1980s are only slightly better than those found in other neighbouring countries suffering industrial decline (Figure 15.2). This is particularly true of the Sava River. This is the axis of Croatia's river network (length 562km in Croatia), on which Zagreb and other cities (Slavonski Brod, Sisak) are situated, and crosses the fertile agricultural lands of Slavonia. It is still categorised as 'heavily polluted' throughout its course through Croatia. This is true especially downstream of Zagreb, but shows a trend to improve upstream when it enters Croatia from Slovenia. This has mainly resulted for the closure of the Trbovlje brown coal-mine in Slovenia, where mine wastes used to be washed into the river. Improved water quality was also recorded in the River Drava (length 305km in Croatia) with its major tributary the Mura (length 67km in Croatia). Nevertheless, they also still remain in Category 3 (i.e. 'heavily polluted'). This improvement is not mainly 'domestic', but the result of industrial emission reductions in Slovenia (Drava) and Austria (Mura). Waste waters from Varaždin, Čakovec, Koprivnica, Virovitica, Valpovo-Belišće and Osijek still provide a heavy burden for water quality in the Drava. The Neretva River with its 'moderate pollution' (Category 2), unchanged since the 1980s, enters Croatia from Bosnia-Hercegovina and crosses its territory only shortly before reaching its mouth on the Adriatic coast (length 20km in Croatia).

In Croatia itself many industrial and communal waste waters still flow without purification into rivers. So far, mechanical or biological cleansing plants have been built mainly in the tourist resorts along the Adriatic coast, i.e. along the western coast of Istria from Umag to Pula, in the Kvarner region including all major islands, on the Dalmatian island of Brač, at the Makarska Riviera and in and around Dubrovnik. The large city of Split surprisingly enough does not yet have a purification plant at its disposal.

In the coastal hinterland and the country's interior a rather limited number of cleansing plants exists: there are two in Istria (Buzet and Labin), two in Dalmatia (Benkovac and Imotski), two in the Gorski kotar (west of Ogulin and

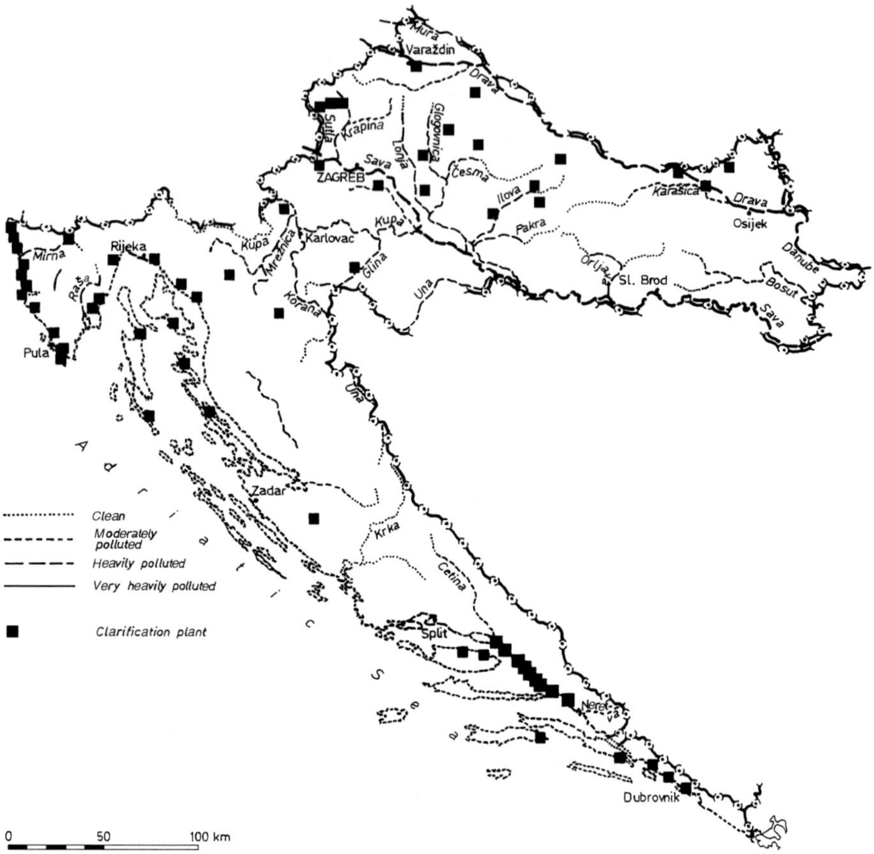

Figure 15.2 Water pollution and clarification plants in Croatia

Source: Salaj 1997.

Plaski), and one each in the Kordun (near Karlovac), the Banja (Topusko), and the Croatian Baranja which drains directly into the Danube (Beli Manastir). There are eight in the wider Zagreb region, five each in the Slavonian Sava catchment area (mainly in its western part along the rivers Ilova and Pakra), and the Drava catchment area (Donji Miholjac, Koprivnica, and Valpovo, Varaždin and Virovitica). Among Croatia's larger urban centres (besides Split and Osijek) and the smaller cities, Sisak, Slavonski Brod and Vinkovci lack any purification plants (Salaj 1997, p.221).

This deplorable state of affairs is especially reflected in the water quality of smaller rivers with their poor run-off (Kozina 1995, p.71; Salaj 1997, p.135). Only in the mountainous regions of the Dinaric Range and the uplands of Slavonia and the Hrvatsko zagorje (Zagreb's northern hinterland), upstream of urban centres are they in general 'clean' or just 'moderately polluted'; in the hilly regions, lowlands and downstream of towns and cities they usually turn

Table 15.1 Small river pollution of the left bank Sava River catchment area

River	*Tributary of*	*Downstream urban centre*	*Industry*
Krapinčica	Krapina	Krapina	pulp/paper
Krapina	Sava	Zabok	leather/footwear
Zelina	"	Zelina	–
Lonja	"	Vrbovec	food production
Glogovnica	"	Križevci	–
Cesma	"	Bjelovar	–
Toplica	"	Daruvar	–
Ilova	"	Grubišno Polje	–
Pakra	"	Lipik	–
Črnec	"	Sesvetski Kraljevci	–
Kutinica	"	Kutina	chemicals
Sumetlica	"	Nova Gradiška	leather/food production
Orljava	"	Požega	food production
Bosut	"	Vinkovci	leather/footwear

Source: Salaj 1997, p.135.

into the 'heavily polluted' category, in exceptional cases even into 'very heavily polluted' (Table 15.1). Along the right bank of the Sava catchment area the following river sections are still heavily polluted: the River Mrežnica almost from its source to the confluence with the Kupa River in Karlovac; the short River Reka downstream of Jastrebarsko town; the River Korana, which crosses the Plitvice Lakes (Plitvička jezera) when it re-enters Croatia from Bosnia; the River Kupa downstream of the towns of Karlovac and Petrinja; the River Glina, Odra and Una along the Bosnian border, downstream the confluence of the River Sana.

The region of the karstic Dinaric mountain range and its coastal fringe directly drain into the Adriatic Sea. The pollution of surface waters can easily affect reserves of drinking-water. Therefore especially dangerous are the waste waters flowing into the ponors (crevices) of the Gorski kotar and the Lika basins. Cases in point are waste waters produced by the towns of Otočac and Gospić. On the Istrian peninsula (Blažević 1996) the largest river is the Mirna, which is heavily polluted downstream of Buzet town with its chemical industry and brewery, right down to its mouth. The water of the River Pazinšćica in the centre of Istria is very heavily polluted (4, the worst category) by the chemical and food industry in Pazin town. The river disappears down a ponor (Pazinska jama) and nourishes a major karst spring in the Bay of Raša (Raški zaljev) as well as fountains in the surroundings of Pula in southern Istria. Also the rivers Boljunšćica and Raša flow into Raša Bay and are partly in Category 4 mainly due to emissions from black coal-mining and the leather industry. However, black coal-mining in Labin is on the decline and will probably be closed down in the near future.

Pollution of coastal waters

The quality of coastal waters has not changed significantly (Figure 15.2). In general it is much better than along the Italian coast of the Adriatic Sea. As tourism is one of Croatia's strongest economic assets along the extended sea coast and has future potential, the environmental image of the coast and the quality of its waters should be a matter of utmost public and economic concern (see Schneider-Jacoby 1996). However, the environmental consciousness of a wider public will still take time to develop.

The ratio between anthropogenic waste discharged from coastal settlements along the coast and natural waters brought in by rivers, submarine springs, underground waters and surface ablution is less than half as unfavourable along the eastern Adriatic coast as it is along its western counterpart in Italy (i.e. 0.22 per cent as compared to 0.52 per cent in the share of waste water to overall discharge into the sea (Sekulić and Sondi 1997, p.99)). Unfortunately, water quality in western Istria is worse than further south, as the Gulf of Venice is flat and fed by the large Italian rivers like the Po, Adige, Brenta, Piave and Tagliamento; all drain into the Adriatic from a highly urbanised and industrialised hinterland which also has intensive agriculture. They all help increase the sea's content of nitrates and phosphates. It is estimated that already 250sq.km of sea bed in the northern Adriatic (Gulf of Venice) is without any animated life (Serman 1995).

Croatian coastal waters are classified into four quality categories. The worst category is found only in the almost sea-locked harbour Pula basin, the Austro-Hungarian Empire's former maritime base. The next category up is frequently found by major cities, ports and industrial zones. From north to south these include Raša Bay on the eastern Istrian coast, where black coal is shipped from the Raša coal-mine; Plomin Bay with its coal-burning thermo-electric power-stations (Blažević 1996, p.136); Rijeka, a city of 168,000 inhabitants and Croatia's largest seaport concentrating on trading in crude oil, coal, iron ore and other harbour industries including a coking plant and oil refineries (Turk 1997); Zadar, a city of 76,000 inhabitants and an important ferry port linking the central Croatian coast with Ancona in Italy; Šibenik with its large aluminium factory; Kaštelanski Bay near to the city of Split (189,000 inhabitants), noted for industrial activity and ferry port facilities; Makarska, a small town and major tourist centre to the southeast of Split; Ploče Bay, the transit port for Bosnia-Hercegovina at the mouth of River Neretva; and Dubrovnik, the major tourist centre in the south with 50,000 inhabitants. This category contains health risks for swimmers.

Most other coastal waters belong to the second category ('moderately polluted'), where human health is not in jeopardy. The best category is 'clean water', where oysters flourish and is confined to sheltered beaches and channels, such as the Limski Canal, an inundated former valley on the west Istrian coast; others include Soline Bay on Krk Island, Lake Prokljan at the mouth of the River Krka near Šibenik, and the Malostonski Canal, a narrow bay between the Pelješac peninsula and the mainland coast near the mouth of the River Neretva.

Soil degradation

In Croatia soil degradation has worsened from the consequences of depopulation and deagrarisation exposing the unprotected soil to the natural forces of water and wind erosion. This compares with the intensive and industrialised agriculture typical of Hungary, the Ukraine or Romania under communist rule. In these countries, mechanisation, intensive use of chemical fertilisers and pesticides, monocultures, large plots, unprofessional melioration measures and so on have all contributed to the deterioration of soil quality.

This is especially true in the thinly populated region of the karstic Dinaric Range and coastal islands with its high energy relief. Rural areas within this region have become depopulated through emigration and migration to urban areas since the end of the nineteenth century. A low agricultural population leaves few people to care for the land. Animal husbandry since prehistoric times and intensive timber exploitation mainly in the Roman and Venetian periods have meant the maritime slopes of the Dinaric Range as well as many islands have only a dispersed forest cover. Thin soils lie on carbonate rocks. These include brown soils, limestone lithosols, limestone rendzinas and terra rossa, mainly in Istria, but also in the Kvarner region and along the Dalmatian coast (Fink *et al.* 1984; Benac and Durn 1997). They lack any shelter and, apart from the Julian Alps in Slovenia and NE Friul, they are exposed to the most intensive precipitation in Central and Southeastern Europe. This reaches up to 3500mm annually at the highest elevations (Steinhauser 1973; Bognar 1996) which frequently occurs as heavy rainfall. On steeper slopes this results in intense denudation including landslides.

On the coast, especially at junctions where the Dinaric Range forms steep slopes, the Bora wind is an additional erosion agent. With a velocity of up to 25m/sec, it sweeps down from the mountain ranges to the sea, also creating a salt water spray; it further contributes to the salinisation of soils, especially on the chain of islands adjacent to the mainland (Yoshino 1976). Water erosion threatens nearly two-fifths (21,755sq.km. or 38.4 per cent) of Croatia (Hebrang 1998, p.539). This places Croatia among the world's most endangered countries in terms of soil erosion. In addition, due to rural depopulation, extended and permanent fallow results in the destruction of the cultural landscape.

Forest damage

Forests and forested land cover over one-third (20,226sq.km. or 35.8 per cent) of Croatia's territory (Hebrang 1998, pp.539f.). Almost half of the mountain forests suffer from the impact of acid rain which is for the most part caused by cross-border air pollution. It is estimated that forested areas are diminishing by about 1000 hectares per year due to the development of various infrastructural facilities and cutting down trees. Further deforestation has been caused by occasional summertime forest fires, which have devasted 7 per cent of the total forested area over the past twenty years. Another destructive factor has been war operations between 1991 and 1995 (see below). Less than one-fifth (19 per cent) of Croatia's forests are privately owned, while the rest belong to the state-owned enterprise 'Croatian Forests' and 2 per cent to scientific institutions as well as to organisations managing protected areas (ibid., p.540).

Waste deposition and waste treatment

Organised waste deposition, treatment and recycling is still in its initial stages in Croatia. It was not until 1991 that Zagreb, Croatia's metropolis and then a city of 707,000 inhabitants, started to plan a comprehensive concept for waste treatment. This was called IVO (Izbjegavanje–Vrednjovanje–Odlaganje, i.e. Avoiding–Utilisation–Deposition). The population was to be trained to avoid waste through public relations exercises and financial measures. Several biocompost plants and five waste collection centres have been established. The city's main waste deposit at Jakuševec is going to be reorganised and more than 300 'wild' waste deposits are to be closed down (Kuduz 1997).

It has been estimated that in Croatia 7.7mln.t of waste (1.4t/pc) annually is disposed of in about 700 unregulated waste deposit sites at the disposal of communal enterprises and about 1300 wild deposits, forming a danger to human health and the environment. A mere one-tenth of the waste is treated according to regulations. Regulated and controlled waste deposits are disposed of only at Buje and Poreč in Istria as well as Garešnica and Sisak in the Sava region do they dispose of regulated and controlled waste deposits together with Rijeka and its INA oil refinery. In Sisak, facilities for special waste disposal also exist (Salaj 1997, p.145). By 2005, it is planned to establish large waste deposits near Rijeka, Split and Osijek in addition to that in Zagreb and smaller ones in each of the twenty counties (ibid., p.222).

Environmental pollution from war

The period of political insecurity and military disturbance in Croatia lasted from 1990 to 1995. It began on 17 August 1990 with road blockades around the Serb-dominated hinterland of Zadar in northern Dalmatia. After heavy fighting from the summer of 1991 to January 1992 the Serbian ethnic minority in Croatia formed *de facto* a separate state which lasted until 1995. Known as the 'Republic of Serbian Krajina', it covered an area of about 15,000sq.km along Croatia's borders with Bosnia and Serbia and a population of 549,083 of whom 52.4 per cent were ethnic Serbs (Klemenčić 1997, p.352) (Figure 15.3). This totalled over a quarter (26.5 per cent) of Croatia's territory and one-tenth (11.5 per cent) of its population (ibid.). In May 1995 the Croatian army reconquered Serbian territory in west Slavonia (the UN 'West' sector); by early August it had added territory along the Bosnian border from the Banja in the north to northern Dalmatia in the south (UN sectors 'North' and 'South'). Eastern Slavonia (UN sector 'East') also regained the status of an integral part of Croatia, but remained under UN forces (UNTAES) control until 15 January 1998.

Damage from warfare and ethnic cleansing was most severe in eastern Slavonia and Syrmia (e.g. Vukovar), as well as parts of the 'Republic of Serbian Krajina' which were originally settled by Croats (e.g. larger pockets in Kordun, Dalmatinska zagora, Posavina). In areas beyond the fringes of the Republic were settlements,

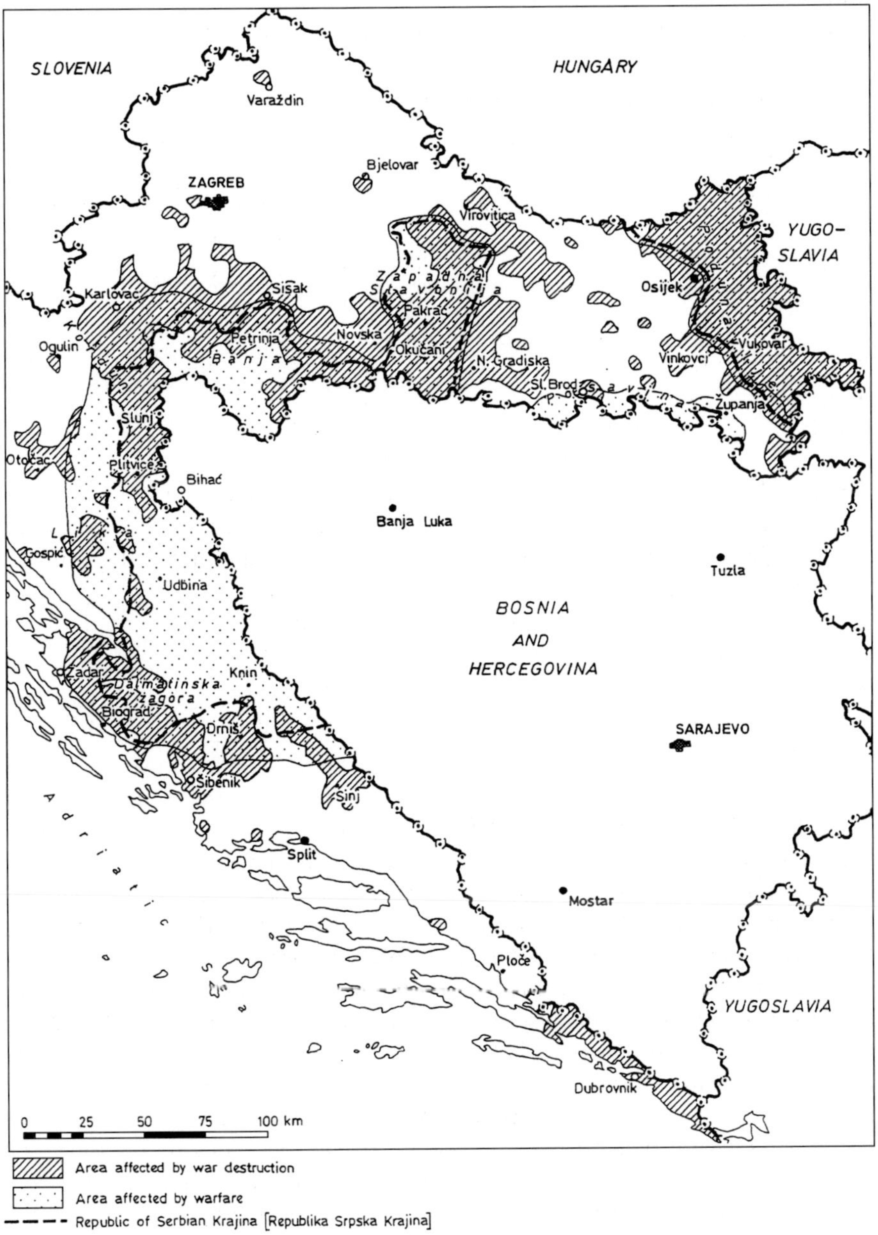

Figure 15.3 War damages in Croatia

Source: Klemenčic 1997.

industrial plants and transportation routes shelled from Serbian territory (e.g. Osijek, Vinkovci, Sisak, Karlovac, Gospić, Maslenica bridge, Zadar, Biograd, Šibenik, Drniš, Dubrovnik) (Figure 15.3). A total of 590 Croatian settlements were damaged or destroyed; damage affected 12 per cent of all housing, 10 per cent of all tourist facilities, 30 per cent of industrial capacity, fifty-seven bridges and 725 cultural monuments, mainly churches and cemeteries (Baletić *et al.* 1994, pp.48ff.).

When environmental damage in the narrower sense is taken into account, forests take first place. Forests were especially hard hit around Dubrovnik, in Slavonia, Plitvice Lakes and Krka National Park; also woods in Lonjsko polje and Kopački rit floodplain. During military operations between 1991 and 1995 damage inflicted on forests (excluding eastern Slavonia) was estimated at about US$75mln (Hebrang 1998, p.540). Another form of environmental damage is the dissemination of land-mines because it prevents the local population from using agricultural land and natural resources. They are most common around the fringes of the former Serbian republic (Figure 15.4), and it will take some years to eliminate them completely.

Land use however has not only declined through mines, but even more so due to rural depopulation in the area. Between 1991 and 1992, ethnic Croats were expelled from 'Serbian' territory. During the Croatian reconquest in 1995 this happened to the Serbian population in western Slavonia and from the Banja down to northern Dalmatia. These territories (excluding eastern Slavonia) were inhabited by about 350,000 people during the population census of March 1991, of whom approximately two-thirds were Serb and one-third Croat. By late August 1995 the area was almost depopulated. In August 1998 125,000 people, of whom two-thirds were Croat and one-third Serb, lived there (Večernji list 1 August 1998, pp.22, 29). This number comprised some former residents entitled to return (about 45,000 Serbs, fewer Croats) and a more numerous 'new population', consisting of Croatian refugees from Bosnia, the Voivodina or eastern Slavonia. This 'new population' very rarely took up agriculture, mainly preferring to settle in urban areas. Thus all over Croatia the extent of agricultural land has decreased, dropped from 2.02 million hectares in 1991 to 1.56 million hectares in 1995 (Salaj 1997, p.109). This resulted not only from the war, but also from growing urbanisation, depopulation and population exchange which, when combined, certainly played a major role. The environmental consequences have led to greater secondary wilderness in the areas in question, which in turn has had harmful effects on the aesthetic appearance of the landscape and increased the possibility of ecological risks.

Legislative environmental measures

Environmental legislation

Laws on the Protection of the Environment and the Protection of Nature were passed in 1994 to regulate the preservation of the environment and natural wildlife habitats. In addition, these laws aim at protecting the environment by enforcing sustainable development, and recognise the need for long-term maintenance of sources of biological diversity in natural habitats and ecological stability (Hebrang 1998). The Air Protection Act, passed in 1995, is the basic law governing the implementation of the policy of protecting and improving air quality in Croatia. Organised monitoring of air pollution started in 1967 and today includes all major cities. An annual survey of air pollution emissions has been systematically carried out since 1993. Water management is regulated by the Water Act and that on the Finance of the Water Supply System, passed in 1995. Croatia has joined the Barcelona Convention on the Protection of the Mediterranean Sea from Pollution and it participates in the Mediterranean Action Plan.

The State Directorate for Environmental Protection is a government body established in 1994 and the highest-ranking executive body for environmental affairs in Croatia. It has the the power to implement laws, deliver regulations for their implementation and perform administrative inspections. A Committee for Urban Planning and Environmental Protection chosen from the House of Representatives in Croatia's National Parliament reinforces the policy for particular environmental protection and planning issues.

Areas under legal protection

Nature protection through the establishment of national parks and other protected areas has a long tradition in former Yugoslavia. Independent Croatia intends to continue this tradition by maintaining existing protected areas and creating new ones (Figure 15.4). National parks are protected against invasion by the tourist industry, but may be regarded as 'parts of a qualitative touristic offer of the wider region' (Salaj 1997, p.148). There already exist seven national parks having an overall area of 827sq.km which were established between 1949 and 1985 (Salaj 1997, p.149; Hebrang 1998, p.540). Plitvice Lakes and the Paklenica national parks were first to appear, both having been established in 1949.

Plitvice Lakes National Park is situated in a wood-covered karstic region in southeastern Kordun and northeastern Lika; it is focused on a natural attraction formed by a water cascade through sixteen small lakes in travertine rocks interconnected by waterfalls. It also contains 80ha of almost virgin forest composed of fir and beech trees up to 50m tall. After later expansion, the park currently covers 195sq.km, and in 1979 was included in the UNESCO world list of natural and cultural heritage. Apart from communal waste water, it is mainly endangered by tourists who have visited this attraction since 1995 in growing numbers. The war events between 1991 and 1995 also did some harm to the park. Paklenica National Park covers a barren and bizarre southeastern section of the Velebit mountain range with peaks rising over 1800m above sea-level, with deeply intersected valleys and canyons. Two of the deep canyons have rock walls 300–400m high. Since 1979, the park has been expanded to cover 102sq.km.

Risnjak National Park was established in 1953, and following expansion now covers 64sq.km of a mountain massif (altitude 1528m) in the Gorski kotar northeast of Rijeka. The main motives for protection were the presence of rare vegetation due to the maritime exposition of the higher mountain massif (fir and beech, mountain beech, bush-like beech, pine, mountain pine) and the

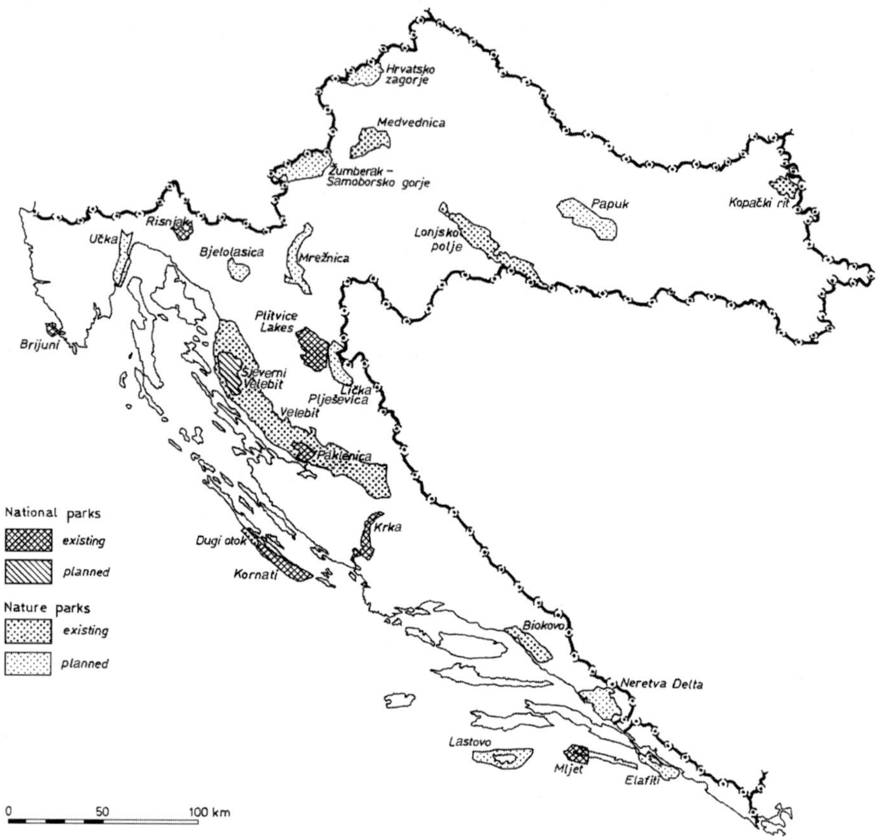

Figure 15.4 Areas under legal protection
Source: Salaj 1997.

attractiveness of the natural landscape in the upper Kupa valley enriched by karstic phenomena. Mljet National Park covers the western third of the Dalmatian island of Mljet and was established in 1960. An extension to include a 500m belt of coastal waters in 1997 has meant that it currently covers 54sq.km. The island is protected mainly for its Mediterranean woodlands including Aleppo pine, evergreen holm oak, and dense evergreen macchia. The diversity of its flora and fauna, two lake-like sea bays and a twelfth-century monastery on a small island in a bay all enhance its attractiveness.

The Kornat Islands National Park contains 234sq.km composed of 51sq.km of firm ground and 183sq.km of a maritime belt at a distance of 200–500m around this barren and picturesque archipelago formed from 101 small islands in central Dalmatia. The park was established in 1980 mainly to protect maritime life and the rocky islands from being colonised by weekend second homes. In none of these respects has environmental protection been really successful. Brijuni National Park facing the Istrian coast near Pula was established in 1983 after Tito's death in 1980. Previously, the islands were the summer residence of the Yugoslavian president, and have been protected for their Mediterranean vegetation and monuments from the Roman and Byzantine eras. Protection extends to 36sq.km including the surface of fourteen islands and the surrounding maritime area. As part of the intensive tourist region of western Istria, the Brijuni Islands are constantly in danger of being threatened by the tourist industry. Since the early 1990s tourist access to the islands has been regulated; a kind of 'soft tourism' has been developed here.

The last of the already existing national parks originated in 1985. Krka National Park is located along the middle/lower course of the River Krka in central Dalmatia, northeast of Šibenik. In 1997 it was extended to include the Krka valley from Knin to Skradin, an area of 142sq.km. The River Krka is deeply incised into the limestone plateau of the Dalmatian zagora and forms deep canyons, lakes and waterfalls. Medieval monuments such as the fourteenth-century monastery on the river island of Visovac are additional reasons for protection. The park is mainly affected by waste water from the towns of Knin and Drniš as well as by industrial and agricultural activities in the region. A new national park is planned in the northern Velebit mountain range around the Veli Alan mountain pass and opposite Rab island. This is mainly to protect the landscape, which is of especial interest for its geomorphological karstic phenomena and botanical richness (Figure 15.4).

Apart from national parks, Croatia has six nature parks (Figure 15.4). These cover an even larger area than the national parks, totalling 3175sq.km (Salaj 1997, p.223; Hebrang 1998, p.541) and include cultivated and settled land. Protection in this category of parks aims at the 'preservation of asthetic, ecological, educational, cultural-historical and touristic-recreational values' (Salaj 1997, p.150) but does not exclude economic activities. Protection of this kind is extended to swamps and floodplains with ornithological values. These include Kopački rit between the Danube and Drava rivers (178sq.km); Lonjsko polje along the Sava River southeast of Zagreb (506sq.km); parts of islands like Telašćica on Dugi otok island (67sq.km) and mountain ranges such as Velebit (2000sq.km), Biokovo (196sq.km) and Medvednica (228sq.km). Ten further nature parks with similar themes are planned. These include the mountain massifs or ranges of Učka, Žumberačko and Samoborsko gorje, Lička Plješevica, Bjelolašica and Papuk, the uplands of the Hrvatsko zagorje, the valley of the karst River Mrežnica, the Neretva delta, the Elafites archipelago near Dubrovnik and Lastovo island (Figure 15.4).

When all these plans have materialised, 14 per cent of Croatia's territory will be under some kind of nature protection (Republic of Croatia 1992). On Istria the idea is being discussed of declaring the central and eastern parts of the peninsula a 'Homeland Park' for the protection of natural and cultural heritage (Bertić 1997, pp.33f.).

References

Baletić, Zv., Esterajher, J., Jajčinović, J., Klemenčić, M., Milardović, A., Nikić, G. and Višnar, F. (1994) *Kroatien zwischen Aggression und Frieden*. AGM: Zagreb.

Benac, Č. and Durn, G. (1997) 'Terra rossa in the Kvarner Area – Geomorphological Conditions of Formation', in *Acta Geographica Croatica* 32, Žagreb, pp.7–19.

Blažević, I. (1996) 'Neki oblici degradacije okoliša u Istraskoj županiji', in Z. Pepeonik (ed.) *I. Hrvatski geografski kongres. Geografija u funkciji razvoja Hrvatske*, Zagreb, 12. i 13, listopada 1995, Zbornik radova. Hrvatsko geogragsko društvo: Zagreb, pp.135–142.

Bognar, A. (1996) 'Croatia – the Land and Natural Features'. *GeoJournal* 38 (4), pp.407–416.

Feletar, D. (1996) 'Industrija u funkciji razvoja Hrvatske', in Z. Pepeonik (ed.) *I. Hrvatski geografski kongres. Geografija u funkciji razvoja Hrvatske*, Zagreb, 12. i 13. listopada 1995, Zbornik radova. Zagreb, pp.74–85.

Feletar, D. (1998) 'Oporavlja se industrijska proizvodnja'. *Hrvatski zemljopis* 34, Zagreb, p.47.

Feletar, D. and Stiperski, Z. (1996) 'The Development and Structure of the Croatian Economy'. *GeoJournal* 38(4), pp.437–444.

Fink, J., Stefanovits, P. and Schappelwein, K. (1984) 'Soils [1: 2,000,000]', in J. Breu (ed.) *Atlas of the Danubian Countries.* Österreichisches Ost- und Südosteuropa Institut: Vienna, Plate 161.

Hartung, A., Jordan, P., Nefedova, T. and Plit, J. (2001) 'State of the Environment in Central and Southeastern Europe around 1995' in P. Jordan (ed.) *Altas of Eastern and Southeastern Europe*, Österreichisches Ost- und Südosteuropa Institut: Vienna 1.7–6.7.

Hebrang, B. (1998) *Hrvatski almanah 1998/99.* Hrvatska izvještajna novinska agencija: Zagreb.

Klemenčić, M. (1997) *Atlas Europe.* Miroslav Krleza Lexicographical Institute: Zagreb.

Kozina, I. (1995) *Zemljopis II.* Mladinska Knjiga: Zagreb.

Kuduz, M. (1997) 'Otpad kao ekološki problem grada'. *Geografski horizont* 43(1), Zagreb, pp.47–57.

Malić, A. (1996) 'Suvremene promjene u agraru Hrvatske', in Z. Pepeonik (ed.) *I. Hrvatski geografski kongres. Geografija u funkciji razvoja Hrvatske*, Zagreb, 12. i 13. listopada 1995, Zbornik radova. Hrvatsko geogragsko društvo: Zagreb, pp.86–92.

Office for European Integration (ed.) (1999) *Action Plan for European Integration of the Republic of Croatia.* OEI: Zagreb.

Republic of Croatia (ed.) (1992) *National Report on Environment and Development.* Republic of Croatia: Zagreb.

Salaj, M. (ed.) (1997) *Strategija prostornog uređenja Republike Hrvatske.* Ministarstvo razvitka i obnove Republike Hrvatske: Zagreb.

Schneider-Jacoby (1996) 'Pogled izvana: Zaštita prirode u Hrvatskoj kao ulog u budućnost zemlje'. *Turizam* 44(1–2), Zagreb, pp.276–292.

Sekulić, B. and Pečar-Ilić, J. (1997) 'Klimatske, morfološke i ekološke značajke istočne i zapadne obale Jadrana'. *Turizam* 45(1–2), pp.3–12.

Sekulić, B. and Sondi, I. (1997) 'Koliko je Jadran doista opterečen antropogenim i prirodnim unosom tvari?' *Hrvatski geografski glasnik* 59, Zagreb, pp.95–105.

Serman, Z. (1995) 'S.O.S. za Srednozemno more'. *Priroda* 815 and 816. Zagreb.

Statistički Zavod Republike Hrvatske (ed.) (1998) *Statistički godišnjak Republike Hrvatske 1997.* SZRH: Zagreb.

Steinhauser, F. (1973) 'Annual Rainfall [1: 2,000,000]', in J. Breu (ed.) *Atlas of the Danubian Countries.* Österreichisches Ost- und Südosteuropa Institut: Vienna, Plate 142.

Tar, D. (1996) 'Industrija u funkciji razvoja Hrvatske', in Z. Pepeonik (ed.) *Hrvatski geografski kongres. Geografija u funkciji razvoja Hrvatske.* Zagreb, pp.74–85.

Taschler, B. (1989) 'Agricultural Structure [1: 6,000,000]', in J. Breu (ed.) *Atlas of the Danubian Countries.* Deuticke: Vienna, Plate 322.

Turk, H. (1997) 'Rijeka – geografska obilježja'. *Geografski horizont* 43(2), Zagreb, pp.35–45.

Yoshino, M. M. (1976) *Local Wind Bora.* University of Tokyo Press: Tokyo.

16 Macedonia

Stefan Bužarovski and Aleksandar Stojmilov

Introduction

Macedonia was the only former Yugoslav republic to acquire its independence without war, and thus it has been hailed as one of the more successful examples of peaceful political transition in Central and Eastern Europe, given the challenging international context in the early years (Figure 16.1). Yet the economic and social legacies of Yugoslav communism, combined with regional political instability, pose an obstacle to more rapid development which could help to resource programmes geared to solving environmental problems. Within the former Yugoslavia, Macedonia was widely seen as an economically underdeveloped area, though this perception does not necessarily reflect historical realities which situated the country at a strategic and relatively prosperous Balkan crossroads, much coveted by the great powers. But its geographic unity was broken in 1913, when Bulgaria, Greece and Serbia partitioned the region into three parts following the withdrawal of the Ottoman Empire.

Serbian Macedonia (of which the present Macedonian state is a direct successor) then became a backward Yugoslav province, with traditional agriculture as the main source of revenue, until increased federal investment after the Second World War accelerated the pace of industrialisation and urbanisation. However, despite the fact that Yugoslavia's post-1945 economic growth rate was among the highest in the world, there was a notable gap between the advanced north and the lagging south in terms of the technological level of development and the industrial structure (Nikolovska 1995, p.6). Hence, when Macedonia declared independence in 1991, it had experienced a high level of rural–urban migration but the infrastructure was only moderately developed and the technologically outdated heavy industries were linked with significant pollution problems. Moreover, independence introduced new economic and political difficulties, so that the environmental issues could not be given the greatest attention. Thus there is still no comprehensive assessment and monitoring of the state of the Macedonian environment, and data in general are scarce and fragmented. But stabilisation and economic recovery are now creating a situation where ecological issues can be more fully considered (Dimitrijevik 1995).

Figure 16.1 Salient features of Macedonia

Physical environment

Situated in the centre of the Balkans between Albania, Bulgaria, Greece and Yugoslavia, with an area of 25,713sq.km, Macedonia occupies the heart of a region with a distinctive combination of natural conditions. About 80 per cent of the country is mountainous and one-third of the territory is higher than 1000m (Stojmilov 1995, p.22), but the diverse relief could be likened to a chessboard because of the alternations between high fault-block mountains and deep basins. The highest mountains are in the west, dominated by the lofty Šar and Korab ranges, whose summits reach 2800m. Several deep grabens lie at the fringes, and two of them – Ohrid and Prespa – are occupied by tectonic lakes. A second area of rugged mountain terrain in the centre is broken only by the flat Pelagonian basin, while the low-lying Vardar valley, one of the Balkans' most important communication corridors, is to be found further east. Finally, another zone of deep valleys and high mountains comprises the easternmost part, near the Bulgarian border.

The climate is equally diverse: 'modified Mediterranean' in the south, 'mild continental' in the centre and north and 'mountainous' in the higher regions. But there are many local variations reflecting the diverse patterns of topography and air circulation. Most of the precipitation drains into the Vardar River and 87 per cent of the hydrographic network falls to the Aegean catchment: the rest of the drainage is oriented to the Adriatic with a symbolic 0.1 per cent for the Black Sea. Many soil types are present and the pattern is like a mosaic in view of the local variations, but there are six distinct interlocking 'life zones' based on altitude and latitude. Apart from the cultural vegetation of the tectonic basins and valleys, there is Mediterranean scrub in the south, replaced consecutively by deciduous oak, beech and coniferous zones (altogether eighty-one forest communities have been identified, covering about 35 per cent of the country) and finally various types of grassland account for exactly a quarter (Kolčakovski 1997, p.113).

Socio-economic processes

The post-1945 development already referred to brought about an eightfold rise in GNP between 1950 and 1990 and the most dramatic increase in the industrial component of all the former Yugoslav republics: from 15 to 47 per cent (Stojmilov 1995, pp.136–9). This was naturally reflected by the occupation structure (Table 16.1) which has changed again since 1991 with the rise of the tertiary sector at the expense of both industry and agriculture. However, the official statistics do not take proper account of the informal sector: the grey or black economy. The industrialisation phase was associated with rapid population growth of 80 per cent over the 1948 census level to reach today's figure of just over two million (Stojmilov 1995, p.79). Immediately after the Second World War, rates of natural increase reached 25/th but today's much lower rate of 6/th is still high by European standards. Indeed, the northern and western areas (Debar, Gostivar, Kičevo, Kumanovo, Tetovo, Skopje and Struga, in descending order) still register very high growth (reaching 15/th at the 1991 census). And while the urban population in Kumanovo and Skopje benefits from in-migration, allowance must be made for the cultural factor through the high birth rate among the Muslim population and the massive in-migration by Kosovar Albanians during the Yugoslav period. Population density is now 81/sq.km on average but there are much higher figures in the basins of Skopje (310), Tetovo (167), Struga (124) and Kumanovo (112) (Figure 16.2). The demographic explosion has meant heavy pressure on natural resources especially in these areas (Hadi Pecova and Dimitrijevik 1996, p.3).

Depopulated mountain regions are complemented by densely inhabited tectonic basins and river valleys to a greater extent than ever before, such has been the retreat from the high ground in favour of both urban – and to some extent rural – settlement in the lowlands (Gelevski 1982, p.11). The mountains in the

Table 16.1 Economic structure of the active population

Sector	*1953*		*1961*		*1971*		*1991*	
	th	*%*	*th*	*%*	*th*	*%*	*th*	*%*
Primary	361	72.3	327	62.5	280	50.0	103	20.0
Secondary	75	15.0	109	20.8	163	29.0	231	44.0
Tertiary	63	12.7	87	16.7	118	21.0	196	36.0

Source: Stojmilov 1995, p.89.

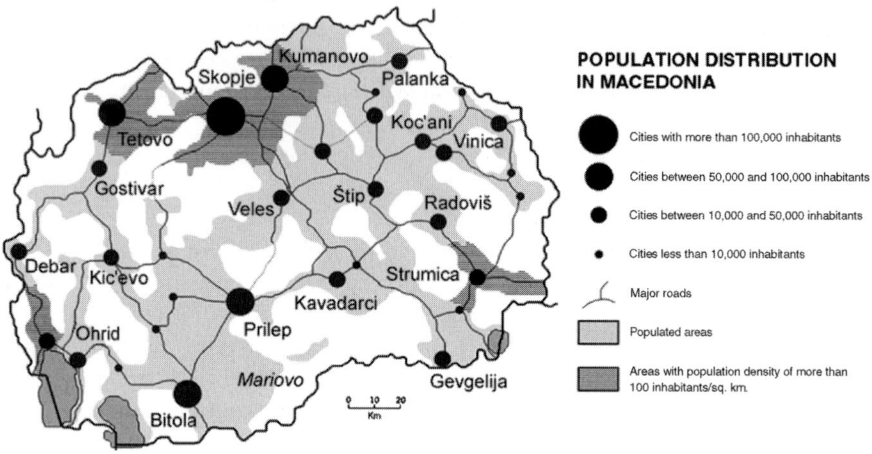

Figure 16.2 Population distribution in Macedonia
Source: Stojmilov 1995; JPPUP 1998.

centre (Poreče, Mariovo and Resen) and border regions have lost population to the urban areas and especially the key centres: 37.8 per cent of the total urban population now lives in Skopje, with a further 22.8 per cent in the four largest provincial cities: Bitola, Gostivar, Kumanovo and Tetovo (Hadji Pecova and Dimitrijevik 1996, p.2). Meanwhile, agriculture's GNP share fell from 57 per cent in 1950 to 15 per cent in 1990 and the number of farmers per hectare of arable land more than halved from 133 to 60 during the same period (Stojmilov 1995, p.83). However, despite the growing urban population, which accounted for 31.3 per cent of the population in 1953 but 48.8 per cent in 1971, 55.2 per cent in 1981 and 58.0 per cent in 1991 (Stojmilov 1995, p.81), the rural population remains considerable in the basins where an explosive growth of villages has occurred around the larger towns. The rapid development of these settlements has had far-reaching consequences upon the quality of the environment, as the municipal governments have not been able to provide for their growing infrastructural needs (Gelevski 1982, p.12). A typical case is the narrow, north–south orientated Polog basin (400sq.km), with a population of 300,000 and a 30 per cent level of urbanisation, where rural settlements along the valley have merged into high-density linear zones along the two local roads linking Tetovo and Gostivar.

Natural resource exploitation

There has been a moderate level of natural resource use because, although the country is relatively rich in raw materials, the intensity of extraction has been quite variable due to the recent history of central planning and the low density of population and inaccessibility of some areas. But exploitation is now intensifying and this is bound to have a negative effect on the fragile physical environment. With a varied geological base, Macedonia has reserves of antimony, chrome, copper, iron, lead and zinc and, while not all reserves are economically attractive, there are eight mines and four smelters located near Kavadarci, Skopje, Tetovo and Veles. None of these installations have adequately solved the problems of liquid effluent and solid waste; thus they pose a serious threat to the immediate environment in each case (Petruševska 1996, p.38). Moreover, there is a particularly disturbing case of iron ore being mined in open pits near to the Mavrovo National Park (Bužarovski 1997, p.45). Meanwhile, the situation is slightly different when it comes to the extraction of non-metallic minerals, because exploitation is widespread across the country with increasing intensity over recent years since corrupt local authorities are failing to exercise proper control. There is much quarrying of marble (18,000cu.m in 1995) and many quarries are situated in the immediate vicinity of towns such as Gostivar and Makedonski Brod, where they create problems of visual and air pollution (Stojmilov 1995, p.230). As in other Mediterranean countries, Macedonia's forests were limited and degraded before the Second World War, but their condition improved dramatically with the marginalisation of the mountainous areas and the measures taken by the Yugoslav government to protect and rebuild (with the reforestation of 9500ha in 1980 alone). According to official data, the current annual rate of cutting is about 1.0mln.cu.m which is only 55 per cent of the growth rate. But experts believe the real figure is much higher, partly because the state-run municipal enterprises do not have adequate management powers (Andonoski 1996, p.42).

With an average annual precipitation of 680mm, Macedonia does not possess extensive water resources: the total water volume is estimated at $6629.6{*}10^6$ cu.m). But increased water exploitation is evident, especially in the central regions (Kumanovo, Štip, Strumica, Valandovo and Veles) where there are frequent water supply problems in summer. The problem is enhanced by the lack of water conservation practices and the poor state of the water mains which supply 99 per cent of households in the towns and 65.4 per cent in rural areas (Hadi Pecova and Dimitrijevik 1996, p.11). In addition, large enterprises are not being encouraged to use water economically: the metallurgical works at Skopje use enormous quantities of high-quality spring water and in 1995 total industrial demand was 270.1mln.cu.m (ibid.). According to the latest estimates, the hydropower potential is 6436mln.kwh and 24 per cent is already exploited at sixteen separate stations. Meanwhile 5235mln.kwh of the total production of 6.6bln is based on five coal-burning stations, but the actual production varies significantly and energy is often imported from neighbouring countries during winter. The thermal plants use the lignite extracted in four open pits (annual production 6.65mln.t) but eight more sites with reserves of 900mln.t could also be used economically (Stojmilov 1995, p.213).

But there is a great need for an effective energy policy, as in other parts of the former Yugoslavia. The case for renewable energy is supported in principle, since Macedonia has a vast potential for geothermal energy use, but there has been only limited application in agricultural production so far (JPPUP 1995). The situation is even less satisfactory in the case of solar energy, although the annual average insolation is 2100–2450 hours (Kitanovski 1996). Meanwhile, planning continues to favour options which threaten the environment: thus the Physical Plan of 1982 (currently under revision) envisaged the construction of one nuclear plant, two new thermal stations and ten dams (six of which directly threaten protected nature reserves) in the period up to 2000. Although the nuclear option has been abandoned as a result of fierce public opposition, construction has already begun on one large dam site, which will flood more than 10sq.km of pristine canyonland south of Skopje. Further plans exist to regulate the whole of the Vardar catchment through nine new dams, including four in the untouched wilderness of the Crna and Radika tributaries.

Environmental problems

Air pollution

This is a significant problem because, despite the low level of industrial and technological development, industrial air purification and emission control systems are inefficient, energy is used irrationally and urban traffic pollution is high. Heavy industry was emphasised under socialism and accounted for some 45 per cent of Macedonia's GNP before independence (Lazarevska and Srbinovski 1997). But most plants are inefficient in their raw material and energy use and pollution is particularly bad in the urban area of Skopje where nearly two-thirds of the country's heavy industry is located (including the steel works, the OHIS chemical plant and the Usje cement factory) with far-reaching effects on the health of the local population (Nikolovska 1995, p.5). In 1995, 381 polluting industrial plants (with 1293 emission points) were registered, and in the context of local meteorological conditions (temperature and humidity, precipitation, wind speed and direction and atmospheric stability) this results in sharply contrasting patterns of contamination. There are notable concentrations of pollutants in specific areas, especially in some basins and valleys (where all the large urban areas are situated) prone to natural radiation inversions in winter (Gorgieva 1998).

Most of the emissions come from metallurgical and electro-metallurgical plants like those of Kavadarci (ferro-nickel), Skopje (steel), Tetovo (chrome) and Veles (lead-zinc). Significant air pollution is also generated by the three coal-burning power-stations of Bitola, Kichevo and Negotino, responsible for more than 1mln.t of dust annually, as well as central heating plants, smaller industrial units and domestic hearths. Traffic, especially in the winter months, is the main cause of the high concentration of heavy metals (especially lead) in city air: 329,218 motor vehicles were registered in Macedonia in 1995 and many are old, with inefficient exhaust systems (Hadji Pecova and Dimitrijevik 1996, p.8). A network of thirty measuring points, set up to sample the situation in the most endangered areas (Table 16.2), confirms the centre of Skopje as

Table 16.2 Maximum annual concentrations of CO_2, SPM (suspended particulate matter) and NO_x (micrograms/m^3/24 hours)

City	*Highest registered*			*Days (per year)*		
	CO_2	*SPM*	NO_x	CO_2	*SPM*	NO_x
Skopje	414	348	–	11	79	–
Gostivar	81	48	48	–	–	1
Bitola	33	124	–	–	18	–
Tetovo	18	182	–	–	29	–
Štip	55	103	–	–	6	–
Berovo	45	84	–	–	2	–
Kumanovo	66	136	–	–	10	–
Prilep	59	153	–	–	24	–
Veles	405	179	–	47	8	–
Kičevo	44	50	–	–	–	–

Source: Republicki hidro-meteoroloski institut (unpublished data).

the most polluted area, with a maximum registered annual concentration of 414 micrograms/day, followed by Veles with 405 micrograms/day. Air pollution in the larger urban areas poses a serious health threat to the population, owing to a dramatic increase in chronic respiratory diseases, allergies and ailments of the internal organs (Kočubovski *et al.* 1995). Moreover, the deposition of contaminating particles around industrial plants has caused a deterioration of livestock health, as well as various degrading processes in the local flora and fauna (Petrovska 1998, p.116).

Water pollution

Eighty-five per cent of Macedonia's rivers originate within the country, comprising as it does the upper part of the watersheds of the Aegean, Adriatic and Black Seas, but since almost all the populated areas gravitate towards the valleys, the water resources (including ground water and lakes) are exposed to contamination from agriculture and the urban-industrial areas. Moreover, the valleys are the natural traffic corridors which comprise a further pollution hazard (Table 16.3). The gravity of the situation is increased by the fact that, except for Dojran, Ohrid and Struga, none of the towns possess waste water treatment facilities while effluent from industrial and mining facilities, livestock farms and landfills is entirely uncontrolled. Hence, apart from severe organic and microbiological contamination, alarming levels of toxic elements (cadmium, chromium, lead and zinc) have been noted in the Vardar River, most critically during the summer when most streams either disappear or are reduced to a trickle (JPPUP 1995).

The Macedonian Hydro-Meteorological Institute (RHMZ) measures the quality of surface water at sixty sampling locations by monitoring the values of more than ten parameters. According to the Act for Classification of Surface

Table 16.3 Major water-polluting industries in Macedonia

Type of industry	*Municipality*	*River*	*Type of pollution*
Chemical (OHIS)	Skopje	Vardar	Organic, inorganic, toxic (mercury)
Leather (Godel)	Skopje	Vardar	Organic, inorganic, toxic (chromium)
Steel mill	Skopje	Vardar	Inorganic
Chromium processing (Jugohrom)	Tetovo	Vardar	Inorganic, toxic (chromium)
Meat	Skopje	Vardar	Organic
Meat	Gostivar	Vardar	Organic
Textiles (Goteks)	Gostivar	Vardar	Inorganic
Meat processing	Negotino	Vardar	Organic
Lead and zinc smelter	Veles	Vardar	Inorganic, toxic (lead, cadmium)
Livestock farm	Negotino	Vardar	Organic
Leather	Kumanovo	Kumanovska Reka	Organic, inorganic, toxic (chromium)
Leather	Bitola	Dragor	Inorganic
Sugar	Bitola	Dragor	Organic
Meat	Štip	Bregalnica	Organic
Meat	Štip	Bregalnica	Organic
Textiles (Astibo)	Štip	Bregalnica	Inorganic
Paper	Kochani	Bregalnica	Organic
Textiles (Tetex)	Tetovo	Pena	Inorganic
Automobile parts (HTD)	Ohrid	Lake Ohrid	Inorganic
Chemicals (Biljana)	Prilep	Stara Reka	Inorganic

Source: Study for the Protection of the Vardar River, Institute of Construction (GIM, 1995, p.48).

Water Flows, the condition of the water is expressed through four categories, ranging from drinking-water (Class 1) to a quality unfit for either agricultural or industrial use (Class 4) (Figure 16.3). According to RHMZ, only two major rivers (Radika and Crni Drim) belong to Class 1, but all the springs and most ground water resources do; a satisfactory situation in that several towns (Debar, Gostivar, Kičevo, Kriva Palanka, Kruševo, Prilep, Skopje and Struga) and most villages are supplied from natural springs. Moreover, the people of Delcevo, Gevgelija, Kočani, Ohrid, Probištip, Radoviš, Štip, Skopje and Vinica obtain drinking-water directly from underground aquifers. On the other hand, several streams (including the Dragor and Kumanovska rivers) have been assessed as 'dead' upon reception of untreated municipal and industrial sewage from Bitola and Kumanovo, respectively. A similar situation exists on the Bregalnica, Crna Reka and Vardar which are contaminated by discharges from such places as Gostivar, Kičevo, Kočani, Prilep, Štip, Skopje, Strumica, Tetovo and Veles; a situation aggravated by inadequate development of the sewerage systems in most populated areas.

Figure 16.3 River pollution in Macedonia

Source: GIM 1995.

Soil pollution and degradation

The twenty major sources of soil degradation include contaminated ground water, suspended particulate matter, industrial and agricultural activities (leaching of heavy metals and organic contaminants), erosion, open pit mining, infrastructure development, urban sprawl and waste disposal. As regards chemical pollution, total pesticide use in 1996 amounted to 556t, with fungicides, herbicides and insecticides taking a dominant share. This level is five to ten times lower than that in Western countries. So far analyses have not registered any contamination of agricultural products, with the exception of rice grown in Kočani and Strumica, and apples cultivated in the Prespa region. The use of chemical fertilisers is also low and decreasing: today it amounts to approximately 180kg/ha for fruit orchards and 160kg/ha for other crops. Chemical fertilisers are not applied to pastures and grazing lands. But the lack of adequate policy instruments for the organisation, management and use of land in the peripheral urban areas has resulted in serious soil pollution problems, further aggravated by contamination from transport routes and populated areas.

Another major soil degradation factor is erosion, as more than 1700 periodically active rills and gullies exist in the country: 1350 are located in the Vardar drainage basin, while the remaining 200 and 150 are to be found respectively in the Crni Drim and Strumica catchments. Only 228 gullies are managed, while the others continue to pose a threat to the physical and human environment. In fact, 96 per cent of Macedonia's territory is subject to erosion and the highest intensity (categories 1–3 out of a total of six) account for 38 per cent (9423sq.km) of the total area affected. The total loss of soil by erosion (685cu.m p.a./sq.km) is a product of topography, soil, vegetation and human intervention (Prostoren Plan – Nacrt 1998, p.145). Finally, reference must be made to the seasonal tourist activity, as the lake shores and skiing centres are saturated during the summer and winter periods respectively, while popular excursion destinations in the urban areas regularly receive a large number of visitors who contribute to soil pollution and environmental degradation generally.

Municipal and industrial waste

The absence of environmentally acceptable methods for the generation, collection, processing and disposal of solid and liquid waste is one of the most serious environmental problems today. This situation is further aggravated by the poor technological level of waste management, which lacks government legislation, civic initiative and public consciousness. Presently, there are no facilities for incineration or physico-chemical treatment of waste, not to mention recycling or composting (Hadji Pecova and Dimitrijevik 1996, p.8). In fact, most of the disposal sites (with the exception of a new landfill near Skopje) do not fulfil even minimal environmental standards and release pollutants into the air, water and soil, even though some hazardous industrial waste is involved (Biljanovski 1996, p.2) (Figure 16.4). Qualitative and quantitative data are very scarce, due to the lack of monitoring systems, but it is clear that paper, plastics and biodegradable residues dominate municipal waste. Furthermore, while most of the larger landfills are located near urban settlements (Figure 16.4), locations have often been selected without any consideration for the physical environment (and the geology in particular). Around several landfills there are ideal conditions for the pollution of both underground and surface waters, while fires

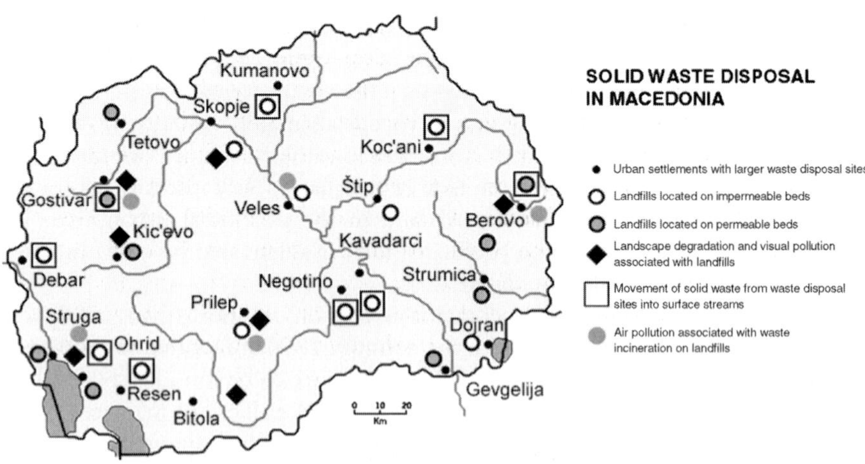

Figure 16.4 Waste disposal sites in Macedonia

Source: Biljanovski 1996 and National Environmental Action Plan.

frequently occur because of inappropriate disposal and compression methods, and in several cases, visual pollution affects scenic panoramic views and recreation sites (ibid., pp.2–10).

One of the most critical regions is the drainage basin of Lake Ohrid, which includes three large disposal sites in direct hydrogeological contact with the underground waters feeding the lake. There are a further eighteen illegal dumps within the nearby Galičica National Park, located directly above porous karstic beds in the lake's catchment area. But the problem of uncontrolled waste disposal is becoming increasingly evident throughout the country, since an illegal dump exists in the vicinity of practically every settlement, usually near a river or roadway (Meandiski 1989). Moreover, almost all the accessible areas, especially urban and recreation areas (beaches and ski slopes), are subject to extensive littering. A special problem concerns hazardous industrial waste containing heavy metals and other toxic substances which is generated by some twenty mining, smelting and energy facilities. Since none of them provide adequate treatment, harmful emissions are released into the environment on a daily basis. Among the most serious problems is the uncontrolled deposition of zinc and lead at the smelter situated only 4km from the town of Veles (Biljanovski 1996, p.6).

Urban problems

Macedonia's cities face growing environmental pressures through rapid development and lack of effective planning policies. Apart from air, water and soil pollution, significant deterioration in the quality of life arises from the inadequacy of infrastructure. During the massive rural–urban population movements in the 1960s, many immigrants were unable to find affordable housing within the city limits and so they built without permits in surrounding urban areas, thereby creating suburbs not very different from Third World slums. Such 'wild settlement' can be clearly seen around Bitola, Kumanovo, Ohrid and Tetovo, but especially around Skopje. Usually the local authorities are unable to meet the infrastructural needs because they lack the necessary planning and policy instruments to monitor and regulate urban growth (Dimitrijevik 1995, p.6). Meanwhile, much green space is being lost due to the building boom of recent years (aggravated by corruption in local government), and there is great concern over the destruction of architectural heritage in old city cores. This process began in the 1950s when urban planners were under political direction to manage vehicular traffic in the old Turkish town centres, but the damage continued through communist 'modernisation' plans applied to Bitola and also to Skopje in the aftermath of the 1963 earthquake when the historic centre was exposed to additional devastation.

Environmental legislation and public participation

Macedonia inherited from the Yugoslav era a plethora of more than 120 legal acts regulating environmental management; an impressive number that did not ensure adequate protection because many provisions were not properly implemented through lack of inspection (Marjanovik 1995, p.20). The lack of any clear determination of law enforcement duties meant that responsible executive branches could exercise discretion and allow themselves to be neutralised; for widespread corruption within state institutions and the courts hindered efficient legal sanctions even where criminal codes were clearly established (Juras *et al.* 1993). What was needed was a unified supervision system for all environmental provisions instead of a system where inspection organs were divided between the various executive branches, but the acts themselves were often incoherent and incomplete, with a low level of coordination. Strategic policy documents might project strikingly different attitudes towards a particular issue, depending on the ministries involved (Marjanovik, 1995).

The situation began to change during the 1990s following the reform of the Ministry for Urban Planning, Construction and Environmental Protection (MUPCEP) in 1992 (Dimovski 1995). One of its main tasks was to create the necessary legislative base for environmental protection, and in 1992 a draft Law for Environmental Protection was passed after four years of public debate and much amendment. For the first time in Macedonia's history, this act attempted to bring some order into environmental protection by taking a unified approach towards all major issues. Most of the new provisions were concerned with inspection and sanctions: a special Inspectorate for Environmental Protection was established within MUPCEP as an independent institution supervising the implementation of the law, with the authority to file lawsuits. In 1998, the Ministry of Environment was separated from MUPCEP and it has since been functioning as an independent government unit.

Several other laws relevant to the environment have also been passed during the past eight years: dealing with local self-government, physical and urban planning, hunting and fishing, and water. But there is still inadequate regulation of protected areas where available legislation was enacted in the communist period and is now in direct conflict with the new constitutional and legislative provisions. The criminal codes need further elaboration, and more efficiency in the implementation of controls and mechanisms is required (JPPUP 1995). Another obstacle to successful environmental regulation is the court system where long delays in dealing with cases imply an element of resistance to social and political change in general. Finally, more efforts will also have to be made to adopt new sub-legal provisions and standards (ibid.), since the regime inherited from the old system is not appropriate in the context of current economic and social realities and creates many problems for the executive organs. Reform of this segment of the legal system, as part of an integrated government strategy, should accord with more stringent EU pollution standards and international environmental standards in industrial production. One of the first steps has been taken with the adoption of the National Environmental Action Plan in 1996 as the basis for governmental and NGO policy in the domain of environmental protection.

Biodiversity and nature conservation

Macedonia's remarkable wildlife diversity reflects the varied relief, geology, natural history and human influence: more than 3500 vascular plants (150 of which are local endemics) have been registered, along with 78 species of mammals, 330 types of birds, 31 reptiles and 55 species of fish (Kolčakovski 1997, p.113). But there has been no systematic research and classification of the flora and fauna over the whole country; thus new endemic species of animals and plants are discovered in isolated mountainous areas on an almost annual basis. Presently, the natural heritage is legally protected in three national parks (IUCN Category 2), three habitat and species management areas (IUCN Category 4) and forty-eight natural monuments (IUCN Category 3). In addition, there are fourteen special nature reserves aimed specifically at the conservation of certain plant communities. The total surface of all protected areas is 1702sq.km, or 6.62 per cent of the national area (Micevski and Anastasovski 1995, p.5). But a major revamping of the nature protection and management institutions is necessary to bring them in line with the current economic and legal realities. More than a hundred different areas are now listed as deserving protection and it is very necessary for action to be taken before it is too late. Another priority is further scientific knowledge of Macedonia's natural heritage, as well as the adoption of international nature conservation agreements and conventions (ibid., p.50) (Figure 16.5).

The largest national park is Mavrovo, established in 1949 in the northwestern part of the country: it includes several mountainous regions rich in rare geomorphic phenomena (karstic and glacial relief, deep gorges), as well as more

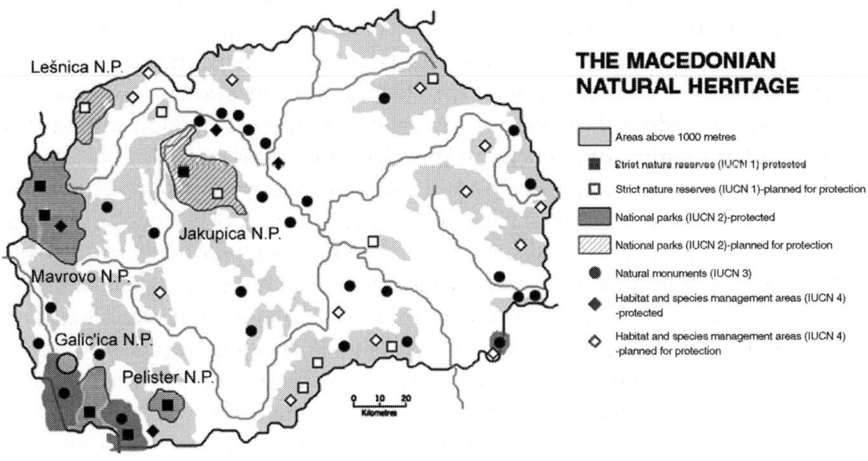

Figure 16.5 Protected areas in Macedonia

Source: Gelevski 1982; Micevski and Anastasovski 1995; Prostoren Plan – Nacrt 1998.

than 1000 plant species, of which 30 per cent are endemic (Bužarovski 1997, p.25). Another park, Galičica, is to be found in the southwest, on the high karstic 'mesa' hovering above lakes Prespa and Ohrid; it too is extremely rich in rare flora and fauna. Nearby is Pelister National Park, the oldest in Macedonia, established in 1948 to protect the rare Macedonian pine (Pinus peuce) communities. The parks have been divided into three zones of land use: strict nature reserves (IUCN Category 1), managed reserves and tourist recreation areas, although this classification does not reflect actual management patterns. There is only one additional 'strict' nature reserve outside the national parks: this comprises 188sq.km of rugged mountain land in the central region (Gelevski 1982, p.20). It has provided for the complete protection of the flora but the same is not true for wildlife, since all types of game are hunted. Furthermore, construction of a large dam in the Treska Gorge started in 1994 and effectively splits the reserve into two parts. Finally, the world-renowned Lake Ohrid is an 'internationally designated area' that was declared a UNESCO World Heritage Site in 1979: it is one of the world's oldest, deepest and clearest lakes, inhabited by 147 types of endemic organism that are relics from the Tertiary epoch. The 1998 Draft Physical Plan provides for two new national parks as well as eight strictly protected nature reserves. Apart from the proposal to establish Jakupica National Park (28,000ha), transfrontier park proposals have been put forward for Šar Mountain (Macedonia and Yugoslavia) over 51,858ha, and for Lake Prespa basin, where three existing national parks (two in Macedonia and one in Greece) could be merged (Schneider-Jacoby 1995, p.20).

However, the current network of protected areas is under heavy pressure and some of their qualities are being destroyed due to external threats and lack of management resources. One of the greatest problems is the cadre structure of the National Park Management Offices, whose role during socialism was reduced very largely to the management of forests and wildlife, i.e. the cushioning (but not the complete termination) of logging and hunting within the national parks (Bužarovski 1997, p.30). Today, these organisations are unable to control unsustainable tourist development, and potentially damaging investments in communal services, infrastructural growth and mineral exploitation. In fact, all the major summer and winter tourist centres are located precisely on the territories of the three national parks, exerting mounting strain on these fragile environments. There are a further fifty rural settlements within the parks, with inadequate sanitation facilities and major communal problems, stemming from uncontrolled waste deposition and housing developments. In addition, mineral resources are intensively exploited on several sites in Mavrovo and Galicica parks, while in Mavrovo there is also a large man-made lake (two more are planned), several chemical and textile factories and a marble quarry. Moreover, to accommodate the growing needs of the population and tourism, more than fifty springs and rivers have been capped without any supervision (Stojmilov 1992); the same is true for electric power lines and other infrastructural projects (Ivanova 1996). Meanwhile, the remaining nature reserves, managed by a special executive branch acting within the Ministry of Environment, face similar problems.

However, this network does not ensure an adequate degree of conservation for Macedonia's entire natural heritage. Only forty-eight bird species are officially protected, while a further 150 vascular plants and 247 animal species have been placed on a waiting list for legal preservation; but Macedonia does not possess an official red data book of vulnerable and endangered species because monitoring systems are lacking (Micevski and Anastasovski 1995, p.25). Even within protected areas rare plants are being lost: a meat-processing plant built in a legally protected nature reserve caused the loss of a unique community of the fern Osmundia regalis, a relict species from the Ice Age (DEM 1990). Animal species are even more threatened by habitat destruction (affecting protected types of eagles and vultures) and uncontrolled exploitation for export: even legally protected species such as bear and lynx have been poached during recent years. Over the past fifty years, several invaluable marshland biotopes have been drained to make way for agricultural expansion and to meet the health requirements of large communities in the river valleys and basins. Even legally protected sites of international importance have been affected in this way, as in the case of the Katlanovo swamp near Skopje (Micevski and Anastasovski 1995, p.25). Meanwhile, more intensive resource use and infrastructural developments continuously threaten large tracts of mountain wilderness. The Draft Physical Plan of 1998 provides for a range of conservation regimes over about one-tenth of the country up to 2020.

The Macedonian environmental movement

The development of civil society during the past ten years has had a significant impact on environmental issues, by establishing a number of new forms of civil participation in the decision-making process. Since the late 1980s, more than 100 ecological NGOs have been founded in Macedonia although a strong 'green' political force is still lacking. Under communism, nearly all facets of civil society, including those dealing with environmental protection and development, were organised through the Socialist Union (Tokarev 1992). But the 'green lobby' was generally restricted to individual experts and intellectuals, who tried to prevent or ameliorate some ecological problems. Then, during the initial years of democracy, several NGOs attracted media publicity, due partly to their highly critical stance over socialist environmental policies. They played a leading role in those cities with serious ecological problems such as air pollution or forest destruction. One of the first groups to emerge was the 'Opstanok' (Survival), founded in Skopje in 1989: it quickly organised massive public protests against air pollution and established a 'green phone' for feedback on urgent environmental problems. The 'Vila Zora' ecological society was formed in Veles later in the same year to reflect strong public opposition to the local lead-zinc smelter. It managed to regulate (and for a time completely halt) the operation of the plant and also managed to stop logging activity near the city.

Following an Opstanok initiative supported by seven similar NGOs, the Ecologists' Movement of Macedonia (DEM) was founded in Veles in 1990 to act as a green lobby nationwide (Meandiski 1990). Despite limited finance and manpower, an extensive organisational structure was quickly built up as more than thirty ecological societies, distributed across the country, came under its umbrella (Juras *et al.* 1993). Frequent protests and actions were organised and many public statements were issued to discuss environmental problems, including a unified declaration against nuclear energy in 1996 which effectively confronted the growing nuclear lobby. However, although funding is no longer a serious problem, there has been increasing inertia and resistance to change within the component societies. Many have been unable to overcome inherited tendencies towards rigid power structures and undemocratic decision-making. The DEM as a whole is beginning to lose its leading role in the Macedonian Green movement as it is being overtaken by new, more specialised NGOs, some of which operate within professional scientific institutions: the Society for Bird Protection and the Macedonian Ecological Society operating within the Institute for Biology; the Society for Nature Protection within the Institute for the Conservation and Development of Nature; and the environmental protection branch of the Mining Institute. The changes have been assisted by a number of foreign foundations working actively on the training of younger NGO staff. One of the most successful NGOs in 1997 and 1998 has been the Peoni Speleological Society, expanding its activities from recreational caving to the conservation of caves and karstic areas in general. The society has already managed to recruit public support for the conservation of a number of caves, especially in the Matka Canyon near Skopje, and a new bill for the permanent protection of eighty-six speleological sites is awaiting adoption (Speleolosko drustvo Peoni 1998).

The existence of many different groups on the one hand, and serious transboundary environmental problems in the Balkans on the other, has contributed to international NGO activity. One of the earliest Macedonian efforts to establish non-governmental cooperation with neighbouring countries involved the 'Last Call for the Salvation of Lake Dojran' (DEM 1992) issued in November 1990 by DEM and the Thessaloniki Ecological Movement from Greece (Ivanova 1990). Several similar steps have been taken in the Prespa area, where scope for transfrontier action resulted in designation as a 'European Ecological Bricks' site following an initiative by WWF-Austria in 1990, though it has not received much scientific recognition in Macedonia (EEA 1995). A joint symposium on the protection of the lake was held in Albania at Gorica (Albanian Korce) in 1997, with the participation of Albanian, Greek and Macedonian NGOs. Unfortunately, none of these initiatives has produced practical results in the Prespa region, due to the lack of governmental action. But with respect to Lake Ohrid, the World Bank donated US$40mln for environmental protection in 1998, partly dedicated to joint NGO actions. Following this decision, the US ambassadors to Albania and Macedonia met with World Bank representatives in Washington later in the year and signed an agreement to protect the lake with joint management of the surrounding area to protect water resources and wildlife (ATSH 1998). Finally, one of the most widely publicised recent NGO activities has been the 'Teddy Project', set up to protect the remaining brown bear (Ursus arctos) population in the Balkan mountains. Coordinated by the Greek organisation 'Arcturos', numerous NGOs and scientific institutions from Albania, Bulgaria, Greece, and Macedonia are now involved (Arcturos 1996).

Initially, the DEM openly expressed the desire to grow into a political party but the interested members eventually transferred to a newly formed Green Party in 1992 (Meandiski 1990), which subsequently merged with a new centre–right Democratic Party due to limited funding and membership (Tokarev 1992). No other Green political structures have emerged and the Green agenda is now divided between several political parties and the NGOs. However, there is a satisfactory degree of cooperation between environmentalists and the government in some aspects of the decision-making process, as in the case of the Act for Environmental Protection and the National Environmental Action Plan. But the Green movement will have to reorganise and become more professional if it wants to operate effectively under current economic and political conditions. A major priority is the education of NGO staff and the formation of new, more specialised organisations. There is also an obvious need to define clear NGO strategies to increase public participation in line with the development of civil society.

Conclusion

Macedonia faces a wide range of environmental problems which must be related to recent social and economic experience at the domestic and international levels. Former Yugoslav policies have combined with specific physical and demographic conditions to create a highly varied pattern of environmental degradation, with very serious problems encountered in a few relatively small and geographically isolated areas. But progress is being made and the National Environmental Action Plan, a strategic policy document adopted in 1996, highlights some of the main priorities for the future enhancement of environmental quality, including the need for comprehensive and continuous pollution monitoring systems with further research and assessment of natural resources, better legal regulation and economic incentives for sustainable development, technology transfer and transformation: for example, to achieve greater energy efficiency and an urgent improvement of waste management. Macedonia's first years of independence have been challenging, but European integration and foreign financial support should provide a new impulse for the establishment of much-needed cross-boundary government cooperation in the Balkans. Significant results have already been achieved in the area of Lake Ohrid, although the same is not true for Lake Dojran (on the Macedonian–Greek border) which has been brought to the verge of extinction as a result of uncontrolled water use. One of the most significant results of international cooperation could be the institution of transfrontier parks as grounds for the development of eco-tourism in passive, wilderness-rich border regions.

References

Andonoski, A. (1996) 'Šumite i šumskite povrsini vo strategijata za zastita na zivotnata sredina vo R. Makedonija', *Životna sredina* 6, p.41.

Arcturos (1996) 'Teddy Project', Arcturos, Thessaloniki.

ATSH News Agency (1998) 'Albania, Macedonia, World Bank to protect Lake Ohrid', 13 August.

Biljanovski, M. (1996) 'Separatna studija od oblasta: Analiza i procenka na sostojbite i upravuvanjeto so cvrstite otpadoci', NEAP, Komunalna Higiena, Skopje.

Bužarovski, S. (1997) Mavrovo National Park, Dvizenje na ekologistite na Makedonija, Skopje.

Dimitrijevik, J. (1995) 'Metodi za ocenka na efikasnosta na zastitata na zivotnata sredina', *Životna sredina* 5, pp.5–7.

Dimovski, M. (1995) 'Pravni i institucionalni aspekti – Institucionalno ureduvanje', *National Environmental Action Plan Tema* 3, pp.1–10.

Dvizenje na ekologistite na Makedonija (DEM) (1990) 'Posleden povik za zpas na Dojranskoto ezero', DEM, Skopje.

DEM (1992) *Save the Macedonian Lakes*, DEM, Skopje.

European Environment Agency (EEA) (1995) *Europe's Environment – the Dobříš Assessment*, EEA, Copenhagen.

Gelevski, B. (1982) 'Kategorizacija na prirodnite celini i prirodnite retkosti od aspekt na zastita na prirodata', in *SR Makedonija Prostoren Plan*, Republicki zavod za zastita na kulturata, Skopje, pp.2–25.

Gorgieva, E. (1998) 'Ušte eden Veles ne ni e potreben', *Eko-Glas* 39, p.2.

Gradežen institut 'Makedonija' (1995) *Studija za zaštita na rekata Vardar od zagaduvanje (1995)*, Gradežen institut 'Makedonija', Skopje.

Hadi Pecova, S. and Dimitrijevik, J. (1996) 'NEAP-Izvadok od sintezniot izveštaj', *Životna sredina* 6, pp.1–37.

Ivanova, M. (1990) 'Ezerski egzodus', *Večer*, 10 November.

Ivanova, M. (1996) 'Sostojbata na nacionalnite parkovi vo zemjata', *Večer*, 10 August.

Javno pretprijatie za prostorno i urbanističko planiranje (JPPUP) (1998) Prostoren plan – Nacrt (Skopje: JPPUP).

Juras, A., Kosterink P., Gjigas, O. and Cochrane, P. (1993) *Report on a Visit to the Former Yugoslav Republic of Macedonia*, Regional Environmental Center for Central and Eastern Europe, Budapest.

Kitanovski, M. (1996) 'Ima li nuklearno lobi vo Makedonija?', *Zelen Veter* 20, pp.4–6.

Kočubovski, M., Filanski, P. and Mancevski, D. (1995) 'Sledenje na vlijanieto na zagadeniot vozduh vrz zdravjeto na naselenieto na grad Skopje vo periodot X. 1989 do III.1993 godina', *Žvotna sredina* 5, pp.19–30.

Kolčakovski, D. (1997) *Biogeografija – interna skripta, Univerzitet 'Sv. Kiril i Metodij'*, Prirodno-Matematički fakultet, Skopje.

Lazarevska, A. and Srbinovski, N. (1997) *Zaštita na okolinata, Nacionalna strategija za ekonomskiot razvoj na Republika Makedonija*, MANU, Skopje.

Marjanovik, G. (1995) 'Makedonskoto ekološko krivichno pravo – Kritichki inventar i predlozi za reforma', raboten materijal, Skopje.

Meandiski, V. (1989) 'Vardar-otvorena deponija', *Nova Makedonija*, 7 October.

Meandiski, V. (1990) 'Zelenite vo ofanziva', *Nova Makedonija*, 31 April.

Micevski, B. and Anastasovski, V. (1995) 'Za biodiverzitetot i upravuvanjeto so zaštitenite podračja vo Republika Makedonija', *National Environmental Action Plan tema* 12, pp.1–37.

Nikolovska, M. (1995) 'NEAP – Geneza na problemite na opkruzuvanjeto vo ramkite na opštestveno-ekonomskite odnosi vo Makedonija vo izminatiot period', Ministerstvo za urbanizam, gradezništvo i zaštita na životnata sredina, Skopje.

Petrovska, A. (1998) 'Zaštita i unapreduvanje na životnata sredina', Javno pretprijatie za prostorno i urbanističko planiranje, Skopje.

Petruševska, L. (1996) 'Zagaduvanje na životnata sredina od industriskoto proizvodstvo vo Republika Makedonija', *Životna sredina* 6, p.38.

Prostoren Plan – Nacrt (1998) *Javno pretprijatie za prostorno i urbanističko planiranje*, Republika Makedonija, Skopje.

Prostoren Plan – Publikacija (1988) *Institut za prostorno planiranje – Ohrid*, Ohrid, pp.3–30.

Schneider-Jacoby, M. (1995) *Project-Lake Prespa*, EURONATUR Društvo za zastita i proucuvanje na pticite na Makedonija, Skopje.

Speleološko drustvo Peoni (1998) *Inicijativa za donesuvanje zakon za zaštita i upravuvanje so pesterite i karstnite podračja*, NACRT, Skopje.

Stojmilov, A. (1992) *Osnovi na turizmot*, Prosvetno delo, Skopje, p.175.

Stojmilov, A. (1995) *Geografija na Republika Makedonija*, Prosvetno delo, Skopje.

Tokarev, M. (1992) 'Introduction Macedonia', in REC (ed.) *Directory for Central and Eastern Europe*, Regional Environment Centre, Budapest.

17 Romania

David Turnock

Like other countries in East Central Europe Romania began the transition with some major environmental problems because legislation failed to deliver effective pollution controls. Although scarcely apparent during the last years of communism, an ecological movement achieved early political success and some of the worst affected areas have now seen substantial improvements. In addition, the slowdown in the economy helped to reduce the scale of environmental damage, although rising unemployment induced a survival mentality which has made the public wary of supporting radical environmental programmes. However, with the help of foreign capital and technology, progress continues to be made on a broad front and the legislative programme of 1995 to 1996 should go a long way towards bringing Romania into line with EU practices. In addition to the key problems of air and water pollution, much is now being done to rationalise the woodlands and other rural land uses; also to develop a sound conservation policy with particular attention to protected areas and to enhance the cultural landscape for the benefit of rural tourism. In these respects this chapter differs substantially from the coverage in the first edition (Turnock 1993) and it also expands a more recent paper on environmental action since 1989 (Dragomirescu *et al.* 1998).

Introduction: the Romanian environment

Although rich in natural resources, Romania is prone to a number of major environmental hazards including avalanches in the mountains, gully erosion, landslides and mudflows in the hills and plateau country, combined with floods and seismic risks in the lowlands (Balteanu 1992). Some changes in the physical environment are inevitable. In recent years the Prut River, which forms the frontier between Romania and Moldova, has eroded the bank on the Romanian side by 2–3m each year in the Stânca–Costeşti area and the island near Horodistea has been reduced from 500 to 200ha (a total loss of 300ha of territory) because the soil is very light and there are no trees to consolidate the bank. No remedial action has been taken since 1989 because of lack of funds: recovery of 1ha costs 350mln.lei. However, the extent of the damage from routine environmental hazards is often increased by excessive human interference (Mac and Ripeanu

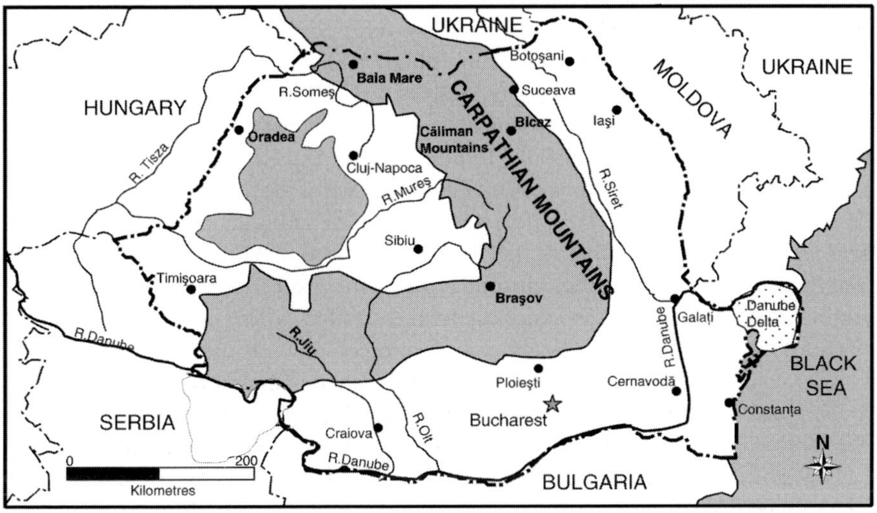

Figure 17.1 Salient features of Romania

1996). For example, there are 3.5mln.ha of potentially flooded lands along the Danube and its tributaries (Argeş, Buzău, Ialomiţa, Jiu and Siret); and similar areas on the Banat–Crişana plain associated with the Criş, Mureş and Someş systems (Figure 17.1). But much of this land has been taken out of circulation; so flooding is accentuated and the consequent reduction in the Danube alluvium supply has meant accelerated erosion along the Black Sea coast. This in turn has required further intervention linked with tourism: protecting beaches by the erection of wave-breaking dams and defences to protect infrastructure along the Danube delta coast south of Sulina (Armas 1991–92). Meanwhile, poor water management has resulted in salinisation in some of the Danube meadows converted for irrigated crop farming, while there has been an increase in marshland through massive infiltrations into phreatic water levels in the Olt valley linked with dams and dikes.

Lignite-producing areas in the southwest (Gorj and Vâlcea) have been desecrated by strip-mining and the tipping of waste. Mountain areas, like the Căliman, Guţâi and Metaliferi ranges, have also been damaged by mining. But much greater problems have arisen through deforestation which has increased run-off and made flooding more likely after very heavy rain. Once flood waters move down the valleys, the closure of natural flood plain reservoirs (noted above) and the blockage of streams by bridges increases the possibility of disaster. The vegetation zones are under climatic control and large areas may be destroyed by gales, such as the 200km/h storms of November 1995 (the worst winter weather for half a century) which led to catastrophic damage over 9000ha in eastern Transylvania (Covasna, Harghita and Mureş). 1.4mln.cu.m of timber was destroyed in 1998 when the June floods damaged 4000ha of woodland in

Bistriţa–Năsăud, Covasna, Harghita and Neamţ. But vulnerability is increased by planting inappropriate species for short-term economic gain, and woodland will be progressively thinned if animals are allowed to destroy young trees (Stoiculescu 1990). In addition, extensive clear felling can result in accelerated run-off during heavy rainfall. Over the longer term, deforestation has contributed to erosion in the Subcarpathians, but in modern times it is inappropriate agricultural methods that have been most significant in aggravating the incidence of gullies, landslides and mudflows in areas with complex structures of sandstone, clays and shales such as the Râmnicu Sărat and Râmna basins (Surdeanu and Ichim 1991). Cumulatively these problems are serious because there are some 2.0mln.ha of degraded land across the country where erosion may remove 300–5000t/ha p.a. The total annual wastage of soil through erosion amounts to 120 mln.t (107mln.t of which come from agricultural land).

The year 2000 proved to be something of a nightmare in ecological terms on account of pollution linked with outdated methods used by the mining industry of Maramureş in the far north of the country. Water with a high concentration of cyanide was released from a lagoon near Baia Mare belonging to Aurul, a joint venture company with capital invested by Esmeralda of Australia. Exceptionally heavy rain caused the dam to burst and pollution worked its way into the Tisza and thence down the Danube to the Black Sea. It was three days before matters were brought under control, by which time 100,000cu.m of cyanide-polluted water had been released. Compared with a maximum permitted cyanide level of 0.1milligrams/litre, actual concentrations of 12.40 were recorded; decreasing downstream to 2.8 in the Szolnok area and 1.0 on the Yugoslav border. Some 1.2th.t fish were killed and many livelihoods threatened, while water supplies for Szolnok were disrupted when abstraction had to stop: 80,000 people were left without water. Aurul lost US$350,000 a week while operations were suspended but Hungary is seeking US$100mln in compensation, although there is no valid international compensation accord between the two countries and The International Court of Justice may be used. The Regional Environment Centre (Szentendre) is calling for a decisive cooperative effort through a 'Tisza River Response Project'. Euro20,000 has been donated by the Dutch government to help NGOs cope with urgent work in connection with what was described as the worst environmental crisis in Europe since Chernobyl. An additional dam will be built with 250,000cu.m capacity to act as an emergency buffer. But two other incidents occurred in Baia Mare in 2000, affecting the installations of the state mining company Remin. First, in July, a faulty pipe at the Herja lead-mine discharged sludge containing zinc and lead into the Tisza basin, and in September a pipeline leak caused 140cu.m of mining waste to pollute the soil and ground water. Finally, another Maramureş mining area – Baia Borşa – suffered a dam burst through pressure of melting snow in February when 20,000t of heavy metal sludge was released into the Vişeu and thence into the Tisza, resulting in lead content double the permissible level and zinc nine times.

The EU has moved to set up a task force which is considering the measures needed to cope with pollution hazards throughout the basin. A survey detailed in Chapter 4 has revealed a cluster of high-risk sites around Baia Mare (including Baia Sprie, Băiuţ, Cavnic, Herja, Ilba, Nistru and Săsar) belonging to the Aurum, Aurul, Baia Mare and Phoenix companies concerned with the mining and processing of ores for non-ferrous and precious metals. The hazards arise from mine waters, process water and hazardous products with pollutants in the form of cyanide, acids and heavy metals. Hazardous waste in store amounts to 61mln.t plus 2.2mln.cu.m. Another cluster of high-risk sites in the Apuseni Mountains – belonging to the Abrud, Baia de Arieş, Coranda, Minvest and Roşa Montană companies – has a similar profile, with the amount of storage again substantial at 52mln.t and 3.0mln.cu.m. The low-risk sites in this area are also concerned with mining, since the Bauxit, Deva, Iara, Minvest and Zlatna companies have hazardous materials in store to the extent of 5.35mln.t and 10.20mln.cu.m. Another significant group covers the three high-risk sites concerned with heavy metals, at Borşa and Rodna where the Borşa and Remin companies store 10.90mln.t. These sites typically consist of dams and tailings ponds with wastes and sludges containing heavy metals and toxic substances. Poor design is a problem but also inadequate maintenance (especially at abandoned sites) increases the risk of leakage and collapse. All sites are to be evaluated and placed under regular surveillance.

Environmental action and public opinion

Pre-1989 approaches to the environment did not show adequate concern for the dynamism of the environment and the potential consequences of human action. There were perceived economic and social benefits linked with resource management (Constantinescu 1980; Herseni 1979) and an attractive landscape was always beneficial for tourism (Vasilescu 1963). There were also notions of efficiency through a judicious balance between specialisation and self-sufficiency. Much interest was shown in specialisation in the 1970s through zoning of production (Andeoca and Chirculescu 1976; Puia 1977, 1979). But the 1980s saw a shift to local food supply and greater self-sufficiency to make better use of rural labour and reduce transport costs (Puia and Soran 1986). Many intellectuals would have liked such an approach strengthened through more production in rural areas and less concentration in large cities. However, the debate between concentration and dispersion did appear to bring a pause to radical schemes of rural settlement regrouping under 'sistematizare' which was launched in the late 1970s (Bold 1974; Defour and Boucher 1977), but then deferred until well into the 1980s after momentum had been lost for the best part of a decade. The programme was then brought to a halt by the revolution of 1989 before too much damage had been done (Turnock 1991).

Despite continued difficulties, Romania has made substantial progress since 1989 and this is in no small measure due to public opinion. Under communism there was no significant environmental protest. The worst problems were not widely appreciated by the population at large and the intelligentsia was disinclined to go beyond professional publishing to underline the importance of the Communist Party's own pronouncements on the importance of environmental policy, sadly compromised by the drive for higher industrial output (Turnock 1982). However, immediately after the revolution conservationists came out of the woodwork and many ecological groups arose. The most significant development has been the formation of political parties. The two Bucharest-based parties, the Romanian Ecological Movement (Miscarea Ecologistă din România (MER)) and the Romanian Ecological Party (Partidul Ecologist Român (PER)), did surprisingly well in the elections of May 1990, securing two senate places (one for each party) and a total of twenty seats in the Chamber of Deputies (twelve for MER and eight for PER). The two parties attracted 1.18mln votes or 8.1 per cent of the total. Exactly half the votes (0.59mln) came from Bucharest and eight other industrial areas where the share of support was more than 20 per cent above the national average and where leaders of local authorities were often members of ecological parties.

However, this early progress has not been maintained. As the scale of economic reconstruction became apparent people became concerned primarily with survival and showed little inclination to sacrifice jobs for a cleaner environment, even in the worst cases of pollution damage (such as Copşa Mică). Action was also limited by an acute capital shortage. Although some finance is obtained from PHARE and other EU sources, the major source remains the state budget and company funds. The major parties also formulated their own environment policies and in the early 1990s the governing coalition did enough to convince most of the electorate that all reasonable measures are being taken. Government has been pragmatic over major public works: for example, moving in favour of smaller-scale hydropower projects which are far less disruptive than the massive artificial reservoirs built under communism. Nuclear policy has also been approved (as noted below), since the chosen Candu technology has an excellent safety record and the extra generating capacity has eased winter power shortages. Meanwhile some influential leaders have been compromised, such as M. Bleahu (of MER) who was the Environment Minister when the Sibiu waste scandal was uncovered (Ciulache and Ionac 1995) and there were tensions within MER in 1996 when the party president exceeded his prerogatives by seeking links with the nationalist Party of National Unity (PUNR) which would have merged MER candidates into PUNR lists for the local elections. Meanwhile, PER is one of the parties in the opposition grouping 'Convenţia Democratică' (currently in government) and cannot be evaluated as an independent entity. MER continues with a separate existence, but failed to gain any parliamentary representation in either 1992 or 1996. Even in local government elections its showing was generally poor, although it topped the poll in Zlatna in 1991 taking 46.2 per cent of the vote, while MER candidates were also elected to the office of mayor in Băile Herculane, Bistriţa and Orăştie (Pop and Bodocan 1991). There has been a grouping of forces with the amalgamation of MER and PER in 1998 and a wider 'Ecological Pole' has been arranged for the 2000 elections, drawing in two other parties: the Ecology Convention Party and the Green Ecology Alternative. Meanwhile, the number of ENGOs has been

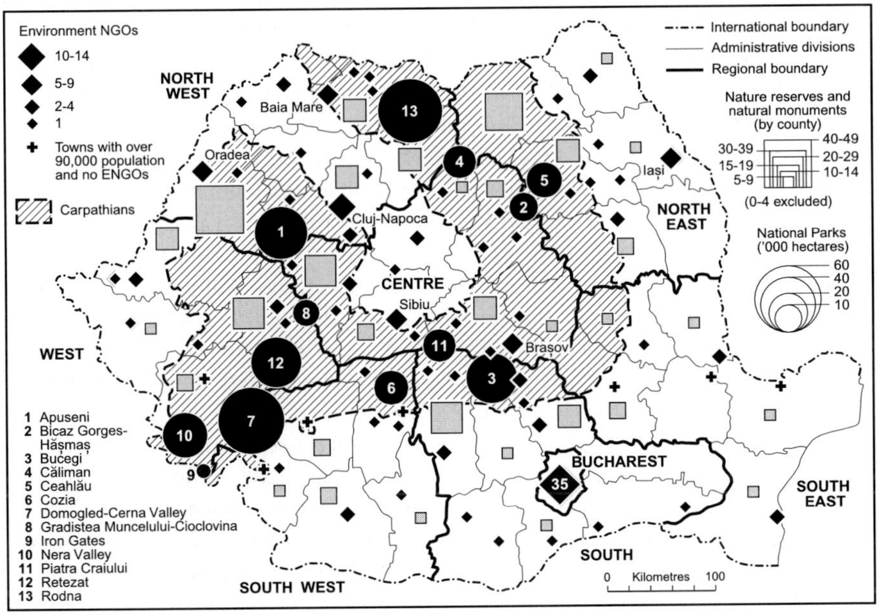

Figure 17.2 Environmental NGOs, nature reserves and national parks in Romania

Source: National Commission for Statistics, Bucharest; European Mountain Forum, Gland, Switzerland.

officially reckoned at around 160, of which thirty-five are based in Bucharest. The rest are based overwhelmingly in the towns, with significant clusters in Baia Mare, Braşov, Iaşi, Oradea, Sibiu and – above all – Cluj-Napoca (Figure 17.2). They cover a wide spectrum ranging from large national groups with regional structures to small local organisations; and in political tone they range from the conventional to the radical (Dinu 1994). There is, however, no tradition of NGOs in Romania and they did not really feature in the opposition to communism in Romania in the late 1980s (in sharp contrast to the situation in other East European countries).

Action during the early transition years

There has been a greater frankness in discussing environment issues (Hancu 1990), supported by major ministerial assessments (MAPPM 1994), whereas glasnost was largely absent prior to 1989. There has been more discussion of public health issues linked with pollution and also with attempts to relieve pollution, for large amounts of chlorine used to make drinking-water safe then created risks of digestive infections. Atmospheric pollution related to industry, heating and traffic affects the normal physical and psychological development of children (Enache 1994, pp.140–1). But there are also problems for young children through respiratory infections, with heavy metals (especially lead) a special hazard (Cucu 1992). Copşa Mică's Carbosin plant with its carbon black production profile aggravated bronchitis and asthma and the town had the highest infant mortality in the country in the 1980s. Health risks for children have also been examined with reference to the Baia Mare chemical industry (Iacob 1994; Ipatiov 1996), alleviated to some extent by protein and vitamins, and the cement works at Bicaz and Taşca (Vasilov and Ichim 1993). However, pollution is only a part of the picture. Health is primarily a reflection of living standards, though poverty is certainly aggravated by pollution and inadequate health services (Dumitrache and Armas 1998). The areas with poorest health are to be found mainly in three clusters: the southeast (Giurgiu, Ialomiţa, Călăraşi and Constanţa), northeast (Botoşani and Iaşi) and the west (Caras–Severin, Hunedoara and Satu Mare), with no obvious correlation with industrial pollution levels. This seems to reflect the picture across Eastern Europe where pollution is one component of a stressful transition which promotes unhealthy life-styles (with excessive alcohol and increased suicide), high rates of cardiovascular diseases and decreased life expectancy (Phillips and Groenewegen 1998).

Priority areas

Fourteen priority areas (Arii Considerate Prioritare în Planul Strategic de Combatare a Poluarii) have been identified where there are serious pollution problems requiring immediate action (Figure 17.3 and Table 17.1). These are the industrial zones of Bacău, Baia Mare, Braşov, Brazi (Ploieşti), Borzeşti–Oneşti, Copşa Mică, Govora (Râmnicu Vâlcea), Işalniţa (Craiova), Piteşti, Suceava, Târgu Mureş, Turnu Măgurele, Tulcea and Zlatna. Plant closures have been generally resisted in preference for investment in modernisation and substantial improvements have now been made following improved monitoring, which includes a project in Bacău with Dutch support. In 1994, 3bln.lei were allocated to the small town of Zlatna in the western Carpathians in order to clean up the Ampellum enterprise concerned with non-ferrous metallurgy and sulphuric acid (Mihailescu and Ciobanu 1990; Serban 1993). The obsolete sulphuric acid plant was a particular problem at the so-called 'death works' but the old works (Uzina Veche) has now been refurbished and production was restarted in 1996: the sulphuric acid needed elsewhere in the complex is now brought in. Pollution is also being eliminated at the non-ferrous metallurgical complex of Baia Mare where the Romplumb enterprise was assisted by the Swiss government to install filters to collect the fine lead powder that used to be pumped into the atmosphere. Furthermore, with help from Finland the Phoenix factory in the same city was able to update its filtering system and install a new furnace to eliminate sulphur dioxide pollution currently dispersed through chimneys 350m high (a European record): though this was partly counterproductive because of the inversion phenomenon which forced the pollution back to ground level. However, the Ministry of Waters, Forests and Environmental Protection was able to fulfil its promise of 1992 that flowers would grow on Romplumb's premises within four

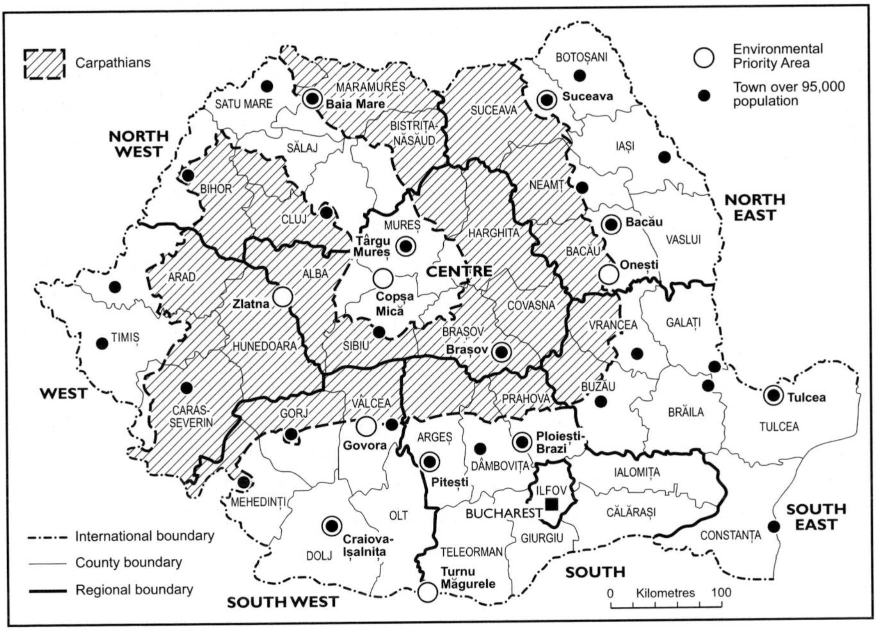

Figure 17.3 Environmental priority areas of Romania

Source: National Commission for Statistics, Bucharest.

years! There has also been action over water pollution in Baia Mare: copper concentrations in the Săsar River (16.4 times higher than the legal maximum) have been reduced by action at the Phoenix works and the Herja/Baia Sprie mining operations, including pipes for the hydro-transport of waste from flotation plants. The Săsar has been transformed through landscaping and flood control measures as part of the programme for a clean environment achieved by the end of 1997. At Iaşi a German company (Mannesmann of Düsseldorf) has refurbished the sewage plant in the city and solved the problem of pollution in the Bahlui (which in turn protects the Prut and the sensitive Danube delta).

Other initiatives

There has been some treatment of tips and strip-mining areas, with 260ha of degraded land reclamation reported in Gorj (Balutescu 1998). After the mining has ceased, lakes in the hollows may be used for recreation and fish farming, but after the first two or three years when cultivation is rewarded with the fertilising effect of the coal dust the land has no agricultural value (Candea 1993). Farming on waste tips in the coalfield now includes the growing of fruit trees at Beterega and Garla and willows have been planted to stop torrential erosion. Prevention of further soil pollution is a priority (Rauta and Carstea) but ecological restoration of polluted soil has been estimated to cost US$25mln (MAPPR

Table 17.1 Pollution in Romanian towns 1993–94

Town[o]	*Main pollutants*			
	1	*2*	*3*	*4*
Alba Iulia	61.6[f]	22.3[n]	0.4[a]	
Arad	10.6[f]			
Bacau	15.7[a]	[p]0.2[j]	[p]0.11	
Baia Mare	83.2[i]	42.0[b]	14.9[n]	21.41
B'le Herculane	[p]38.9[f]			
Bals	[p]10.7[f]			
Barlad	15.5[f]			
Bicaz	57.2[f]	[p]25.0[g]		
Bocsa	[p]10.7[f]			
Botosani	21.8[f]	20.7[n]	2.9[a]	
Braila	17.1[f]			
Brasov	[p]57.1[f]	[p]25.0[j]	[p]8.3[a]	
Bucharest	17.9[n]	[p]1.6[j]	[p]1.0[a]	
Calan	52.1[n]	41.8[f]		
Calarasi	[p]40.0[f]	[p]0.3[b]	[p]0.3[k]	
Campia Turzii	[p]42.0[f]	[p]37.0[m]		
Campina	[p]51.1[m]	[p]4.2[h]		
Caracal	[p]25.8[a]	[p]21.4[f]	0.2[c]	
Caransebes	[p]31.2[f]	[p]0.6[a]		
Cernavoda	31.2[f]			
Cluj-Napoca	[p]83.3[f]			
Codlea	[p]11.1[f]			
Constanta	[p]11.5[f]	3.7[a]		
Corabia	[p]45.5[f]			
Cugir	[p]33.0[f]			
Dej	[p]75.0[h]			
Deva	11.3[f]	[p]0.1[j]		
Dorohoi	[p]33.6[n]	10.2[f]	7.1[a]	
Dragasani	[p]50.0[f]			
Dbta-T.Severin	10.2[f]	2.4[a]	2.4[h]	[p]0.2[j]
Eforie Nord	[p]25.0[f]			
Fagaras	[p]16.7[f]			
Fieni	69.1[f]			
Focsani	29.4[h]	1.2[m]	[p]0.3[h]	[p]0.3[j]
Galati	67.0[f]	1.8[a]	[p]0.4[k]	[p]0.1m
Gheorgheni	18.3[f]	2.1[n]	[p]1.6[a]	
Hateg	[p]50.0[f]			
Hunedoara	66.2[f]	59.7[n]	0.3[j]	
Husi	[p]57.0[f]			
Iasi	39.9[f]	[p]0.1[a]		
Motru	18.7[f]			
Navodari	[p]62.5[f]			
Ocna Mures	[p]88.2[f]	[p]36.7[m]		
Odorhei	14.7[f]			
Oltenita	[p]25.0[f]			
Petrosani	32.9[f]			
Piatra Neamt	24.0[f]	23.5[a]	14.9[j]	
Ploiesti	[p]100.0[f]	[p]28.0[n]	[p]15.4[a]	[p]14.0[n]
	[p]6.4[c]	[p]3.6[j]	[p]2.2[h]	[p]2.0[g]
Ramnicu Sarat	17.7[f]			
Ramnicu Valcea	[p]15.0[n]	3.6[a]	3.8[e]	[p]1.2[h]
Resita	[p]34.2[f]	6.7[m]	1.2[a]	[p]0.1[j]
Roman	[p]33.0[a]	[p]14.3[g]	[p]14.31	[p]11.1[f]
Rovinari	75.5[f]	[p]3.7[m]		
Sebes	[p]83.0[f]			
Sf.Gheorghe	34.0[f]	1.3[n]		
Sibiu	10.6[f]			
Simleu Silv.	18.2[f]			
Slatina	22.6[f]	4.6[a]	[p]3.4[n]	[p]1.2[j]
Targoviste	10.3[f]	[p]8.0[j]		
Targu Bujor	33.3[f]			
Targu Jiu	32.3[f]			
Targu Neamt	29.4[f]	[p]1.7[a]		
T'gu Secuiesc	[p]10.0[f]			
Tarnaveni	[p]20[f]	[p]16.0[n]		
Tasnad	[p]10.5[f]			
Tecuci	37.5[f]			
Timisoara	53.1[f]	[p]3.2[c]	0.7[n]	[p]0.3[a]
Tulcea	13.2[f]	4.5[a]	2.1[n]	[p]1.0[j]
Turda	19.1[f]			
Turnu Magurele	25.0[f]	[p]5.9[a]	[p]3.1[j]	[p]1.6[l]
Vaslui	59.0[f]			
Zalau	[p]14.4[f]	2.0[m]	0.3[a]	0.2[j]
Zlatna	63.8[f]	64.7[n]	31.4[l]	

Source: *Anuarul Statistic al României* 1994, pp.62–81; 1995, pp.50–63.

Notes

Pollutants: a Ammonia; b Cadmium; c Carbon Oxide; d Chlorine; e Chlorine Hydride; f Deposited powder; g Formaldehyde; h Hydrogen sulphide; i Lead/lead compounds; j Nitrogen dioxide; k Phenols; 1 Sulphur dioxide; m Sulphuric acid; n Suspended powder.

o Towns are included where the percentage of readings for one of the observed pollutants exceeds the authorised limits on more than 10 per cent of occasions (figures for Piatra Neamt include Savinesti and for Ploiesti, Brazi).

p Indicates that the data refer to only one year (otherwise the figures are averages of the percentages for two years).

1994) and UN FAO has supported a scheme for the 'Rehabilitation of Polluted Soils' worth US$264,000. This was a two-year project to use the most efficient means of improvement on farmland polluted by oil waste, mining operations, pesticides and chemical fertilisers. Implementation is in the hands of the Institute for Research in Pedology and Agrochemistry and the Academy of Agricultural Sciences and Forestry.

There is a need for zoning for different crops in the context of their susceptibility for irrigation and constraints of soil deterioraion including acidification and pollution by heavy metals (Ianos and Florea 1996). It may be that intensification can be reduced most effectively by working out the optimum crop combinations in 'natural micro- and meso-regions' so that the cropping that is retained will generate maximum output (consistent with appropriate rotation and avoidance of monoculture – and also, presumably, commensurate with the logic of mixed farming and social considerations regarding rural employment where cropping is substantially reduced).

Elsewhere a range of significant initiatives are being taken. In Bucharest a start is being made on the problems of sanitation and garbage collection. An improved garbage collection system is operating in Sector 5 of the city, thanks to the Romanian–Italian joint venture Ecosal. Garbage is being collected in high-capacity containers with separate compartments for wood, paper, plastic or glass waste. Residents are rewarded for their cooperation by tickets that can be exchanged for cash at the council office. Ecosal is also commissioning an installation to convert human waste into agricultural fertiliser. Meanwhile, the French company Sater–Parachini is providing for the ecological storage of waste in Bragadiru commune and another ecological dump for household waste has opened at Mioveni (Argeş). The latter, with a lifespan of some thirty years, has a capacity of 117,800cu.m and a stockpile for reuseable materials. Finally, on the Black Sea coast, Năvodari and Ovidiu have joined forces with Constanţa in a joint venture with Sater–Parachini. The Consulab company is making a US$5mln investment, using the French partner's financial and technical resources in establishing an ecological waste dump comprising ten garbage pits over 32ha. Constanta is also undertaking a major water supply upgrade costing nearly Euro100mln, with three-quarters coming from the EU and the rest from EBRD credit and local authority contributions.

Energy conservation

When the Agency for Energy Conservation was created in 1991 as a public body within the Ministry of Industry, Romania was the first of the former socialist countries to create a government organisation specifically for energy conservation. Emerging out of the former Energy Inspectorate, the Agency can help industrial enterprises both technically and financially in energy efficiency, with scope for expansion into agriculture as well as the commercial sector (including offices and public buildings) and residential areas. The aim is to achieve the efficient use of energy in all fields with the provision of financial and technical assistance (Patterson 1994). The way was opened for international collaboration to pioneer new/alternative energy sources for energy-efficient production and transportation/distribution with replacement of hydrocarbons. The Agency can supply up to 30 per cent of the cost of energy efficiency projects. But it was necessary to get around the public perception of energy savings as a 'sacrifice', because the country consumes three times as much power as the EU per unit of GNP and also produces power inefficiently in large-capacity stations working way below full capacity. To help reduce consumption and so moderate the polluting emissions which cause the greenhouse effect, a 'competition' has been set up whereby the local authorities of Constanţa, Craiova and Ploieşti submit organisational frameworks, with the cost of US$7mln being met by US and French Funds for Environmental Protection, the EU PHARE programme and the Romanian government. The project will include the development of local power generation strategies, improvement of fuel-burning efficiency, recovery of heat in industry and better management in the power industry. However, Romania is not yet included in the network of Organisations for Promoting Energy Technologies emerging across Europe with the support of the European Commission.

Although energy consumption is falling, fuel is still being imported, while domestic hydrocarbon resources could be conserved if present demand were to be reduced further. 'Upgrading the electricity, natural gas and heat supply systems will be costly and time consuming; such expense and effort will make little sense if the electricity, natural gas and heat supplied continue to be wasted' (Patterson 1994, p.42). Conservation measures usually pay for themselves within a few years, but Romania does not yet have an 'efficiency culture': in the communist period energy saving meant simply denying power to domestic consumers and therefore energy conservation is greeted with suspicion. Moreover, in the energy industry there are wasteful attitudes.

The old Energy Inspectorate functioned as a policing system during the last two decades of communism, ensuring that industry used the energy stipulated under the plan: it did not concern itself with efficiency, as had been the case during the 1960s, and tended to become an agent of repression under the self-sufficiency drive between 1984 and 1989. This has not made things easy for its successor to gain public confidence, especially given the shortage of qualified personnel. This is not to mention the passivity of the population under communism and the difficulty in appreciating that everyone can make a contribution in using plugs to check waste. Despite the sophistication of the Romanian oil-refining industry, including the system of terminals and pipelines and the integration of refining with petrochemistry, 'the sector is beset by poor operational performance, extended downtime, extensive losses and disregard for energy conservation and the local environment' (ibid., p.72). Therefore winning public support and providing trained energy managers will be a lengthy process. However, there has been a powerful stimulus from rising energy costs since 1995 with the scaling down of subsidies: electricity went up 4.6 times and heat from district stations nine times.

Nuclear power

Romania's first nuclear power project dates back to 1978 when the decision was taken to proceed after some ten years of negotiation with potential foreign partners. However, during the 1980s there were major difficulties due to the shortage of hard currency which resulted in delays in obtaining essential equipment, despite the large number of workers massed on the site. There was a major re-evaluation in 1990 linked with problems over poor building standards and equipment performance: many welded joints failed stress tests. But Ansaldo (Italy) and AECL (Canada) agreed in 1991 to provide experts to oversee the completion of Unit 1 working to a December 1994 deadline. US$400mln was received in credits in 1992 while the Romanian government allocated a total of 670bln.lei from the budget between 1994 and 1996. The CANDU (Canadian Deuterium Uranium) reactors operate with uranium and heavy water produced in Romania. Cernavoda Unit 1 700MW Candu reactor was inaugurated in April 1996 in the presence of the Romanian and Canadian prime ministers and the presidents of AECL, Ansaldo, General Electric and the International Agency for Atomic Energy. Nuclear fission began and the station was connected with the national electricity network by the end of the year. However, the National Nuclear Programme aims at the completion of a further four units at Cernavoda. Nuclear power will ultimately lead to reduced imports of hydrocarbons: 1.4mln.t per unit. It is important that uncertainties surrounding the other units should be removed and that the loss of specialist staff should be halted. Finance is now assured for Unit 2 which should be finished by 2005. Ultimately there should be five units of 700MW: this will involve enormous cost but experts claim that it is the only rational option if demand for energy (including a margin for export) is to be met and environmental pollution contained.

There is a strong argument for further nuclear power-stations (specifically Cernavoda Units 3–5) on the grounds that pollution will be much smaller than for the thermal equivalent. But this creates a dilemma because retaining the national nuclear programme will involve some substantial investments: Zorcom (for titanium and zirconium ingots), Grantmetal (titanium and alloy pipes) and a fourth heavy water module at the Halinga (Drobeta–Turnu Severin) factory. Continuing the programme will also require subsidies for some enterprises that are central, in varying degrees, to the country's nuclear programme. Four echelons of companies have been identified in terms of priority for state support; for example, through conditioned privatisation (especially Echelons 1 and 2) to maintain a strategic element of nuclear activity: (1) crucial know-how and facilities; (2) major importance in manufacturing; (3) benefit to the state through recovery and profitable operating; and (4) other enterprises connected with the nuclear programme which are unprofitable and have worn-out equipment (Anon. 1998). If the programme does go ahead much capital will have to be borrowed, although some co-funding might be secured. The idea of finance through a Black Sea Power Ring might realise US$10bln over ten to fifteen years in respect of 2100MW of capacity in Units 3–5 placed in the hands of an international company owned by riparian states adhering to the Black Sea Economic Cooperation Treaty. Another option is international bidding.

The new environmental legislation of 1995 to 1996

Further environmental legislation has now been enacted following the early transition years when survival was the most pressing priority. There was clearly a need for much better environmental management including a clear statement of responsibilities to resolve confusion arising from previous directives (Rascu 1993): a decree on pesticides and fungicides issued in 1993 did not specify that the local authorities had an obligation to enforce it. The new legislation (published at the end of 1995 and adopted in 1996) seeks to conserve natural resources and maintain biodiversity by protecting air, soil and water and to safeguard designated protected areas (nature reserves and national parks) and natural monuments as well as settlements. Responsibility for environmental protection, including enforcement of pollution limits, will now rest with the central authority (the Ministry of Waters, Forests and Environmental Protection) and the environmental agencies in each county which will record observations on air and water quality and monitor the situation in the nature reserves. There are other responsible institutions subordinated to the Ministry, such as the Institute for Silviculture (ICAS), while the Romanian Academy and the National Committee of UNESCO will help to decide the criteria for the designation of protected areas.

As well as clarifying the responsibilities of the central administration and local authorities with regard to environmental protection, the new law provides codes for handing all dangerous substances, and it applies the 'polluter pays' principle (PPP): compliance programmes will cover all polluters over a period of five years: thereafter failure to meet standards could result in closure. Provision has also been made for land management because new developments will require an impact study, which will have special significance in sensitive areas. In the case of a large project an *accord de mediu* will have to be drawn up by the Ministry or the local agency, with scope for public debate and local authority resolution of conflicts. Finally, it should be pointed out that the general law is an umbrella under which more specific legislation will shelter. These other laws will deal with such matters as water management, the forestry code and nuclear protection following public consultation, plus a backlog of other issues of environmental importance delayed by shortage of parliamentary time. Action over motor vehicles will soon be necessary in a country where lead-free petrol is currently little used. Substantial improvements may be expected as a result of the new legislative programme, with eventual EU membership providing the stimulus to harmonise with European legislation by the end of 1998.

Implementation

The national programme for action includes 296 projects: 34 per cent short term needing 288bln.lei and 66 per cent medium term costing 1780bln. Some finance will be from PHARE and other EU sources, but the Ministry of Waters, Forests and Environmental Protection strategy envisages an annual investment of 0.6 per cent of GDP (similar to 1994 when 270bln.lei was spent on environmental protection). An alternative variant would double the investment rate and allow major projects to be completed in a shorter period, though this is not realistic in political terms. However, the existence of legislation is no guarantee of enforcement and it may be some time before pollution limits are effective and planning procedures are followed everywhere in an objective fashion. It is too easy to blame inconsistency on corruption: rather there is an ingrained culture of localism which legitimises accommodations which enable local authorities to coexist with their communities at the expense of the national interest. As pluralism advances, even rural societies may become more fragmented and in this context authorities will need to be more neutral in their approach and the more traditional methods of problem solving may disappear in favour of greater standardisation. On the other hand, the intense competition for investment is likely to enhance the discretion shown by local authorities to existing industrial operations and the planning applications of others.

The need for more education and research is an important contextual issue (Jordan and Tomasi 1994). Before the revolution, Romanian schools and universities were covering environmental issues, and for geography in particular this orientation was particularly important in strengthening the links between the human and physical branches of the discipline (Turnock 1982). The environmental emphasis in the schools has become stronger since the revolution and some private companies are operating incentive schemes: a recent Swissair initiative to encourage cooperation over the maintenance of Bucharest's Herastrău Park by teams of schoolchildren offered a trip to Zurich as the first prize. Universities, both state and private, have multiplied since 1989 and environmental teaching is quite common, sometimes in connection with tourism which is another growth area in education.

Energy is being included in the curriculum of some special industrial schools and energy centres are included in the higher education system. A chair in energy and the environment has been created at the Polytechnic University of Bucharest with the support of UNESCO. But mentalities will also have to be changed in the enterprises where managers lack practical experience in adapting strategies to meet realities. A new generation is required to confront problems inherited from the past as well as the new challenges which relate to international business and to the domestic scene, both economic and environmental. Management training programmes concerned with human resources and innovation in restructuring will have implications for an improved R&D performance that will be particularly important for the environmental programme. The Urban Sociology Centre is also involved in perfecting the R&D programme and improved laboratories are being provided.

A coordinated research programme is needed (Jelev 1992) and World Bank funding is being absorbed by the Ministry of Water, Forests and Environment Protection (MWF&EP) through a Biodiversity Conservation Strategy and Action

Plan. The Ministry of Research and Technology has drawn up a national programme (Orizont 2000) for scientific research concerned with technology and environment which has now received government approval. There is a strong environmental component covering environmental quality and global change with interdisciplinary research in the various ecological regions (Vadineanu 1992), including the Danube valley, Danube delta and the Black Sea. Other work is concerned with the protection of river basins with land prone to erosion (Baloi and Ionescu 1986; Traci and Ivana 1990) and the negative effects of irrigation (Varlan 1993). Much research is connected with the Black Sea coast. The extreme south is considered important because the human impact in the past was only slight. In 1980 plans were announced to protect the natural habitat for scientific and recreational use, but no formal arrangements were made, and since then the expansion of the port of Mangalia has caused some deterioration. Costineşti-Olimp has been selected as a location for experimental mariculture: a base will be established for the culture of mussels and other molluscs and fish in cages. In Mamaia Bay there is reduced species diversity as a result of tourist pressure. Research is trying to find methods to repopulate the sandy bottoms. Research is also concerned with the breeding of turbot and flounder. Research on species diversity has been enhanced by exploration of the Movila sulphurous cave discovered near Mangalia in 1986: a small part of an underground sulphurous complex extending under the town. Thirty new species have been identified which get their energy not by photosynthesis but by chemosynthesis with the help of micro-organisms which use sulphuretted hydrogen as an energy source. It is a unique ecosystem isolated from open air life.

There is also a widespread appreciation that Western technology is essential for the solution of pollution problems. In the past, foreign assistance was largely geared to industrial production. This meant that the organisation responsible for pollution monitoring and control equipment (set up in 1978 in Bucharest, with branches in Cluj-Napoca, Galaţi, Ploieşti, Suceava and Timişoara) was thrown back on domestic resources and forced to use outdated technology to produce various types of detection equipment. But in particular Romanian academics have kept a close eye on Western literature, despite foreign currency shortages which have greatly complicated the supply of books. There are several valuable texts which draw heavily on non-Romanian sources (Rosu 1987). There has been involvement with UNESCO programmes (linked with the three biosphere reserves established in 1979) and with the World Heritage Convention. Since the revolution, Romania has rejoined the World Conservation Union (IUCN) following a lapse through most of the 1980s at a time when pressures for great commitment to conservation in the Danube delta were being resisted. Now IUCN's East European Programme is lending strong support to the Romanian initiative in setting up the Danube Delta Biosphere Reserve. Foreign business is also playing a stronger role. There have been discussions with the German firm Lurgi regarding the most serious problems at chemical factories. In addition, Kvaerner, an internationally diversified engineering and construction group registered in Norway, now controls the Bucharest enterprises IMGB and FECNE, and specialises in technologies for environmentally friendly solutions in processing natural resources (woodland, hydrocarbons, hydropower and minerals).

International agreements in Southeastern Europe have seen Romania linked with her neighbours over Black Sea economic cooperation to secure more effective pollution control, preserve natural beauty and encourage tourism. Since 1990 there has been the Bucharest Convention (1992) and the Odessa Declaration (1993) to deal with pollution and environmental protection. Measures are under consideration to protect fishing and vines. International assistance has been involved in the Danube delta (referred to below), and water quality in the Danube is also subject to international agreement (Equipe Cousteau 1993). Romsilva (the Romanian silvicultural authority) has collaborated with Moldova and there is a Romanian–Ukrainian joint geo-ecological survey investigating the northwestern part of the Black Sea. Monitoring is to be extended to the Danube delta, since the main polluting agent of the marine environment in the Black Sea continental shelf is Danube water and sediment which provokes eutrophication (Panin 1992). There is also an intergovernmental commission for the Black Sea operating under UNESCO auspices on which Romania is represented by specialists from the Institute of Marine Research. The international Greenglobe programme seeks to harmonise tourism with the natural environment and there are specific programmes for ecotourism in the delta and the wider Black Sea region.

Water management and forestry

The new Water Law passed in 1995 established principles for water management. Rational water use must involve consideration both of environmental protection and economic development. But there is scope for public participation and the water authorities can only implement their management plans after consultation (except for emergency measures to deal with flood or drought). There will be separate management for each basin through a committee set up under the public corporation Romanian Waters. Each committee will include representatives from local government and local consumers as well as NGOs. The management plan will be implemented through a regime of bonuses and penalties in order to minimise pollution on the basis of PPP. Despite the resources of the natural lakes (1.0bln.cu.m in an average year) and the underground water supplies (9.6), the key resource comprises the national rivers (39.6), rising in the Carpathians, because the massive volume in the Danube (53.2) is peripheral and relatively impure. Hence, the mountains are Romania's natural 'water tower' and to safeguard supplies there should be further research and better management to bring about a reduction in pollution and controlled development of pastoralism, tourism and forestry operations (Zavoianu 1993). The water industry should be restructured 'so as to assure the circulation of first-degree quality water through the mountain space by means of purification stations and by self-purification' (Gastescu 1990, p.92).

Reservoirs are essential to cope with seasonal variations in run-off, but care will be needed to avoid the sedimentation which has been a problem in the

Subcarpathian zone (Tecuci 1993). Small storage should also be encouraged to avoid the disruptive effects of very large projects. At present there are some 400 water storages with a total capacity of 13.0bln.cu.m. Further provision will be linked with an extra 930MW of hydro-capacity (currently 5700MW in total) (Michael 1993). In view of the central position of the mountains there should be gravitational supplies to the adjacent lowlands by underground pipes. At the same time care should be taken to minimise waste, and distribution losses are to be reduced from 40–45 per cent to 15 per cent. Irrigation projects are to be managed so that waterlogged soil, degradation and salting are avoided. Ultimately, the aim is to increase Grade 1 water quality from 54.3 to 90.0 per cent by improved purification and re-equipment of factories (Gastescu 1998a). The share of the urban population receiving piped water and sewerage should increase from 87 to 99 per cent; and in the case of the rural population from 19 to 85 per cent (Ion-Tudor 1996). This will require substantial pipelines such as the one proposed across the Transylvanian Plateau between the lakes of Rastoliţa and Someşu Cald (Zoltan 1996). Other aspects of the water programme include flood protection for 440,000ha in addition to the 2.13mln.ha covered at present; and rehabilitation of irrigation systems (total area 3.2mln.ha) with expansion of irrigation over a further 300,000ha. Irrigation systems should be managed so that waterlogging, soil degradation and salting is avoided.

Forest policy

In 1996 a new legal framework (Codul Silvic) was drawn up for the long-term protection and rational exploitation of the forests through better management (backed by computing and remote sensing) to reduce losses and achieve a lower level of annual cutting. The legislation is quite comprehensive and covers forest strips, trees planted along streams and roads and on grazings; also degraded land plantations, parks and dwarf pine areas in the mountains. The code includes procedures for examining the environmental impact of economic and social activities and anticipates preparation of a law for land management (amenajarea teritoriului). Cutting is being restricted to 13.5–15mln.cu.m (a level reached in 1994) compared with 23–24 in the communist era, but it is hoped that an improved age distribution will not only ensure continuity of wood production but also allow an increase in annual cutting to 16–17mln.cu.m by 2000 to 2005. It is also considered necessary to return to the natural woodland structure (where it has been artificially modified) in order to increase the proportion of more valuable timber. It was common in Europe in the 1950s and 1960s to concentrate on the most productive species and plant pure stands outside the natural zones. The policy of species change has proved counterproductive in the face of severe storms (Stanescu 1996b). Research also showed evidence of disease and die-back, leading to a decision in Romania in 1976 to grow trees appropriate to the natural zones (Chirita 1981), as in Maramureş (Mac and Tudoran 1990). In future species will correspond more closely to natural vegetation. In the plains, where exotics have been relatively unstable, 4000ha will have to be reforested at a cost of 10bln.lei. This effort will require 18mln saplings and more than 1000kg of spruce tree seeds. The degrading process has been stopped (Ureche 1996) and work has started on ecological reconstruction (Simionescu 1998; Stanescu 1996a) including conservation of genetic resources (Enescu 1997).

Work must also be done to restore mixed woodlands which are both ecologically stable and productive. 'The ecological and wood production study of these forests is of considerable value to the afforestation and forest reconstruction planning at regional level' (Abrudan 1997, p.19). There is enhanced soil protection 'because of the different root system of the three species (shallow for spruce, very deep for fir and medium for common beech) the soil resources are better used by the beech–conifer mixtures than by the pure stands. The resistance to windthrow is also higher in mixed stands than in pure ones' (ibid., p.25). Moreover, 'common beech litter reduces surface drainage so the quantity of water infiltrated in the soil is greater than in pure stands of spruce or fir' (ibid.). Romsilva collaborates with research institutes in France and Moldova and research has been carried out on the damaging impact of grazing on woodland (Stoiculescu 1990) which has had a bearing on the new law greatly restricting this practice. Grazing is now forbidden in publicly owned forests (including afforested degraded land and protective forest strips) and stiff penalties are imposed when damage occurs. However, exceptions will be made for flocks in transit; also to provide shelter and allow some routine grazing in well-established parts of the forest. Carts must keep to forest roads. Dwarf pine brushwoods (jnepenişuri) are strictly protected. Private woodland must be managed according to principles laid down by the local forestry office (Ocol Silvic). Silvicultural diseases will be tackled more energetically. Drought, accompanied by intense pollution, calls for a programme of recovery for lowland forests from drying and erosion: Program de Regenerare a Arboreţelor de Salcâm afectate de Uscare. Fast-growing species crucial for the cellulose and paper industries will be encouraged. The hunting of wild animals will be managed on a sustainable basis and conserving biodiversity will call for protected areas amounting to 8–10 per cent of forest stock, including many new protected zones in the hill and plain regions.

Forest replanting amounted to only 100–300ha p.a. after the revolution. Regeneration has increased from 18,854ha in 1993 and 24,010 in 1994 to an estimated 25,500ha in 1996: 13,500ha natural regeneration and 12,000ha plantation (Salagean 1996) and, following the removal of fallen timber uprooted by the storms of the winter of 1995 to 1996, plans have been drawn up for regeneration, with allowance for snow breaks and priority for trees which are relatively resistant to windblow and can therefore protect other species in vulnerable areas (beech, larch, maple, rowan and sycamore). It is necessary for rational exploitation to bring all forests into use except where protection is the overriding consideration. There was interest in the communist period in 'whole tree harvesting', but the transport costs were too high and policy therefore seeks sustainable cutting in all commercial stands. Only 65 per cent of the forest stock is accessible, which means than 2.2mn.ha of forest cannot be effectively exploited. At present Romania has only 39,200km of forest roads: 6.1m/ha, compared with the norm of 20–30 in the EU (45 in Germany, 44 in Switzerland, 26 in France). It was appreciated before the revolution that road access was unsatisfactory (Bereziuc 1987). The optimum density for Romania is considered to be 12–15m/ha, which means that another 40,000km of roads are needed (Madaras 1998). This will have be arranged through better organisation by each Ocol Silvic and by a special fund for new roads. Meanwhile, there will continue to be damage as logs are dragged overland and along streams. Moreover, inadequate roads encourage clear felling in the more accessible areas and this practice is being challenged in the Căliman Mountains by the Rhododendron Environmental Association on the grounds that this has not been sanctioned by the Environment Ministry.

In addition to large forest stands there will be provision for degraded land plantations, blocks of woodland on grazing land, and belts of trees along streams, railways and roads (with action to ensure adequate management of those belts already in existence). For many years there have been programmes to protect eroded land (Baloi and Ionescu 1986) and planning has extended across whole river basins prone to erosion. Regulation of torrential streams (50km in 1993 and 81km in 1994) is following traditional methods (Traci and Ivana 1990). Such work enhances production because soil losses and torrential floods are reduced and employment maximised (Traci and Ivana 1990). There are also improvements regarding tourism and aesthetics (Clinciu and Lazar 1997). But now the foresters have been collaborating with the Agriculture Ministry to explore ways of taking over eroded lands that cannot be efficiently farmed. Meanwhile, there is also greater activity in treating both degraded lands (1506ha in 1993 and 1530 in 1994) – ultimately 3mln.ha of degraded land (2mln without any vegetation). Moreover, in 1994 4000ha of agricultural land were identified as being suitable for planting: 780ha were planted in 1994 and 900ha approved for 1995. Private woodland (which now accounts for 5 per cent of the total) must be managed according to 'Amenajamentul Ocolului Silvic' in each district. Private owners requesting help will be given free seed. But in critical areas protective strips will be considered as public works (lucrare de utilitate publică). However, more generous restitution was approved in 1999 and this will provide woodland holdings of up to 10ha for former owners and society members. Hence there is a danger that without rigorous land-use planning and enforcement procedures more forest will be lost through owners seeking to realise the value of their assets.

Approximating the EU environmental *acquis*

Adopting European directives is now a dominant issue in environmental policy. Romania's progress in assimilating the environmental *acquis* is charted through a series of regular reports published by the European Commission since 1998, when two new directorates were set up under the Ministry of Waters, Forests and Environmental Protection (MWF&EP): for the preservation of biological diversity and management of protected areas; and for European integration and international relations. However, comprehensive reorganisation was considered necessary, particularly with regard to the organisational structure for enforcement, since the monitoring of air and water was far behind EU requirements for the most part. In 1999 limited progress was reported, though there were still serious problems of air protection, water management and waste management. It was also considered problematic that Romania 'still lacks an overall environmental strategy that includes directive specific cost assessments implementation and financing plans'. There was no comprehensive policy approach and the MWF&EP lacked capacity at a time when greater EU funding was becoming available through ISPA and SAPARD: this was thought to be the result of its limited consultative role over environmental matters involving other ministries. However, a committee combining the ministries responsible for agriculture, environment, industry, physical planning and transport was helping to integrate environmental protection requirements into other sector policies. It was also conceded that priority for the accession process was evident in the National Programme for the Adoption of the Acquis (adopted in 2000 for the period 2000 to 2003) and shows that further transposition of the *acquis* is a key priority for all sectoral ministries.

Administrative capacity is still considered low in view of the accidents in Maramureş during 2000. While the MWF&EP needs restructuring, the protection agencies (EPAs) which operate in forty-two counties have been strengthened by wider regional responsibilities in eight instances. And by charging for their services the EPAs can obtain additional funding to use on monitoring and laboratory equipment to enhance their effectiveness over inspection and enforcement. An environmental fund has been set up to sustain public investment in the environment although there is not sufficient legislation for it to become operational in terms of the PPP. Furthermore, there is a lack of information on the cost of full compliance with the environmental *acquis* – especially in sectors like air, water and waste. Costs in these sectors have been estimated by the EU at Euro20bln, although ISPA funding – worth E208–270mln annually between 2000 and 2006 – will help with drinking-water, waste water treatment, solid waste management and air pollution, including the Piatra Neamţ Waste Management Programme and Constanţa Waste Water Treatment Plan. However, the National Action Plan for Environmental Protection has been updated and most sectoral approximation strategies have been finalised. Moreover, several reforms have been introduced during 2000, including the ratification of the Århus Convention on public rights to information, justice and decision making in environmental matters: secondary legislation is needed to ensure implementation. And several acts have been issued but financial plans for implementation remain weak. These cover the conservation of natural habitats, including ratification of international conventions on the conservation of endangered species; industrial pollution control and risk management; genetically modified organisms; and chemical substances that deplete the ozone layer. Also an emergency ordinance regulating the handling of waste oils, used batteries and packaging materials – as well as waste incineration, shipment and landfill – should accelerate the assmilation of EU waste legislation which has so far been neglected. But action is awaited on air and water quality as well as nuclear safety and radiation protection.

Protected areas: nature reserves and national parks

'Protection of rare species and representative vegetation types is essential at a time of increasing anthropic pressure. Such action must be combined with measures to maintain landscape diversity and safeguard wildlife habitats' (Muica and Popova-Cucu 1993, p.15). Another important feature of the legislative programme of 1995 to 1996 is the codification of the responsibilities of local government for environmental protection including nature reserves and national parks (Figure 17.3). A National Network of Natural Protected Areas includes 810 separate areas covering 1.13mln.ha or 4.8 per cent of Romanian territory. Romania has participated in UNESCO's Man and Biosphere programme (Soran and Borcea 1985) with increased enthusiasm since 1989 and the list is now headed by three internationally recognised biosphere reserves: Danube Delta (also a World Heritage and Ramsar site), Pietrosul Rodnei and Retezat (Table 17.2). The latter is a long-established national park comprising the Gemenele scientific reserve (1630ha) surrounded by an extensive buffer which has been progressively extended to its present size of 42,476ha (Figure 17.4). Fifty-two per cent of the area is forested while much of the remainder consists of an Alpine zone with dwarf pine, meadows and scree slopes. There are more than eighty lakes, mostly glacial (the largest being Bucura, with an 8.9ha surface and a depth of 14.7m) (Pascu 1984). Since 1990 restitution means that some woodland in the buffer zone will become privately owned but not that in the core. Tourism has increased (regulated by NGOs) while the local councils (who manage the grazing areas) have been allowing excessive numbers of sheep to enter the park. The World Bank is supporting the conservation effort in Retezat and also in two other areas (the Piatra Craiului mountains and the forest park of Vânători Neamţ) through a US$6.6mln grant provided under a joint funding arrangement for the conservation of biodiversity. In 1997 Geomatics International of Canada began

Table 17.2 Protected areas 1996

Type of area	*Number*	*Area (th.ha)*
Biosphere reservations	3	583.6
National parks[a]	14	376.4
Scientific reservations	40	53.9
Reservations of nature preservation	573	118.2
Natural monuments	180	2.7
Total	810	1134.8

Source: *Anuarul Statistic al Romaniei* 1997.

Note
a National parks do not yet have management regimes.

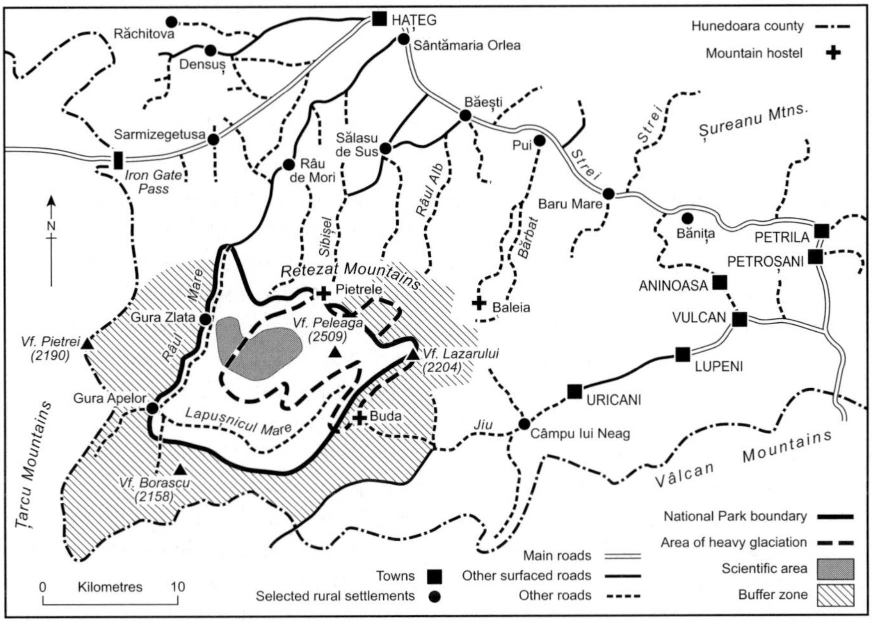

Figure 17.4 Retezat National Park, Romania

Source: Romsilva, Deva.

work on a Global Environmental Fund (GEF) project for Integrated Protected Areas and Conservation Management: one aim is a proper administrative structure for Retezat National Park and support for a management plan (Stanciu 1998). This will be very necessary if the road link between Băile Herculane and Petroşani is completed across the Cerna–Jiu watershed.

Romania's principal lowland reserve: the Danube Delta (Figure 17.5)

The conservation value of the delta arises from the fact that the river and wetlands constitute a major habitat which should be preserved in the interests of biodiversity. Twenty-nine separate ecosystems (Gastescu and Oltean 1997) involve a range of habitats and wealth of flora and fauna. Studies between 1991 and 1994 have revealed some 5200 plant and animal species, many noted for the first time and some new to Romania. The delta is a real paradise for fish and birds. There are more than ninety species of fresh, brackish and salt-water fish, both sedentary and migratory (including several species of sturgeon): 'of the 64 fish species found over the last three years two were new to Romania and one was new to science' (Petranu 1997, p.164). There are also over 300 species of birds, 176 of which breed in the delta, involving colonies running into tens of thousands, notably terns and herons. Several globally threatened species are found: white pelican (Pelicanus onocrotalus); pygmy cormorants (Phalacrocorax

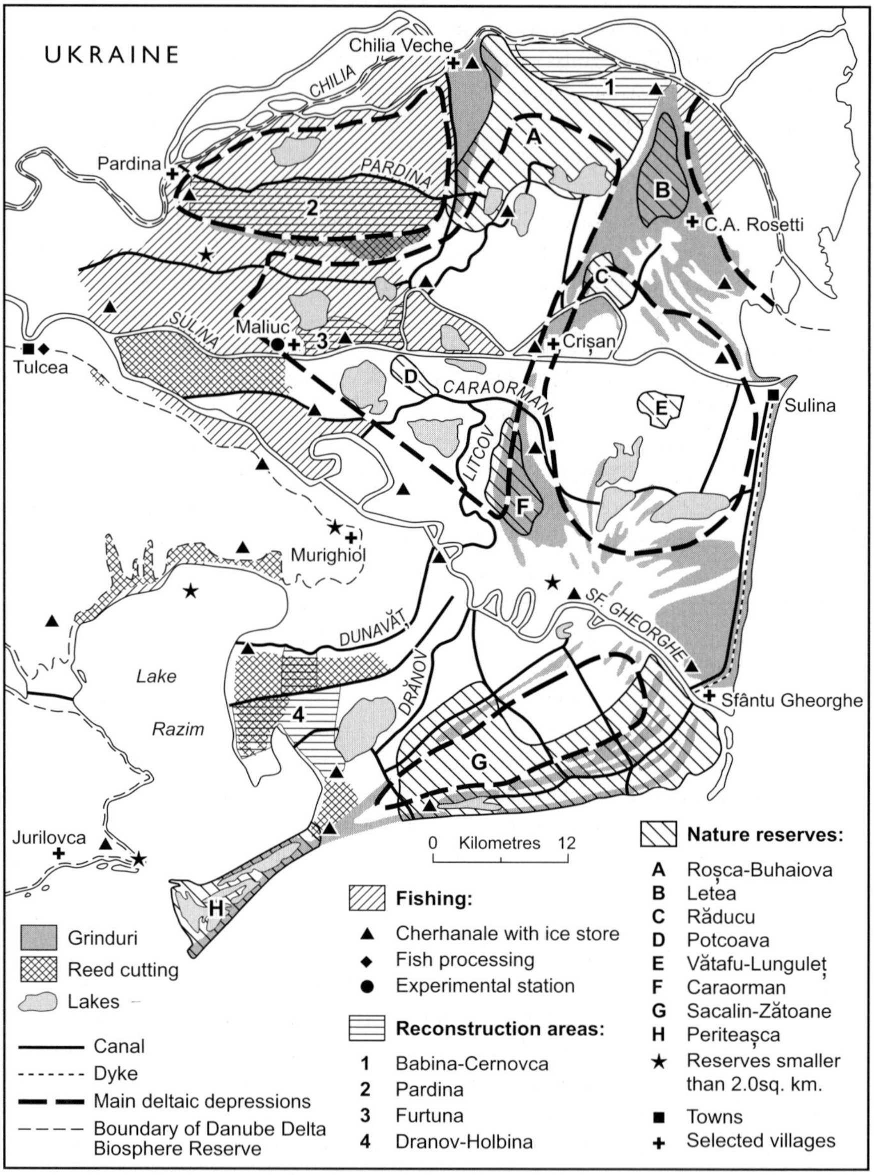

Figure 17.5 Danube Delta Biosphere Reserve, Romania

Source: Gastescu 1992, 1996.

pygmeus); red-breasted geese (Branta ruficollis) and slender-billed curlews (Numenuis tenvirostris). The delta is also important in the Eurasian context for bird migration and nesting. Some rare birds are present at certain times of the year (common pelican, red-necked duck and white egret) and there are colonies of mink, otter and pheasant which are in danger of extinction. Wild boars are an important component of the fauna although they are not so vulnerable: they cause considerable damage but the figure of thirty-eight animals actually hunted in 1994 to 1995 was way below the quota of 245: hunters were therefore encouraged to hunt these animals during the 1995 to 1996 season.

Several reserves were established in 1940 (Letea forest and Roşca-Buhaiova) but with few restrictions. Further reserves were created after 1956 as a result of the establishment of a Natural Monuments Commission within the Romanian Academy. The total protected area increased to 40,000ha. But these areas were affected by the changes in ecological equilibrium arising from development measures to extend agriculture, fish farming and forestry (Gastescu 1991) and better protection was therefore required. Hence the formation of the Danube Delta Biosphere Reserve (DDBR) by Act of Parliament in 1991, confirmed by UNESCO in 1991 (Gastescu 1992, 1993) and an administrative regime for the Danube Delta Biosphere Reserve Authority (DDBRA) was put in place in 1993. In addition to becoming part of the UNESCO Man and Biosphere Programme, parts of the reserve were also listed under the Ramsar Convention on Wetlands of International Importance in 1991, and over half of its area was placed on the World Heritage List later in the same year. The administration has set out to assess the ecological status of natural resources, designate functional zones and restore damaged ecosystems. It is also acting as a regional environmental agency and collaborates with local authorities to protect the interests of the local population, and it grants concessions for use of renewable resources. It is cooperating with national and international bodies to promote research and exchange of information. Other responsibilities include ecological education, the monitoring of waste disposal and the regulation of navigation in cooperation with the Ministry of Tourism. Local authorities will monitor the species inhabiting the reserve and a database, using a geographic information system, will be set up for the drafting of management projects.

A management structure has been set up (following Goriup 1995) along with a Research Scientific Council (which is preparing plans for ecological reconstruction) and an environment quality survey unit. Management includes notions of long-term qualitative/quantitative capacity with planning procedures; integrated research/monitoring systems; educational materials for local population and visitor awareness; and international support. A regime of strict protection applies to eighteen zones (totalling 506sq.km or 8.7 per cent of the DDBR) with little-disturbed ecosystems: most are wetlands but there are some woodlands. Then there are thirteen buffer zones for the protection area totalling 2233sq.km (38.5 per cent) and 3061sq.km of economic zones comprising lakes and channels, bars and embankments, agriculture-forestry-fishing areas and settlements. The emphasis throughout is on sustainable management (Vadineanu 1997) which means farming without fertilisers and clearly specifying hunting and fishing areas in the interests of conserving biodiversity. But traditional activities are supported in the interests of improved living standards for the local inhabitants. Three per cent of the delta is set aside for ecological restoration (Gastescu 1996). Management is restoring the polders of Babina and Cernovca for wildlife. These areas were earmarked for rice cultivation but only dikes had been constructed by the end of 1989. There is also restoration of over 3000ha at Dranov-Holbina which was developed for fishing but is now abandoned. Furtuna is a further reconstruction area and the southern part of Pardina has been proposed by Petranu (1997, p.166), although this is not official. Water preservation strategies are another essential component of the DDBR management plan (Pascariu 1997).

It is also important to consider the wider issue of the Danube River (Gastescu 1998b) which has been greatly modified between 1965 and 1987 when about 80 per cent of the Romanian Danube floodplain was taken out of the natural flood regime. There was a decrease in diversity because channels and lakes, with their great abundance of aquatic plants, reed beds and floating mats (plaur), willow and poplar forest, were badly affected by draining and diking and the breaking up of natural meadows. This complicated the work of flood relief, while the building of dams contributed to a lowering of the water-table which affected ground water resources. Management of the DDBR should be reinforced by measures upstream to control the nutrient sources in the river catchment and rehabilitate floodplain wetlands. In 1998 a project was launched for Culoarul Verde al Dunării based on cooperation between Bulgaria, Hungary, Moldova, Romania and Ukraine; with input from research institutes in Austria (Vienna), Germany (Rastatt) and The Netherlands (Lelystad). The World Wide Fund for Nature (WWF) has convened a symposium on floodplain drainage and is considering the possibility of converting some drained areas that are not well used for agriculture. This would involve the demolition of dikes and the restoration of green land utilisation. The current protected area of 773.2th.ha is to be extended by 300.00ha of wetland, lakes, meadows and forest which will provide business opportunities for fishing and tourism. Action has now been taken over water quality (Equipe Cousteau 1993) and it is recommended that construction of further dams should be prohibited; also that bypasses at existing dams should guarantee the passage of minimum quantities of water required for ecological reasons. A Danube Basin Ecological Convention is currently in draft form and attention is also focusing on the need to upgrade Black Sea water quality (Kharitonov 1997; Shilov 1997).

Conclusion

In comparison with the extreme pressures of public opinion over issues such as the Gabčikovo–Nagymaros hydropower project in Hungary and the nuclear power debate in Poland, it may seem that Romanians are relatively indifferent to environmental issues. It may be that further education is needed to stimulate greater public interest and supervision. For it is interesting to see that the Danube delta, which attracted a great deal of national and international attention, has been the subject of legislation and is now modestly funded. By contrast, the national parks (although strongly supported by groups of professionals, including biologists, silviculturists, geographers and speleologists) have made relatively modest progress since the original intentions were noted. Progress has been made along a fairly broad front and pollution levels are rarely excessive outside the blackspots, where much work has been done over the past four years. However, the complexity of the Romanian transition means that careful consideration must always be given to job security.

The high cost of an improved environment in East Central Europe generally (ECU300bln over fifteen years) could put a brake on progress towards EU membership. EU environmental legislation is complex and specialised, making it difficult for all candidate countries to fall into line. And since the population is tired of shock therapy after the traumas of the Ceauşescu era, the government has read the public mood correctly in maintaining a gradual reform process. Limited resourcing for environmental action has however enabled policy action to move in the right direction with international cooperation and some environmental education to stimulate public awareness. However, much more needs to be done, since the European Parliament's 'Resolution on Romania' in October 2000 emphasised the need for enforcement of the 'strictest European criteria' in the environmental domain when the reopening of enterprises in the Baia Mare area is considered. It also asks the Romanian authorities to remove obstacles in the field of environmental protection in general.

References

Abrudan, I. V. 1997, 'Ecological and wood-production investigations of the semi-natural beech-conifer mixtures in the Upper Draganul watershed (northwest Romania)' (High Wycombe: Forest Products Research Centre, Buckinghamshire College/Brunel University Ph.D. thesis).

Andeoca, C. and Chirculescu, I. 1976, *Economie agrară* (Bucharest: Editura Didactică si Pedagogică).

Anon. 1998, 'Romania's nuclear programme should continue'. *Romanian Business Journal* 5(51–52), pp.8–9.

Armas, I. 1991–92, 'Aspects of the anthropic landforms in Romania and its mapping'. *Studia Universitatis Babeş-Bolyai: Geographia* 40–1, pp.91–7.

Baloi, V. and Ionescu, V. 1986, *Apararea terenurilor agricole impotriva eroziunii alunecarilor si inundatiilor* (Bucharest: Editura Ceres).

Balteanu, D. 1992, 'Natural hazards in Romania'. *Revue Roumaine de Geographie* 36, pp.47–55.

Balutescu, O. 1998, 'Consideratii geografice privind organizarea spatiului agrar in perimetrul bazinului minier Rovinari', in D. Buga and C. Talanga (eds) *Gorjul geografic I* (Targu Jiu: Societatea de Geografie din Romania Filiala Gorj), pp.83–5.

Bereziuc, R. 1987, *Retele de drumuri forestiere* (Bucharest: Editura Ceres).

Bleahu, M. 1998, 'Turismul si protectie peisayului', in M. Bleahu, *Ecologie nature om* (Bucharest: Editura Metropol), pp.125–36.

Bold, I. 1974, *Sistematizarea rurala* (Bucharest: Ed.Tehnica).

Candea, M. 1993, 'The coal exploitation influence upon the geographic environment in the Motru coalfield', in C. Muica and D. Turnock (eds) *Geography and Conservation: Proceedings of the second Romanian–British Seminar* (Bucharest: Geography Institute, International Geographical Seminars 1), pp.99–102.

Chirita, C. 1981, *Pădurile româniei* (Bucharest: Editura Academiei RSR).

Ciulache, S. and Ionac, N. 1995, 'Public control of the rural environment: a case study', in D. Turnock (ed.) *Rural Change in Romania* (Leicester: Leicester University Geography Department Occasional Paper 33), pp.78–81.

Clinciu, I. and Lazar, N. 1997, 'Efectele lucrărilor de amenajare a bazinelor hidrografice torentiale', *Revista de Silvicultura a Sud-Estului Transilvaniei* 2(2), pp.23–5.

Constantinescu, N. N. 1980, 'Protecţia mediului natural şi productivitatea muncii sociale', *Ocrotirea Naturii şi Mediul Înconjurător* 24(1), pp.9–12.

Cucu, M. 1992, 'Poluarea mediului şi sanatatea in Romania', *Mediul Înconjurător* 3(1), pp.23–5.

Defour, D. and Boucher, I. 1977, *Sistematizarea localitaţilor rurale* (Bucharest: Ed.Ceres).

Dinu, A. 1994, 'Country profile: Romania', *IUCN/World Conservation Union East Europe Newsletter* 13, pp.6–8.

Dragomirescu, S., Muica, C. and Turnock, D. 1998, 'Envionmental action during Romania's early transition years', *Environmental Politics* 7, pp.162–82.

Dumitrache, L. and Armas, I. 1998, 'The health state of the Romanian population during the transition period', *GeoJournal* 44, pp.151–60.

Enache, L. 1994, 'The quality of environment in Romania', in P. Jordan and E. Tomasi (eds) *Zustand und Perspecktiven der Unwelt im Ostlichen Europa* (Vienna: Peter Lang Europaischer Verlag/Wiener Osteuropa Studien), pp.131–57.

Enescu, V. 1997, *Conservarea biodiversitatii şi a resurselor genetic forestiere* (Bucharest: SC Agris).

Equipe Cousteau 1993, 'Environmental programme for the Danube river basin', *Mediul Inconjurator* 4(2), pp.21–32.

Gastescu, P. 1990, 'Water resources in the Romanian Carpathians and their management', *Revue Roumaine de Geographie* 34, 85–92.

Gastescu, P. 1991, 'Les transformations morpho-hydrographiques et l'equilibre écologique dans le Delta du Danube' *Revue Roumaine de Géographie* 35, 37–43.

Gastescu, P. 1992, 'The Danube delta: biosphere reserve', *Revue Roumaine de Géographie* 36, pp.71–80.

Gastescu, P. 1993, 'The Danube delta – a biosphere reserve: principles of organization legal framework boundaries and structure', in C. Muica and D. Turnock (eds) *Geography and Conservation: Proceedings of the Second Romanian–British Seminar* (Bucharest: Geography Institute, International Geographical Seminars 1), pp.109–18.

Gastescu, P. 1996, 'The Danube delta biosphere reserve: present state and management', *Revue Roumaine de Géographie* 40, pp.27–33.

Gastescu, P. 1998a, 'Bucharest municipality: its periurban zone water supply situation', in I. Iordan *et al.* (eds) *Socio-economic Changes in the Suburban Areas of Large Cities in Romania and Poland* (Bucharest: Academia Română Institul de Geografie), pp.41–5.

Gastescu, P. 1998b, 'Danube River', in R. W. Fairbridge (ed.) *Encyclopaedia of Hydrology and Water Resources* (Dordrecht: Kluwer), pp.159–61.

Gastescu, P. and Oltean, M. 1997, *Ecosystems Map of the DDBR* (Bucharest: Academia Romana).

Goriup, P. (ed.) 1995, *Management Objectives for Biodiversity Conservation and Sustainable Development in the Danube Delta Biosphere Reserve, Romania* (Newbury: Nature Conservation Bureau for the DDBR Authority).

Hancu, S. 1990, 'Calitatea mediului înconjurător in Romania: perspective de îmbunătăţire', *Mediul Înconjurător* 1(1), 5–9.

Harjoaba, I. and Conachi, M. 1993, 'Trotuşul: un râu sacrifiat', *Terra* 25(1–4), pp.38–42.

Herseni, T. 1979, 'Ecologia socială cultură mentală', in M. Berneasi and A. Calciu (eds) *Ecologia umană* (Bucharest: Meridiane), pp.325–90.

Iacob, G. 1994, 'Regiunea metalurgică Baia Mare: continuitate diversitate şi implicare în mediul înconjurător', in P. Urdea *et al.* (eds) *The First Regional Geography Conference: Geographical Researches in the Carpathian–Danube Space* (Timişoara: Universitatea de Vest din Timişoara, Deparamentul de Geografie), pp.269–74.

Ianos, G. and Florea, N. 1996, 'Aspecte privind sustenabilitatea terenurilor din zona periurbană a municipiului Timişoara in vederea dezvoltării durabile a teritoriului', *Analele Universităţii de Vest din Timişoara* 6, pp.67–76.

Ion-Tudor, C. 1996, 'Development strategies for waters, forests and environment protection', *Romanian Business Journal* 3(4), p.11.

Ipatiov, F. 1996, 'Influenţa poluarii industriale asupra starii de sanitate populaţiei în oraşul Baia Mare', *Studia Universitatis Babeş-Bolyai: Seria Geographia* 41, pp.175–82.

Jelev, I. 1992, 'Programul unificat de cercetare în domeniul protecţiei mediului în România pentru perioadă 1991–1995', *Mediul Înconjurător* 3(2), pp.3–10.

Jordan, P. and Tomasi, E. 1994, 'Environmental management, education and policy in Romania', in P. Jordan and E. Tomasi (eds) *Zustand und Perspecktiven der Unwelt im Ostlichen Europa* (Vienna: Peter Lang Europaischer Verlag/Wiener Osteuropa Studien), pp.159–69.

Kharitonor, I. A. 1997, 'Integrated coastal zone management around the Black Sea', in U. Graine (ed.) *Sustainable Development for Central and Eastern Europe* (Berlin: Springer), pp.209–32.

Mac, I. and Ripeanu, M. 1996, 'Changes in the Romanian landscape as a result of voluntarist activities', *Studia Universitatis Babeş-Bolyai: Seria Geographia* 41, pp.32–7.

Mac, I. and Tudoran, P. 1990, 'Etajul supraforestier din Munţii vulcanici Oaş-Guţâi-Ţibleş', *Studia Universitatis Babeş-Bolyai: Seria Geographia* 35(2), pp.30–3.

Madaras, I. 1998, 'Optimizarea reţelei de drumuri auto forestiere din România: criterii ecologice şi parametri economico-sociale', *Revista Pâdurilor* 113(1), pp.42–5.

Michael, I. 1993, 'Aspecte ale corelarii amenajarilor hidroenergetice cu mediul inconjurator', *Mediul Înconjurător* 4(1), pp.67–9.

Mihailescu, I. and Ciobanu, C. 1990, 'Poluarea industrială a solurilor şi a vegetaţiei forestiere în Zona Zlatna', *Revista Pădurilor* 105, pp.129–35.

Ministerul Apelor Padurilor si Protectiei Mediului (MAPPM) 1994, *Starea factorilor de mediu în România* (Bucharest: MAPPM).

Muica, C. and Popora-Cucu, A. 1993 'The composition and conservation of Romania's plant cover', *GeoJournal* 29, 9–18.'

Panin, N. 1992, 'Situaţia ecologică în partea de nord vest a Marii Negre', *Mediul Înconjurător* 3(1), pp.27–34.

Pascariu, S. 1997, 'Water preservation strategies and the regional plan for the DDBR' in U. Graute (ed.) *Sustainable Development for Central and Eastern Europe* (Berlin: Springer Verlag), pp.145–62.

Pascu, S. 1984, *Recherches ecologiques dans la parc national de Retezat* (Cluj-Napoca: Academie de la RSR).

Patterson, W. 1994, *Rebuilding Romania: Energy Efficiency and the Economic Transition* (London: Earthscan Publications).

Petranu, A. 1997, *Black Sea Biological Diversity: Romania* (New York: UN Publications GEF Black Sea Environmental Programme).

Phillips, D. and Groenewegen, P. 1998, 'Health environment and development: issues in developing and transitional countries', *GeoJournal* 44, pp.97–102.

Pop, G. P. and Bodocan, V. 1991, 'Opţiuni electorale pentru allegerea primarilor in Banţat, Crişana-Maramureş şi Transilvania'. Studia Universitatis Babeş-Bolyai: *Geographia* 36(2), pp.66–75.

Puia, I. 1977, 'Probleme ale agriculturii ecologice', in I. Ceauşescu and I. Ionescu (eds) *Probleme ale agriculturii contemporane* (Bucharest: Editura Ceres), pp.23–37.

Puia, I. 1979, *Agroecosistemul bioiproductivitatea şi alimentaţia* (Bucharest: Editura Academiei RSR).

Puia, I. and Soran, V. 1986, 'Conceptul de agrosistem şi agricultura ecologică', *Forum* 28(10), pp.55–68.

Rascu, P. E. 1993, 'Aspecte şi propuneri în legatura cu degradarea patrimoniului natural şi cultural-istoric în România', *Mediul Înconjurător* 4(1), pp.19–24.

Rosu, A. 1987, *Terra-Geosistemul-Vieţii* (Bucharest: Editura Ştiinţifică şi Enciclopedică).

Salagean, L. 1996, 'Forest Code: fundamental forest law in Romania', *Romanian Business Journal* 3(27), p.11.

Serban, R. 1993, 'Impactul uzinei "Ampellum" Zlatna asupra calitaţii aerului: prezent si perspective', *Mediul Înconjurător* 4(1), pp.41–8.

Shilor, A. S. 1997, 'Ecological prospects for the Black Sea region', in U. Graute (ed.) *Sustainable Development for Central and Eastern Europe* (Berlin: Springer Verlag), pp.201–8.

Simionescu, A. 1998, 'Starea de sanatate a pădurilor din Romania 1995 si 1996', *Revista Pădurilor* 113(1), pp.26–31.

Soran, V. and Borcea, M. 1985. *Omul şi biosfera: etica protejarii natura* (Bucharest: Ştiinţifică şi Enciclopedică).

Stanciu, E. 1998, *Retezat National Park: Short Presentation* (Deva: Romsilva).

Stanescu, V. 1996a, 'Din nou despre "reconstrucţie ecologică" a pădurilor', *Revista Pădurilor* 111(4), pp.21–6.

Stanescu, V. 1996b, 'Observaţii si propuneri privind doboraturile de vânt din pădurile judeţului Covasna', *Revista de Silvicultura a Sud-Estului Transilvaniei* 1(1), pp.4–6.

Stoiculescu, C. D. 1990, 'Influenţa exercitata de păsunat asupra radacinilor fine de gorun: factor de deterioare a echilibrului ecologic', *Revista Pădurilor* 105, pp.127–8.

Surdeanu, V. and Ichim, I. 1991, 'Alunecarile de teren din bazinele subcarpatice ale rîurilor Rîmnicu Sărat şi Rîmna ca surse de aluviuni', *Terra* 23(2–4), pp.33–6.

Tecuci, I. 1993, 'Soluţii constructive şi masuri pentru atenuarea fenomenului de colmatare a lacurilor de acumulare', *Mediul Înconjurător* 4(2), pp.51–3.

Traci, C. and Ivana, S. 1990, 'Efectele lucrărilor de amenajare în bazinul hidrografic al torentului Valea lui Bogdan (Prahova)', *Revista Pădurilor* 105, pp.136–42.

Turnock, D. 1982, 'Romanian geography reunited: the integrative approach demonstrated by the conservation movement', *GeoJournal* 6, pp.419–31.

Turnock, D. 1991, 'Romanian villages: rural planning under communism', *Rural History: Economy Society Culture* 2, pp.81–112.

Turnock, D. 1993, 'Romania', in F. W. Carter and D. Turnock (eds) *Environmental Problems in Eastern Europe* (London: Routledge), pp.135–63.

Ureche, M. 1996, 'Valorificarea prin impădurire a terenurilor cu fenomene de degradare din Ocolul Silvic Sălişte', *Revista de Silvicultura a Sud-Estului Transilvaniei* 1(3), pp.37–9.

Vadineanu, A. 1992, 'Priorities for environmental protection in Romania', in A. Vadineanu 1997, 'Changes and opportunities for integrated management of the Razim-Sinoe lagoon system', *International Journal of Salt Lake Research* 6(2), pp.135–44.

Varlan, M. 1993, 'An example of modified geosystem in the Moldavian Plain: the irrigated perimeter Tansa-Belcesti', in C. Muica and D. Turnock (eds) *Geography and Conservation* (Bucharest: Geography Institute, International Geographical Seminars 1), pp.33–8.

Vasilescu, A. 1963, *Gurghiu–Harghita: calauza turistului* (Bucharest: Ed. Uniunii de Cultura Fizica şi Sport).

Vasilov, M. and Ichim, R. I. 1993, 'Efecte ale poluarii aerului cu pulberi de ciment asupra sanitatii scolarilor expusi', *Mediul Înconjurător* 4(1), pp.15–17.

Zavoianu, I. 1993, 'Romania's water resources and their use', *GeoJournal* 29, pp.19–30.

Zoltan, I. 1996, 'Sistemele de alimentare cu apă din Câmpia Transilvaniei: prezent şi perspective', in C. Vert (ed.) *Proceedings of the Second Regional Geography Conference: Geographic Researches in the Carpathian–Danube Space* (Timisoara: Universitatea de Vest din Timişoara Departamental de Geographie), pp.173–8.

18 Yugoslavia

Richard Clarke

Introduction

Present-day Yugoslavia covers the territory of what was left of the Socialist Federal Republic of Yugoslavia *Socialistička Federativna Republika Jugoslavija* (SFRJ) following the secession, from late 1991, first of Slovenia, then, successively, of Croatia, Bosnia and Hercegovina, and finally, Macedonia. This 'rump' – the Federal Republic of Yugoslavia *Savezna Republika Jugoslavija* (SRJ) – consists constitutionally of two sovereign republics, Serbia and Montenegro (Figure 18.1). Each has a separate government, legal and administrative system within the Federal constitution. They are often separately represented at international fora: within SRJ their relationship is uneasy and its future uncertain. Serbia today includes the former (SFRJ) 'autonomous provinces' of Vojvodina to the north and Kosovo to the south. Since 1987, both were progressively assimilated – administratively and politically – into the Republic of Serbia and were formally stripped of their autonomy under a new constitution adopted by Serbia in September 1990. Both SRJ and its constituent entities have uncertain status in international law. The declaration in April 1992 by Serbia and Montenegro that SRJ was the legal successor of the SFRJ was a *de facto* recognition of the secession of the other four republics. However, the United Nations ruled in September of that year that this could not automatically be the case and excluded SRJ from the General Assembly; subsequently the recognition of SRJ by other nations has been uncertain. Kosovo is presently under military control of NATO (and Russian) armed forces (KFOR), its administration in the hands of a United Nations mission (UNMIK); its future can only be a matter of conjecture.

Examination of environmental issues in Yugoslavia must be informed by two principal considerations: (1) the physical and ecological characteristics of the region, and its social and economic development up to and including the collapse of the former Yugoslavia in 1991; and (2) events since 1991, including socio-economic changes, the effect of external sanctions consequent on Yugoslavia's involvement in the civil war in neighbouring Bosnia and Hercegovina (1992–95) and, most recently, the civil war in Kosovo and the intervention of NATO. The latter, in particular, cast a shadow over any analysis of Yugoslavia and its future, including the matters dealt with in this chapter, which therefore includes

Figure 18.1 Salient features of Yugoslavia

an assessment of environmental damage and prospects for environmental remediation against the backcloth of an analysis of the pre-1999 situation in the region.

Physical and ecological features

At a little over 102,000km^2, SRJ covers less than half the size of the former Yugoslavia. Serbia covers 88,361km^2, is the larger of the two states and is landlocked; the Danube (whose catchment covers almost 80 per cent of the country) provides navigable access through Romania to the Black Sea. Montenegro covers 13,812km^2 and possesses some 200km of coastline, including the Bay of Kotor, on the Adriatic. SRJ's terrestrial borders to the west are with Bosnia-Hercegovina and Croatia, Hungary to the north, Romania (northwest), Bulgaria (southwest) and with Macedonia and Albania to the south.

Much of the north of the country comprises rich and fertile plains, derived to significant degree from anthropogenic drainage, and characterised by a continental climatic regime. Limestone ranges characterise the east and central parts of the area. Older mountains rise from coastal plains in the southeast, dominated by Mediterranean climate. Vegetation patterns are based on this geomorphological and climatic variation and fall broadly into three major biogeographical regions: Pannonian, montane and coastal. The Balkans as a whole have long been recognised as of exceptional value for biodiversity; within this SRJ is one of the richest components, home to over one-third of European angiosperm species, two-thirds of its birds and mammals and half of its fish species. Endemic species comprise over 9 per cent of the flora.

Economic and social background

The total population of SRJ in the 1991 census was 10.6 million. Some two-thirds of the people of Serbia (9.8 million) have Serb ancestry, the remainder reporting Croat, Hungarian, Montenegrin, Romanian and Slovak nationality. Some two-thirds of the population of Montenegro (0.8 million) are Montenegrins, with other groups including Albanians, Muslim Slavs and Serbs. Differential birth rates and the departure of Kosovan Serbs mean that the ethnic Albanian population of Kosovo has grown from under 60 per cent in 1945 to between 70 to 80 per cent just prior to NATO bombing. Kosovo's population remains the youngest (with around 50 per cent below the age of 20) and most rapidly growing (with a crude birth rate of twenty per thousand) in Europe. Vojvodina comprises a mixture of all ethnic groups with Hungarians and Serbs being in the majority, As in other parts of Yugoslavia, nationalism has for the past two centuries been an ever present undercurrent of political life. However, it was largely quiescent within the SFRJ with the exception of Kosovo, where periodic eruptions of Albanian nationalism and Serb nationalist reaction in the late 1980s led to events which precipitated the collapse of socialist Yugoslavia.

As with other former republics, the territory of SRJ prior to 1945 was underdeveloped and primarily agricultural. From the 1950s to 1970s enormous development and economic growth produced rates of industrialisation and urbanisation among the highest in the world. This growth was based on energy and materials intensive industrialisation, accompanied in low-lying areas by the rationalisation and intensification of agricultural production. During the 1980s economic stagnation led to reductions in growth rates (but also to severe limitations on environmental maintenance); as elsewhere in SFRJ such economic difficulties were critical in precipitating the radical institutional reforms of

1989 which were followed by the disintegration of SFRJ itself. Loss of markets elsewhere in former SFRJ led to dramatic reductions in SRJ economic performance: this was considerably exacerbated by UN sanctions imposed on SRJ in 1992 for its role in the war in Bosnia-Hercegovina. A partial recovery consequent upon a stabilisation programme in 1994 proved short-lived. Further sanctions followed by the NATO attack in 1999 have led economic activity to plummet further. In general such economic decline has led to reduced emissions of industrial and agricultural effluent to air and water. However, any consequent environmental improvements have at least to some degree been offset by increased use of low-quality fuels and by reduced investment in environmental protection (REC 1999a, 1999b). In any such general account the existence of great regional variations should be noted. For example, Kosovo, which occupies some 11 per cent (10,887km^2) of SRJ's land area but has over 20 per cent (2.2 million) of its people received, within SFRJ, disproportionately high inward investment. This was partly focused on exploitation of its mineral resources, partly on the modernisation of highly traditional and backward social structures, and partly on relieving the relative poverty which remains a consequence of the latter.

Sanctions, civil war and NATO bombing

This chapter is not the place to examine the origins of the conflict in Kosovo. These go back well before pre-1991 Yugoslavia to the troubled relations of the Serbs with the Ottoman Empire. Their antiquity is symbolised by Kosovo Polje, the 'field of blackbirds' on which, in 1389, Serb armies were defeated and virtually the whole of present-day Yugoslavia passed into Turkish control, under which it remained until the Balkan wars of 1912 to 1913. The date, 28 June, remains the most celebrated anniversary in Serb history. From the end of the Second World War (during which Kosovo was incorporated by the Italians into a greater Albania, and Serb and Montenegrin civilians were massacred by Albanian fascists) it has continued to be a focus of civil conflict. In 1989 Serb suppression of Kosovo Albanian demands for autonomy and secession precipitated the breakup of SFRJ. A strategy of provocation by ethnic Albanian paramilitaries and Yugoslav army retaliation led, a decade later, to foreign intervention. For seventy-eight days, between 24 March and 9 June 1999, Yugoslavia was subject to intense aerial bombardment. In 10,484 sorties, NATO forces dropped 23,614 bombs (of which 1011 were British, including 531 cluster bombs dropped by the RAF). Most attacks were directed against Serbia (including Vojvodina and Kosovo) although a number of sorties also involved targets in Montenegro. Initially, only military targets were attacked, but sorties were soon launched against any targets seen as having a strategic importance, including power-stations, agricultural installations, a tobacco factory, bridges and housing estates.

The consequences of the war – political and social as well as environmental, and for the whole of SRJ, not just Kosovo – have yet to be evaluated and in many cases will remain (like the political future of the region) unpredictable. Little monitoring is taking place: what factual data are available are often unreliable and objectivity has been one of the 'casualties' of the conflict. Two independent studies on the environmental impact of the bombing of Yugoslavia provide the only objective assessment to date. The first, in June 1999, was conducted by a team of expert staff from the Regional Environmental Centre for Central and Eastern Europe (REC) assisted by a variety of specialist contracted experts (REC 1999a). The second, by a United Nations Balkan Task Force (BTF), was initiated in May 1999 to carry out an assessment of the impact of the Kosovo conflict on the whole Balkan region, under the auspices of the UN Environment Programme (UNEP) and the UN Centre for Human Settlements (UNHCS).

The BTF exercise involved a series of sectoral visits by specialist teams in August and September 1999 (Balkans Task Force 1999). One independent scientific team studied the environmental impact on human settlements in Kosovo itself. Another, comprising sixteen experts and two mobile laboratories, studied the impact on targeted industrial sites in Serbia and Vojvodina. A third team examined the Danube River and associated pollution sources and a fourth studied the effects of the conflict for biodiversity in protected areas. In addition to these field missions, a fifth team (comprising representatives from the World Health Organisation, the International Atomic Energy Authority and the Swedish Radiation Protection Institute, as well as UNEP), reported on possible consequences for the environment and human health, of depleted uranium (DU) weapons used in the conflict.

The conclusion of both the REC and BTF studies is that predictions made during the conflict of a major environmental catastrophe affecting the Balkans region (or indeed, Yugoslavia) as a whole have proved unfounded, but that locally there are serious threats to environmental quality and human health. 'At these sites . . . immediate clean-up action, which should be seen as a part of humanitarian assistance to the region, is called for' (United Nations Environment Programme 1999, p.1).

It should be noted, however, that these analyses have themselves been criticised for defining the 'environment' in narrow and reductionist terms (Staddon 1999). In this 'Cartesian' view (Merchant 1992) the environment is seen as an objective physical and biological externality rather than (as Staddon would argue) a complex of biophysical and social relations. Irrespective of these criticisms, it should be noted that both the REC and the BTF studies, and this chapter, exclude the geographically wider impacts of the conflict. Such impacts include the unknown ecological consequences of more than a hundred bombs jettisoned into the Adriatic Sea by NATO aircraft returning to bases in Italy (some ninety-three of which have been located and detonated by NATO forces; the remainder still lie in deep water) and the dumping by KFOR of unexploded ordnance and other waste into Lake Prespa in FYR Macedonia. They also include the environmental effects caused by the influx of huge numbers of refugees from Kosovo into both Macedonia and Albania.

Legislation, policy and planning

As noted above, SRJ regards itself as the legitimate successor to the former, larger SFRJ; both Serbia and Montenegro adopted their current 'new' (1990) constitutions before the secession of the other former Yugoslav republics and have therefore not subsequently seen as great a need for formal constitutional changes. One of the features of socialist Yugoslavia was its unique system of workers' self-management, with enterprise autonomy (and a degree of market discipline) superimposed on an essentially socialist economic base. This was coupled with a high degree of decentralisation, a high level of political awareness and debate, and openness to 'Western' ideas, influence (and investment).

In all but the most essentially federal matters (such as defence), legislation and policy-making was devolved to the constituent republics. Within each republic, implementation and decision-making was often further devolved to a local level. Day-to-day environmental regulation and management was generally the responsibility of municipalities or individual enterprises, subject to general policy guidance (and funding) from the responsible ministries, together with advice and/or regulation via a parallel system of semi-autonomous institutes. Under the broad direction of the responsible ministry, these institutes maintained a policy and regulatory oversight of matters such as water management, wildlife protection, archaeological and cultural heritage protection and urban planning. The April 1992 constitution of SRJ left the republic structures of legislation and administration largely intact. More than any other former Yugoslav republic therefore (possibly excepting Bosnia-Hercegovina, where continuing territorial divisions have inhibited environmental legislation), environmental legislation, policy and planning is derived from that of pre-1991 Yugoslavia.

One of the unfortunate consequences of this situation is that no systematic and internationally comparable State of the Environment reporting exists. Neighbouring countries (with the exception of Croatia) currently participate in a European Environment Agency (EEA) and UNEP EU Phare funded programme to establish accessible, up-to-date and comparable data regarding the state of the environment. In SRJ however, such information must be gleaned from a number of official and unofficial government reports and from statistical data derived from individual studies and organisations. There is however an awareness of the growing importance of publicly available information and participation. An integrated environmental reporting system for Yugoslavia has been proposed by the Federal Government (Federal Ministry for Development 1997b) but is not yet in place.

At federal level there is constitutional recognition of the right to a healthy environment, manifest in Resolutions on Environmental Protection and on Saving Biodiversity. However, the SRJ Constitution prescribes that only basic principles of environmental protection lie within the competence of the federation; as a consequence a draft Law on the Principles of Environmental Protection has not yet been adopted. Both states (as well as SRJ itself) have their own Ministries of Environment. At the 1992 Rio conference, Montenegro declared itself the world's first 'environmental state', pledged to develop a harmonious relationship to nature. However, at the formal level legislation in Serbia is probably the more advanced. Both Serbia (1991) and Montenegro (1996) have framework environmental legislation which is manifest in statutes on air, water, soil, chemical and toxic waste, natural and cultural heritage protection, and spatial planning; in Serbia the last includes a requirement for environmental impact assessment of certain categories of development.

In both countries the major deficiencies concern the degree to which such policy instruments are implemented and enforced (Balkans Task Force 1999). This is not just a matter of administrative efficiency but of public perception. Within SFRJ all natural resources with the exception of private holdings of up to 10ha were defined as social property. Whatever the advantages of social ownership in principle, its practical consequences were often that 'what is everyone's is no one's' and financial mechanisms for securing environmental objectives could not be applied. Moreover, environmental costs (for example, of air and water pollution) were, even when recognised, perceived as 'externalities' and were often disregarded in policy decisions.

Kosovo faces particular environmental problems, especially in respect of industry and mineral extraction, which pre-date the recent armed conflicts. As elsewhere in SRJ, large-scale exploitation of mineral resources after 1945 has produced massive landscape destruction. Related industrial development has compounded this with significant air and water pollution. Particular problem areas are Kosovska Mitrovica (lead and zinc mines) and Glogovc (ferro-nickel mines), both areas with major metallurgical plant. Large-scale energy production at Obilić is associated with open-cast lignite mines; General Janković/Elez Han is characterised by limestone quarries and cement production. Production in all these areas has virtually ceased subsequent to NATO bombing, but the evidence of environmental damage remains. Future developments are contingent on the resolution of political conflicts. For example, along the boundary between Albanian-held Kosova and Serb-dominated northern Kosovo, the main shaft of the Trepča mining complex is in Albanian hands but Serbs control the processing facilities.

In respect of the urban environment it is clear that beyond the enormous direct damage to infrastructure and housing caused by the NATO attack and the subsequent systematic emptying and destruction of towns and villages, first by Serbian forces and then by returning ethnic Albanian paramilitaries, the major problem is the collapse of municipal administrative systems (including the destruction or loss of all cadastral records). The inability to re-establish in the province any effective legislative or policy framework on housing and property (which is an essential precondition for any significant improvement in the urban infrastructure including the operation of utilities) is the biggest problem faced in Kosovo. The one environmental consequence of the conflict that Kosovo is free from, but which exists elsewhere in SRJ (and even more so in neighbouring Macedonia and Albania), are the problems associated with local concentrations of masses of displaced people. These include the pressure on sanitation and drinking-water supplies in overcrowded refugee camps.

Within the service sector most environmental issues pre-1991 were associated with tourism. In general these were less significant in the territory of SRJ than in other areas of the former Yugoslavia, but impacts were nevertheless significant in some locations particularly in coastal areas of Montenegro. Tourism in SRJ plummeted following the disintegration of SFRJ and fell even further following the imposition of sanctions in 1992 (Savezno Ministarstvo za Razvoj Nauke i Životne Sredine 1996). Following NATO bombing, external tourism in SRJ is now non-existent and any attempts to regenerate income from this source await a resolution of the current political situation.

The ambient environment

Environmental quality is related to the level of economic activity in complex ways. It seems likely that as the latter has declined, levels of ambient pollution may have improved (REC 1999a); however, financial restrictions and institutional and infrastructural disruption have reduced capacities for environmental protection and monitoring. The consequences of the war have been both mixed and geographically uneven. In some places damage to processing and storage facilities has produced pockets of intense local pollution. The BTF study identified at least four environmental 'hot spots' where urgent action is necessary. One of the most seriously affected areas is Pančevo, an industrial complex near Belgrade. Here, NATO targeted first the Lola Utva aircraft factory, then the HIP Petrohemija petrochemical complex, then the HIP Azotara fertiliser plant and finally launched seven separate attacks on the NIS oil refinery. The consequences at one stage were so serious that most of the town's inhabitants had to be evacuated.

The consequences of this bombing and of similar action in other areas have been serious. At Pančevo, over 2000t of 1,2 dichloroethane (EDC) and 8t of metallic mercury were released, producing serious contamination of a waste water canal which is leaking into the Danube. This was in addition to the release of 460t of vinyl chloride monomer (VCM), 80,000t of oil and oil products, and about 250t of liquid ammonia. Destruction of the Zastava car plant at Kragujevac has resulted in major ground contamination with polychlorinated biphenyls (PCB) and dioxins. The attack on the oil refinery at Novi Sad released very large quantities of oil, much of which found its way directly into the Danube, but which has also contaminated ground waters and may cause hazards to drinking-water supplies. At the Bor ironworks, very large quantities of sulphur dioxide gas were released into the atmosphere, and major stocks of PCB oils remain unsecured. It is likely that such documented effects have been replicated on a smaller scale in other areas such as in Čačak, following the bombing of the 'Sloboda' household appliance factory (28 March).

With regard to water pollution, waterways and water cycle management, little general data are available. It is clear that in addition to 'domestic' pollution from industry, agriculture and human settlements, a significant amount of pollution is brought in upstream by rivers, so that FRY is part of a surface water system that extends well beyond its own borders. Since the late 1960s river

Table 18.1 Atmospheric emissions in SRJ (10^3 tonnes p.a.)

	1985	*1990*	*1992*	*1994*	*2000 (projected)*
SO_2 emissions	239 (478)	254 (508)	198 (395)	212 (401)	(680)
NO_2 emissions	58 (58)	66 (66)	49 (49)	52 (55)	(88)

Source: Savezno Ministarstvo za Razvoj Nauke i Životne Sredine, 1996: *Izveštaj o Stanju Životne Sredinu y Saveznoj Republici Jugoslaviji za 1994 Godinu* Belgrade: Savezno Ministarstvo za Razvoj Nauke i Zivotne Sredine: 36 (Figures in parentheses from Economic Commission for Europe, 1995: *Strategies and Policies for Air Pollution Abatement.* Geneva: United Nations).

water quality has fallen significantly and most rivers have dropped by one or two classes in the four-class water quality system adopted in 1968. However, recent improvements have been recorded subsequent to the decline in economic activity caused by UN sanctions (Balkans Task Force 1999, p.24). Most monitoring focuses on the Danube and its tributaries.

Air pollution in SFRJ prior to 1989 was considerable but mainly concentrated in urban areas, and has since tracked the fortunes of the economy. Principal sources include thermal power-stations, industrial processes, domestic heating and automobile exhausts. As elsewhere in former SFRJ, inefficient energy utilisation and low technical efficiency of manufacturing and other plant have made the situation worse, and have offset the otherwise positive consequences (for air quality) of economic decline. A major problem in SRJ (as elsewhere) is incompatibility in datasets. Table 18.1 shows major disparities between SO_2 emission data released by the Ministry of Environment and estimates prepared for the UN under a 1994 major review under the Convention on Long-range Transboundary Air Pollution (Economic Commission for Europe 1995; it is likely that, following NATO action, projections for 2000 need to be revised downwards). Data for ambient concentrations, by contrast, are more reliable.

Well-documented consequences of attacks on chemical plants, most particularly the petrochemical and nitrate fertiliser plants in Pančevo, have included dense clouds of toxic chemicals including vinyl chloride monomer, ethylene dichloride, chlorine, phosgene and ammonia. Bombing of electricity transformer stations has resulted in leaks of pyralen and other compounds. The health and environmental consequences of such emissions have yet to be evaluated, but all these compounds are extremely toxic. Several are environmentally persistent and accumulate in the food chain; several are carcinogenic and mutagenic. It is certain that at least some effects extend beyond Yugoslavia's borders. Atmospheric dioxins recorded in Skopje (Macedonia), downwind of Pančevo, immediately following the NATO bombardment, are said to have reached seven times the internationally agreed 'safe' levels. Dioxins and other airborne toxins resulting from NATO bombing have also been recorded in Poland, Hungary and Greece. Again, further systematic monitoring is required to see whether these emissions, severe but localised in time and space, will lead to significant long-term effects.

Major fears during the NATO bombardment related to the Danube. The location on its banks and tributaries of industrial works, many of which were targeted by NATO, led to fears of a major ecological catastrophe, a fear fuelled at the height of the bombing by reports of 20km surface oil-slicks floating downstream. After the bombing of chemical process works in Pančevo on 18 April, some 1000 tonnes of ethylene dichloride, a similar volume of 33 per cent hydrochloric acid and 3000 tonnes of sodium chloride were released into the Danube. In addition, eutrophication and high levels of coliform bacteria were reported in the Danube, Sava and other watercourses following damage to sewage works in Novi Sad and elsewhere. The BTF study found no evidence of an ecological disaster for the Danube as a whole. However, chemical analysis of river sediments and biotic indices revealed significant chronic pollution, both upstream and downstream, of sites directly affected by the conflict, with the likelihood of deleterious effect on aquatic ecosystems and their associated species beyond, as well as within, Yugoslavia's borders. Subsequent to NATO bombing and overshadowing it, in terms of impact on water quality is the February 2000 cyanide spill from the Aurul gold-mine in Baia Mare, Romania. This contaminated the Somes, Tizsa and Danube rivers and has had catastrophic consequences for the Danube ecosystem in SRJ, as well as in Romania and Hungary. It will be some time before the long-term effects of this are known, but it seems likely that they will be considerable.

As with the environmental 'hot spots' identified above, the BTF team found it difficult in some cases to distinguish the consequences of NATO action from those arising from years of environmental neglect. In the case of Danube pollution sediment, analysis showed a record of accumulation of toxic pollutants from the 1960s, 1970s and 1980s (Balkans Task Force 1999, p.10). There is a critical need for systematic monitoring of water quality, and for investment in clean production and waste management processes. It is clear that international sanctions have significantly held back efforts to integrate SRJ into international and European agreements for water quality monitoring, pollution reduction and emergency response. Ironically, in other areas it seems likely that economic collapse brought about by sanctions and intensified by the bombing has meant that aquatic pollution has been (temporarily, at least) much reduced. At Pančevo, the most heavily targeted area, no data on preconflict contamination levels were available, but the BTF team was supplied by a local NGO with a list of accidents within the industrial complex over the past twenty-five years. Local councillors reported a high incidence of 'Pančevo cancer' in the area, and the BTF team concluded that this was most likely to be angiosarcoma of the liver, resulting from exposure to high levels of VCM (Balkans Task Force 1999, p.35). The environmental consequences of dioxins and other toxins released as a consequence of the burning of 460 tonnes of VCM are impossible to estimate; it seems likely that low-level releases of VCM have been going on for a considerable period of time, with direct human health effects. One local resident declared, on meeting with the BTF team, 'finally you came, we have been waiting for you for ten years'.[1]

One issue specific to the Danube concerns through traffic. Although, due to the UN embargo, practically no international traffic has been registered for Yugoslavia since 1995, the volume of international traffic through the area was considerable (Schneidewind 1998). All through traffic on the Danube has been brought to a halt by the bombing, although a canal system around Novi Sad (built by the Austro-Hungarian engineers who drained the lakes of Vojvodina) allows 'friendly' (principally Russian and Ukranian) vessels to bypass Novi Sad and thus complete their transit through Djerdap and the Iron Gates to the Black Sea. Serb authorities see repair or replacement of the bridges as the responsibility of the 'international coalition' that destroyed them and have placed additional conditions on permitting the work to start. These include readmission of Yugoslavia into international organisations including the IMF. The lack of haste no doubt relates at least in some degree to the fact that next to Serbia itself, the principal casualties of the bombing are upstream states such as Hungary that allowed NATO to use their airspace for the attacks. A joint report by Hungary and Austria in February 2000 proposed clearance of the ruins of three destroyed bridges in Novi Sad at a cost of $24 million: it seems unlikely that Serbia will agree to any move which excludes reconstruction as a minimum requirement. Destruction of the bridges is associated with a number of particular environmental problems, real or hypothetical. In the latter category is the possibility that when the Danube freezes, ice-floes might be trapped by the still-standing bridge pillars and huge blocks of concrete roadway lying just below the water surface. This could cause water to back up distances up to 100km and overtop river banks in Hungary and Croatia as well as in Vojvodina itself.

Fears that NATO would bomb the Vinča Nuclear Research Institute, near Belgrade, which it had declared a 'legitimate' target, were, happily, unrealised. The Institute includes an operating research reactor and a second reactor sealed since 1984 but with a still active core, and waste stores of various levels of radioactivity. Bombing would certainly have released radioactive nucleotides. The worst-case scenario of a 'Chernobyl'-like incident in the centre of Europe has receded. However, the plant (and several other lesser installations) is still considered to present an environmental hazard, particularly given the uncertain levels of monitoring.

Munitions and mines

A new environmental problem, which, since the NATO attack, Yugoslavia now shares (uniquely in Europe) with Bosnia and Hercegovina, is the hazard of land-mines. It is estimated that there are upwards of 6000 minefields, most of which are in Kosovo, particularly in border areas. Some other fields were laid in other border areas, particularly in Serbia and Vojvodina, as part of the preparations to frustrate a NATO invasion with ground troops. However, these present less of a threat partly because their extent is less, and partly because their locations are known to a greater degree of precision, than those minefields laid by both contending parties in Kosovo.

To the threat of land-mines must be added the very real hazard (not restricted to Kosovo) of largely unrecorded military wastes together with the legacy of unexploded mines and munitions dropped by NATO forces. In addition, Yugoslav munitions attacked by NATO were often incompletely destroyed. Much remains as environmental hazards with a similar effect. A particular hazard arises from NATO's use of anti-personnel cluster bombs, in defiance of the Hague and Geneva Conventions. These continue to have consequences beyond the immediate death and suffering they caused. It is estimated that between 7 and 13 per cent of bomblets fail to explode immediately and that as a result some 14,000 cluster bombs lie unexploded. When disturbed, their consequences are identical to that of land-mines (which have also been outlawed by the Geneva Convention). Ironically, the first NATO fatalities in the Kosovo conflict occurred in June, after the NATO attack had ceased, when two Ghurka guardsmen were killed, not by Yugoslav mines or bullets, but by unexploded cluster bomblets that they had collected in a clearance operation. Subsequently, over a hundred individuals (mainly ethnic Albanians, with children disproportionately represented) have been killed or seriously injured by such ordnance.

Other environmental effects of NATO bombing include the production of secondary toxins. The most serious of these include the use of depleted uranium (DU) munitions. This is used not merely as a substitute for lead in bullets, but also as armour-piercing enhancement in shells. In this latter use it explodes on impact, releasing a ceramic aerosol which is directly toxic when inhaled and which becomes dispersed as a radioactive dust into the environment. After months of denial, NATO eventually admitted that DU armour-piercing shells were routinely carried and had been fired by A-10 ('Warthog') aircraft. More munitions were used in the attack on Yugoslavia than in the whole of the Gulf War, when almost one million 30mm uranium tipped bullets and over 14,000 large calibre shells were fired, resulting in between 300 and 800 tonnes of DU dust scattered over soil and watercourses. A major problem in assessing levels of immediate and long-term environmental risk from DU is that NATO has refused to release any data about the type, quantity and location of use of DU weapons, claiming that no scientific study has ever yet proved a link with cancer (Balkans Task Force Desk Assessment Group 1999). KFOR and humanitarian staff have been privately advised to stay away from areas where DU weapons were used, particularly around Djakovica, Mitrovica, Priština and Uroševac as well as in Serbia proper. However, as in Iraq (where there is clear evidence of cancer clusters around sites of DU weapon use), NATO has refused to cooperate with the BTF DU team. Indeed the BTF from the start found great difficulty in securing cooperation from Western governments, which refused to release satellite data which would assist its work so that these had eventually to be obtained from the Indian government. This lack of cooperation meant that unlike the four field missions of the UNEP/UNHCS Balkan Task Force, the DU team's work had to be primarily a desk assessment. The most immediate need identified by BTF is for NATO to make available data regarding how much DU was used during the conflict, and where. Without such information, initial risk assessments and the implementation of precautionary measures are impossible. In addition, there has still not been any proper study of the health and environmental risks of long- and medium-term exposure to DU, which urgently needs to be undertaken under the auspices of the WHO.

The rural landscape

Yugoslav agriculture is mixed with large enterprises characterising the northern plains and small family farmsteads in the upland and marginal areas. One recent feature – particularly in intensively cultivated regions such as Vojvodina – has been the impact of sanctions on farming practices. Serbia in particular has reached virtual self-sufficiency in agricultural products since sanctions began. This has led, in some areas at least, to a more intensive use of land, including the cultivation of previously marginal areas. However, the embargo on the supply of chemical products, including fertilisers, together with the loss of export markets, has led to the strong growth in organic and low-input farming. This trend has been accentuated further in 1999 by NATO bombing of fertiliser factories. The probable consequence has been a lessening of agricultural impacts on the environment.

In Kosovo, too, human pressures in rural areas have dropped, but as a consequence of the displacement of people and the destruction of villages. Despite the return of (ethnic Albanian) refugees, the inaccessibility of large areas of the countryside because of mines seems likely to delay the re-establishment of agricultural practices in rural areas, many of which are following those of Bosnia into 'decay', with scrub encroaching on former pasture and arable land.

Total forest cover in SRJ has gradually increased from its low of under a quarter of the land surface in the early 1960s to some 28 per cent in the early 1990s. This is around the European average and higher than in Croatia, but much less than in Slovenia or Bosnia-Hercegovina. Since 1960 the forest area has increased by about 4 per cent. Cover varies greatly in different regions, from under 5 per cent in Vojvodina to almost 40 per cent in Kosovo and Montenegro (Savezno Ministarstvo za Životnu Sredinu 1994). Fears of NATO's possible use of defoliants (such as those used in Vietnam) in preparation for a ground war were fortunately unrealised. In the event, the most direct effects were limited in extent, such as the major fire in the Lipovica Forest following bombing of fuel storage on 26 March 1999. Much more significant is likely to be the consequences of reduced or absent management and illegal logging by local inhabitants (Table 18.2). However, the extent of this is unknown, since (as in other areas) little monitoring is taking place.

Natural and cultural heritage protection

SRJ is located at the centre of one of the most important biogeographical regions in Europe. The high level of scientific education and university expertise has for many years been combined with a well-developed awareness – and

Table 18.2 Damage to forests in SRJ

	Illegal cutting	*Other damage*
1975	19,462	1,809
1985	29,804	6,907
1990	45,804	6,425
1991	63,963	5,611
1992	79,726	2,816

Source: Savezno Ministarstvo za Životnu Sredinu, 1994: *Izveštaj o Stanju Životne Sredine u SR Jugoslaviji u 1992* Belgrade: Savezno Ministarstvo za Životnu Sredinu.

concern – to protect the natural resources (Stevanović and Vasić 1995; Federal Ministry for Development 1997a). As with the ambient environment, legislation and administration for nature conservation in each of the constituent republics broadly corresponds to the situation as it was before 1991 (Jančar 1987; Fisher 1990; European Parliament Directorate-General for Research 1991; IUCN 1991; Terselić and Juras 1991). Despite a technical understanding of the region's importance, species protection in both Serbia and Montenegro is generally recognised to be weak. Outside protected areas, this is based on scheduling of vulnerable species with some 215 plant species protected in Serbia, though only fifty-seven are protected in Montenegro (Federal Ministry for Development 1997a).

Conservation sites and protected landscapes in SRJ cover an area of approximately 6000km^2 or 6 per cent of the territory (over 5 per cent of Serbia and almost 8 per cent of Montenegro). The nine national parks (Biogradska Gora, Djerdap/Iron Gates, Durmitor, Fruška Gora, Kopaonik, Lovćen, Šar Planina, Skadarsko Jezero and Tara) themselves cover a total of approximately 400,000ha (equivalent to some 4 per cent of the land surface) and represent the range of habitats and landscape types in the region (Table 18.3). All these areas were in existence prior to 1991. However, the Landscape Management Plan for the Republic of Serbia proposes to declare some 10 per cent of the land surface as protected by 2010. Prior to NATO bombing, several areas were being prepared for national park status, including Prokletije. In Serbia, management of national parks is a matter for their five separate autonomous public authorities, while in Montenegro the four national parks are managed under the auspices of a single national parks authority. Day-to-day management of lesser-protected sites (such as nature reserves and nature parks) remains (in both states, as in the former SFRJ), delegated at a local level (to municipalities and often to individual enterprises). For example, in Serbia, most forested protected areas outside national parks are managed by Srbijašume, the state forestry enterprise. Overall control is exercised through a tripartite relationship between local authorities, the responsible ministries and the quasi-autonomous institutes.

Transboundary issues are particularly important. Four out of Serbia's five parks (including the proposed Prokletije National Park which abuts Albania) are situated in boundary areas; Kopaonik National Park is also situated partly in Kosovo. Two of Montenegro's four parks are also on the frontier with Albania.

Table 18.3 Protected areas in SRJ

Category	*SRJ*	*Serbia*	*Montenegro*
National park	9	5	4
Nature park	19	19	–
Nature reserve	89	85	4
Landscape monuments	3	3	–
Natural spaces around immovable cultural monuments	45	45	–
Natural monuments	316	261	51
Natural collections	2	2	–
Woodland/forest parks (and other specific areas)	4		4
Internationally protected areas	6	3	3

Sources: Savezno Ministarstvo za Razvoj Nauku i Zivotnu Sredinu, 1996: *Izvestaj o Stanju Zivotne Sredinu y Saveznoj Republici Jugoslaviji za 1994 Godinu* Belgrade: Savezno Ministarstvo za Razvoj Nauku i Životnu Sredinu: 46 and Institute of Nature Protection (pers com).

In some areas, most particularly those (such as Lake Skadar) which abut Albania, there is little or no cooperation with the adjoining conservation authorities. By contrast collaboration with Hungary in respect of the Selevenj sands (which adjoins Kiskunsag National Park) is cited as a model for transboundary collaboration with a joint management plan and proposals to facilitate border crossings (Brunner 1998). SRJ is signatory to all the international conventions involving designated areas and a number of its protected areas are designated as World Heritage Sites (e.g. Durmitor National Park, Tara Canyon, Kotor and Risan Bays), Ramsar sites (Obedska bara, Skadarsko Jezero) or Biosphere Reserves (Tara and Durmitor).

The degree of protection afforded by designation is variable, however. Most national parks are seen as major recreational venues and tourist attractions. To the threats from development and agricultural intensification must be added hunting and the generalised effects of pollution. Eutrophication and drainage are a particular threat to wetland areas; Obedska Bara, the first protected area to be designated in Yugoslavia (1874) and now a Ramsar site, is just one of many such areas to have lost much of its biological interest, though rehabilitation programmes have begun.

Because NATO bombing has been most concentrated in residential and commercial areas, direct damage to species and ecosystems seems to have been relatively light. Nevertheless, in Serbia, four out of five national parks (Kopaonik, Fruška Gora, Šar Planina and Tara) as well as Prokletije (which was to have been declared a national park in 2000) have received direct missile damage. Inevitably, there is an absence of empirical data on the effects of the war on wildlife. The BTF study found that direct physical damage to biodiversity from air strikes was significant within limited areas but of relatively minor importance in relation to the overall size of protected areas and ecosystems around the sites that were hit. There was significant local damage during attacks on communications facilities, to forest ecosystems, and to meadow habitat particularly in Mount Kopaonik, where the Baćište Hotel and meteorological station were targeted, and in Fruška Gora National Park. However, the BTF concluded that there appeared to have been no significant long-term genetic, species or ecosystem effects of the bombing (Balkans Task Force Biodiversity Mission 1999), though this conclusion was based on short visits to only a limited number of sites. Much more significant are the institutional effects, including the disruption to management and monitoring regimes, as a consequence of financial restrictions and administrative disruption. Existing initiatives in environmental protection and conservation have been set back many years by the bombing.

In Kosovo itself a major new problem is the lack of expertise and staff. Most qualified staff – some hundred urban planners and foresters, twenty Ministry of Environment inspectors, plus the thirty staff of the Šar Planina National Park, left during the conflict and have not returned (Balkans Task Force Biodiversity Mission 1999). The Institute for Nature Conservation office in Priština (Kosovo) was abandoned. Although it has nominally reopened it seems unlikely that any significant work is proceeding (KFOR have stated that the legitimate staff are those who worked there previous to the conflict). The same situation would appear to exist with regard to other offices of environmental agencies. Conservation programmes in Kosovo (such as the reintroduction of the Kosovo peony, Paeonia decora) have ceased. Efforts by Yugoslavia to implement international conventions such as the Berne and Biodiversity Conventions have been frustrated as has Yugoslav participation in the activities of international bodies such as IUCN and EUROPARC. In addition, unexploded ordnance is an immediate safety issue, posing a risk to staff working in protected areas. All this also presents a barrier to the regeneration of tourism on which much of the income of these areas depends. However, tourism in some areas, including development of skiing infrastructure in Kopaonik, had for some time been regarded as a threat by conservation authorities. Effective monitoring of species and ecosystems has yet to recommence.

Outside as well as inside protected areas, the enormous damage to property and infrastructure and the need to prioritise humanitarian and economic recovery will inevitably, for a considerable period to come, consume resources that could otherwise be devoted to environmental conservation and to natural and cultural heritage protection. Moreover, the heavy materials demands of reconstruction (particularly on minerals, wood products, water and energy) will themselves pose particular environmental threats. Moreover, as Stritih (1999) points out, consequent to the destruction of rural communities and national economies, the economic value of ecosystem resources is also undergoing a transformation. Such values are close to zero for those who have had to flee their land, while the subsistence value of land can only be realised if there is sufficient security to farm (Stritih 1999).

Cultural monuments and archaeological heritage whose level of protection may be said to be good prior to 1991 have suffered significantly from 'collateral' damage from NATO bombing in addition, in Kosovo, to the deliberate destruction of mosques (by Serbs) and of orthodox monuments by ethnic Albanians. There is some evidence also that Serbian cultural monuments in Kosovo and elsewhere in Serbia (such as Gračanica and other monasteries) have been deliberately targeted. If true, such damage is in contravention of international law. Cultural monuments include relatively modern as well as older entities. For example, at Kragujevac (the location of the bombed Zastava car plant) most of the male population of the city was machine-gunned in the Šumarice Forest by Nazis in 1941 in reprisal for Partisan attacks. Hundreds of monuments commemorating such events are protected throughout SRJ and are at the centre of a category of protected area less common in other parts of Europe (Table 18.3).

Civil society and non-governmental organisations

The strong tradition of civil action and of non-governmental organisation (NGO) activity in SFRJ has, inevitably, been adversely affected by the internal political situation and by international isolation. However, a 1996 REC study of the state of Yugoslavia's environmental movement found that

> despite the embargo, the isolation and the repression, quite a lively NGO movement has continued to exist in Yugoslavia. Even without external support, Yugoslav NGOs have pursued their basic activities, and it appears that, unlike in Bosnia and Hercegovina, where the environmental movement all but disappeared, there are few obstacles to hamper cooperation among them.
>
> (Gjigas 1997, p.13)

Many current forms of civic self-organisation developed in the 1980s with the growth of political pluralisation in SFRJ. The 1996 REC study identified some 1000 alternative, non-profit NGOs in SRJ of which around a hundred had an environmental focus, and were equally divided between Serbia, Vojvodina, Montenegro and Kosovo. Over 80 per cent of these were registered during the 1990s, of which a large number claim to have over 1000 active members. However, membership data are unreliable since only a small proportion of NGOs require payment of dues; levels of activity (and in some cases the existence, other than on paper) of some NGOs may be doubtful (Gjigas 1997). Many NGOs also face economic difficulties and declines in membership. In addition, the bombing has had direct consequences for NGOs. In May 1998, a new Regional Environmental Centre (REC) office was opened in Belgrade, its work funded by the US Environmental Protection Agency. The REC had already held a series of training seminars for NGO activists (funded by the Netherlands Embassy) and had begun the development of a Local Environmental Action Plan (LEAP) process for Yugoslavia. All this work ceased during the conflict and is only now, a year after it ended, being painfully restarted: for example, the REC office in Belgrade has reopened, and is working principally on issues related to the Århus Convention.

Many NGOs continue to be active however, and can be judged to 'have considerable experience and are important and capable sources of information' (Balkans Task Force 1999, p.27). Most active are those scientifically based organisations, which occupied a 'semi-official' niche in the environmental establishment prior to 1991. Examples of such 'establishment' non-profit organisations include Ecolibri-Bionet and Mustella, both of which are involved in the elaboration of the National Biodiversity Strategy. Since 1991 however, new NGO structures have developed. For example, rural areas have seen the recent growth of new and different forms of agricultural cooperatives. An example is the NGO Terra, formed in Subotica (Vojvodina) with the aim of encouraging organic farming, processing and marketing both in the local region and in the whole of Yugoslavia. The initiative is associated with the Subotica-based Open University (formerly the Yugoslav Workers' University) created in 1992 to foster civil society and deliver adult education (Csagoly 1999).

Conclusions: the future

Environmental futures in Yugoslavia, as in the other southern republics which once belonged to the much larger socialist federation which bore the same name, are fragile and unpredictable. Post-fragmentation recession, aggravated by economic sanctions, international isolation and the NATO bombardment, has considerably reduced economic activity and hence industrial pollution, but these problems have also greatly reduced the impetus for environmental protection. Although predictions of a widespread environmental disaster seem confounded by subsequent studies, direct local effects are severe and, beyond the direct effects of the bombing, the generalised environmental consequences arising from the economic, political and humanitarian consequences of NATO action have indeed been catastrophic. These have impacted most heavily on Kosovo itself, where an effective form of civil administration, an essential precondition for environmental protection, has yet to be established.

Now that the bombing is over, economic reconstruction should be a priority second only to humanitarian aid to those who suffered – from the ethnic conflicts which led to the NATO intervention, from the NATO attacks themselves, and from the ethnic cleansing it precipitated. However, in both Kosovo and the rest of Yugoslavia, environmental remediation and protection must be an essential component of this process and one for which the NATO states bear prime responsibility. It is unfortunate that current reconstruction policies of the 'international community' say little about environmental protection apart from local infrastructural projects (such as improving water and waste management systems) and are almost entirely focused on securing a dynamic market economy backed by a strong public institutional framework (European Commission and World Bank 1999).

Some commentators contend that far from being unintended consequences of the conflict, or 'collateral' damage, its human and environmental costs were both predictable and deliberately inflicted. Staddon, in particular, insists that 'environmental assessments of the NATO campaign in FRY must proceed from a broader understanding of the complex relations between contemporary Great Power militarism and the environment . . . environmental destruction, pace current official viewpoints [is] not a "regrettable byproduct" of armed conflict, but one of its key means' (Staddon 1999, p.1). Without necessarily accepting this argument, it seems clear that remediation measures will only be encouraged by the NATO states that participated in the campaign to the extent that they further its objectives of influencing the trajectory of Yugoslavia's economic and political development. The environmental prognosis is not good. UNHCR has already complained that, having 'won' the air war, NATO countries are turning their backs on their own commitment to help clean up the mess they have created. The United Nations High Commission for Refugees complained in August 1999 that it had received less than 50 per cent of the money long pledged to it for emergency relief (and that of all countries, Britain had contributed the least, only $800,000, rather less than the cost of one cruise missile). Most of the money contributed to date is for humanitarian aid in Kosovo; NATO countries have declared that they will refuse aid of any kind for the rest of Yugoslavia until changes in political leadership occur. Inevitably, the environment is well down the list of donor priorities. Ironically, some of the positive developments have derived from initiatives taken in respect of other environmental disasters, such as the Baia Mare cyanide spill in February 2000, where Yugoslavia has received at least some assistance from international funds which have been directed not only at physical clean-up operations but also at information and public participation initiatives.

International support (including technical assistance and collaboration from state, scientific and university bodies outside Yugoslavia) for programmes to repair damaged natural and cultural heritage and to protect the ambient environment is vital. These programmes should include individuals from all ethnic groupings within Yugoslavia. They should include measures designed to reintegrate Yugoslavia with the wider European community of which it is a part. They should also include encouragement to progressive trends within SRJ; for example, the development of local community initiatives as well as political encouragement to adopt and implement international and European environmental conventions. The Århus Convention (formally the Convention on Access to Information, Public Participation in Decision-making and Access to Justice in Environmental Matters) is particularly significant because it links humanitarian concerns and civic rights (including freedom of information) to environmental protection. This was signed in June 1998 by the EU and thirty-five governments including Yugoslavia, although like the US, SRJ has yet to ratify it. The UNEP/UNHCS Balkans Task Force report concluded: 'the suspension of international co-operation has had an undisputedly negative impact on environmental management and institutional development in FRY' (Balkans Task Force 1999, p.24).

As Yugoslavia's economy painfully recovers and as civil society is slowly re-established, environmental protection must not be seen as an optional 'add-on'. Stabilisation will be accompanied by international investment and both processes will provide opportunities for environmental remediation, for the protection of nature and natural resources, and for the strengthening of conservation institutions, including NGOs. Yugoslavia is not yet party to or a beneficiary of the Regional Environmental Reconstruction Programme (RERP, which has been incorporated into the Stability Pact agenda for the area and which will become the basis for the Environmental Action Plan for South East Europe and for world environmental policy in the environment in the region) (REC 2000). It is important that reconstruction and environmental improvement is based on Yugoslav environmental authorities, enterprises and professionals – not consultants flown in from outside (REC 1999a); they must also have regard to the transboundary nature of many environmental issues. Investment in these areas of a relatively small fraction of the donor funding that will be spent in reconstruction in future years will yield multiple returns, both for the local environment and its peoples and for Europe as a whole (Stritih 1999). A quality environment is an essential complement to human rights and social justice as the foundation for a stable society, in which all citizens can begin again to live together in peace – with each other and with nature.

Acknowledgements

Acknowledgement and thanks are due for help and information provided by colleagues at the Federal Ministry for Development, Science and Environment, and at the Serbian Ministry of Environment and to Institute of Nature Conservation staff in Belgrade and Niš. Much of the research for this chapter was carried out during study visits conducted with the assistance of the British Council, Belgrade, and of Birkbeck College, London University.

References

Balkans Task Force. 1999: *The Kosovo Conflict: Consequences for the Environment and Human Settlements.* Nairobi: United Nations Environment Programme; United Nations Centre for Human Settlements.

Balkans Task Force Biodiversity Mission. 1999: *Assessment of the Damage to Biodiversity in Protected Areas of the Federal Republic of Yugoslavia.* Geneva: United Nations Environment Programme; United Nations Centre for Human Settlements.

Balkans Task Force Desk Assessment Group. 1999: *The Potential Effects on Human Health and the Environment Arising from Possible Use of Depleted Uranium During the 1999 Kosovo Conflict: A Preliminary Assessment.* Geneva: United Nations Environment Programme; United Nations Centre for Human Settlements.

Brunner, R. 1998: *Transborder Protected Area Cooperation in Yugoslavia – A Technical Assessment.* Vienna: Report on a visit to Serbia, December 1997.

Csagoly, P. 1999: 'Norther Serbia's organic necessity'. *REC Bulletin*, 8 (2), pp.18–19.

Economic Commission for Europe. 1995: *Strategies and Policies for Air Pollution Abatement.* Geneva: United Nations; 1994 major review prepared under the Convention on Long-range Transboundary Air Pollution.

European Commission and World Bank. 1999: *Toward Stability and Prosperity – A Program for Reconstruction and Recovery in Kosovo.* Brussels: European Commission, World Bank. Prepared for the United Nations Mission in Kosovo.

European Parliament Directorate-General for Research. 1991: *Nature Conservation in Austria, Finland, Norway, Sweden, Switzerland, Bulgaria, Czechoslovakia, Hungary, Poland, Romania, Yugoslavia and the Soviet Union.* Environment, Public Health, and Consumer Protection Series 17. Luxemburg: European Parliament.

Federal Ministry for Development, S and E. 1997a: *Biological Diversity of FR Yugoslavia: Assessments, Threats and Policies.* Belgrade: Ecolibri-Bionet.

Federal Ministry for Development, S and E. 1997b: *Summary Report on the State of the Environment in FR Yugoslavia for 1998.* Belgrade: Ecolibri-Bionet.

Fisher, D. 1990: 'Environmental politics in Yugoslavia'. *Environmental Policy Review*, 4 (2), July, p. all. The Soviet Union and Yugoslavia.

Gjigas, O. (ed.) 1997: *NGO Directory: A Directory of Environmental Nongovernmental Organisations in Central and Eastern Europe* (3rd edn). Szentendre, Hungary: Regional Environmental Centre. See also http://www.rec.org/REC/Publications

IUCN. 1991: *Environmental Status Reports No. 2: Albania, Bulgaria, Romania, Yugoslavia.* Cambridge: International Union for the Conservation of Nature.

Jančar, B. 1987: *Environmental Management in the Soviet Union and Yugoslavia: Structure and Regulation in Federal Communist States.* Durham, NC: Duke University Press.

Lauševic, R. 1999: 'General environmental effects of NATO aggression'. *Tehnokratia*, 2, pp.16–21.

Merchant, C. 1992: *Radical Ecology – The Search for a Liveable World.* London: Routledge.

REC. 1999a: *Assessment of the Environmental Impact of Military Activities During the Yugoslavia Conflict.* Szentendre, Hungary: Regional Environmental Centre for Central and Eastern Europe. For European Commission DG-XI.

—— 1999b: *Sourcebook for Economic Instruments for Environmental Policy in Central and Eastern Europe.* Szentendre, Hungary: Regional Environmental Centre for Central and Eastern Europe.

—— 2000: *Regional Environmental Reconstruction Programme for South Eastern Europe.* Szentendre, Hungary: Regional Environmental Centre for Central and Eastern Europe.

Savezno Ministarstvo za Razvoj Nauku i Životnu Sredinu. 1996: *Izveštaj o Stanju Životne Sredine u Saveznoj Republici Jugoslaviji za 1994 Godinu.* Belgrade: Savezno Ministarstvo za Razvoj Nauku i Životnu Sredinu. (In Serbian Cyrillic).

Savezno Ministarstvo za Životnu Sredinu. 1994: *Izveštaj o Stanju Životne Sredine u SR Jugoslaviji u 1992.* Belgrade: Savezno Ministarstvo za Životnu Sredinu.

Schneidewind, P. 1998: 'The Danube Waterway as a key European transport resource', in V. Braute *Sustainable Development for Central and Eastern Europe – Spatial Development in the European Context.* Berlin: Springer, pp.127–43.

Staddon, C. 1999: 'Militarism and nature at the millennium: the environmental consquences of the 1999 NATO campaign in FRY', unpublished. Faculty of the Built Environment, University of the West of England.

Stevanović, V. and Vasić, V. 1995: *Biodiverzitet Jugoslavije sa Pregledom Vrsta od Međunarodnog Značaja.* Belgrade: Ecolibri i Biološki Facultet.

Stritih, J. 1999: 'Nature in the Balkans: an economic resource'. *European Nature*, 3, p.31.

Terselic, V. and Juras, A. 1991: 'Nature conservation in Yugoslavia', in Directorate General for Research (ed.) *Nature Conservation in Austria, Finland, Norway, Sweden, Switzerland, Bulgaria, Czechoslovakia, Hungary, Poland, Romania, Yugoslavia and the Soviet Union.* Environment, Public Health, and Consumer Protection Series, 17. Luxemburg: European Parliament, Directorate-General for Research, pp.117–37.

United Nations Environment Programme. 1999: 'Balkans Task Force Recommends Immediate Environmental Action as part of Humanitarian Aid', Press Release, 14 October.

Part IV

Conclusion

19 Conclusion

F. W. Carter and David Turnock

A concern for the well-being of the environment in the ECECs began to grow during the 1980s. In many of these countries environmental legislation was passed during the central planning period, but its practical implementation remained very feeble. Often a single administrative authority was a major pollutant yet at the same time responsible for its prevention. Only after 1989 did significant changes in environmental conditions begin to take place, as recorded in the Kyoto Protocol (Table 19.1). Unfortunately, the Kyoto Protocol has been described as a sketchy agreement by rich countries to cut greenhouse gases to below 1990 levels. In Kyoto, in the eagerness to frame an accord, the Protocol failed to answer the question as to which countries should make the cuts to everyone's satisfaction. Only developed countries agreed to binding targets for emission reductions while the others agreed to nothing. Thus with no agreement over the question as to which countries should be involved, the UN conference on global warming at Buenos Aires in November 1998 was able to disregard the major question of how to make the cuts (Anon., 1998a).

Environmental specialists have pointed out that many environmental improvements in the region have resulted from the collapse of communist power. One should remember, however, that at least part of this process across the former socialist states resulted from a cumulative fall in the Gross Domestic Product (GDP) and the collapse of COMECON (Baker and Jehlička, 1998). These factors led to significant reductions in previous energy consumption levels. Moreover, many countries in the region eliminated energy subsidies; these had previously made a significant contribution to the wasteful utilisation and over-employment of energy resources, particularly in state enterprise and domestic households. After 1989, energy prices were allowed to rise to world levels; in turn this helped to reduce energy consumption. Inefficient former state-owned enterprises now had to discover how to survive in this new economic climate. This was due to world market price exposure, not only in the energy sector, but more generally in the cumulative privatisation process across the region. The choice was stark; either close down the economic unit or develop new restructuring methods to enable more efficient production. In turn, this often meant reducing energy consumption. There have been positive environmental improvements in spite of the

Table 19.1 1997: Kyoto Protocol emissions (as % of 1990)

Country	*Annex I parties*	*Annex B parties*
Bulgaria	92	–
Croatia	–	95
Czech Republic	92	–
Hungary	94	–
Poland	94	–
Romania	92	–
Slovakia	92	–
Slovenia	–	92

Source: Anon., 'The Kyoto Protocol', *Science for Democratic Action*, 6(3) (Institute for Energy and Environment Research), Merrifield, VA, 1998, p.8.

Note
Annex I parties included in UN, *Framework Convention on Climatic Change*, New York 9 May 1992; Annex B parties included in the Kyoto Protocol December 1997.

economic difficulties these post-communist countries have had to face during the transition process. They directly result from new environmental policies adopted in these former centrally planned economies after 1989, together with financial/know-how contributions by many EU countries and the USA towards reconditioning environmental conditions there. While this is the generally accepted scenario, closer analysis reveals this was not always the case.

One of the greatest environmental risks identified in the region has been airborne pollution. In 1995, countries from the region still dominated the top ten European emitters list for SO_2; Bulgaria was top with 176kg p.a., followed by the Czech Republic (106), with Hungary (69) in fifth place, Poland (68) in sixth, Slovenia (60) seventh, and Macedonia (50) tenth. Only the Czech Republic figured in a similar NO_x list, coming seventh (OECD, 1997; Anon., 1998b). It is around major urban centres and during summer periods that vehicle movement provides most air pollution. In this region several countries have identified a growing number of so-called air pollution 'hot spots' which have been recognised and catalogued; included in the list are twenty-five in Romania and nine in Bulgaria. Others have been identified in northern Estonia, and along the Czech–Polish border which forms part of the 'Black Triangle' region. Many of these so-called 'hot spots' include urban areas which suffer from the close proximity of industrial pollution, the burning of fossil fuels for heating domestic residencies and the ever growing degree of traffic (Golubić *et al.*, 1999). Besides increasing levels of vehicle use, other associated factors include the intensive exploitation of diesel oil in Bulgaria, the burning of low-grade fuels in Lithuania, deteriorating public transport services in Hungary, and mounting financial costs in Romania and Slovakia.

Links between air pollution and acid rain continue in the region. Plagued by pollution during the Soviet era, the ECECs, in contrast to North America and Western Europe, are at an earlier stage in addressing acid rain. However, analysts

Table 19.2 East Central Europe: acidification of forests 1995 ('000 ha)

Country	*Forest area*	*Acidification*	%
Belarus	7,865	2,045	26
Czech Republic	2,630	2,499	95
Hungary	1,811	1,195	66
Poland	8,942	3,845	43
Russia	163,308	1,633	1
Slovakia	2,016	825	31
Ukraine	9,458	473	5
Others (12)	23,695	23	0

Source: E. Sollander (1999), 'Europe: Exporters of acid identified', *Acid News*, 1 (March), pp.14–15.

Table 19.3 East Central Europe: forest damage 1996–98 (%) (above 25% defoliation)

Country	*1996*	*1997*	*1998*
Belarus	40	36	30
Bulgaria	39	50	60
Croatia	30	33	26
Czech Republic	72	69	49
Hungary	19	19	19
Latvia	21	19	17
Poland	40	37	35
Slovakia	34	31	32
Slovenia	19	26	28
Ukraine	46	31	52

Source: P. Elvingson (1999), 'Europe's forests: general trend for worse', *Acid News*, 4 (December), pp.6–7.

believe that as the region emerges from economic crisis, efficiency measures and the adoption of EU standards will result in a decline in sulphur emissions and a more modest drop in nitrogen emissions from 1990 levels by 2010 (Cofala *et al.*, 1997). Recent research into acidification of forests on a European scale has identified a handful of countries as the chief sinners, including some ECECs (Table 19.2). During the annual European forest survey in summer 1998, the most extensively damaged forests were found in the Czech Republic, Slovakia, the southern parts of Poland and Belarus (Table 19.3). The main concerns over human health have mainly been linked with sulphur dioxide (SO_2) and particulate matter (PM), both by-products of various types of older industrial facilities, including coal-burning power plants. East Central European countries like Poland, Hungary and the Czech Republic have all begun improving their coal sectors. This is proving a painful exercise since it involves abolishing jobs in societies still emerging from the old communist system (World Bank, 1997). Wide use in the factories/domestic households of fossil fuels and brown coal, which have very low heating value, has resulted in annual carbon dioxide (CO_2) and sulphur dioxide emissions being well above world limits.

Rapid changes in air pollution are now being experienced in nearly all ECECs. Some countries do suffer from transboundary air pollution, as, for example, in Romania and Slovakia. While there has been a marked decline in emissions of major industrial pollutants, especially SO_2, the quality of urban air has worsened. This has resulted from rising levels of smaller particulates, NO_x and other greenhouse gases, thanks to the growing amount of motorised traffic. Such growth has led to greater noise and safety problems, as well as a reduction in physical exercise which, in turn, has had a considerable bearing on the need for medical and health care resources.

Recent OECD scrutiny of the Czech Republic (a member since 1995) is of interest here. Air quality has improved considerably thanks to a switch away from using brown coal and an extensive retrofitting of larger power plants with desulphurisation equipment. Even so, per capita emissions in the Czech Republic are still the highest in the OECD. Its report dutifully records how extensive investments have been made in environmental improvements; it also notes problems with pollution charges; these were originally either too low, or have been worn away by inflation. The report also castigates continued secrecy adopted by the government towards its public on environmental matters. In future, the OECD report urges the government, NGOs, industry and local authorities to maintain 'closer and more sustained' relations between themselves (OECD, 1999, p.12; Anon., 1999a).

Evidence from the country studies

As expected, these have revealed both positive benefits and negative drawbacks in the environmental situation of the region during the 1990s. The positive environmental benefits have varied considerably between countries. However, in most countries the full/partial closure of heavily polluting and uneconomic plants has meant that environmental pollution in the worst hot spots has been eradicated, leading to evidence of some improvements in both air and water quality. Even in the FSU, the post-1990 transition period has benefited the environment through reduced industrial output. The question remains however as to whether these changes will be permanent. Certainly such improvements have resulted from economic decline post-1989, the impact of anti-pollution measures in the 1980s, new environmental policies in the 1990s, and the gradual economic restructuring away from heavy industry and the desulphurisation of major power plants. Nevertheless, small-scale environmental problems have remained numerous in some countries, thanks to the communist heritage. Even so, in general, pollution levels have been found to be rarely excessive outside known blackspots.

The pursuit of sustainable development has experienced substantial progress. Since 1990 there has been stronger environmental emphasis on sustainability principles; this has been especially so in tourism which has much to gain from such attitudes. Positive environmental gains have led to the permanent conservation of protected areas through the establishment of new national parks, for example, and enhanced countryside management, but all this can still be limited by insufficient financial state support. Improved farming methods also play their part but are often, as in East Germany, more environmentally friendly in marginal rather than more fertile areas, while in Albania the imposition of exclusion zones for environmentally sensitive areas and prohibition of fishing in others add their weight to sustainable development.

The problem of war has also influenced the environment. In Bosnia and Hercegovina the environment provided a focus for the most positive post-conflict development represented by the recent growth of NGO activity, environmental legislation and greater awareness of the quality of natural resources. In Yugoslavia, post-war stabilisation has helped stimulate international investment which in turn assists in remedying environmental problems through nature protection and the strengthening of conservation institutes, including NGOs. It is essential that environmental improvement is founded on home-based Yugoslav authorities, enterprises and professional practitioners, as well as an awareness of their transboundary environmental issues. Meanwhile, increased consumerism has led to tourist growth and the need for infrastructural investments. Some of these could help in the revival of traditional industries that have low production costs as, for example, in Croatia. Such investments could also help provide greater integration between national and regional/local plans with respect to environmental protection at present much needed in Poland.

Negative drawbacks in the region's environment after 1989 resulted from both economic and political changes. These have included greater use of motor vehicles especially in urban areas, leading to increased atmospheric pollution/congestion as well as increased plastic packaging/waste. In East Germany, for example, negative environmental changes have been represented by a general increase in consumerism with higher car ownership and household waste production. Changes in political circumstances have tended to squeeze out environmental problems from the public arena and more or less eliminated popular social movements. The popular environmental politics of 1989 have now faded from national life in many countries, leaving the environmental movements in disarray. Politically, no consensus exists in some countries about the need to develop an efficient system of environmental management and integrate economic and environmental problems. In addition, potentially dangerous environmental developments have arisen due to capitalism through what may be called 'dirty' Western investment. For example, in Bulgaria, the role of international agencies, corporations and foreign government bodies has become surprisingly influential in the formulation of public environmental policy. In contrast, independent environmental NGOs such as those in the Czech Republic has become weaker, because the government ignores such groups as there is little public support for them. In Slovenia, the post-1990 environmental situation has remained neglected and there is a clear need to appreciate the role of sustainable development. The social crisis there in the early 1990s meant that environmental protection was neglected; this was not helped by the belief among politicians and the general public that environmental protection hindered economic growth and secure employment. Similarly in Poland, environmental topics have declined to be of interest in government circles.

The results of war have also taken their toll on the environment. In Bosnia and Hercegovina the Balkan wars had an indirect impact on the environment, both in the natural and cultural sense. It is most probable that the consequences of this war will dominate that country for many years to come. In Croatia, the Balkan wars have resulted in additional environmental problems including damage to woodland, a decline in land use and the presence of land-mines. These wars also distracted people from environmental problems. Both among the general public and in professional circles, environmental consciousness is at a low ebb. In Yugoslavia, the impact of war has meant that an effective form of civil administration has yet to be established, especially in Kosovo, yet this is an essential precondition for environmental protection. In addition, suspension of international cooperation has had a negative impact on environmental management and institutional development. Environmental protection is only seen as an additional option in any package for the re-creation of civil society and economic development.

The key to the future appears to be dependent on the environmental attitudes of each country's inhabitants and is therefore complex. In comparison with their West European counterparts, the present generation in the region have a lack of 'environmental/green' education. Many people seem to have become indifferent to environmental issues and there is a genuine need for further education to stimulate greater public interest. In Slovakia, for example, there is a demand for more and better information on the environment, although there are limited financial and labour resources available, especially in the FSU.

Finally, the country studies revealed the significance and potential impact of the EU on the region. Its standards are seen as the yardstick to which they must all aspire if membership is to become a reality. Formal relations with the EU are bringing about the greatest change in each country's environmental policies and practices. In some countries, such as the Czech Republic, environmental reconstruction has been driven by these EU requirements, including rigorous pollution controls. Poland's environmental efforts have been encouraged by the EU's positive response to its endeavours, but the country still needs firm government especially in the environmental field. For example, Poland's increased transport use has meant the country still exceeds EU levels for gases and dust and must improve its environmental standards. The high financial cost of environmental improvements can be disheartening for some countries like Romania. Furthermore, the complexities associated with EU environmental legislation can act as a deterrent, but have to be mastered. Unless they are met they will impose a curb on EU membership. Time is also a critical factor. In Slovenia, for example, it is becoming rapidly apparent that adaption to EU environmental standards will only be achieved in the medium term; EU directives in all environmental fields there, such as urban waste water, waste disposal and drinking-water standards, will only be reached over a longer time period. In the FSU, the EU's imposing environmental standards are obviously less significant.

Economic influences

There appears to be no straightforward answer to the type of influence the free market has had on the environment of the region. However, certain factors may be noted. First, there was no repetition of the obvious excesses found during the communist period; the most polluting industrial plants were closed and others submitted to state regulation. Some time did elapse before these laws could be fully realised, but indulgences were not tolerated. An important impact was also the attention given to intense pollution cases by the foreign media sources. Initially, foreign aid was mainly given for research (Lofstedt and Sjostedt, 1996, p.215), while at home political priority was given to cleaning up the worst pollution problems, such as the withdrawal of subsidies for certain polluting practices like fertiliser use and heavy 'smokestack' industries. Recent experience has also involved implementing pollution charges and the use of environmental permits (Bluffstone and Larson, 1997).

Second, privatisation for private companies meant competition and striving for profit, which enhanced short-term thinking but left few financial resources for environmental security. Raw material prices soared to international levels but were not accompanied by any immediate attempt to reduce energy use. Much of the new legislation/environmental regulations were simply not enacted, because production was vital for the struggling economies despite some of the major employers being heavy polluters. Furthermore, many foreign investors did not wish completely to modernise production processes, preferring instead to use filters, scrubbers, etc. to reduce pollution in the production process. Unfortunately this proved no basic solution to the problem. In addition, foreign aid proved insufficient for implementing rapid environmental measures because of inadequate funds (Manser, 1993). Fortunately, many of these impediments were only transient and it is hoped over time the privatisation process will force companies to accept new regulatory standards, while higher energy prices will reduce utilisation. With more stability in the economies of ECECs, money will hopefully be available for answers to environmental problems and eventually the free market will encourage amelioration in the environmental situation (Klarer and Francis, 1997).

Finally, the environment became exposed to new threats. Prior to the collapse of communism many people in the ECECs looked enviously at Western life-styles and prosperity. When they themselves became democratic/capitalist societies, people with money quickly embraced the Western consumerist life-style, setting an example to the others. Moreover, with the increased use of plastic, production became less sustainable. Such consumerism also augmented the quantity of waste and amount of raw material utilisation (Klarer and Francis, 1997). Even worse, foreign companies began constructing polluting plants in the region. The attractions were more lenient, poorer controlled legislation than in the West and the fact that pollution of a foreign country generated much less opposition from environmentalists back home. For example, the nuclear industry in the West lobbied for more power plants in the region, though this stimulated very different reactions among the various states. Nevertheless, the nuclear power industry, despite the evidence presented by the Chernobyl disaster of April 1986 in the Ukraine, continues to minimise the potential possibility for catastrophic reactor accidents in future (Makhijani, 1998). Another incentive for relocation to ECE was the lower wages paid for parts of the goods production process, which in turn leads to increased transport and hence pollution. Moreover, most FDI was targeted at resource extraction and utilisation without much concern for the impact on environmental degradation (Tickle and Welsh, 1998). Since the beginning of the 1990s exploration for various metals has been undertaken by seventy nations in a vast area combining ECE with Latin America, Africa, Asia and the Pacific region. Countries in these regions have modified their mining codes in order to attract investors (Symonds, 1997) while others have abolished mining taxes and introduced generous economic incentives (Dorian, 1997).

Governments cannot always plan for unknown natural disasters. In July 1997 severe flooding in Central Europe pushed rivers such as the Oder to record levels and economic losses were estimated at US$5.3bln (Koenig, 1997). A more recent example occurred three times within a period of seven weeks. In early 2000, East Central Europe suffered environmental disasters of the highest magnitude, all originating in Romania. In January, a cyanide spill at the Aurul gold-mine near Baia Mare eventually leaked into the River Someş, a tributary of the Tisza (Prentice and Hillen, 2000). The river carried the toxic material into the Danube, polluting areas of the Vojvodina in Serbia, and parts of Romania and Bulgaria, before discharging its dangerous load into the Black Sea. About six weeks later, heavy rains and melting snow led to a dam burst at the Baia Borsa lead/zinc mine in northern Romania. About 20,000 tons of heavy metal sludge again entered the Tisza, this time via the Vişeu River. Lead concentration in the Tisza River became twice the permissible level, and zinc nine times. Four days later another heavy metal spill again occurred at Borşa. These incidents clearly illustrate that environmental disasters have no known frontiers; they also illustrate how one country's shortcomings regarding environmental issues can have serious knock-on effects as a result of transboundary pollution.

Future help for the region's environment may lie with the multilateral development banks (Sandberg, 1998; Gutner, 1999). The three major actors have been the World Bank, the European Bank for Reconstruction and Development (EBRD), and the European Investment Bank (EIB) in descending order of financial involvement. The World Bank has been most active in supporting environmental reform, especially 'green' projects in the region, thus tending to refute NGO criticism of its worldwide role in this field. The EIB has rendered the least environmental help for the region, while the EBRD has played more of a rather median role. The environmental activities of these three financial sources have been mainly related to energy and water problems. The World Bank has concentrated on 'green' projects largely related to energy efficiency, the amelioration of waste water collection and treatment, and quality of potable water. The EBRD has chiefly focused on credit facilities for energy efficiency projects and the improvement/treatment of waste water. While lacking the breadth and range of the World Bank's commitment, it remains more impressive than that of the EIB. Of the latter's ninety-six projects in the region, only six have been concerned with environmental 'green' projects. Of these, three-fifths of total investment has gone into one project in the Czech Republic, namely the purchase of desulphurisation equipment for several thermal power plants in the heavily polluted area of northern Bohemia. Other projects were shared by the EIB with either the World Bank or EBRD, and included sewage treatment/water supply for Riga, the Latvian capital, improving district heating facilities in Estonia, and funding a Polish forestry project and a Bulgarian energy scheme. Whilst all these 'green' projects are very laudable, there does appear to be a different approach adopted by each financial organisation. The EIB, and to a lesser extent the EBRD, is more private sector driven with a stronger banking mentality; in contrast the World Bank is more policy orientated and willing to boost environmental targets.

Political influences

While the process of democratisation is well under way in the region, one persistent problem relates to those remnants of the former state structure and establishment personalities still present in the political sphere. An important section of the political elite still contains some of the former nomenklature, who retain a conservative mentality resistant to change *per se*. They form a large barrier to those wanting environmental reform, which demands widescale structural change (Tickle and Welsh, 1998). This obstacle is particularly evident at the regional level where old communist arrangements are still common. In contrast, state level structures received plenty of attention during the democratisation process but, unfortunately, at the regional level political structures remained as they were, with the same people holding significant prominence. This was a particular problem for environmental policy, because it is highly dependent on administration at the regional and local level (Baker and Jehlička, 1998). Ecological issues are best dealt with by regional governments because of their practical make-up; thus executive power at the local and regional levels becomes very important because environmental problems from a national perspective are often driven down the political agenda.

Over the past decade as part of the globalisation process, there has been a decay in the power of the nation-state. It became very apparent that the nation-state was not the ideal medium for environmental protection. It was clear that supranational governments, regional administrations and NGOs were much better adapted for this task. However, the opposite effect has been experienced during the post-communist transition. Under communism the nation-state was quite weak but has since gained strength. After 1989, the nation suddenly became a major factor exemplified by the spread of unfettered nationalism. As a result the political structure became less capable of dealing with environmental problems (Tickle and Welsh, 1998). However, this may only be a transitory problem where countries seem destined to become EU members and will therefore be assimilated into a supranational structure.

The legacy of environmental legislation from the communist period was inadequate, yet it did not receive immediate precedence after 1989 and has also been gaining greater priority recently. As already noted, significant legal progress has resulted from prospective membership of the EU which is demanding a complete rethink and is creating opportunities for the adoption of international standards (e.g. from UNESCO). Of course EU membership does not justify unlimited environmental optimism, even though environmental degradation in the European periphery has been tackled by several large, infrastructural projects funded by the EU in the past. Membership entails some form of economic development which will place further strain on the environment (Baker and Jehlička, 1998) and this is giving rise to concern, for example, in the Czech government and Poland, over the future of the environment (Fagin and Jehlička, 1998; Wódż and Wódż, 1998). This is why there was some jubilation at the formal adoption in December 1999 of a 'Protocol to Abate Acidification, Eutrophication and Ground-level Ozone' agreed by twenty-seven nations in Göteborg, Sweden; including the signatures of Bulgaria, Croatia, the Czech Republic, Hungary, Romania, Slovakia and Slovenia. The protocol assigns national emission limits for SO_2, NO_x, NH_3 (ammonia) and volatile organic compounds (VOCs). It also established emission limits for particular pollution sources (e.g. large combustion plants and vehicles). This comes into operation either a year after the protocol is effective, or by December 2007, whichever comes first, but there are some experts who believe that reduction of the four named pollutants will still be insufficient for a cleaner environment (Ågren, 2000).

Social change and environmental movements

With the transfer from a communist regime to that of a free market democracy, two basic ideological options remained. A minority of people dissatisfied with the changes, including unemployment and economic recession, preferred a return to communism. The majority, however, attracted by Western life-styles, desired a capitalist society. As no alternative system surfaced, it left little support for environmentalists (Jancar-Webster, 1998), as there was quite a radical acceptance of the democractic/free market. In contrast to the West, which had accepted the restrictions on individual freedom in relation to issues of gender, sexuality, race and environmental degradation, civil societies in the ECECs of the 1990s found it difficult to handle these issues (Tickle and Welsh, 1998). Basic ground rules such as compromise and self-limitation were harder for the general public to accept, given their minimal training in a free market economy.

The deep negative influence from the communist period remained, as illustrated by opposition to hydro-electric schemes like the Gabčikovo–Nagymáros Dam Project (Deets, 1999). Such relatively environmentally friendly energy projects were linked to the enormous dam-building plans, typical of the communist period. Similarly, it became difficult to convince the population to trust the scientific research findings on pollution and environmental damage; this attitude reflected former communist scientific research which used such results to justify projects that have since been proved to have been damaging to the environment.

Environmental issues did not really concern the communist leaders of the region. However, the 1980s saw the emergence of newly formed protest groups, linked with the growing menace of pollution, and they became part of a much larger swell of rapidly growing and popular protest movements against communist regimes in general. For example, in Slovakia environmental dissent became part of an emerging nationalist movement. Moreover, protests against the environmental conditions became a relatively safe way to express one's own displeasure with the political regime. Open recruitment to such environmental groups fared best in the freer political climate of Poland and Hungary (Jancar-Webster, 1998). In Poland, environmental issues were fundamental in the ideology of the Solidarity movement, whilst environmental abuse had profound significance for Czech dissidents (Tickle and Welsh, 1998). Such anti-government movements were to exercise an ever-increasing function in organising the downfall of communism (Baker and Jehlička, 1998).

Many of these movements lost their significance after the demise of communism. Once the main political aspirations had been attained, prominent environmentalist members merged into mainstream politics because there was no longer a stimulus to join such a protest movement (Baker and Jehlička, 1998). Nevertheless, some Green parties did experience excellent results in the first free elections, but by the end of the 1990s many have gradually disappeared from the political scene partly because environmentalists had the apparent stigma of being both anti-market and anti-nation. Organisations that have remained true to their faith have now diverted their attention to influencing plans for corporatist policy-making, a task requiring greater expertise and professionalism in contrast to the popular enthusiasm of the anti-communist movements (Jancar-Webster, 1998). These organisations now form part of international NGO networks, inspired by new social movements on a global scale (Tickle and Welsh, 1998; Tickle, 2000). Following the demise of communism, perhaps another reason why environmentalist organisations have failed to obtain greater overall support is related to the isolated nature of their political programmes. For ordinary citizens, they appear to lack any appreciation of the serious economic and social problems encountered in everyday life (Klarer and Francis, 1997).

Unfortunately, regard for environmentalism in the various countries by both governments and electorates has proved to be a poor indicator in the transition process from communism to democracy. Furthermore, environmental liability was neglected during the early privatisation period in most countries. Environmental protection laws were entering the statute books, but failure to integrate economic questions into these decrees made for excessively idealistic legislative measures and pollution targets. Environment ministers' liberty to express individual viewpoints were soon curtailed and ministers who maintained contact with NGOs were quite exceptional.

While beneficial results have resulted from international initiatives, they could be criticised for emphasis on theory at the expense of financial provision through cash payments to fund more practical measures. For example, in the case of nuclear power, the EU emitted conflicting signals; this encouraged financial investors from some countries to become involved in this sector at the expense of other areas. Bulgaria and Slovakia in particular were able to evade any EU rulings in this field. The distant prospect of EU membership blunted the impact of pressure from Brussels to terminate their use of the ageing nuclear power-stations. More generally, while environmental policy has been increasing in sophistication since 1995, it might be asked if a crucial opportunity to affect the long-term future of environmentalism in the ECECs was lost in the first years of transition. Given the high profile and prestige of the environmentalists in 1989, should they not have been more strongly assisted to engage public interest and transform popular concern about the state of the environment into individual activism? More funding could have helped NGOs in the area to strengthen their links with the public and project environmental activity as an expression of civil liberty and the individual's right to choose.

Conclusion

It was becoming clear after 1989 that a new society was emerging in the region, following the collapse of communism. The ensuing transition period was perceived in the West and by environmentalists in the East as a rare chance to combine environmental elements and economic factors in fields such as agriculture, energy, retailing, tourism and transport. Unfortunately, this did not transpire immediately, because the new post-communist societies could not ignore the impact of a wider international dimension. There was some improvement in the environmental situation, because it was impossible for a democratic society to tolerate pollution levels recorded during the communist period. Conversely, some negative polluting factors emerged, associated with greater mobility and introduction of contaminating foreign industries. The conflicts over nuclear power and other projects demonstrate that environmental issues are more complicated than they seemed in the communist era when it was enough just to oppose the government. The fiasco of the Gabčikovo–Nagymaros scheme shows that while simplisitic arguments may raise the profile of the issue, the populist approach does not necessarily result in the best outcome. Sadly, environmental conditions did not improve markedly in spite of the changed scenario. The environment failed to rise very high on the political agenda; other preferences overtook it in the minds of entrepreneurs, bureaucrats and the general public. Tragically, at the end of the 1990s only a very basic appreciation of environmental awareness was found amongst the citizens of the region.

This fact suggests the critical need for more environmental education in all branches of the community, rather than leaving it all to the school curriculum. Crucial here is the educational role of NGOs. As soon as the populus learns that environmentalism and economic development are not mutually exclusive, it will become impossible for politicians to interpret environmentalists' demands as based on trivial dreams. Here modest but forceful environmental organisations could play a vital role in ensuring that greater attention is paid to improving the environment of the future. Fortunately, some constructive evidence is emerging thanks to the potential integration of some ECECs into supranational organisations like the European Union. Candidate countries have delayed the adoption of the EU environmental policy, thus making it more difficult to arrange an accession timetable. However, short and well-defined transitional periods for environmental improvement after membership could provide welcome flexibility. Issues most likely to complicate accession dialogue relate to environmental dumping and cross-border emissions; along with the regular enforcement of environmental legislation, and higher energy and raw material costs to stimulate greater efficiency (Anon., 1999a, 1999b). Let us hope there will be some improvement in the region's environmental situation once the problems of the early and most exacting years of the transformation period have been overcome.

References

Ågren, C. (2000) 'Effects of the new protocol', *Acid News*, 1 (March), pp.8–10.

Anon. (1998a) 'Environmental policy: Kyoto 1997', *The Economist*, 7 November, p.109.

Anon. (1998b) 'Emissions – top ten emitters in 1995', *Acid News*, 2 (June), p.20.

Anon. (1999a) 'Mixed from Czechs', *Acid News*, 2 (June), p.9.

Anon. (1999b) 'Eastern Europe's environment: clean up or clear out', *The Economist*, 11 December, p.46.

Baker, B. and Jehlička, P. (1998) 'Dilemmas of transition: the environment, democracy and economic reform in East Central Europe – an introduction', in S. Baker and P. Jehlička (eds) *Dilemmas of Transition: The Environment, Democracy and Economic Reform in East Central Europe*, London: Frank Cass, pp.1–26.

Bluffstone, R. and Larson, B. A. (eds) (1997) *Controlling Pollution in Transition Economies: Theories and Methods*, Cheltenham/Northampton, MA: Elgar.

Cofala, J., Kurz, R. and Amann, M. (1997) *Application of the Current EU Air Emissions Standards to the Central and Eastern European Countries: An Integrated Assessment of the Environmental Effect*, Final Report to the European Environmental Agency, IIASA, Luxenburg.

Deets, S. (1999) 'Solving the Gabičkovo-Nagymaros dam conflict', *East European Studies* (March–April), Woodrow Wilson International Center for Scholars, Washington, DC, pp.3, 7.

Dorian, J. P. (1997), 'Mining – changing picture in transitional economies', *Mining Engineering*, January, p.13.

Fagin, A. and Jehlička, P. (1998) 'Sustainable development in the Czech Republic: a doomed process?' in S. Baker and P. Jehlicka (eds) *Dilemmas of Transition; The Environment, Democracy and Economic Reform in East Central Europe*, London: Frank Cass, pp.113–28.

Golubić, J., Štrumberger, N. and Jurum-Kipke, J. (1999) 'Traffic and ecological environment of Croatia as integration factor in 21st century Europe', *Promet-Traffic-Traffico*, 11(2–3), Portorož-Trieste-Zagreb, pp.173–8.

Gutner, T. (1999) 'Banking on the environment in Central and Eastern Europe', *East European Studies*, March–April, Woodrow Wilson International Center for Scholars, Washington, DC, pp.5–6.

Jancar-Webster, B. (1998) 'Environmental movements and social change in the transition countries', in S. Baker and P. Jehlička (eds) *Dilemmas of Transition; The Environment, Democracy and Economic Reform in East Central Europe*, London: Frank Cass, pp.69–90.

Klarer, J. and Francis, P. (1997) 'Regional review', in J. Klarer and B. Moldan (eds) *The Environmental Challenge for Central European Economies in Transition*, Chichester: John Wiley and Sons, pp.1–15.

Koenig, R. (1997) 'Floods wreak destruction in Eastern Europe', *Journal of Commerce*, 23 July, pp.10–13.

Lofstedt, R. and Sjostedt, G. (eds) (1996) *Environmental Aid Programmes in Eastern Europe: Area Studies and Theoretical Applications*, Aldershot: Avebury Press.

Makhijani, A. (1998) 'Nuclear power: no solution to global climate change', *Science for Democratic Action*, 6(3) (Institute for Energy and Environmental Research), Merrifield, VA, pp.1, 14–16.

Manser, R. (1993) *The Squandered Dividend: The Free Market and the Environment in Eastern Europe*, London: Earthscan Publications, pp.72–94.

OECD (1997) *Environmental Data 1997*, EMEP Report January, for non-OECD countries, Paris.

OECD (1999) *Environmental Performance Reviews: Czech Republic*, Paris, p.12.

Prentice, E-A. and Hillen, S. (2000) 'Cyanide threatens Balkan farms', *The Times*, 15 February, p.14.

Sandberg, M. (1998) *Green Post-Communism: Environmental Aid, Innovation and Evolutionary Political Economics*, London: Routledge.

Symonds, W. C. (1997) 'All that glitters is not Bre-X', *Business Week*, 19 May, p.10.

Tickle, A. (2000) 'Eastern Europe: the place of NGOs, then and now', *Acid News*, 2 (June), p.18.

Tickle, A. and Welsh, I. (1998) 'Environmental politics, civil society and post-communism', in A. Tickle and I. Welsh (eds) *Environment and Society in Eastern Europe*, Harlow: Addison-Wesley, Longman.

Wódż, J. and Wódż, K. (1998) 'Environmental sociology in Poland and the ecological consciousness of Polish people', in A. Tickle and I. Welsh (eds) *Environment and Society in Eastern Europe*, Harlow: Addison-Wesley, Longman, pp.97–113.

World Bank (1997) *Expanding the Measure of Wealth: Indicators of Environmentally Sustainable Development* (Environment Department), Washington, DC.

Index

Page numbers in italics refers to maps/tables/figures.